Mass Spectrometry

Mass Spectrometry

edited by

Charles A. McDowell

Department of Chemistry
The University of British Columbia
Vancouver, B.C., Canada

McGraw-Hill Book Company, Inc.

New York   San Francisco   Toronto   London

Mass Spectrometry

Library of Congress Catalog Card Number 62-22201

44940

List of Contributors

A. J. H. Boerboom — *Laboratory for Mass Separation (F.O.M.), Amsterdam, The Netherlands*

W. A. Bryce — *Department of Chemistry, The University of British Columbia, Vancouver, B.C., Canada*

V. H. Dibeler — *Mass Spectrometry Section, National Bureau of Standards, Washington, D.C.*

H. E. Duckworth — *Department of Physics, McMaster University, Hamilton, Ont., Canada*

R. M. Elliott — *Associated Electrical Industries Limited, Instrumentation Division, Trafford Park, Manchester, England*

J. B. Farmer — *Department of Chemistry, The University of British Columbia, Vancouver, B.C., Canada*

D. C. Frost — *Department of Chemistry, The University of British Columbia, Vancouver, B.C., Canada*

S. N. Ghoshal — *Department of Physics, Presidency College, Calcutta*

L. Kerwin — *Department of Physics, Laval University, Quebec, P.Q., Canada*

J. Kistemaker — *Laboratory for Mass Separation (F.O.M.), Amsterdam, The Netherlands*

F. P. Lossing — *Division of Pure Chemistry, National Research Council, Ottawa, Ont., Canada*

C. A. McDowell — *Department of Chemistry, The University of British Columbia, Vancouver, B.C., Canada*

C. C. McMullen — *Department of Physics, McMaster University, Hamilton, Ont., Canada*

D. P. Stevenson — *Research Department, Shell Development Co., Emeryville, California*

H. A. Tasman — *Laboratory for Mass Separation (F.O.M.), Amsterdam, The Netherlands*

H. G. Thode — *Department of Chemistry, McMaster University, Hamilton, Ont., Canada*

v

Preface

The subject of mass spectrometry has, in the past decade or so, developed very rapidly, and by now mass spectrometers and mass spectrographs are used in practically every branch of science. Naturally, as the subject developed there have appeared, from time to time, books dealing with particular aspects of the then-known knowledge in the field. Though some of the volumes available are most useful, it seemed that a fairly comprehensive account of the present status of the subject would be welcomed by many scientists. In the introductory chapter of the present volume, we have outlined the way this literature has developed. We have also indicated, to some extent, the reasons for the production of the present volume as a cooperative effort. It seemed to us that the whole subject of mass spectrometry has become so diversified that no one person could possibly hope to summarize it all adequately. It was the view of the editor that a more satisfactory book would result from the coordinated contributions of several persons, each with a specialized or detailed knowledge of a particular aspect of the subject. Only by this means could there be produced a volume which would be comprehensive and reasonably authoritative on the diverse topics to be covered in the different chapters. The aim throughout has been to stress the subject of mass spectrometry. Certain of the more fundamental scientific aspects of the results which have been obtained by mass-spectrometric studies have, of course, been discussed; but an attempt was always made to keep the importance of the mass-spectrometric aspects clearly in mind.

There are, of course, certain disadvantages in this method of assembling a book of this type. Had the whole work been written by one author, it is possible that the style would have been more uniform. However, it is the editor's view that the advantages of more comprehensive and authoritative treatment gained by multiple authorship more than compensate for such minor flaws. In the writing of this volume, the editor's view was also that the authors of the various chapters were best left free to treat the subject matter in the way they thought would be the most useful for the potential reader.   They were, however, requested to keep in mind the relation of their own chapter to others in the book.   Because the editor felt that the reader should be able to follow certain rather detailed and involved portions of some of the chapters without constant reference back to earlier chapters, in a few places it was decided that the reproduction of essentially the same diagram or figure was an allowable luxury.   At the same time, every attempt has been made to avoid excessive repetition of the same textual material, and much has been done to give ready cross references to accounts of related phenomena in the different chapters.

We may perhaps illustrate the above points by mentioning a few examples.   Because there is given what is judged to be an adequate account of radio-frequency and quadrupole mass spectrometers in Chapter 2, these subjects are treated in less detail in the chapters on ion optics, ion sources, and high-resolution mass spectroscopes.   It was also felt that because of the very full account of the theoretical aspects of ion optics given in Chapter 5, and also to some extent in Chapter 7, most of the mathematical aspects of the theory of ion optics could reasonably be omitted from Chapter 4, which deals with the other main aspects of the design, construction, and operation of ion sources.   For the same reasons, Chapter 5 on ion optics and Chapter 7 on high-resolution mass spectroscopes are complementary in many respects.   Here again, every effort was made to prevent excessive repetition of the subject matter, though at the same time the chapters were integrated as intimately as possible.   Similar remarks apply to the subject matter treated in Chapters 9, 10, 11, 12, and 13.   In many cases the material is such that these could easily have become repetitious, but every attempt was made to avoid this while at the same time providing a readable and comprehensive account of those aspects of the topics allotted to a particular chapter.

Finally, the editor would like to make it clear that he alone is responsible for the overall plan of the book.   The credit for the excellence of any of the individual chapters must be given to the author or authors.   Thanks are due to the many scientists, too numerous to be mentioned here by name, who supplied reports of their work and who helped in many ways.   The granting of permission to use illustrations published elsewhere is also gratefully acknowledged, and the origin of those used is generally indicated in the captions.

C. A. McDowell

Contents

Mass Spectrometry

1

Introduction

C. A. McDowell

Department of Chemistry
The University of British Columbia
Vancouver, B.C.

As is well known, the whole field of mass spectrometry developed from the original parabolic method of analysis of positive rays introduced by J. J. Thomson [1]. The early work of Goldstein [2], Wein [3], and others is admirably described in Thomson's book "Rays of Positive Electricity and Their Application to Chemical Analysis," published in 1913. The prototypes of modern mass spectrometers are, of course, those forms of apparatus described by Dempster [4] and Aston [5]. An interesting account of the early days of mass spectrometry and its application to isotopic analysis is given in Aston's first book, "Isotopes" [6], published in 1922. A later book by Aston, entitled "Mass Spectra and Isotopes" [7], first published in 1933, gives an excellent account of the rapid developments in the field which occurred during the preceding decade. At that time the few mass spectrographs and spectrometers in existence were almost wholly used for isotopic studies. Since then there has been an amazingly rapid increase in the numbers and types of mass spectrometers and mass spectrographs, and the range of applications has broadened so greatly that these instruments are now used in practically all branches of physics and chemistry.

With this great development in the range and scope of mass spectrometry, there has, of course, been the need, from time to time, to have conferences and review articles describing much of the new knowledge and its applications. Though a second edition of Aston's book "Mass Spectra and Isotopes" appeared in 1942, it was a decade or so before other books or substantial review articles appeared to record the very rapid modern developments.

Two of the earliest of these volumes were "Mass Spectrometry" [8], published by the Institute of Petroleum, London, in 1952, and "Mass Spectrometry in Physics Research" [9], published by the National Bureau of Standards, Washington, D.C., in 1953. The former book was a report of a conference at which papers were presented dealing with various applications of mass spectrometry in certain areas of chemistry and physics. The range of applications discussed was not comprehensive, but did include some quite important applications in chemistry. The latter book gave a comprehensive account of the known developments in the field of mass spectrometry, particularly in so far as physics was concerned. It is particularly noteworthy that this gave an excellent coverage of the then recent developments in precision high-resolution mass spectrometers and certain aspects of the theory of ion dispersion and ion optics in general. Certain new types of mass spectrometers such as nonmagnetic radio-frequency and time-of-flight instruments were brought to the general attention of workers in the fields of physics and physical chemistry for, perhaps, the first time. This book also contains many articles describing researches which it turns out were to be most important in defining how the subject was to develop during the ensuing decade.

Almost simultaneously with the publication of the above two books there appeared the monograph by Barnard entitled "Modern Mass Spectrometry" [10]. This book was the first of several modern books which were to appear in fairly rapid succession. Barnard's book attempted to give an account not only of the instrumental side of mass spectrometry, but also of some of the more important fundamental applications in the fields of physics and chemistry. In this it succeeded admirably and has remained a standard reference work. It is, however, lacking a little in that it failed to include a sufficiently detailed account of high-resolution mass spectrometers and certain other aspects of mass spectrometry which have since become of considerable importance in both physics and chemistry. In many cases, of course, this was inevitable, as one cannot always foresee what are to be important areas in the future, for frequently, little expected technical advances completely change the whole picture.

During October, 1953, a conference was held in London, sponsored by the Institute of Petroleum and called Applied Mass Spectrometry. The papers which were presented at that meeting were published in book form in 1954 in a volume entitled "Applied Mass Spectrometry" [11]. These papers dealt with various aspects of mass spectrometry but were largely concerned with applications of interest in different branches of chemistry. In some ways this volume is important, not only because it gives a very good indication of the greatly increased number of areas of chemistry in which mass spectrometry was having an important impact, but also because, in many instances, one can again see, as has been noted above for the National Bureau of Standards Washington Conference of 1951, that certain areas of study, which were then only in a rather preliminary state, were nevertheless to point the ways in which advances were to be expected in the years ahead.

In the same year (1953) Ewald and Hintenberger published their book "Methoden und Anwendungen der Massenspektroskopie" [12]. This book is a particularly important source of information on the design and construction of high-resolution mass spectrometers and gives an excellent account of the applications of mass spectrometers and mass spectrographs to isotopic studies. A somewhat less detailed account is given of other areas, such as the application of mass spectrometry to geology and chemistry. The work will, however, remain a standard reference on those topics to which the authors themselves have contributed so much to advance our knowledge, namely, ion optics and the accurate determination of atomic masses.

A somewhat different type of book made its appearance in 1954. This was the little book by Inghram and Hayden [13] which described, in an economy of detail, the salient features of mass spectrometers at the then known stage of development of the instruments. This booklet, in our opinion, has become a classic and was one of the most valuable publications in the field of mass spectrometry in the past decade.

Since the applications of mass spectrometry in chemistry were becoming extremely important in the mid-1950s, it was not surprising that several books should appear dealing with this material. Robertson's little book "Mass Spectrometry" [14], published in 1955, will for long remain a somewhat interesting, if in some ways uniquely quaint, account of certain chemical aspects of this subject. Field and Franklin's book, entitled "Electron Impact Phenomena" [15], was the first work to give in considerable detail an authoritative account of the study of the ionization and dissociation of molecules and molecular ions by modern mass-spectrometric methods. It gives an excellent comprehensive survey of this most extensive field of chemical physics and has become a classic to which workers in the field make constant reference. As is so frequently the case with a work of this nature, the authors seemed to foresee the areas in which important advances were to be anticipated, and their discussions of heats of formation of molecular ions, of ion-molecule reactions, and of like topics must have stimulated many future workers in these areas. Of a somewhat different nature is the volume entitled "Electromagnetically Enriched Isotopes and Mass Spectrometry" [16]. This is based on papers presented at a conference in London and is a useful source of information on isotopic mass spectrometry at that time.

One of the few attempts to survey the whole field of mass spectrometry in recent years resulted in the excellent book written by Duckworth, entitled "Mass Spectroscopy" [17]. This work, which is of wide general appeal, gives a very good account of the principles governing the operation of mass spectrometers and mass spectrographs. A particularly good treatment is given of the methods used to obtain mass resolution and dispersion and for the production and detection of ions. Excellent, though of necessity brief, discussions are given of the applications of mass spectrometry to physics, geophysics, chemical physics, etc. It thus forms a very good modern introduction to the whole subject of mass spectrometry.

Since the publication of Duckworth's book, no other work has appeared which attempts to deal with the subject of mass spectrometry as a whole. However, an excellent volume, "Advances in Mass Spectrometry" [18] (J. Waldron, editor), based on papers presented at a conference in London in 1958, jointly organized by the Hydrocarbons Research Group, Institute of Petroleum, and the ASTM Committee E-14, gives an account of developments in many areas of mass spectrometry and its main applications to physics and chemistry. More recently, the volume *Proceedings of the International Conference on Nuclidic Masses* [19] (H. E. Duckworth, editor) has provided a comprehensive account of developments in high-resolution mass spectrometry and their application to the determination of atomic masses with very high precision.

It is easy to understand that, as the range of applications of a subject like mass spectrometry becomes wider and wider, more specialized books will tend to appear. Thus Benyon's outstanding recent book, "Mass Spectrometry and Its Applications to Organic Chemistry" [20], reflects this trend. This work gives a unique account of the applications of mass spectrometry to organic-chemical problems. It is particularly noteworthy for its account of the important recent work on the use of high-resolution mass spectrometers in the elucidation of the structure of organic compounds. Another recent book covering similar topics is Biemann's "Mass Spectrometry: Organic Chemical Applications" [21].

The present volume represents the result of an attempt to produce a book which aims at giving a comprehensive account of the whole field of mass spectrometry. The view taken at the outset was that the subject matter has become so widespread, in a sense consisting of a group of specializations, that no one person can hope to summarize it all in an adequately satisfactory manner. It was therefore the view of the editor that it would be best to proceed by persuading several colleagues to write chapters on topics of which they had a special knowledge. By this means, it was thought, a volume could be produced which would be comprehensive and at the same time authoritative on the various widely chosen topics covered in the different chapters.

It is emphasized that though the aim of the editor was to realize a volume which would be both comprehensive and yet authoritative in treatment, covering the whole field of mass spectrometry, there was no attempt at making the work encyclopedic. As has been indicated above, the impact of mass spectrometry has been so great and its influence so widespread in so many branches of physics and chemistry that any attempt even to mention all the material in the published literature would surely produce a work of tremendous dimensions. Perhaps that is something which should be done; the present work is not, however, offered in any way as an encyclopedia of mass spectrometry.

A further word might be said about the method of compiling this book. Certainly, if one author had written the entire book, the volume that would have resulted would probably have been more uniform in style and perhaps even in outlook. Those advantages are, in the editor's opinion, greatly outweighed by one possible great disadvantage which has been mentioned earlier, namely, that, in the editor's belief, the whole field of mass spectrometry has in the past decade become so extensive that no one person can hope to be able to write comprehensively and at the same time authoritatively on all the important aspects of the subject. The same sort of problem, of course, presented itself in choosing the material to be included in this book. Our aim has been to stress the subject of mass spectrometry rather than to be comprehensive about the important scientific implications of the vast amount of data which has arisen from the study of different phenomena by mass spectrometers and mass spectrographs. It has, of course, frequently been necessary to consider in some detail certain of the fundamental implications of some of the results, but where possible this has been done in such a way as to keep the importance of the mass-spectrometric approach at all times clearly in mind. Again, this viewpoint to some extent dictated the choice of subject matter. Many areas of the field had in any course to be included in a book on mass spectrometry, but in attempting to provide a volume which would guide the novice in this important scientific area as well as give the more experienced scientist already working in the subject a book for ready reference was not an easy undertaking. It was, however, our view that a volume such as the present one is appropriate, and we have endeavored to ensure that the needs of a wide range of scientists engaged in the numerous areas of mass spectrometry will be met.

REFERENCES

1. J. J. Thomson: "Rays of Positive Electricity and Their Application to Chemical Analysis," Longmans, Green & Co., Ltd., London, 1913.
2. E. Goldstein: *Berl. Ber.*, **39**, 691 (1886).
3. W. Wein: *Verhandl. Physik. Ges.*, 17 (1898).
4. A. Dempster: *Phys. Rev.*, **11**, 316 (1918).
5. F. W. Aston: *Phil. Mag.*, **38**, 709 (1919).

6. F. W. Aston: "Isotopes," Edward Arnold (Publishers) Ltd., London, 1922.
7. F. W. Aston: "Mass Spectra and Isotopes," Edward Arnold (Publishers) Ltd., London, 1933.
8. "Mass Spectrometry," The Institute of Petroleum, London, 1952.
9. Mass Spectrometry in Physics Research, *Natl. Bur. Std. Circ.* 522, 1953.
10. G. P. Barnard: "Modern Mass Spectrometry," The Institute of Physics, London, 1953.
11. "Applied Mass Spectrometry," The Institute of Petroleum, London, 1954.
12. H. Ewald and H. Hintenberger: "Methoden und Anwendungen der Massenspektroskopie," Verlag Chemie GmbH, Weinheim, Germany, 1953.
13. M. G. Ingram and R. J. Hayden: A Handbook of Mass Spectrometry, *Natl. Acad. Sci.-Natl. Res. Council, Nucl. Sci. Ser., Rept.* 14, 1954.
14. A. J. B. Robertson: "Mass Spectrometry," Methuen & Co., Ltd., London, 1955.
15. F. H. Field and J. L. Franklin: "Electron Impact Phenomena," Academic Press Inc., New York, 1957.
16. M. L. Smith: "Electromagnetically Enriched Isotopes and Mass Spectrometry," Butterworth & Co. (Publishers), Ltd., London, 1956.
17. H. E. Duckworth: "Mass Spectroscopy," Cambridge University Press, London, 1958.
18. "Advances in Mass Spectrometry" (J. Waldron, ed.), Pergamon Press, New York, 1959.
19. *Proc. Intern. Conf. Nuclidic Masses* (H. E. Duckworth, ed.), University of Toronto Press, Toronto, Canada, 1960.
20. J. H. Beynon: "Mass Spectrometry and Its Applications to Organic Chemistry," Elsevier Publishing Company, Amsterdam, 1960.
21. K. Biemann: "Mass Spectrometry: Organic Chemical Applications," McGraw-Hill Book Company, Inc., New York, 1962.

2

Types of Mass Spectrometers

J. B. Farmer

Department of Chemistry
The University of British Columbia
Vancouver, B.C.

2.1. Introduction

The term mass spectroscope is applied to any device which has the ability to separate gaseous ions according to mass-charge ratio. Although mass-spectroscopic techniques have been used to study free ions formed in electric discharges and flames, and even ions existing in the upper atmosphere, it is more commonly desired to examine neutral molecules. Such molecules are, first of all, converted into gaseous ions in an ion source. Then the ions, which may be singly or multiply charged and have masses corresponding to the original molecules or fragments thereof, pass from the source to a mass analyzer, where they are separated into the appropriate mass-charge groups. This is achieved by the application of an electric or a magnetic field or a combination of the two. Finally, the separated ions produce an indication of their presence at a detector. In the ideal mass spectroscope an ion travels from source to detector without suffering a collision with another ion or molecule. In practice, this situation is approximated by working at gas pressures below 10^{-5} torr, where the mean free path is long compared with the ion trajectory.

The uses of mass spectroscopy are so diverse that no one instrument is suitable in all instances. Consequently, one finds in the literature numerous descriptions of ion sources, mass analyzers, and ion detectors, some differing in principle, some merely in detail. In certain instruments, usually referred to as mass spectrographs, the ion detector is a photographic plate on which the separated ion beams form images. The positions of the images depend in a predictable way on the mass-charge ratios of the ions. Hence, if the respective ionic charges are known, the masses can be compared. The relative abundances of the ions can be estimated from the plate by photometric means, but this procedure is subject to considerable error. The primary, although not the exclusive, task of mass spectrographs is therefore mass comparison. Instruments featuring electrical detection are usually referred to as mass spectrometers. The currents carried by the ion beams leaving the analyzer are measured. It follows that mass spectrometers are primarily useful for work involving abundance determination. However, some mass spectrometers have been designed to function as mass comparators in competition with the mass spectrographs.

In this chapter a survey is made of the various types of mass spectrographs and mass spectrometers. The principles of operation are discussed, and the outstanding properties of each type are indicated. It is convenient to classify the instruments according to the field arrangement used for mass analysis. Instruments with static fields were the first to be developed and are still the most common. Dynamic instruments, using a periodic electric field, possess virtues of their own, including simplicity of construction and, in one version, high speed of response. Magnetodynamic instruments, using a static magnetic field and a periodic electric field, have found special applications in the areas of mass comparison and high-vacuum gas analysis.

Several symbols which appear frequently in this chapter are defined below in accordance with the mksa system of units:

m = mass of an ion, in kilograms
M = mass of an ion, in atomic mass units
N = Avogadro's number, in (kilogram mole)$^{-1}$
e = electric charge of an ion, in coulombs
v = velocity of an ion, in meters per second
W = kinetic energy of an ion, in joules
V = voltage applied to an electrode
E = electric field, in volts per meter
H = magnetic field, in amperes per meter
$B = \mu_0 H$ = magnetic induction, in webers per square meter
μ_0 = permeability of free space = $4\pi \times 10^{-7}$ henry per meter (by definition)

2.2. Static Instruments

A. History

The forerunner of all mass spectroscopes, built by Thomson [1] at the beginning of the century, is depicted in Fig. 2.1. A beam of positive ions from a gas-discharge tube is collimated in a narrow channel (Thomson used a hypodermic needle for this purpose) and then projected into a mass analyzer, where parallel electric and magnetic fields are simultaneously applied.   After deflection by the fields, the ions impinge on a fluorescent screen.   With this arrangement ions of a given m/e ratio are deflected so that they reach the screen at points along a segment of a parabola, while those of other m/e ratios fall on different parabolas. The length of each parabolic segment depends on the spread of velocities of the ions responsible for it.   Ions from a discharge have widely varying velocities, so long segments are observed.   If all ions of the same m/e ratio had the same velocity, point images would result.

The original parabola apparatus and several later variants provided much important information concerning the nature of the ions formed

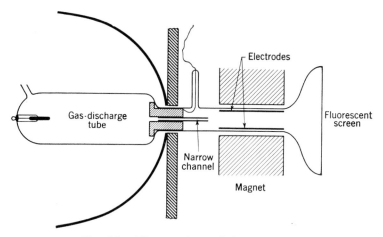

Fig. 2.1.   Thomson's parabola apparatus.

in discharges and gave the first demonstration of the existence of isotopes of nonradioactive elements.   Thomson summarized his pioneering researches in his book "Rays of Positive Electricity" [2], published in 1913, and in the preface he made the following statement: "I feel sure that there are many problems in chemistry which could be solved with far greater ease by this than by any other method.   The method is surprisingly sensitive, more so even than that of spectrum analysis, requires an infinitesimal amount of material and does not require this to be specially purified.   The technique is not difficult if appliances for producing high vacua are available."   Thomson's confidence in the value of mass spectroscopy has been amply justified.

In spite of its early triumphs, the parabola apparatus was eventually superseded by improved devices.   Aston [3, 4] constructed his mass spectrograph, with which he was destined to make a distinguished contribution to our knowledge of atomic structure.   The principle is

illustrated in Fig. 2.2.   Positive ions from a discharge tube are collimated by means of two slits, and then, in a mass analyzer, they are deflected in turn by an electric and a magnetic field.   Although ions having the
same m/e ratio enter the analyzer with a large range of velocities and
therefore follow different paths, they converge to a line image at the
surface of a photographic plate located as shown.   This action is called
velocity focusing.   Ions having different m/e ratios come into focus at
different positions on the plate, and a mass spectrum of parallel lines
results.   The spectrograph had a mass resolution of 130, or, in other

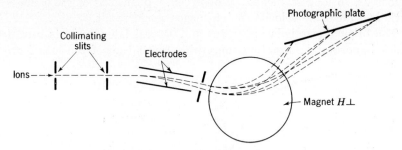

Fig. 2.2.   Aston's mass spectrograph.

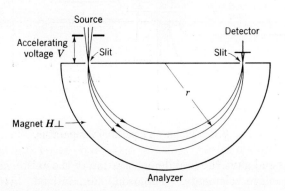

Fig. 2.3.   Dempster's mass spectrometer.

words, it could separate ions having m/e ratios differing by as little as
1 part in 130.   Aston was hence able to conduct a systematic search
for the isotopes of over 50 of the lighter elements.

Working concurrently with Aston, Dempster [5] carried out a similar
search for isotopes with an apparatus of entirely different design.   Dempster's apparatus, the first mass spectrometer, is shown in Fig. 2.3.   Ions
are produced with low kinetic energy by heating a salt on a metallic
filament or by bombarding a salt with electrons.   They are electrically
accelerated, and then proceed from the source via a slit to a mass analyzer,
where they are acted upon by a transverse magnetic field.   During

acceleration through a voltage V, an ion acquires kinetic energy eV. Provided that the energy increment is large compared with the initial energy,

$$eV = \frac{mv^2}{2} \tag{2.1}$$

where v = velocity of the ion after acceleration. All the ions carrying the same charge are accelerated to the same energy. If the magnetic induction in the analyzer is B, the ion experiences a force evB, which is normal to the direction of the induction and to the direction of motion. The result is a circular orbit such that the centrifugal force balances the deflecting force. If the orbital radius is r, it follows that

$$evB = \frac{mv^2}{r}$$

or
$$r = \frac{mv}{eB} \tag{2.2}$$

Eliminating v between Eq. (2.1) and Eq. (2.2),

$$r = \left(\frac{2mV}{eB^2}\right)^{\frac{1}{2}} \tag{2.3}$$

From the foregoing it is clear that the ions in Dempster's mass spectrometer are deflected according to their m/e ratio. For values of V and B satisfying Eq. (2.3), ions of a given m/e ratio are deflected through 180° along a circular path of radius r to a slit. The ions pass through the slit to a collector, and the collector current is measured with a quadrant electrometer. When V or B is varied, different m/e groups arrive successively at the collector, and a record of the collector current constitutes a mass spectrum. By itself, a magnetic analyzer is sensitive to the momentum of an ion rather than to its mass alone, as shown by Eq. (2.2). However, it becomes mass-sensitive when coupled with a source delivering ions of uniform energy. One important feature introduced by Dempster was direction focusing. This focusing is indicated in Fig. 2.3. Even though ions enter the analyzer in a divergent beam, their tracks through the magnetic field converge on the collector slit. Thus the need for collimation, with its attendant current attenuation, is much reduced.

By the mid-1920s it had been recognized that a great number of the lighter elements possess isotopes. The scrutiny of the remaining elements awaited the development of mass-spectroscopic equipment capable of increased resolution. Furthermore, it had become apparent that the atomic masses of isotopes deviate slightly from integral values, contrary to what had previously been supposed. Researchers therefore began to concentrate on the problems of accurate mass determination, and here again high-resolution equipment was necessary. Improved versions of the velocity-focusing mass spectrograph were set up by Aston [6–8] and by Costa [9].   By enlarging the analyzer and narrowing the collimating slits, a resolution of 2,000 was finally attained.

Bainbridge [10] carried out some significant isotope studies with the mass spectrograph depicted in Fig. 2.4.   A parallel ion beam from a discharge tube enters a velocity filter, where it is subjected to crossed electric and magnetic fields at right angles to the direction of motion. The emerging beam consists of those ions which have suffered no deflection and is homogeneous in velocity, a condition which ensures mass

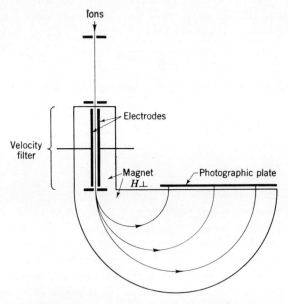

Fig. 2.4.   Bainbridge's mass spectrograph.

separation in a subsequent 180° magnetic analyzer.   The radius of the path of an ion in the analyzer is proportional to the ionic mass, as concluded from Eq. (2.2), so the scale of the spectrum is linear.

B. Direction-focusing mass spectrometers

For many mass-spectroscopic investigations it is essential to be able to determine the relative abundances of ionic species.   Aston made abundance estimates by comparing the densities of the photographic images produced by his spectrograph, but this routine is extremely tedious and is complicated by the nonlinear behavior which is characteristic of the photographic process.   Generally speaking, a mass spectrometer, with its electrical detector, is more satisfactory for this job.   Almost all the early mass-spectrometric work was performed with instruments equipped with monoenergetic ion sources and 180° magnetic analyzers, in the manner of Dempster. Descriptions were given by Bleakney [11], Tate, Smith, and Vaughan [12], and Nier [13]. The ions were commonly formed by bombarding gases with electrons of controlled energy. At first, a quadrant electrometer served as the current-measuring element, but a vacuum-tube electrometer later became standard.

In 1940, a milestone was reached with the introduction by Nier [14] of the sector magnetic analyzer. The ion-optical properties of the sector magnetic analyzer had been known for some years and had been exploited in mass spectrography. Nier's arrangement is shown schematically in Fig. 2.5. It can be seen that deflection takes place in a wedge-shaped magnetic field and that direction focusing is present. The ion beam enters and leaves the field at right angles to the boundary, so the deflection angle is equal to the wedge angle, namely, 60°. The geometry is symmetrical in that the source and the detector are equidistant from the magnet. As with the 180° analyzer, the radius r of the ion path in the

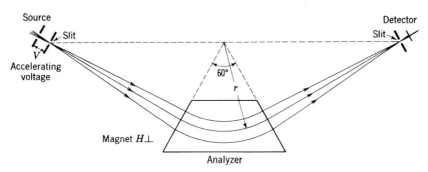

Fig. 2.5. Sector magnetic analyzer.

field, the accelerating voltage V, the magnetic induction B, and the m/e ratio are linked by Eq. (2.3). In the plane of the detector slit the ions are laterally dispersed according to the m/e ratio. Successive m/e groups are brought into focus at the slit by sweeping V or B. Ion-optical theory proves that mass separation and focusing can be achieved in a sector analyzer with any deflection angle. Provided the ion beam crosses the field boundary at right angles, focusing occurs when the source and detector slits fall along a line passing through the apex of the field. The 180° analyzer corresponds to an extreme case of this generalization. A 90° sector configuration was adopted by Hipple [15]. An improved 60° spectrometer was built by Nier [16]. Accounts of the design and construction of general-purpose sector spectrometers have been published by Graham, Harkness, and Thode [17] and Pelchowitch [18].

Hagstrum [19] has described in detail a series of four sector mass spectrometers intended for specific research problems. Schematic and perspective views of one of these instruments are shown in Figs. 2.6

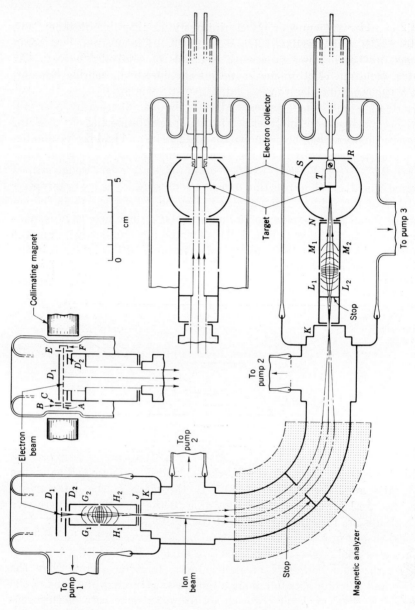

Fig. 2.6. Schematic view of a sector mass spectrometer. *(From Hagstrum [19].)*

and 2.7. This version is designed for the study of electron ejection from metal surfaces by positive ions, and the layout is typical of contemporary practice. The ions are produced in an electron-bombardment source and separated in a 90° magnetic analyzer. Special facilities necessary for electron-ejection measurements are located at the exit

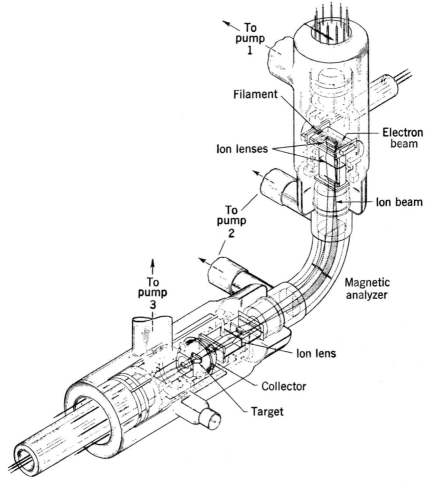

Fig. 2.7. Perspective view of a sector mass spectrometer. (*From Hagstrum* [19].)

of the analyzer. In a standard instrument an ion detector, usually a simple collector, would be placed behind slit K. With the exception of the magnet, the ion-optical components of the spectrometer are enclosed in an evacuated tube fabricated from glass and nonmagnetic metal.

Both 180° and sector mass spectrometers are in common use at the present time. Numerous commercial models are available. The sector instrument has the advantage that its source and detector are comparatively free from the mass-discriminating influence of the analyzer field. This is important when absolute isotope ratios are being determined, because here it usually must be assumed that the isotopes are ionized with equal probability and that the mass spectrometer is equally sensitive to each of the resulting ionic species. For most other measurements in chemical analysis, however, mass discrimination is a secondary consideration, because the spectrometer can be calibrated with standard substances. Unfortunately, even in the sector spectrometer the ion source is not completely field-free. For its correct operation the customary type of electron-bombardment source needs a collimating magnetic field parallel to the analyzer field, and when the source is remote from the analyzer, the field must be derived from an auxiliary magnet, as shown in Fig. 2.6. The source field amounts to perhaps one-tenth of the analyzer field.

For equal performance the magnet of a sector spectrometer is more compact than its 180° counterpart, and if an electromagnet is used, the demand for current is less. One problem in designing a sector spectrometer springs from the fact that the magnetic field does not possess a sharp boundary, as is assumed in the simple, ion-optical concept. The ions must cross a fringe region, where a field gradient exists. Their trajectories are modified, and to maintain the focusing action, the source and detector must be displaced slightly from the positions indicated by theory. The displacement is difficult to compute. Normally, provision is made for shifting the magnet with respect to the spectrometer tube, and the focusing condition is found empirically.

As concluded by Thomson from his early experiments, it is possible to obtain a mass spectrum from "an infinitesimal amount of material." Samples consisting of 1 ml of gas at STP are routinely handled. Reynolds [20] has described a high-sensitivity spectrometer for the isotopic analysis of minute samples of the inert gases. The spectrometer tube is pumped to 5×10^{-11} torr in order to reduce the background spectrum from residual gases. During a run the tube is isolated from the pumping system, contrary to the usual procedure in mass spectrometry, and the sample is introduced batchwise rather than continuously. With this method, 5×10^5 atoms of any isotope of xenon can be detected. The limit is set by the noise level of the detector, which is an ion-electron multiplier. For some m/e ratios the limit is set by the magnitude of background currents.

Sometimes, as in geochemical investigations, it is necessary to determine the relative abundances of two isotopes with high precision so that small deviations in a series of samples can be detected. McKinney et al. [21] attained a precision of 0.01% with a mass spectrometer in which the ion beams from two isotopes were measured simultaneously with twin collectors. This technique compensates for fluctuations of the sample concentration in the ion source and for drifts in the electronic circuitry.

In a magnetic-deflection mass spectrometer an ion image is superposed on the detector slit. For the symmetrical analyzers generally employed, the width of the image is equal to the width of the source slit together with any aberrations which may be present. Two important aberrations result from the spread in direction and velocity of the ions leaving the source. If the ion beam emerges from the source with a half angle of divergence of α radians, where α is small compared with unity, the corresponding direction aberration is proportional to α^2. Since a term involving the first power of α is eliminated by the focusing, the analyzer is said to be first-order direction focusing. The magnetic analyzer provides no velocity focusing. In the simple theory it is assumed that all ions of a given m/e ratio leave the source with the same energy. However, an actual source may have an energy spread which produces an appreciable aberration of the ion image. The detector slit is made somewhat wider than the ion image, so that all ions of one m/e ratio can pass through at the same time. Then, when a spectrum is scanned, flat-topped peaks with heights strictly proportional to the corresponding beams are obtained.

Ions of two slightly different m/e ratios are completely resolved provided that the combined width of the detector slit and an ion image is equal to or less than the dispersion of the ions in the plane of the detector slit. The dispersion is proportional to the radius of the beam path in the magnetic field. For a sharp ion image the spectral peak is trapezoidal and an appropriate measure of the resolution of the spectrometer is the ratio $M/\Delta M$, where ΔM is the width at the base of the peak for singly charged ions of mass M amu. In fact, the ion image is frequently diffuse at the edges, so the peak does not terminate abruptly but tails into the base line. There then arises the problem of how best to define resolution. There is no universally accepted convention, but the peak width is often taken at an arbitrary 1% of the height. The practical limit to the resolution of a direction-focusing mass spectrometer is about 1,500. For many applications this is more than enough and resolution is often sacrificed in the interest of sensitivity.

Factors which contribute to the diffuseness of an ion image are the energy spread of the beam, small-angle elastic scattering of the ions by molecules in the analyzer, and space-charge repulsion. A tail caused by image diffuseness may obscure a small adjacent peak, or at least reduce the precision of its measurement. The ability of a spectrometer to determine the relative abundances of ions of similar m/e ratio is signified by its abundance sensitivity. This is defined as the reciprocal of the fractional number of ions of mass M which are detected at a distance of one atomic mass unit on the mass scale. The abundance sensitivity is usually mass-dependent, increasing with decreasing mass, but it bears only a secondary relationship to resolution as defined by the 1% criterion. The best abundance sensitivity obtainable with a direction-focusing mass spectrometer is about 10^5. When the abundance sensitivity is limited by scattering, performance can be improved by operating two mass analyzers in series. Ions which are scattered into an adjacent beam in the first are removed in the second. Inghram and Hayden [22] and White and Collins [23] have described two-stage spectrometers. White and Collins reported an abundance sensitivity of 7.5×10^4 at 40 amu for a single analyzer and one of 10^7 for a two-stage combination. The tail of the peak is attenuated by this technique, but the main part is not greatly affected.

C. Double-focusing mass spectrographs and mass spectrometers

In the double-focusing mass analyzer advantage is taken of the ion-optical properties of a radial electric field established between the plates of a cylindrical condenser, as shown in Fig. 2.8. If an ion is projected into a radial field E at right angles to the boundary, it experiences a force eE normal to the direction of motion. The result is a circular orbit of which the radius r is simply calculated:

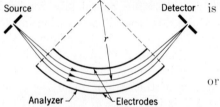

$$eE = \frac{mv^2}{r}$$

or

$$r = \frac{mv^2}{eE} \tag{2.4}$$

Fig. 2.8. Radial electric analyzer.

A radial electric field therefore deflects ions according to their kinetic energy. Direction focusing is also obtained, as indicated in Fig. 2.8. Ions from a slit placed at one field boundary focus at the other boundary if the deflection angle is $127°17'$ ($\pi/\sqrt{2}$ radians). For smaller deflection angles the conjugate foci lie outside the field. A radial electric analyzer can be combined with a magnetic analyzer in such a way that the velocity dispersion is opposite and approximately equal in the two analyzers. Hence the velocity aberration of the ion image is substantially eliminated. Direction focusing is at the same time preserved.

Mass spectrographs embodying double-focusing analyzers of this sort were devised in the 1930s to provide the high resolution necessary for accurate atomic-mass determinations. The first models were built by Mattauch and Herzog [24, 25], Dempster [26], and Bainbridge and Jordan [27]. The arrangement used by Mattauch and Herzog, shown schematically in Fig. 2.9, has the distinction of being double-focusing simultaneously for all m/e ratios. The other versions are double-focusing for one chosen m/e ratio and approximately so for nearby ratios. The deflection angles in the electric and magnetic analyzers are 31°50′ ($\pi/4 \sqrt{2}$ radians) and 90°, respectively. The ions leaving the electric analyzer are dispersed according to energy and focused at infinity. Ions having energies within a selected range are admitted to the magnetic analyzer, where mass separation takes place. Double focusing can also be obtained when the electric and magnetic fields are coincident rather than consecutive. A coincident-field spectrometer was reported by Bondy, Johannsen, and Popper [28]. In order to meet the demand for

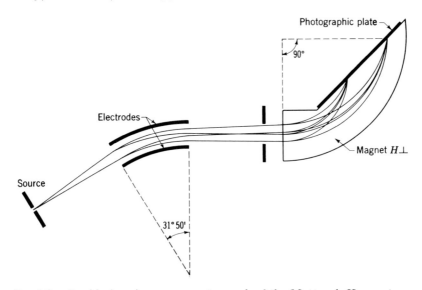

Fig. 2.9. Double-focusing mass spectrograph of the Mattauch-Herzog type.

nuclidic-mass data, the technical development of double-focusing spectrographs and spectrometers has continued to the present day.

In recent years double-focusing instruments have been increasingly applied to problems of chemical analysis. Their good resolution permits the investigation of materials of high molecular weight such as oils and waxes. Even in the low mass range benefits accrue from the use of double focusing. Molecular and structural formulas of molecules can be deduced from precise mass measurements on the ions formed by the dissociative and nondissociative ionization of the molecules. Because of the velocity focusing, the measurements on fragment ions are not affected by the presence of any kinetic energy of formation. In modern technology there is a frequent need to analyze solids for trace quantities of elements. Mass spectroscopy has been found highly satisfactory

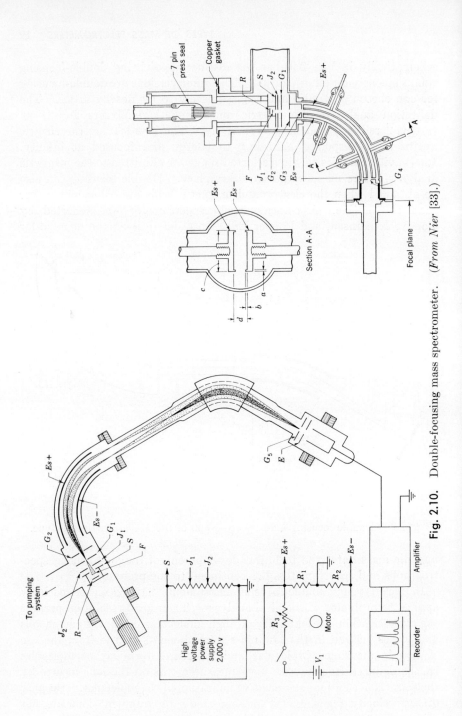

Fig. 2.10. Double-focusing mass spectrometer. (*From Nier* [33].)

for this purpose. Samples are generally ionized in an r-f spark source. This type of source is preferred because it has about the same sensitivity for all elements. However, it produces ions with a large energy spread, so double focusing is mandatory even in the low mass range.

Hannay [29] has described a double-focusing mass spectrograph of the Mattauch-Herzog type equipped with a spark source and intended for trace analysis. A photographic plate is used as the detector because it integrates the fluctuating ion currents produced by the spark source. Craig and Errock [30] have given a detailed account of a double-focusing mass spectrometer designed to identify organic compounds by measuring

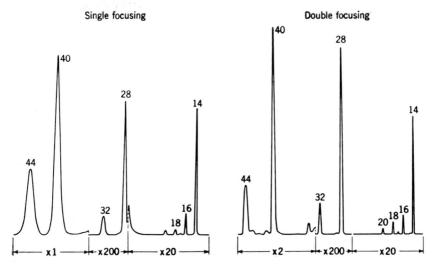

Fig. 2.11. Mass spectra for air taken with the double-focusing spectrometer of Fig. 2.10 and with a single-focusing instrument having a magnetic analyzer of similar dimensions. (*From Nier* [33].)

the masses of their ionization products with high precision. Mass measurements on an unknown ion are made by determining the voltage across the electric analyzer necessary, first, to focus the unknown ion beam and, second, to focus any known ion beam having a mass number within 10% of the unknown. Use is then made of the relationship that mass is inversely proportional to analyzer voltage. A precision of 1 part in 10^5 is possible by this technique. Voorhies et al. [31] have constructed several experimental double-focusing spectrometers and have discussed the importance of the many design parameters. Robinson, Perkins, and Bell [32] have developed a multipurpose Mattauch-Herzog system. Several types of ion sources may be installed, and both photographic and electrical detection are available. The resolution is about 2,500. With the narrow lines resulting from this high resolution a visible photographic image is produced with as few as 3,000 ions.

Nier [33] has published a description of a small double-focusing spectrometer suitable for performing gas and isotope analyses. Its sensitivity and resolution make it competitive with instruments several times as large employing only a magnetic analyzer. If a permanent magnet is used, the weight and power consumption of the spectrometer are both low, making it attractive for applications where these factors may be of importance. Drawings of the spectrometer are given in Fig. 2.10. Two mass spectra for air are shown in Fig. 2.11. One was taken with the double-focusing instrument, and the other was taken under identical conditions with a single-focusing instrument with a magnetic analyzer of similar dimensions. Since voltage scanning was used, the improvement in performance brought about by the double focusing is more noticeable at the high-mass peaks, which are produced by low-energy ions.

White, Rourke, and Sheffield [34, 35] have constructed a three-stage spectrometer with two magnetic analyzers and an electric analyzer in series. In this way high resolution and high abundance sensitivity (10^{10} at 24 amu) are obtained. White et al. [36] have used a large double-focusing analyzer to produce samples of highly enriched isotopes. The sample is deposited on a metallic target at the usual detector position. White et al. suggest that this technique could be employed for doping semiconducting materials.

D. Cycloidal-focusing mass spectrometers

Figure 2.12 illustrates the mode of operation of a cycloidal-focusing mass analyzer. An ion beam diverging from a source slit is deflected

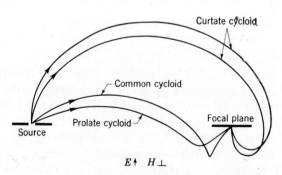

Fig. 2.12. Cycloidal-focusing mass analyzer.

by crossed electric and magnetic fields. Ions of a given m/e ratio pursue various cycloidal paths to a focus. The image is a replica of the source slit. There are no aberrations arising from the spread in direction and velocity of the ions leaving the source, and in this sense the focusing is said to be perfect. The ions can be visualized as performing circular motions in a coordinate system moving from left to right with velocity E/B. The cycloidal pitch, which is equivalent to the focal distance, is $2\pi m E/eB^2$, so linear mass dispersion is obtained. Although all paths, common cycloidal, prolate cycloidal, and curtate cycloidal, are permissible in theory, the curtate geometry is generally used. The common-cycloidal path is objectionable because of space-charge defocusing at the cusp. The analyzer can be used in conjunction with a photographic plate located in the focal plane or with a slit followed by an electrical detector. The latter arrangement is more common. An isometric view of a cycloidal-focusing mass spectrometer with curtate geometry is shown in Fig. 2.13. The electric field is established in the analyzer by applying a voltage to a pair of parallel plates, and the field is homogenized by feeding a stack of guard rings from a voltage divider placed between the plates. The ratio of the beam accelerating voltage to the analyzer

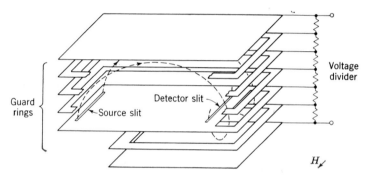

Fig. 2.13. Isometric view of a cycloidal-focusing mass spectrometer.

voltage is kept constant to ensure that the focused ion beam passes through a series of apertures cut in the guard rings.

The cycloidal-focusing principle was originated by Bleakney and Hipple [37], and these workers built two experimental spectrometers based on its use. Mariner and Bleakney [38] have published details of a high-resolution spectrograph intended as a mass comparator, but the actual instrument was never completed. Monk and Werner [39] have presented curves, monographs, and equations which simplify the computation of cycloidal paths. Robinson and Hall [40] have discussed the design and performance of a compact, general-purpose spectrometer. This instrument is designed for gas analysis in the mass range of 12 to 150 amu. If the electric field is switched off, masses lower than 12 amu can be focused on an auxiliary detector via a 180° path. Satisfactory focusing in the cycloidal analyzer is obtained with ion beams having a 10% energy spread and a 6° divergence from the median path. Under normal operating conditions the peaks are free of tails. Robinson [41] has shown theoretically that nonuniformities in the electric and magnetic fields of a cycloidal-focusing mass spectrometer may cause a deterioration of the image quality or merely shift the focal point. Space charge shifts the focal point out of the plane of the detector slit. These effects can be compensated by adjusting the guard-ring voltages. Hall, Hines, and Slay [42] have described a cycloidal-focusing mass spectrometer capable of producing up to 10 spectra per second. The detector is an ion-electron multiplier located outside the analyzer. After being focused at the detector slit in the usual way, the ion beam is channeled toward the multiplier by means of an electric field applied between a pair of specially shaped electrodes. Voorhies et al. [31] have constructed a cycloidal-focusing mass spectrometer with a resolution of 2,500. To accomplish this performance the fields must be uniform to a high degree. The electric field is homogenized with 68 guard rings, and to reduce surface-charge effects the rings are heated to 220°C.

2.3. Dynamic Instruments

A. Radio-frequency mass spectrometers

Radio-frequency techniques were first applied to mass spectrometry by Smythe and Mattauch [43, 44]. These workers built several versions of a mass spectrometer in which an ion beam was subjected to a series of lateral r-f electric fields. Ions traveling with one preferred velocity emerged from the final field with no net deviation or displacement and continued through a slit system while other ions were deflected away. The r-f analyzer itself was velocity-sensitive, and mass separation was accomplished by adding a radial, electrostatic analyzer. In practice, the design had several serious defects. Fringing fields were troublesome, and ions moving with certain velocities other than the preferred one gave rise to spurious peaks, which were difficult to suppress. The latter problem was discussed in detail by Hintenberger and Mattauch [45]. The Smythe-Mattauch method of mass analysis was ultimately abandoned and is now remembered only for its historical interest.

Most of the modern r-f mass spectrometers have made use of axial fields rather than lateral fields, and the analyzers have generally taken the form shown in Fig. 2.14. An ion beam, homogeneous in energy, passes through a succession of r-f modulator stages separated by field-free drift spaces. Ions having one certain velocity are preferentially accelerated and can overcome a static retarding field, while ions having other velocities are rejected. Because of the homogeneity in energy of the incoming beam, the velocity of an ion is characteristic of its mass. Hence the velocity analysis achieved in the spectrometer is equivalent to mass analysis. Each modulator stage of an r-f analyzer consists of an odd number of plane, parallel grids divided by equal gaps. Alternate grids are electrically connected. The first, third, etc., grids are maintained at a constant potential, and an r-f voltage is applied to the intermediate grids. The kinetic energy which an ion acquires during passage through a modulator stage depends on its initial velocity and on the phase of the r-f cycle at the time of entry. A maximum energy increment is obtained at one particular ionic velocity and entry phase. Operating conditions are usually chosen so that the incremental energy is small compared with the initial energy. The drift spaces are of such lengths that accelerated ions emerging from one stage reach the following stage at the optimum phase for further acceleration. This requirement is met if the ions travel between the first grids of two adjacent stages during an integral number of r-f cycles. Although these synchronous ions gain the maximum possible energy in the analyzer, ions with certain other velocities are harmonically synchronous and can gain a sizable fraction of the maximum energy. The cycle numbers of the drift spaces are selected so as to dephase harmonic ions between

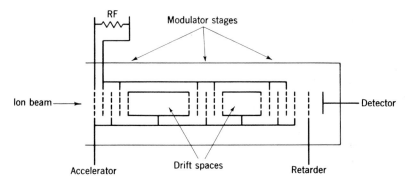

Fig. 2.14. Radio-frequency mass analyzer.

stages. The retarder voltage is set at a level which allows an ion to proceed from the analyzer to a detector only if it has received more than a specific fraction of the maximum incremental energy. Harmonically synchronous ions are rejected, together with ions of untuned mass and those of tuned mass entering the analyzer at a phase sufficiently removed from the optimum one so that they fail to gain the requisite energy. The mass resolution can be enhanced by raising the retarding level, but only at the expense of reducing the current efficiency of the analyzer, since the entry phase for an ion of tuned mass must then fall within a more restricted interval if the ion is to reach the detector. The mass range is scanned by varying the energy of the ion beam incident on the analyzer or by varying the frequency of the r-f voltage. The former way is usually more satisfactory.

Bennett [46] has described three axial-field, r-f mass spectrometers, which have served as models for subsequent designs. They contain one, two, and three modulator stages, respectively, and each stage has a simple three-grid structure. Wherry and Karasek [47] have achieved better mass resolution with a five-stage version. Redhead and Crowell [48, 49] have reported a single-stage spectrometer having 21 modulator grids, and they have derived expressions for its mass resolution and current efficiency. Dekleva and Peterlin [50] have demonstrated that the mass resolution of the three-stage Bennett spectrometer can be markedly improved by adding a harmonic component to the r-f voltage. They illustrate their results with mass spectrograms obtained with the simple, sinusoidal waveform $V_0 \sin \omega t$ and with the modified waveform $V_0(\sin \omega t - \sin 2\omega t)$. Cannon and Testerman [51] have devised a method for calculating the energy gained by an ion in a multistage, three-grid analyzer, based on the use of an analog computer. Shcherbakova [52] has carried out a mathematical analysis of the Bennett three-stage analyzer, tracing the relationship between the choice of the drift-space cycle numbers and the resulting mass resolution and harmonic suppression.

Vorsin et al. [53] have published a comprehensive treatment of the axial-field r-f mass spectrometer which permits a detailed evaluation of any proposed electrode configuration. These authors consider a generalized analyzer having $p + 1$ stages with p drift spaces. The entry grids of adjacent stages are separated by distances $l_1, l_2, \ldots, l_p$. Each stage consists of $q + 1$ grids with q gaps, where q is an even number, and the gap width is s. A voltage of the form $V_0 \sin (\omega t + \theta)$ is applied to the appropriate grids. The incremental energy $\Delta W (\gamma, \theta)$ received by an ion entering the analyzer at $t = 0$, phase θ, with velocity v, is shown to be

$$\Delta W(\gamma, \theta) = -2eV_0 \frac{\tan \frac{1}{2}\gamma \sin \frac{q}{2} \gamma}{\gamma} \left\{ \cos \left(\theta + \frac{q}{2}\gamma \right) + \cos \right.$$
$$\left. \left[\theta + \left(\alpha_1 + \frac{q}{2} \right) \gamma \right] + \cdots + \cos \left[\theta + \left(\alpha_1 + \cdots \alpha_p + \frac{q}{2} \right) \gamma \right] \right\} \quad (2.5)$$

where $\gamma = s\omega/v =$ tuning parameter
$\quad \alpha_1 = l_1/s, \alpha_2 = l_2/s, \ldots, \alpha_p = l_p/s$
For ΔW to attain a maximum value it is necessary that

$$\frac{d}{d\gamma} \frac{\tan \frac{1}{2}\gamma \sin \frac{q}{2} \gamma}{\gamma} = 0 \quad (2.6)$$

This condition is met when

$$\sin \gamma \sin \frac{q}{2} \gamma - \gamma \left(\sin \frac{q}{2} \gamma + \frac{q}{2} \sin \gamma \cos \frac{q}{2} \gamma \right) = 0 \quad (2.7)$$

The maximum value of ΔW corresponds to a tuning parameter γ_0 given by the principal solution of Eq. (2.7). If $q = 2$, $\gamma_0 = 2.34$; if $q = 4$, $\gamma_0 = 2.93$; etc.

Further conditions required for a maximum are

$$\theta_0 = h\pi - \frac{q}{2}\gamma_0 \tag{2.8}$$

$$\alpha_1\gamma_0 = 2\pi n_1 \qquad \alpha_2\gamma_0 = 2\pi n_2 \qquad \cdots \qquad \alpha_p\gamma_0 = 2\pi n_p \tag{2.9}$$

where h, n_1, n_2, . . . , n_p are integers. θ_0 defines the optimum entry phase for a synchronous ion and n_1, n_2, . . . , n_p are equivalent to the cycle numbers of the drift spaces. The velocity of a synchronous ion may be written

$$v = \frac{s\omega}{\gamma_0} = \left(\frac{2eV}{m}\right)^{1/2} \tag{2.10}$$

where V = beam voltage of the ions entering the analyzer. From Eq. (2.10) it is easy to deduce the tuning equation for the spectrometer:

$$M = \frac{2NeV\gamma_0^2}{s^2\omega^2} \tag{2.11}$$

The energy gained by any ion, synchronous or not, entering the analyzer at the phase corresponding to maximum acceleration is given by

$$
\begin{aligned}
\Delta W(\gamma) = 2eV_0 &\left| \frac{\tan \frac{1}{2}\gamma \sin \frac{q}{2}\gamma}{\gamma} \right| \{(p+1) + 2[\cos \alpha_1\gamma + \cos \alpha_2\gamma \\
&+ \cdots + \cos \alpha_p\gamma + \cos (\alpha_1 + \alpha_2)\gamma + \cos (\alpha_2 + \alpha_3)\gamma \\
&+ \cdots + \cos (\alpha_{p-1} + \alpha_p)\gamma + \cos (\alpha_1 + \alpha_2 + \alpha_3)\gamma \\
&+ \cos (\alpha_2 + \alpha_3 + \alpha_4)\gamma \\
&+ \cdots + \cos (\alpha_{p-2} + \alpha_{p-1} + \alpha_p)\gamma \\
&\qquad + \cdots + \cos (\alpha_1 + \alpha_2 + \cdots + \alpha_p)\gamma]\}^{1/2} \tag{2.12}
\end{aligned}
$$

Figure 2.15 shows a graph of ΔW versus γ for the simplest possible analyzer, that having a single stage with three grids. The maximum energy ΔW_{max} is gained by ions specified by the tuning parameter $\gamma_0 = 2.34$. If the retarder voltage is set to reject those ions which have received energy less than $k\,\Delta W_{max}$, where k is a fraction, the ions specified by values of the tuning parameter varying from γ' to γ'' will continue to the detector. As the retarder constant k approaches unity, $\gamma'' - \gamma'$ decreases, and consequently the mass resolution of the analyzer increases. The subsidiary maximum in the graph of Fig. 2.15 refers to the energy gained by harmonically synchronous ions corresponding to the tuning parameter γ_1. The height of this maximum places a lower limit on k if harmonic peaks are to be excluded from the mass spectrum. From Eq. (2.12) expressions can be derived for the theoretical mass resolution $M/\Delta M$ and current efficiency η of an r-f analyzer:

$$\frac{M}{\Delta M} = \frac{\gamma_0}{4\sqrt{2}}\left[\frac{q^2}{12(1-k)} + \left(\frac{2\pi}{\gamma_0}\right)^2\right.$$

$$\left.\frac{n_1^2 + n_2^2 + \cdots + n_p^2 + (n_1 + n_2)^2 + \cdots + (n_{p-1} + n_p)^2}{+ \cdots + (n_1 + \cdots + n_p)^2}\right]^{1/2} \quad (2.13)$$

$$\eta = \frac{\sqrt{2(1-k)}}{\pi} \quad (2.14)$$

Equation (2.14) is based on the assumption that the grids of the analyzer are perfectly transparent to ions.

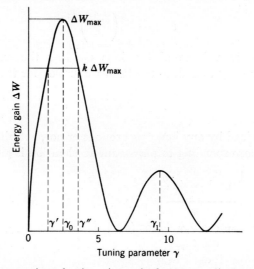

Fig. 2.15.   Energy gained by ions in a single-stage radio-frequency analyzer.

If the Bennett three-stage analyzer is appraised in the light of Eq. (2.13), it is seen that any increase in n_1 or n_2, the cycle numbers of the drift spaces, tends to improve the mass resolution. However, the choice of combinations of n_1 and n_2 is restricted because the heights of the subsidiary maxima of Eq. (2.12) must not exceed $k\,\Delta W_{\max}$. If k is set at the typical value of 0.9, a maximum resolution of 24 is obtained when the cycle numbers are 3 and 13. For perfectly transparent grids, the current efficiency of the analyzer, calculated from Eq. (2.14), would be 0.14. In practice, each of the modulator and retarder grids transmits about 0.95 of the ions, so the current efficiency becomes about 0.14 × 0.95^{10}, or 0.084. For comparison, a single-stage analyzer of the Redhead-Crowell type, with $k = 0.9$ and a mass resolution of 24, would require 49 modulator grids and would have a current efficiency of only 0.14 × 0.95^{50}, or 0.011.

Vorsin et al. assess the effect of supplying various nonsinusoidal voltages to Bennett's analyzer. They find that, for a voltage of the form $V_0(\sin \omega t - \sin 2\omega t)$, used by Dekleva and Peterlin, and a retarder constant of 0.9, the cycle numbers can be raised to 6 and 22 before harmonic peaks appear, and the resulting resolution is 42. The current efficiency is still given by Eq. (2.14), so remains unchanged. Kuchkov [54] has evolved an equation for the shape of the mass-spectral line produced by an r-f spectrometer.

An obvious advantage of the r-f mass spectrometer is that it requires no magnet. The analyzer is compact, and its construction is simplified by the fact that there are no defining slits to be aligned. The absence of slits also means that ion beams of large cross section can be handled, and this provides compensation for the low current efficiency of the analyzer. The instrument has been found ideal for the analysis of gases in the upper atmosphere. In this application there is a premium on small size and weight and the ability to operate under mechanical strain and over a wide range of temperatures. Pavlenko et al. [55] have given details of a rocket-borne spectrometer having a five-stage analyzer with a mass resolution of 50. Neutral molecules are ionized in an electron-bombardment source, and naturally occurring ions are accelerated directly into the analyzer. The total weight is only 3.3 kg, and the power consumption is only 6 watts. Boyd and Morris [56] have investigated the distribution of ions in a plasma with a spectrometer built in the form of a probe.

Donner [57] has reported a mass spectrometer in which the ions receiving the greatest deceleration in an r-f analyzer are selected in a subsequent, electrostatic analyzer and passed to a collector. A drawback to this design is that defining slits are necessary.

Tretner [58] has devised a type of r-f mass spectrometer in which ions execute oscillations in an electrostatic potential well maintained between two parallel electrodes. The oscillation frequency is mass-dependent, so ionic species can be selectively accelerated to an electrode by applying the appropriate frequency. The mass range can be scanned by sweeping the frequency. The spectrometer has a resolution of only 20, and its performance is marred by the presence of harmonic peaks and by a nonlinear dependence of peak height on concentration. However, it has been found useful for monitoring residual gases in vacuum systems.

B. Time-of-flight mass spectrometers

In the time-of-flight mass spectrometer a bunch of ions is accelerated and then projected through a field-free space toward a detector. The flight time for each ion is a function of its mass, so the detector receives a series of separate ion bunches, one for each mass present. A mass spectrum is obtained by recording the detector current versus time.

Mass spectrometers working on the time-of-flight principle have been reported by Cameron and Eggers [59], Keller [60], Glenn [61], Ionov and Mamyrin [62], Wolff and Stephens [63], Katzenstein and Friedland [64], Wiley and McLaren [65], and Agishev and Ionov [66, 67]. The mass resolution of the earlier models was very limited, but this defect was remedied by further development. In their paper Wiley and McLaren present a detailed account of the design of time-of-flight mass spectrometers and discuss, in particular, the improved version which they introduced. The salient features of this are shown in Fig. 2.16. An electron gun is switched on for a few hundred nanoseconds by the application of a voltage pulse to a control electrode, and the electron beam passes through an ionization chamber, which is field-free at this time. A bunch of ions is thereby generated. After the gun is switched off, an electric field is applied across the ionization chamber for a few microseconds, during

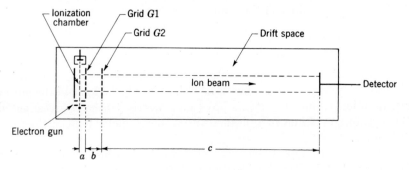

Fig. 2.16.   Time-of-flight mass spectrometer.

which time the ion bunch is accelerated through grid G1. It is further accelerated by a field applied continuously between grid G1 and grid G2. It enters a drift space and proceeds to a detector. The operation can be repeated as frequently as 50,000 times per second.

The mass resolution achieved by the spectrometer depends on the spread of flight times for ions of the same mass. Since the ionizing electron beam has a finite thickness and since ion diffusion occurs before the first accelerating field is applied, the length of the flight path varies from ion to ion. However, the farther an ion travels in the accelerating fields, the greater the velocity it acquires. Hence ions which have to travel different distances in the fields tend to converge later. The accelerating fields can be chosen so that the ions are focused in the plane of the detector. Ions begin their flight, not at rest, but with some thermal motion and, especially in the case of fragment ions, with some motion resulting from the process of ion formation. Ions of a given mass therefore have a range of velocity components in the direction of the flight

path.   In the interest of minimizing the spread of flight times due to this
cause, the acceleration period is kept as brief as practicable so that ions
move at their terminal velocity during most of the flight.

For a mathematical analysis of the time-of-flight mass spectrometer
it is helpful to define the following quantities:

a = distance from point of ionization to grid G1
b = distance from grid G1 to grid G2
c = length of drift space
E_a = first accelerating field
E_b = second accelerating field
W_0 = initial kinetic energy of an ion
W = kinetic energy after acceleration

The flight times t_a, t_b, and t_c for the segments a, b, and c of the path are
then

$$t_a = \frac{(2m)^{\frac{1}{2}}}{eE_a} [(W_0 + eaE_a)^{\frac{1}{2}} \pm W_0^{\frac{1}{2}}] \qquad (2.15)$$

$$t_b = \frac{(2m)^{\frac{1}{2}}}{eE_b} [W^{\frac{1}{2}} - (W_0 + eaE_a)^{\frac{1}{2}}] \qquad (2.16)$$

$$t_c = \frac{(2m)^{\frac{1}{2}}c}{2W^{\frac{1}{2}}} \qquad (2.17)$$

The total flight time t is obtained by summation:

$$t = t_a + t_b + t_c \qquad (2.18)$$

The positive and negative signs in the equation for t_a correspond to
initial velocity components directed, respectively, away from and toward
the detector.   In the ideal case, where all ions begin their flight in a
plane located at a distance a_0 from grid G1 with zero kinetic energy,
$a = a_0$ and $W_0 = 0$.   In this situation the energy W' after acceleration
is given by

$$W' = ea_0E_a + ebE_b \qquad (2.19)$$

Let the constant k_0 be defined as follows:

$$k_0 = \frac{a_0E_a + bE_b}{a_0E_a} \qquad (2.20)$$

Then the flight time t' for ideal ions can be derived from Eq. (2.15) to
Eq. (2.20) and written in the form

$$t' = \left(\frac{m}{2W'}\right)^{\frac{1}{2}} \left(2k_0^{\frac{1}{2}}a_0 + \frac{2k_0^{\frac{1}{2}}b}{k_0^{\frac{1}{2}} + 1} + c\right) \qquad (2.21)$$

Ions originating with zero kinetic energy in planes displaced slightly
from the ideal one will focus with the ideal ions at the detector if $dt/da = 0$

at $a = a_0$. This happens when

$$c = 2k_0^{3/2}a_0 \left(1 - \frac{1}{k_0 + k_0^{1/2}} \frac{b}{a_0} \right) \tag{2.22}$$

Equation (2.22) specifies a maximum, a minimum, or an inflectional point for t versus a, according to the values of the instrumental parameters. For an inflectional point a further requirement is that $d^2t/da^2 = 0$ at $a = a_0$. This happens when

$$\frac{b}{a_0} = \frac{k_0 - 3}{k_0} \frac{c}{2a_0} \tag{2.23}$$

If b/a_0 is smaller than the value set by Eq. (2.23), t reaches a maximum at $a = a_0$, and if b/a_0 is larger, t reaches a minimum. In practical designs a maximum is usual. Once the dimensions a, b, and c are fixed, the ratio E_b/E_a is uniquely determined by Eq. (2.22). This focus condition is independent of the ionic mass and the kinetic energy after acceleration. The time separation of adjacent atomic masses M and $M + 1$ can be obtained from Eq. (2.21):

$$t_{M+1} - t_M = \left[\left(1 + \frac{1}{M} \right)^{1/2} - 1 \right] t_M \simeq \frac{t_M}{2M} \tag{2.24}$$

The spectrometer completely resolves adjacent masses if the spread of flight times for ions of a single mass does not exceed $(t_{M+1} - t_M)$. If M_a is the maximum mass for which adjacent-mass resolution is achieved for focused ions originating with zero kinetic energy over a range $\Delta a/2$ on each side of a_0, it can be shown that

$$M_a \simeq 16k_0 \left(\frac{a_0}{\Delta a} \right)^2 \tag{2.25}$$

Ions originating in the ideal plane but with some kinetic energy are considered next. If two ions start their flights with axial velocities of equal magnitude but opposite direction, one will move immediately toward the detector while the other will move away, slow down to zero velocity, and then follow its partner. The second ion will pass the plane of origin with the same velocity as the first, but will be delayed by the turnaround time. This delay will persist all the way to the detector. It follows that the spread of flight times Δt_w attributable to initial kinetic energy from zero to W_0 is equal to the turnaround time for an ion of energy W_0, and so

$$\Delta t_W = \frac{2(2mW_0)^{1/2}}{eE_a} \tag{2.26}$$

The spread of ions of a single atomic mass must not exceed $(t_{M+1} - t_M)$ if adjacent masses are to be resolved. If M_W is the maximum mass for which adjacent-mass resolution is achieved for ions originating at $a = a_0$ with maximum kinetic energy W_0, and if the focus condition of Eq. (2.22) is retained, it can be shown that

$$M_W = \frac{1}{4}\left(\frac{W'}{W_0}\right)^{\frac{1}{2}}\left(\frac{k_0+1}{k_0^{\frac{1}{2}}} - \frac{k_0^{\frac{1}{2}}-1}{k_0+k_0^{\frac{1}{2}}}\frac{b}{a_0}\right) \qquad (2.27)$$

In an actual spectrometer the ions vary both in initial position and in initial energy, but an analysis of this case is prohibitively difficult. Therefore electrode configurations and applied fields are chosen with the aid of the simplified expressions Eqs. (2.22), (2.25), and (2.27). Several of the requirements, however, are mutually incompatible, and compromises have to be made in order to bring about the best overall performance. In practice, adjacent-mass resolution up to about 200 amu is feasible.

The ion bunches received at the detector of a time-of-flight mass spectrometer may have a duration of only 10 nanosec, so wide-band amplification techniques are necessary. Goodrich and Wiley [68] have devised a magnetic electron multiplier which is especially suitable for this purpose. It uses a strip of semiconducting material for the multiplying surface instead of the usual series of metal dynodes. No surface activation is required, and a gain of 10^7 is obtained with a cathode dark current of only 3×10^{-21} amp, or 0.02 electron per second. The output signal can be displayed on a cathode-ray oscilloscope the time base of which is triggered by the spectrometer pulse generator. The mass spectrum thus obtained can be recorded photographically. Betts, Paufve, and Wiley [69] have described alternative procedures. By use of a gating system, the electron beam from the multiplier is directed toward an anode during the period of each cycle when one particular mass is being received at the detector. During the rest of the time the beam is deflected away. The anode signal is fed to a recorder. By insertion of a circuit of appropriate time constant between the anode and the recorder, the signal is averaged over a number of cycles. In this way the effect of random ion-current fluctuations is reduced. If the delay of the gate pulse is varied continuously, the recorder traces out a mass spectrum. When it is necessary to eliminate drifts in peak heights caused by changes in the sample pressure or in the ionizing electron current, two anodes are gated, one with a fixed delay and one with a variable delay. The first gate pulse is set on a reference mass, and a feedback system controls the gain of the multiplier so that the anode signal is maintained at a constant level. A stabilized mass spectrum is obtained from the second anode by varying the delay of its gate pulse. Even when the average number of ions per mass per cycle is considerably less than 1, measurements are still practicable with the gating technique. Individual ions produce pulses at a multiplier anode if its gate is open, and the pulses are totaled over many cycles with an electronic counter. Ratios of peak heights are determined satisfactorily, even in the presence of drifts, by gating two anodes and feeding the outputs simultaneously to two counters. An attractive feature of the gating method is that peak heights are independent of peak widths provided that the duration of the gate pulse is at least equal to that of the ion bunches received by the detector.

Within its limit of resolution the time-of-flight mass spectrometer is useful for many tasks involving abundance determination. A precision of about 1% can be expected. The instrument is particularly valuable for studying the growth and decay of substances taking part in fast, gas-phase chemical reactions. These include explosions, flash photochemical reactions, and reactions initiated by shock waves. The sampling time corresponds to the duration of the ionizing pulse, a matter of hundreds of nanoseconds, and complete mass spectra are available at intervals of a few tens of microseconds. For this type of measurement it is necessary that a specimen of the reacting mixture be pumped through the ionization region at a high rate. Here the open, gridded structure is a great asset, and pumping speeds exceeding 100 liters/sec can be obtained. During the ionizing period the space traversed by the electrons in the time-of-flight spectrometer is field-free, and this is an advantage when electron-impact phenomena are being investigated by the monoenergetic-electron method. The rates of ion-dissociation processes can be studied by introducing a lag between the ionizing pulse and the accelerating pulse and observing the variations in the mass spectrum as the lag time is changed. However, this technique is applicable only for ion half-lives in the rather narrow range of 5 to 20 μsecs.

Although the ionizing period in the time-of-flight spectrometer covers only a small fraction of a cycle, virtually all the ions which are produced reach the detector. By contrast, in the magnetic-deflection spectrometer, where ionization is continuous, the ion beam is restricted geometrically and, at a given moment, ions of only one mass fall on the detector. The result is that the two types of instruments do not differ substantially in overall sensitivity.

C. Quadrupole mass spectrometers

Paul and coworkers [70–73] have developed a type of mass spectrometer in which mass separation takes place in a quadrupole, r-f electric field. Its mode of operation may be explained by referring to Fig. 2.17, which is an end-on view of the analyzer. A two-dimensional quadrupole field is established between four rod-shaped electrodes of hyperbolic cross section. Opposite pairs of electrodes are electrically connected. The voltage applied to the quadrupole array comprises a constant component U and an r-f component $V_0 \cos \omega t$. The potential ϕ at any point in the field may be written in terms of the rectilinear coordinates shown in the diagram

$$\phi = \frac{(U + V_0 \cos \omega t)(x^2 - y^2)}{r_0^2} \qquad (2.28)$$

where $2r_0 =$ electrode spacing.

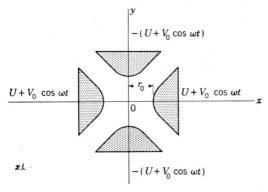

Fig. 2.17. End-on view of a quadrupole mass analyzer.

The motion of an ion injected into the field in the z direction is described by the equations

$$m\ddot{x} + \frac{2e(U + V_0 \cos \omega t)x}{r_0^2} = 0 \qquad (2.29)$$

$$m\ddot{y} - \frac{2e(U + V_0 \cos \omega t)y}{r_0^2} = 0 \qquad (2.30)$$

$$m\ddot{z} = 0 \qquad (2.31)$$

It is now convenient to define the dimensionless parameters A, Q, and ξ:

$$A = \frac{8eU}{mr_0^2\omega^2} \qquad (2.32)$$

$$Q = \frac{4eV_0}{mr_0^2\omega^2} \qquad (2.33)$$

$$\xi = \frac{\omega t}{2} \qquad (2.34)$$

In this problem attention is confined to values of A and Q less than unity. On rewriting the equations of motion for the x and y axes, the following Mathieu equations are obtained:

$$\frac{d^2x}{d\xi^2} + (A + 2Q \cos 2\xi)x = 0 \qquad (2.35)$$

$$\frac{d^2y}{d\xi^2} - (A + 2Q \cos 2\xi)y = 0 \qquad (2.36)$$

The solutions of these equations designate oscillations performed by an ion in the x and y directions, respectively. Meanwhile, the ion proceeds with constant velocity in the z direction. For certain values of A and Q the oscillations are stable, or, in other words, the amplitude remains finite at any time t, but for other values of A and Q the oscillations are unstable and the amplitude increases exponentially to infinity. If an ion is to oscillate stably in both x and y directions, the solutions of the two Mathieu equations must be stable simultaneously. Figure 2.18 indicates the range of values of A and Q for which this is so. The conditions are independent of the initial position and velocity of the ion and of the r-f phase at the time of injection. If the ratio A/Q is kept constant, the representative points for different ionic masses will fall on a straight line passing through the origin of the stability diagram. For masses such as m_1, where the point falls inside the region of stable oscillation, ions continue on an oscillatory path through the quadrupole field, whereas for

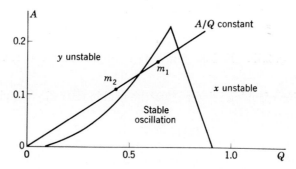

Fig. 2.18. Stability diagram for a quadrupole analyzer.

other masses such as m_2, the ions are eventually lost by lateral deflection. Hence a mass separation is achieved. If A and Q are varied in such a way that the constant A/Q ratio is maintained, the range of masses which continue through the field is also varied. The mass resolution increases as the A/Q ratio is increased and theoretically becomes infinite when the representative line passes through the apex of the stability region, $A = 0.237$, $Q = 0.706$.

The foregoing account has been somewhat idealized. For an actual quadrupole spectrometer, mass separation does not depend on whether ionic oscillation is stable or unstable, but on whether an ion is able to travel through an analyzer of finite length without striking an electrode. An important consequence of this practical requirement is that the mass resolution becomes dependent on the lateral positions and radial velocities of the ions entering the analyzer. Hence the ion beam is usually restricted in cross section and angular divergence by circular apertures placed at the entrance to the analyzer. An ion detector is located at the other end of the analyzer, but no defining aperture is necessary. The mass spectrum is scanned by varying ω or by varying U and V_0, keeping a constant U/V_0 ratio. At low mass resolution most of the ions are transmitted through the analyzer at the appropriate field condition, and trapezoidal spectral peaks are produced. At high resolution, corresponding to operating points near the apex of the stability region, a high proportion of the ions of the selected mass oscillate far enough to strike an electrode so that the transmission is reduced. Also, the peaks are triangular in shape. Köhler et al. [72] have obtained a mass resolution of 8,000 at 85 amu with an instrument having the specifications listed in Table 2.1. The

TABLE 2.1

Beam voltage	24	volts
Diameter of entrance aperture	0.5	mm
Radial energy of ions	0.018	ev
Electrode spacing $(2r_0)$	8	cm
Electrode length	834	cm
Direct voltage (U)	366	volts
Radio-frequency voltage (V_0)	2,180	volts
Frequency $(\omega/2\pi)$	471	kc/sec
Radio-frequency load	140	watts

main experimental difficulty in the construction of a high-resolution quadrupole spectrometer stems from the fact that the electrode rods must be uniform in cross section and parallel over a length of several meters with a tolerance of only a few microns. It is not essential that the cross section be strictly hyperbolic, and a satisfactory field pattern can be produced with properly spaced circular rods.

The high transmission of the quadrupole analyzer at low resolution and its freedom from restrictions on the geometry and energy distribution of the incoming ion beam make it potentially valuable as an isotope separator. Paul's group [71] have exploited this feature. In the interest of providing a large collector current, the resolution is lowered to the extent that ions of several adjacent masses are transmitted. An unwanted isotope can be rejected by a special technique, which may be explained as follows. When ions perform stable oscillations in the quadrupole field, the x component has the same fundamental frequency for all ions of a given mass while the frequency is different for other masses. Therefore a chosen isotopic mass can be rejected by applying a voltage at its fundamental frequency to the x electrodes. The resonant ions then oscillate with increased amplitude in the x direction and strike an electrode. Collector currents of several milliamperes can be obtained when the quadrupole analyzer is operated in this manner.

Brubaker [74] has pointed out that the quadrupole mass spectrometer is very suitable for studying the composition of the atmosphere in the vicinity of an artificial satellite because it is insensitive to the velocities of the incoming molecules or ions. He has shown that quadrupole spectrometers can be designed to accept atmospheric molecules or ions moving with satellite velocities ($\sim$8 km/sec) at angles up to 30° to the axis of the analyzer.

2.4. Magnetodynamic Instruments

A. Omegatrons

An ion traveling perpendicularly to a magnetic field follows a circular orbit, and the orbital angular velocity ω_c in radians per second can be

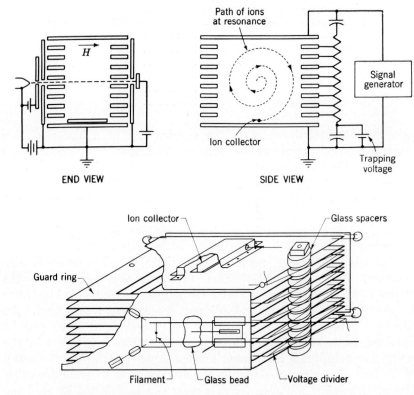

Fig. 2.19. Omegatron. (*From Sommer, Thomas, and Hipple* [75].)

calculated from Eq. (2.2).

$$\omega_c = \frac{v}{r} = \frac{eB}{m} \qquad (2\text{-}37)$$

ω_c, known as the cyclotron velocity, thus depends on m, e, and B, but not v or r. Sommer, Thomas, and Hipple [75] devised an instrument, appropriately named the omegatron, for the determination of ω_c. The instrument is, in effect, a mass spectrometer, and its principal features are shown in Fig. 2.19. An ionizing electron beam of circular cross section is directed along a magnetic field. Ions formed in the beam move under the combined influence of the magnetic field and a perpendicular r-f electric field, $E = E_0 \sin \omega t$. The r-f voltage is applied between a pair of planar electrodes, and the field is homogenized by feeding a stack of guard rings from a voltage divider. If $\omega = \omega_c$, the path of an ion initially at rest is a continuously expanding spiral in a plane normal to the direction of the magnetic field. If ω differs from ω_c, the path is a spiral the radius of which alternately increases to a maximum and

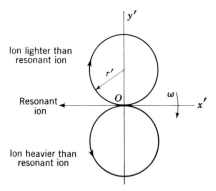

Fig. 2.20. Ion paths in an omegatron. The coordinate system rotates in a clockwise direction with angular velocity ω.

decreases to zero. A resonant ion eventually reaches a collector, as also does a nonresonant ion which achieves a sufficiently large orbit.

A mathematical treatment of the trajectory of an ion in an omegatron has been carried out by Berry [76], and this treatment has been extended by Brubaker and Perkins [77]. These authors refer the position of the ion to a rotating coordinate system. Figure 2.20 illustrates the situation where the rotating x' axis is parallel to the r-f field at $t = 0$ and an ion is formed at the origin at this moment with zero velocity. The coordinate system rotates clockwise with angular velocity ω. Analysis shows that a resonant ion proceeds along the negative x' axis with constant velocity $E_0/2B$. A nonresonant ion describes a circle tangential to the x' axis at the origin with angular velocity $\epsilon = |\omega - \omega_c|$. The circle lies above the axis for an ion which is lighter than a resonant ion and below for one which is heavier. Its radius r' is given by

$$r' = \frac{eE_0\omega}{m(2\omega + \epsilon)(\omega + \epsilon)\epsilon} \qquad (2.38)$$

The maximum orbital radius r_{max} of a nonresonant ion is clearly $2r'$. For conditions near resonance, where $\epsilon \ll \omega$ and $\omega \simeq \omega_c$,

$$r_{max} = 2r' = \frac{eE_0}{m\epsilon\omega_c} \qquad (2.39)$$

Although this relationship has been derived for a special case where an ion is formed at $t = 0$, with zero velocity, it is also valid for a near-resonant ion formed at an arbitrary r-f phase with a thermal-velocity component in the orbital plane. An ion reaches a collector placed at a distance r_0 from the point of formation if r_{max} is equal to or greater than r_0. If ω is varied and ions of a given mass, resonant and nonresonant, are collected from $\omega_c - \epsilon$ to $\omega_c + \epsilon$, the value of ϵ can be deduced by substituting in Eq. (2.39):

$$\epsilon = \frac{eE_0}{mr_0\omega_c} \qquad (2.40)$$

The range of collection $\Delta\omega$ is equal to 2ϵ. If the radius of the electron beam is negligible with respect to r_0, the mass resolution of the omegatron is given by

$$\frac{M}{\Delta M} = \frac{\omega_c}{\Delta\omega} = \frac{Mr_0\omega_c^2}{2NeE_0} = \frac{Ner_0B^2}{2ME_0} \qquad (2.41)$$

It can be seen that resolution is inversely proportional to mass.

A mass spectrum is obtained by sweeping the frequency or the magnetic field and recording the collector current. The center of a mass-spectral line corresponds to the resonance condition. It is noteworthy that the omegatron produces no harmonic peaks, unlike the cyclotron accelerator, to which it bears a formal resemblance. This arises from the fact that an ion in the omegatron is accelerated continuously rather than at discrete intervals.

Since ions have thermal-velocity components in the direction of the magnetic field, axial drifts are superposed on the spiral motions already considered. For nonresonant ions this is beneficial because the ions are intercepted by the guard-ring stack, and thus the axial space charge is reduced. However, to prevent excessive loss of resonant ions, it is usually necessary to have recourse to some method of restricting axial movement. Sommer, Thomas, and Hipple applied a positive trapping voltage to the guard-ring stack. The trapping voltage and the space charge give rise to a radial electric field, and this causes the observed orbital frequency to differ slightly from ω_c. In their experiments Sommer, Thomas, and Hipple made a correction to their data to compensate for this shift. They determined ω_c for the proton, and in the same magnetic field they determined the proton-magnetic-resonance frequency. From these results they computed the magnetic moment of the proton in terms of the nuclear magneton.   The mass resolution at the proton peak was 10^4.

A further effect of a radial electric field in the omegatron is a modification of the path of an ion in the orbital plane.   The guiding center of the spiral is no longer stationary at the point of formation, but moves outward.   By controlling the field, it can be brought about that resonant ions are guided toward the collector while nonresonant ions are lost to some other electrode.   Brubaker and Perkins achieved control of the field by applying direct voltages between the guard rings.   Klopfer and

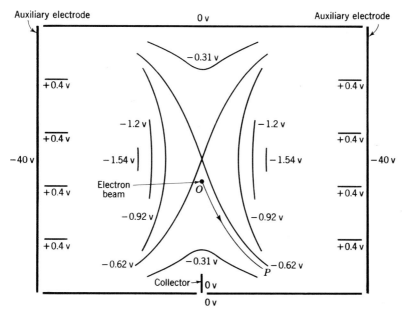

Fig. 2.21. Field distribution in a Klopfer-Schmidt omegatron. The equipotential lines represent the electrostatic field in the median plane normal to the magnetic field. The guiding center of the ion orbit drifts along the equipotential line *OP*. (*From Schuchhardt* [79].)

Schmidt [78] have used an alternative method.   They apply a voltage to a pair of auxiliary electrodes, and the resulting field distribution is indicated in Fig. 2.21.   The equipotential lines represent the electrostatic field in the median plane normal to the magnetic field.   The electron beam is intentionally displaced from the electrical center in the direction of the collector.   Schuchhardt [79] has analyzed the operation of this version of the omegatron and has shown that the guiding center of the ion orbit drifts along an equipotential line (*OP* in Fig. 2.21) at a rate which is independent of ionic mass.   The field is adjusted so that resonant ions spiral out from the moving guiding center to the collector, while other ions, with their limited spiral radii, strike the adjacent electrode. The field can also be modified so that all ions reach the collector, and the omegatron can then be operated as an ionization gauge.

Omegatrons are mainly used in vacuum technology for the analysis of residual gases. They are ideal for this application because they can be made in a compact form and are easily degassed. The Klopfer-Schmidt version has the advantage that it indicates total or partial pressure, according to its mode of operation. Since all resonant ions reach the collector, absolute partial pressures can be calculated from the geometry of the electron beam and the known ionization cross sections of the gases.

B. Other magnetodynamic instruments

The orbital period t_c of an ion circling in a magnetic field is obtained from Eq. (2.37):

$$t_c = \frac{2\pi}{\omega_c} = \frac{2\pi m}{eB} \tag{2.42}$$

Since t_c is proportional to m, the masses of ions moving in the same field can readily be compared by measuring their respective orbital periods. Hays, Richards, and Goudsmit [80] have measured the periods of ion bunches orbiting in a magnetic field by using an electronic timing apparatus. Smith [81–83] has developed another technique. At a certain stage in the orbit an ion beam is velocity-modulated at radio frequency, and the orbital period is measured in terms of the modulation frequency.

REFERENCES

1. J. J. Thomson: *Phil. Mag.*, **13**, 561 (1907).
2. J. J. Thomson: "Rays of Positive Electricity and Their Application to Chemical Analyses," Longmans, Green & Co., Ltd., London, 1913.
3. F. W. Aston: *Phil. Mag.*, **38**, 707 (1919).
4. F. W. Aston: *Phil. Mag.*, **39**, 611 (1920).
5. A. J. Dempster: *Phys. Rev.*, **11**, 316 (1918).
6. F. W. Aston: *Proc. Roy. Soc. (London)*, **A115**, 487 (1927).
7. F. W. Aston: *Proc. Roy. Soc. (London)*, **A163**, 391 (1937).
8. F. W. Aston: "Mass Spectra and Isotopes," 2d ed., Edward Arnold (Publishers) Ltd., London, 1942.
9. J. L. Costa: *Ann. Phys.*, **4**, 425 (1925).
10. K. T. Bainbridge: *J. Franklin Inst.*, **215**, 509 (1933).
11. W. Bleakney: *Phys. Rev.*, **40**, 496 (1932).
12. J. T. Tate, P. T. Smith, and A. L. Vaughan: *Phys. Rev.*, **48**, 525 (1935).
13. A. O. Nier: *Phys. Rev.*, **50**, 1041 (1936).
14. A. O. Nier: *Rev. Sci. Instr.*, **11**, 212 (1940).
15. J. A. Hipple: *J. Appl. Phys.*, **13**, 551 (1942).
16. A. O. Nier: *Rev. Sci. Instr.*, **18**, 398 (1947).
17. R. L. Graham, A. L. Harkness, and H. G. Thode: *J. Sci. Instr.*, **24**, 119 (1947).
18. I. Pelchowitch: *Philips Res. Rept.*, **9**, 1 (1954).
19. H. D. Hagstrum: *Rev. Sci. Instr.*, **24**, 1122 (1953).
20. J. H. Reynolds: *Rev. Sci. Instr.*, **27**, 928 (1956).

21. C. R. McKinney, J. M. McCrea, S. Epstein, H. A. Allen, and H. C. Urey: *Rev. Sci. Instr.*, **21**, 724 (1950).
22. M. G. Inghram and R. J. Hayden: "Mass Spectroscopy," p. 23, National Research Council, Washington, D.C., 1954.
23. F. A. White and T. L. Collins: *Appl. Spectry.*, **8**, 169 (1954).
24. J. Mattauch and R. Herzog: *Z. Physik*, **89**, 786 (1934).
25. J. Mattauch: *Phys. Rev.*, **50**, 617 (1936).
26. A. J. Dempster: *Proc. Am. Phil. Soc.*, **75**, 755 (1935).
27. K. T. Bainbridge and E. B. Jordan: *Phys. Rev.*, **50**, 282 (1936).
28. H. Bondy, G. Johannsen, and K. Popper: *Z. Physik*, **95**, 46 (1935).
29. N. B. Hannay: *Rev. Sci. Instr.*, **25**, 644 (1954).
30. R. D. Craig and G. A. Errock: "Advances in Mass Spectrometry" (J. Waldron, ed.), p. 66, Pergamon Press, New York, 1959.
31. H. G. Voorhies, C. F. Robinson, L. G. Hall, W. M. Brubaker, and C. E. Berry: "Advances in Mass Spectrometry" (J. Waldron, ed.), p. 44, Pergamon Press, New York, 1959.
32. C. F. Robinson, G. D. Perkins, and N. W. Bell: *Fifth Intern. Instr. Meas. Conf. Proc.*, Stockholm, 1960.
33. A. O. Nier: *Rev. Sci. Instr.*, **31**, 1127 (1960).
34. F. A. White, F. M. Rourke, and J. C. Sheffield: *Rev. Sci. Instr.*, **29**, 182 (1958).
35. F. A. White, F. M. Rourke, and J. C. Sheffield: *Appl. Spectry.*, **12**, 46 (1958).
36. F. A. White, F. M. Rourke, J. C. Sheffield, and L. A. Dietz: IRE, *Trans. Nucl. Sci.*, **NS-8**, 13 (1961).
37. W. Bleakney and J. A. Hipple: *Phys. Rev.*, **53**, 521 (1938).
38. T. Mariner and W. Bleakney: *Rev. Sci. Instr.*, **20**, 297 (1949).
39. G. W. Monk and G. K. Werner: *Rev. Sci. Instr.*, **20**, 93 (1949).
40. C. F. Robinson and L. G. Hall: *Rev. Sci. Instr.*, **27**, 504 (1956).
41. C. F. Robinson: *Rev. Sci. Instr.*, **27**, 509, 512 (1956).
42. L. G. Hall, C. K. Hines, and J. E. Slay: "Advances in Mass Spectrometry" (J. Waldron, ed.), p. 266, Pergamon Press, New York, 1959.
43. W. R. Smythe and J. Mattauch: *Phys. Rev.*, **40**, 429 (1932).
44. W. R. Smythe: *Phys. Rev.*, **45**, 299 (1934).
45. H. Hintenberger and J. Mattauch: *Z. Physik*, **106**, 279 (1937).
46. W. H. Bennett: *J. Appl. Phys.*, **21**, 143 (1950).
47. T. C. Wherry and F. W. Karasek: *J. Appl. Phys.*, **26**, 682 (1955).
48. P. A. Redhead: *Can. J. Phys.*, **30**, 1 (1952).
49. P. A. Redhead and C. R. Crowell: *J. Appl. Phys.*, **24**, 331 (1953).
50. J. Dekleva and A. Peterlin: *Rev. Sci. Instr.*, **26**, 399 (1955).
51. W. W. Cannon and M. K. Testerman: *J. Appl. Phys.*, **27**, 1283 (1956).
52. M. Ya. Shcherbakova: *Zh. Tekhn. Fiz.*, **27**, 599 (1957).
53. A. N. Vorsin, E. F. Doil'nitsyn, A. I. Trubetskoi, and M. Ya. Shcherbakova: "Radiochastotnii Mass Spectrometr," Academy of Sciences of the U.S.S.R., Moscow, 1959.
54. E. M. Kuchkov: *Zh. Tekhn. Fiz.*, **30**, 568 (1960).
55. V. A. Pavlenko, A. E. Rafal'son, M. E. Slutskii, G. A. Tsveiman, and M. D. Shutov: *Pribory Tekhn. Eksperim.*, no. 6, p. 89, (1960).
56. R. L. Boyd and D. Morris: *Proc. Phys. Soc. (London)*, **A68**, 1 (1955).
57. W. Donner: *Appl. Spectry.*, **8**, 157 (1954).
58. W. Tretner: *Z. Angew. Phys.*, **11**, 395 (1959).
59. A. E. Cameron and D. F. Eggers: *Rev. Sci. Instr.*, **19**, 605 (1948).
60. R. Keller: *Helv. Phys. Acta*, **22**, 386 (1949).
61. W. E. Glenn: *U. S. At. Energy Comm. Rept.* AECD-3337, 1952.
62. N. I. Ionov and B. A. Mamyrin: *Zh. Tekhn. Fiz.*, **23**, 2101 (1953).

63. M. M. Wolff and W. E. Stephens: *Rev. Sci. Instr.*, **24**, 616 (1953).
64. H. S. Katzenstein and S. S. Friedland: *Rev. Sci. Instr.*, **26**, 324 (1955).
65. W. C. Wiley and I. H. McLaren: *Rev. Sci. Instr.*, **26**, 1150 (1955).
66. E. I. Agishev and N. I. Ionov: *Zh. Tekhn. Fiz.*, **26**, 203 (1956).
67. E. I. Agishev and N. I. Ionov: *Zh. Tekhn. Fiz.*, **28**, 1775 (1958).
68. G. W Goodrich and W. C. Wiley: *Rev. Sci. Instr.*, **32**, 846 (1961).
69. J. F. Betts, E. H. Paufve, and W. C. Wiley: *Appl. Spectry.*, **14**, 119 (1960).
70. W. Paul and M. Raether: *Z. Physik*, **140**, 262 (1955).
71. W. Paul, H. P. Reinhard, and U. Von Zahn: *Z. Physik*, **152**, 143 (1958).
72. R. Köhler, W. Paul, K. Schmidt, and U. Von Zahn: *Proc. Intern. Conf. Nuclidic Masses*, University of Toronto Press, Toronto, 1960, p. 507.
73. F. Von Busch and W. Paul: *Z. Physik*, **164**, 581, 588 (1961).
74. W. M. Brubaker: *Neuvième Colloq. Spectroscopicum Intern.*, Lyon, France, 1961.
75. H. Sommer, H. A. Thomas, and J. A. Hipple: *Phys. Rev.*, **82**, 697 (1951).
76. C. E. Berry: *J. Appl. Phys.*, **25**, 28 (1954).
77. W. M. Brubaker and G. D. Perkins: *Rev. Sci. Instr.*, **27**, 720 (1956).
78. A. Klopfer and W. Schmidt: *Vacuum*, **10**, 363 (1960).
79. G. Schuchhardt: *Vacuum*, **10**, 373 (1960).
80. E. E. Hays, P. I. Richards, and S. A. Goudsmit: *Phys. Rev.*, **84**, 824 (1951).
81. L. G. Smith: *Natl. Bur. Std. (U.S.) Circ.* 522, p. 117, 1953.
82. L. G. Smith and C. C. Damm: *Rev. Sci. Instr.*, **27**, 638 (1956).
83. L. G. Smith: *Proc. Intern. Conf. Nuclidic Masses*, University of Toronto Press, Toronto, 1960, p. 418.

3

Mass Spectrometry in Research

W. A. Bryce

Department of Chemistry
The University of British Columbia
Vancouver, B.C.

3.1. Introduction

Mass-spectral analysis, from its inception by J. J. Thomson in 1913 [1] and with the advances made by Aston, Bainbridge, Dempster, Bleakney, Nier, and others, has proved to be one of the most powerful and versatile research techniques ever developed in the physical sciences. It has been applied, principally as mass spectrometry, to a wide variety of problems in physics, chemistry, and the geological and biological sciences. Some of these applications will have been apparent from the descriptions of mass spectrometers in the preceding chapter. It is the purpose of this chapter to show the diversity of these applications and to indicate in a general way the nature of the results obtained.

3.2. Determination of Isotopic Abundances

The discovery of isotopes and the pioneering work of Aston and Dempster were a direct consequence of the development of the techniques of mass analysis. Their work, extended by others, has led to a complete elucidation of the isotopic distribution of the elements, and hence to a more precise evaluation of atomic weights. Two techniques of mass spectroscopy have developed: the first, *mass spectrography*, involves the simultaneous recording of focused ion beams on a photographic plate and is used primarily for the accurate determination of atomic masses. In the second, *mass spectrometry*, the ion beams are recorded separately by means of an electrometer and related electronic devices.   Until recently mass spectrometers have been more reliable for the measurement of isotopic abundances and have been used extensively for this purpose.

The first mass spectrograph constructed by Aston [2] in 1919 was for the identification of the two isotopes of helium of masses 20 and 22. This was followed quickly by Dempster's mass spectrometer [3], in which the ion beams were recorded electrometrically and by means of which the isotopes of magnesium were discovered.   Interest in this field developed rapidly during the next twenty years, so that the identification of all naturally occurring isotopes was virtually complete by 1939.   Bainbridge [4] gave impetus to the determination of accurate atomic masses by the development of a Dempster-type mass analyzer with a velocity filter.   The resulting linear mass scale greatly facilitated the measurement of mass differences.   One of the chief contributors to the study of isotopic abundance in the past two decades has been Professor A. O. Nier of Minnesota.   A number of new stable isotopes, including those of calcium and sulfur, were reported by him in 1938 [5], and extensive studies of relative isotopic abundances have been made by his research group since that time.   He has developed new techniques, including a double-focusing mass spectrometer in which the ratio of two isotopic ion currents is measured precisely.   Standards such as an Ar^{36}-Ar^{40} doublet of accurately known proportions have been used to correct the results for mass discrimination effects.   The values of isotopic ratios obtained have an absolute precision of approximately 0.1% [6].   A general review of the determination of isotopic abundances was given by Nier in 1955 [7].

One of the more interesting aspects of the study of isotopic abundances has been the variation in natural abundance that occurs because of chemical, geological, and biological processes.   This has been of particular interest in the fields of geochemistry and geophysics and will be considered in more detail later.   Systematic studies of isotopic ratios in minerals, for example, have been of great value in providing reliable information on the ages of geological formations.   In this work the samples being studied must be compared with laboratory standards such as those prepared by the National Bureau of Standards in Washington [8].

3.3. Determination of Atomic Masses

One of the major contributions made by mass spectrography to physics and chemistry has been the accurate evaluation of atomic masses.   The techniques involved and the results obtained have been described by Ewald [9] and Mattauch [10] and are described in detail in Chap. 7.   The usual procedure is to measure accurately the separation of the lines of a mass *doublet*.   The mass of one of the two contributory ions must be known and can therefore serve as a standard. The dispersion of the instrument must also have been determined. The precision obtained in mass measurements by this means is of the order of 1 part in 10^7.

An accurate knowledge of atomic masses is of great importance in nuclear physics in such areas as the evaluation of binding energies and the study of nuclear shell structure and in gaining an understanding of the processes involved in the release of nuclear energy. Of particular interest has been the evaluation of the masses of the so-called "secondary standards" H^1, D^2, and C^{12}, as these can be used in doublet measurements with a variety of nuclides for which a direct comparison with O^{16} is not possible. Duckworth and his coworkers [11] have used a large single-focusing mass spectrometer for the determination of the masses of a number of isotopes of krypton, xenon, and mercury to a precision of 1 part in 2 million. Other developments in precise mass measurement have been discussed in detail by Hintenberger [12] and by Duckworth [20].

3.4. Electromagnetic Separations

The use of large-scale mass spectrometers for the separation of isotopes electromagnetically is illustrated by the work of Keim [13] at Oak Ridge, Kistemaker and Zilverschoon [14] in Holland, Koch [15] at Copenhagen, and the group at Harwell. Developments in this field have been reviewed by Dawton and Smith [16, 17].

The element whose isotopes are to be separated is introduced into a specially designed high-temperature source in the form of a volatile compound, usually a halide. The ion receiver contains a number of collector pockets which catch the separated isotopes, after which they are recovered chemically.

Electromagnetic separation of U^{235} from U^{238} in the years 1942 to 1944 was a major technique in the development of the United States atomic energy project. By 1945 enriched isotopes were being produced in gram quantities, and by 1951 a publication of the United States Atomic Energy Commission showed that most of the 61 polyisotopic elements in the periodic table had been separated and were available for research purposes. These enriched isotopes have been widely used in many areas of research [17]. Information on nuclear transformation, nuclear shell structure, and isotopic cross sections has been obtained by using enriched targets in cyclotrons, van der Graff machines, and other accelerators. The measurement of spins and moments of nuclides and their correlation with other spectroscopic data has been possible. The investigation of natural radioactivity has been greatly facilitated, and the availability of enriched isotopic starting materials has made possible the preparation of a large variety of compounds containing either radioactive or stable tracer elements for metabolic or mechanistic studies in biology and chemistry.

3.5. Nuclear Physics and Chemistry

The application of the mass spectrometer to the physics and chemistry of nuclear processes has been reviewed by Thode [18] (Chap. 10), by Inghram [19], and by Duckworth [20]. Work in these fields has been extended very considerably with the development of more powerful electromagnetic separation methods. The sensitivity of mass spectrometers has also increased markedly with the development of new electronic techniques, so that it is possible, in some cases, to detect as little as 10^{-15} g of sample.

It was mentioned earlier that the rapid growth of the Manhattan Project was greatly facilitated by the mass-spectrometric identification of the uranium nuclide responsible for the fission process [21]. Nier [22] had shown in 1939 that the relative-abundance ratios for the uranium isotopes were $U^{238}/U^{235} = 139\ (\pm 1\%)$ and $U^{238}/U^{234} = 17,000 \pm 10\%$. Subsequent neutron bombardment of microgram quantities of the separated isotopes showed that U^{235} was fissile with slow neutrons, but that U^{238} underwent fast neutron fission only [21, 23]. Values for the natural abundance of uranium isotopes have been reported by Lounsbury [24], and extensive data for the isotopes of lead have been given by Russell and Farquhar [25].

The identification of a considerable number of naturally occurring radioactive nuclides has been made by mass spectrometry. They are illustrated by K^{40} [26], In^{115} [27], and La^{138} [28]. The identification of fission-product nuclides has been one of the major contributions of mass spectrometry to nuclear studies. Of particular interest in work done at the close of World War II were the studies on inert gases. Although they are produced in very small amounts—10^{-3} ml at STP from uranium samples of approximately 30 g—accurate analyses were obtained following preliminary separation. The results [29, 30] showed that the isotopic distributions in fission-product krypton and xenon, for example, are markedly different from those of the normal elements. The analysis of these distributions has provided valuable information on fission yields and fission chains. Cumulative fission-yield studies have been made on a variety of fissile nuclides. The studies include the relative yields of stable and long-lived isotopes of Cs, Rb, and Sr from U^{235} [31], Xe, Cs, and Kr from U^{233} [32], and Xe and Kr from neutron fission of U^{235} and U^{238} [33]. The results obtained have been important in relating the asymmetry of fission to the neutron configuration.

Accurate studies of the branching ratios of certain unstable nuclei have been done mass-spectrometrically. An odd-odd (proton-neutron) nuclide such as Cu^{64} is unstable with respect to the isobaric pair Ni^{64}-Zn^{64} and will tend to decay to these nuclides by β^+ and β^- emission, respectively. The product analysis, as determined by Reynolds [34], led to a value of 1.62 ± 0.11. Similar results for decay processes, in which two β particles are lost simultaneously with a change in nuclear charge of 2, have also been obtained (Ref. 20, p. 128).

A direct determination of the half-life of a radioactive nuclide can be obtained by a mass-spectrometric study of its rate of decay or of the rate of formation of a product nucleus. Nier [35] determined the half-life of U^{235} to be 7.13×10^8 years by measuring the growth of Pb^{207} in a uranium mineral. A seven-year study of the disappearance of Kr^{84} has yielded a value of 10.27 ± 0.18 years for its half-life [36]. For long-lived isotopes the daughter-growth method is more accurate than the parent-decay method, as a large change in a smaller amount of isotope is easier to measure than a small change in a large amount.

Neutron-capture cross sections have been determined by mass-spectrometric studies of the effect of neutron irradiation on the isotopic distribution of a target element. The integrated neutron flux must be known accurately.

3.6. Geochemistry and Cosmochemistry

The isotopic distribution of elements in nature can be altered significantly by a variety of geological, biological, and chemical processes. These variations can be studied accurately by mass spectrometry, and valuable information can be obtained concerning the conditions under which the processes occurred. The technique has proved to be of particular value in geological studies since it has led to reliable methods being developed for the measurement of geological time.

The principal age-determining methods involve the analysis of radiogenic lead [25]. Three of the four stable isotopes of lead, 206, 207, and 208, are end products of the decay of U^{238}, U^{235}, and Th^{232}, respectively. From a knowledge of the half-lives of these latter three nuclides and the observed abundance ratios, the age of the uranium-containing rocks can readily be calculated. The maximum time of formation of the elements has been set at 5.5×10^9 years [37], and the time of formation of the earth's crust, at 3.5×10^9 years [37, 38].

The potassium isotope K^{40}, which has a natural abundance of 0.16%, decays to Ar^{40}. From a knowledge of the decay constants and the K^{40}/Ar^{40} ratio, the age of the potassium-containing mineral can be obtained. The widespread occurrence of potassium in minerals makes this relatively new method very useful. Pliopleistocene intrusive rocks from Sacramento, Calif., have been shown by this method to have an age of approximately 1.6×10^6 years [39]. The same method gave the age of a particular sample of microcline as 1.03×10^9 years [40]. The age of the earth's atmosphere has been calculated from its Ar^{40} content to be 3.3×10^9 years [41], a value very similar to that obtained for the earth's crust.

Age determinations have also been made based on the Sr^{87} content of minerals. The Sr^{87} is formed radiogenically from Rb^{87} (Ref. 20, p. 160). Herzog and Pinson [42] have computed the age of certain chondritic meteorites to be approximately 4×10^9 years by this method, i.e., roughly the age of the earth.

Mass spectrometry has been applied to a variety of problems in cosmochemistry by Urey and his colleagues. Analysis of Pb^{206}/Pb^{204} ratios in meteorites has shown the latter to have been formed from 4×10^9 to 4.6×10^9 years ago, assuming that no contamination by other lead has occurred during that time. The age of certain meteors based on Ar^{40} extraction has been placed at 4.3×10^9 years [43]. The main age-determining methods thus provide results which are in remarkably good agreement.

Isotopic fractionation of structurally similar molecules containing different isotopes of a particular element may occur because of the effect of mass differences on the moment of inertia and vibrational modes, and hence on the zero-point energy of the molecule. The differences in zero-point energy may cause fractionation in equilibrium-exchange processes in which these molecules are involved [44, 45]. The fractionation would be expected to be most extensive for the isotopes of hydrogen because of the large mass difference. The H/D ratio in hydrogen compounds has been studied extensively by Friedman [46], who has observed variations ranging from 10.6% less than the normal value of 6,700 to 5.6% above the normal value. From the observed values the temperature at which a particular H-D exchange process occurred can be estimated.

Variations over a range of 4.5% from the normal value of 89.3 have been observed in the C^{12}/C^{13} ratio in nature [47]. Studies of O^{16}/O^{18} variation have been used in an attempt to gain information about the origin of the earth's atmosphere and of the temperatures existing at the time of deposition of marine carbonates [48].

The variation in the isotopic abundance of sulfur from the normal value of 22.7 for the S^{32}/S^{34} ratio has been studied extensively by Thode and his coworkers. Variations in S^{33} and S^{34} abundance as much as 4 and 8%, respectively, have been reported [49]. Sulfates in general are enriched in S^{34}, and sulfides are depleted. Sulfur in minerals of igneous origin have intermediate S^{34} contents, and organic sulfur generally is deficient in this isotope. Meteoritic sulfur has values close to the overall terrestrial average. Macnamara and Thode [50] have suggested that meteoritic and terrestrial sulfur were initially identical but that isotopic fractionation has occurred in the latter as a consequence of biological and geological processes. Certain sulfate-reducing bacteria produce H_2S, which is deficient in S^{34} [51]. Free sulfur produced by bacterial action at the bottom of certain lakes contains as much as 3.2% less S^{34} than the sulfate from which it is formed [52]. From the S^{34} content of sedimentary sulfides and sulfates of different ages it has been calculated that the first significant participation of sulfur bacteria in terrestrial processes began about 800 million years ago [53].

3.7. Analysis by Mass Spectrometry

The use of mass spectrometry in analysis was first suggested by J. J. Thomson in his book "Rays of Positive Electricity and Their Application to Chemical Analysis," published in 1913 [1]. The technique has developed extensively during the past twenty years in parallel with the improved performance of electronic components and the availability of commercial instruments (see Chap. 9).

Mass spectrometry has been applied most extensively to the analysis of hydrocarbons [54–59]. The method is based on the fact that, when a polyatomic molecule is bombarded in the ionization chamber by electrons of sufficiently high energy, all possible dissociation processes occur. The fragments formed are all converted into positive ions and are analyzed by the magnetic field. The relative intensities of the peaks in the mass spectrum thus produced are characteristic of the particular molecule and can be used for identification purposes.

The analysis of a multicomponent mixture requires a knowledge of the mass spectrum of each of the components, since the spectrum of the mixture is the sum of the individual spectra. The analysis is based also on the fact that when gaseous flow from the sample reservoir through the leak into the ionization chamber is molecular, the partial pressure of each component in the ion source is proportional to the partial pressure of that component in the reservoir. Hence the peak intensities for each component are directly proportional to the partial pressure of that component in the reservoir. From a knowledge of the sensitivity, i.e., peak height per micron of reservoir pressure, of the particular instrument for each component, it is possible to compute the composition of a multi-component mixture from the overall mass spectrum. The sensitivity of the method is indicated by the fact that a complete analysis of a 10-20 component mixture can be done with 1 ml of gas at STP. The method is also capable of distinguishing between isomeric hydrocarbons.

The identification of organic molecules from their mass spectra has been greatly facilitated by the publication periodically by the American Petroleum Institute and the National Bureau of Standards of a Catalogue of Mass Spectral Data. The spectra of over 2,000 compounds are listed.

The success achieved in the analysis of light hydrocarbons has led to the development of techniques applicable to high-boiling hydrocarbons and waxes.   O'Neal [60] has done a detailed study of the mass spectra of polyaromatic hydrocarbons with parent masses up to 600.   Analysis of such high-boiling compounds has been greatly facilitated through the use of heated inlet systems [59].

The interpretation of the mass spectra of mixtures of compounds is greatly simplified if extensive fragmentation is avoided by the use of low-energy electrons.   This technique is particularly useful in the analysis of saturated hydrocarbons [61, 62].   It has also been applied with considerable success to the analysis of high-molecular-weight substances [63, 64], including aromatic concentrates from refinery operation and organic reaction products, as well as substances relating to the geochemistry and origin of petroleum itself [65].

Mass spectrometry has been particularly useful in the qualitative identification of compounds and in the determination of their structures [66–69].   The development of high-resolution mass spectrometers with resolving powers of the order of one part in many thousands of mass units will assist greatly in these identifications [70, 71].   It is now possible to obtain complete resolution of multiplets such as C_3H_7O, $C^{13}C_2H_6O$, $C_2H_3O_2$ (mass 57) and C_2H_4, CH_2N, N_2, CO (mass 28).

Complex organic substances, including polymeric materials, can often be identified by the mass-spectrometric analysis of products obtained from them in thermal decomposition [72].   This technique has been applied successfully to the identification of polystyrene, polyisoprene, polybutadiene, polyethylene [73, 74], organosilicons [75], acrylates [76], and Teflon [77].

The analysis of isotopically labeled molecules by mass spectrometry has been of great assistance in the elucidation of reaction mechanisms. The analyses are very straightforward with simple molecules such as H_2O^{18}, $C^{13}O_2$, HD, etc.   The spectra of a variety of deuterated hydrocarbons of interest in pyrolytic and combustion studies have been reported, but unfortunately there is still a shortage of data in this area, because of the unavailability of pure standard compounds.

Other isotopes commonly used in tracing out chemical and biological processes include C^{13}, N^{15}, and O^{18}.   These estimations are usually done with an isotope-ratio mass spectrometer, in which, for example, the $N^{15}H_3/N^{14}H_3$ ratio in the sample is compared directly with the normal value under conditions for which instrumental variations cancel out.

Tracer elements present in compounds in very low abundance can be determined accurately by the technique of isotope dilution [78–80].   The sample to be analyzed is dissolved in a suitable solvent, and a known quantity of a second isotope of the element to be determined is added and allowed to reach equilibrium in the solution.   The element is then extracted chemically in a form suitable for isotopic abundance measurements.   The quantity of the element originally present in the material to be analyzed can be computed from the isotope ratio in the final sample without a quantitative recovery in the extraction step. The sensitivity of the method is such that uranium in concentrations as low as 4 parts in 10^{12} can be determined [81].

3.8. Solid-state Studies

The techniques necessary for the study of inorganic solids by mass spectrometry are more difficult than those for gases and liquids because of the extremely low volatility of most solid materials. The usual procedure [82, 83] involves heating the solid in a block-metal furnace or tungsten crucible, located in the ion source of the spectrometer, to a temperature sufficiently high to volatilize some of the substance. The technique of depositing the sample to be analyzed directly on the filament has also been used [84]. Palmer [82] has described a high-sensitivity source in which the material under investigation is painted as a solution or slurry on each of several tungsten or tantalum filaments. When the filaments are heated at a low temperature, the material evaporates on to a central high-temperature filament, from which it reevaporates in the form of positive ions. Spark sources are possible in some instances, but the large energy spread in the ions produced plus the difficulty in measuring the ion-beam intensity has made this technique rather difficult. It has been applied successfully, however, to the determination of impurities in low-alloy steels, magnesium, aluminum, copper, graphite, and silicon [85].

The analysis of stainless steels for nickel and copper has been done [89] using a steady spark source with a high-frequency discharge between the rod-shaped sample and a tantalum disk. Ion currents of the order of 10^{-11} amp were recorded, and excellent agreement with chemical analysis was obtained. Carbon, oxygen, and sulfur have been determined in a vacuum fusion apparatus in which a mass spectrometer replaced the normal vacuum train [87]. The method is capable of extension to concentrations as low as 0.001 %. Semiquantitative results for trace amounts of arsenic, phosphorus, antimony, lead, iron, zinc, and cadmium in germanium samples of interest in transistor studies have been obtained by the direct evaporation of samples from a furnace into the ion source [88]. The heat of vaporization of aluminum and the first-order-phase transition temperatures for aluminum, praesodymium, and neodymium have also been determined [89].

A study of the equilibrium vaporization of copper, silver, and gold from carbon and alumina crucibles gave the diatomic-monatomic ratios for these metals as a function of temperature [90]. From these results the dissociation energies of Cu_2, Ag_2, and Au_2 have been calculated. Studies have also been made of the sublimation of silver [91, 92].

A great deal of interest has been centered on the determination of the correct value of the heat of sublimation of graphite because of its direct relationship to the determination of bond-dissociation energies in organic compounds. The values reported in the literature [93], based principally on the analysis of band spectra, range from 125 to 170 kcal. Direct determinations of the heat of sublimation of carbon have been made in recent years by mass-spectrometric methods. Chupka and Inghram [94] volatilized graphite from a tantalum Knudsen cell, heating the cell by electron bombardment. Only electrically neutral particles were allowed to enter the ionization chamber. The absolute pressure of each vapor component was obtained, and from the observed variation in carbon-atom pressure with temperature, a value of 170.4 kcal was calculated for the heat of sublimation. The ionic species detected in the vapor from graphite include C_1^+, C_2^+, C_3^+, C_4^+, and C_5^+ [203].

Honig [95] evaporated graphite from small carbon filaments in the ion source and found that, at 2600°K, the neutral particles were C_3, C_2, and C_1; the graphite was found to emit negative ions containing from 1 to 8 atoms, with C_2^-, C_4^-, and C_3^- the most abundant in that order. The heats of sublimation of C_1 and C_3 were both found to be 178 ± 10 kcal/mole at 2400°K. The value for C_2 was 199 ± 20 kcal/mole. Dissociation energies for the C_2 and C_3 molecules were estimated.

The value of 170.9 kcal/mole, which now appears to be well established for the heat of sublimation of graphite, has necessitated a recalculation of a great many bond-dissociation energies which had been based on the previously accepted lower value of around 135 kcal/mole. A detailed discussion of the evidence favoring the higher value has been presented by Cottrell [96].

The mass spectrum of sulfur vapor has been determined by evaporating sulfur from a heated tube directly into the ionization chamber of a mass spectrometer [97]. Ions of the type S_x^+, with x ranging from 1 to 8, were observed, S_2^+ being the most abundant. Similar studies have been done with phosphorus and arsenic [98]. The tetramer appears to be the most abundant vapor form of these elements, with As_6^+ and As_8^+ having an abundance 10^{-4} that of As_4^+. The heats of activation of sublimation were found to agree with earlier values based on weight-loss studies.

A number of interesting studies have been made on the mass spectra of metallic oxides. Aldrich [99] determined the evaporation products of BaO from heated ribbon filaments of various metals. The mass spectra show the formation of Ba^+, BaO^+, Ba_2O^+, Ba_2O^{++}, and $Ba_2O_2^+$ at high temperatures. A measurable amount of Sr was found from SrO evaporated from platinum. A study of BaO vapor from a Knudsen cell [100] in the temperature range 1500 to 1800°K led to the identification of the same species plus $Ba_2O_3^+$, with BaO^+ the most abundant. In the Al-Al_2O_3 system [101], Al_2O^+ was the most abundant particle. From the appearance potential of Al_2O^+ from Al_2O_3, the heat of dissociation for the process

$$Al_2O(g) = 2Al(g) + O(g)$$

was found to be 256 ± 7 kcal/mole at $298°K$.

In the $B-B_2O_3$ system, B^+, BO^+, $B_2O_2^+$, and $B_2O_3^+$ were identified [102]. A mixture of B and B_2O_3 was found to evaporate mainly as B_2O_2, the heat of evaporation of B_2O_2 from the condensed phase being 94 ± 8 kcal at $1400°K$. From the concentrations of $BO(g)$ in the gaseous phase, the dissociation energy of BO was estimated to be in excess of 100 kcal/mole.

Recent studies [103, 104] have shown that WO_3 evaporates chiefly as W_3O_9, W_4O_{12}, and W_5O_{15}, and MoO_3 as Mo_3O_9, Mo_4O_{12}, and Mo_5O_{15}. Values of ΔH, $\Delta F°$, and $\Delta S°$ for these processes have been estimated. The $Zr-ZrO_2$ system has also been studied [105], and values obtained for $D(Zr-O)$ and $D(Zr-O_2)$.

The dimer $K_2(OH)_2$ has been identified as the principal form in which KOH vaporizes [106] in the temperature range 300 to $450°C$. The heat of vaporization was calculated to be 36 ± 2 kcal/mole of dimer, and the entropy of dimerization of $KOH(g)$ at $353°C$ estimated to be greater than 38 kcal/mole of dimer. Dimerization also occurs in NaCN vapor [106].

Goldfinger has reported [107] the extension of studies conducted in his laboratory on the evaporation of Group III–V and II–IV compounds to include InAs, GaSb, and ZnS. The energies of dissociation of the tetramer to dimer of the Group V elements have been found, and a number of interesting properties of sulfur and selenium vapors have been investigated. These topics are further discussed in Chap. 12.

3.9. Application to Chemical Kinetics

The mass spectrometer has proven to be an extraordinarily useful tool for the kineticist because of its high sensitivity and resolving power (Chap. 11). One of the first kinetic studies in which mass-spectrometric analysis was used was the decomposition of dimethyl ether by Leifer and Urey in 1942 [108]. The reaction mixture was sampled continuously by a molecular leak leading to the ionization chamber, and the partial pressures of the products were measured during the course of the reaction. The existence of formaldehyde as an intermediate was confirmed.

A very large number of investigations have been reported in which detailed analysis of the stable components of gaseous reaction mixtures has been made possible by mass spectrometry. These are illustrated by work on pyrolysis and photolysis [110, 111] of deuterated and undeuterated acetaldehyde and the pyrolysis of specifically labeled hydrocarbons such as 1-butene-4-d_3 [112], deuterated butanes [113], and cyclopentanes [114]. Reactions of deuterium atoms with alkenes have also been investigated [115, 116] with the identifications of the various deuteroisomers formed.   Evidence for the occurrence of free radicals in such systems was provided by the extensive deuterium mixing observed.

The mechanisms of a wide variety of reactions in solution have also been studied by the use of stable isotopes.   These studies are illustrated by the use of H_2O^{18} in working out the mechanism of hydrolysis of lactones [117] and of esters [118].   The rate of exchange of O^{18} with solvent molecules has been of importance in solution reactions involving O^{18}-labeled species [119].   One such reactant used in oxidations studied in solution has been the MnO_4^{18} ion [120, 121].   It has been possible to determine directly the extent to which the oxygen introduced into the oxidized substance comes from the oxidizing agent.   The applications here are fairly straightforward, involving only the measurement of isotopic abundance ratios to determine the distribution of the tracer element among the reaction products.

The study of free radicals by mass spectrometry was first carried out by Eltenton [122, 123] in a flow system.   Methyl, methylene, ethyl, and allyl radicals and hydrogen atoms were detected in the pyrolysis of hydrocarbons and metal alkyls.   Eltenton made use of the fact that the ionization potentials of radicals are lower than the appearance potentials of the corresponding ions from the parent molecules and selected values of electron energy at which the ion currents were due only to radicals produced thermally.   In this way the growth and decay of radical concentrations could be studied as a function of time.   He investigated the kinetics of the reaction of methyl with hydrocarbons over a range of temperature and also studied the radical intermediates formed during the combustion of methane, propane, and carbon monoxide in low-pressure flames.   He was the first to obtain direct evidence of the existence of the HO_2 radical, long postulated as a key intermediate in combustion processes.

The radical studies have been pursued in a number of laboratories in recent years.   Lossing and his colleagues in Ottawa have made extensive contributions to the field [124–126].   In much of their work a flow reactor mounted above the ionization chamber has been used with a quartz-thimble leak of approximately 20μ thickness.   A movable furnace capable of temperatures up to 1200°C has been used in such studies as the formation and recombination of methyl radicals, the pyrolysis of ethylene oxide, propylene oxide, dimethyl ether, and dioxane.   Radicals produced in the pyrolysis of a variety of benzene derivatives have also been studied in Lossing's group, principally with a view to determining ionization potentials, from which bond-dissociation energies for the formation of these radicals can be calculated.

It was mentioned earlier that the HO_2 radical has for many years been of great interest as a postulated intermediate in the oxidation of organic compounds.   It has been studied by Robertson [127] in the reaction of H atoms with molecular oxygen, by Foner and Hudson in the $H + O_2$ reaction [128] and reactions involving H_2O_2 [129], and by Ingold and Bryce [130] in the hydrogen-oxygen reaction and in the reaction of methyl radical with oxygen.   It has also been studied in the methane-oxygen reaction at low pressures [131].

Other studies on radical intermediates include the detection of free OH in the decomposition of water vapor [132] and in the hydrogen-oxygen reaction [130].   Active nitrogen, formed when an electrodeless discharge is passed through N_2 gas under certain conditions, has been studied mass-spectrometrically.   Jackson and Schiff [133] suggest that excited nitrogen atoms may be the active species.   Significant contributions to the understanding of this interesting system have been made by Kistiakowsky and his coworkers [134].   An important application of mass spectrometry to kinetics has been the development of a rapid-scanning mass spectrometer for the study of combustion processes [135, 136]. The reaction mixture is sampled through a quartz leak, and the mass spectrum of the stable species is recorded continuously on an oscilloscope at a scanning rate of the order of 40 units per 5 msec.   A clearer understanding of the mechanism of oxidation of acetaldehyde has been obtained by this means.   A study [137] of reactions induced in ketene and nitrogen dioxide by flash photolysis has been made, using a Bendix time-of-flight mass spectrometer which scans a complete mass spectrum every 50 μsec, with a resolving power of 1 in 40.   The sensitivity of the instrument is such that a component of the reaction mixture present at a concentration of 0.5% can be observed in a single spectrum with reasonable certainty. The energy of the flash was 500 joules at 10,000 volts.   Evidence was found for the formation of a short-lived intermediate, C_3H_4O, which disappeared from the reaction cell after several hundred microseconds.

Photoionization sources have been used in several studies [138, 139, 204] and appear to be very promising for certain types of work because of the comparative simplicity of the spectra obtained.

An interesting new development in the field of combustion has involved the magnetic analysis of ions formed in flames.   Knewstubb and Sugden [140, 141] directed a flame at atmospheric pressure against a platinum disk with a leak 50 μ in diameter and 50 μ deep leading into the magnetic analyzer through a high-speed pumping system.   The ions were detected by a multiplier with a sensitivity such that concentrations of this order of 10^7 ions per cubic centimeter could be detected over a mass range from 1 to 400.   Acetylene, ethylene, ethane, methane, and acetone flames were studied.   Similar work, but with low-pressure flames and a much larger entrance slit to the analyzer, has been reported by Deckers and Van Tiggelen [142, 143].

A variety of equilibrium and exchange processes have been investigated with the assistance of mass-spectral analysis. They include the study of the rates of H-D exchange in the isomerization and hydrogenation of normal butanes [144], isomerization of propane-C^{13} [145], N^{15}-N^{14} exchange in nitrogen and H_2-D_2 exchange on iron catalysts [146], H-D exchange in ethylene on nickel [147], CH_4-D_2 and NH_3-D_2 exchange on nickel [148], the H_2-T_2 equilibrium [149], exchange and hydrogenation reactions of C_2H_2 and C_2H_4 [150], H-D and O^{16}-O^{18} exchange in water [151, 119], and isotopic exchange between NH_3 and D_2 on a variety of metal films [152] and between deuterium and hydrocarbons on Ni-Si catalysts [153]. A great many exchange processes have also been studied in solution. Analyses are achieved by converting the isotopically labeled species to stable gases like N_2 or CO_2.

Valuable information on the reactions of radicals on surfaces can be obtained by mass spectrometry. LeGoff [154, 155] has studied the rates of formation and decomposition of radicals such as CH_3 on tungsten, platinum, and nickel filaments located in the ion source. He has also studied the effect of the carbon content of tungsten filaments on the rate of radical decomposition [156]. The work has been extended to include a study of the formation of CS, SO, and CCl_2 in heterogeneous reactions [157]. The dehydrogenation of ethane at low pressures on incandescent platinum [158] and the sensitivity of propyl radicals to surface collisions [159] have also been investigated. Other mass-spectrometric studies of heterogeneous reaction include the formation of CO on carbon filaments [160] and the evolution of oxygen during the electron bombardment of alkaline-earth oxides [161].

It is well known that the replacement of a light isotope like H^1 or C^{12} by the corresponding heavier isotope H^2 or C^{13} can cause a significant difference in the reactivity of the molecule concerned. This isotope effect can be attributed largely to the lowering of the zero-point energy of the molecule by the introduction of the heavier atom [45]. A great many examples of this phenomenon have been studied with the aid of mass analysis. They are illustrated by study of the thermal and electron-impact-induced dissociation of propane-1-C^{13} [145], the decomposition of C^{13}-labeled oxalic acid [162, 163], the O^{18}-O^{16} effect in the decomposition of H_2O_2 [164], and the N^{15} effect in the deammonation of phthalamide [165]. The deuterium isotope effect has been reviewed by Wiberg [166]. It has proved to be of great value in establishing the mechanisms of reactions in which the rupture of bonds to specific hydrogen atoms is rate-determining.

A theory of isotope effects proposed by Bigeleisen [45] and others has been tested by a number of studies, including a detailed survey of the mechanism of decomposition of C^{13}- and C^{14}-labeled malonic acids. An extended controversy concerning the correct values for the relative rates of the various decomposition modes was finally resolved when radioactive-counting techniques were replaced by the more definitive mass-spectrometric analysis [167, 168].

3.10. Ionization and Dissociation

The dissociation energy of a bond R—X is defined as the heat of the reaction

$$RX(g) = R(g) + X(g)$$

at 0°K. If R—X is bombarded with electrons of gradually increasing energy until the following process occurs,

$$RX + e^- = R^+ + X + 2e^-$$

the appearance potential of the R^+ ion, $V(R^+)$, is given by the energy equation

$$V(R^+) = D(R—X) + I(R) + \text{K.E.} + E_{ex}$$

where $D(R—X)$ is the dissociation energy of the R—X bond, $I(R)$ is the ionization potential of the radical or atom R, K.E. is the kinetic energy of the particles produced, and E_{ex} is the excitation energy (electronic, vibrational, and rotational) of the fragments. For many processes the kinetic-energy term is less than 0.5 ev and the particles will be found in their ground states. Hence the relationship reduces to

$$D(R—X) = V(R^+) - I(R)$$

$V(R^+)$ can be determined readily by electron bombardment of R—X in a mass spectrometer. $I(R)$ can be obtained for a great many radicals by producing them pyrolitically in a flow reactor and measuring the ionization potentials of the radicals themselves.

A great many dissociation energies have been obtained by means of this technique, principally by Stevenson, Lossing, Hagstrum, and McDowell (see Chap. 12). Its importance in providing reliable values has been emphasized by Cottrell [96]. The use of this method in studies on the solid state has been referred to earlier.

Some work by Foner and Hudson [129] illustrates the application of this method. They determined the appearance potential of the much discussed HO_2 radical by the electron bombardment of H_2O_2. They were then able to produce a sufficiently high concentration of HO_2 radicals through the reaction of OH and H_2O_2 to permit them to determine the ionization potential of HO_2 itself. Knowing the heat of formation of H_2O_2, they calculated the value of 47.1 kcal for the endothermicity of the process

$$HO_2 = H + O_2$$

Other studies of electron-impact phenomena have included the determination of ionization cross sections of atoms and molecules [169, 170] and the study of meta-stable ions [171, 172]. These are ions which are sufficiently stable to be accelerated out of the ion source but which decompose during their passage through the mass-spectrometer tube. New charged fragments thus formed appear, usually at nonintegral mass numbers.

The study of negative ions in the gas phase has attracted a good deal of attention in recent years. These ions may be formed by electron-resonance capture or by the production of an ion pair. Such ions are formed efficiently by substances such as O_2, CCl_4, SF_6, HCl, and the interhalogen compounds. The most profitable work appears to be that done with a source providing an essentially monoenergetic beam of electrons. A detailed study has recently been reported on negative-ion potential and electric breakdown in a variety of halogenated gases [173]. It is found that the relative production of negative ions in the particular gas can be correlated with dielectric strength.

Studies on negative-ion formation by Frost and McDowell [174] show that it is possible to obtain a more accurate approximation to the true shape of a resonance-capture peak by using the Fox method for producing essentially monoenergetic electrons. The application of this technique also enables the occurrence of various ion-pair dissociation processes to be identified. It has been particularly successful in the study of the ionization and dissociation of the halogens and the halogen halides.

A further area of mass spectrometry which has been of interest in recent years is the study of gas-phase ion-molecule reactions illustrated by the processes

$$CH_4^+ + CH_4 = C_2H_5^+ + CH_3$$
$$X^+ + H_2 = HX^+ + H$$

Such reactions (Chap. 13) may be quite extensive in ion sources under certain conditions and are therefore of importance in considerations of the origin of mass spectra. Rate constants for some of these processes have been evaluated, as well as charge-transfer cross sections [175, 176].

Another type of reaction of interest in mass spectrometry is electron-induced rearrangement, which can readily occur in the ion source of the instrument with certain types of molecules. For example, rearrangement involving skeletal atoms has been observed prior to dissociation of the parent ion in the mass spectrum of neopentane terminally and centrally labeled with C^{13} [177, 178] and has been studied in detail by Rylander and Myerson for a series of benzene derivatives [179–181]. One of the interesting deductions drawn from this work is that the dissociation of the alkyl benzenes involves a rearrangement to produce not the benzyl ion, but the isomeric seven-membered tropyllium ion. Extensive intramolecular hydrogen-deuterium exchange has been observed in a variety of deuterated unsaturated hydrocarbons under electron impact [182, 183]. The exchange is less marked with deuterated paraffins [184, 185], presumably because of the more local zed character of the bonding in the parent molecule ion.

An exact understanding of the way in which polyatomic ions fragment is difficult to obtain, but substantial advances have been made in this field, notably in theories in which the dissociation probability is approached from a statistical point of view [186]. The impact of the ionizing electron is assumed to create the ion in a highly excited state, with the excitation energy tending to be distributed randomly among its various internal degrees of freedom. The parent ion then undergoes unimolecular decomposition into various initial product ions, each of which may decompose further. The rate constants for these various decompositions are determined by the energy-level densities of the corresponding transition states.

An extensive study of the interpretation of mass spectra of organic molecules in terms of modern theories of physical-organic chemistry has been made by McLafferty [187], who has shown that it is sometimes possible to predict the effect a given substance will have on the mass spectrum of a particular molecule from a knowledge of its effect in more conventional organic reactions.

Ionization and dissociation processes induced by electron impact have been reviewed by Field and Franklin [175] and are discussed in detail in Chap. 12.

3.11. Leak Detection

Because of its extreme specificity and sensitivity, the mass spectrometer can be adapted very readily to serve as a leak detector for metal vacuum systems [188–190]. A number of such instruments are available commercially. They are usually prefocused to record helium, which is then sprayed over the area suspected to contain a leak. The helium entering the leak is pumped into the ion source of the detector, where it can be observed at concentrations as low as 1 part in 200,000. A number of holes can be detected simultaneously without stopping to repair them, thus effecting a considerable saving in time.

3.12. Miscellaneous Applications

A problem of great importance to which mass-spectrometric analyses have been applied is the study of atmospheric pollution [191, 192]. The method requires very small samples and is both sensitive and rapid. It has been applied successfully to the estimation of a wide variety of organic materials [193] and to NO_2, NO, CO_2, CO, etc.   Analysis of the upper atmosphere has been used in connection with weather forecasting [194]. The analysis of engine exhaust gases has been done in combustion studies [195].   Clinical applications are illustrated by the study of respiratory metabolism [196] and the analysis of the gases dissolved in blood [197].

The activation of barium oxide cathodes has been followed by monitoring the evolution of CO and CO_2 [198].   The diffusion of helium through glass has also been investigated in a study of glass porosity [199].

Accurate measurements of the vapor pressures of low-boiling organic compounds have been made using a specially designed low-temperature thermostat in which the partial pressure of the compound concerned could be measured by mass spectrometry over a wide range of temperature [200].   The method has the distinct advantage over other methods in that the results obtained are not affected by the presence of impurities. The vapor pressures of substantial numbers of compounds were measured accurately at temperatures down to the boiling point of liquid nitrogen.

Gas chromatography has developed in recent years as a powerful tool for the analysis of gaseous samples.   The successive elution of the separated components from the chromatographic column is commonly detected by means of a thermal-conductivity cell.   A mass spectrometer can be used directly as a monitoring device, the chromatogram being displayed and recorded oscillographically [201].   The usefulness of gas chromatography as an analytical tool has been greatly enhanced by combining it with mass-spectrometric identification of the separated compounds.   This involves trapping the sample components as they emerge from the chromatographic column and transferring the trapped component to the mass spectrometer [202].   The technique has proved to be of great value in elucidating the mechanism of gas-phase reactions by the use of labeled molecules [112, 114].   The time-of-flight mass spectrometer, with its extremely high scanning rate, can be used directly as a detector for the identification and quantitative estimation of substances eluted from gas chromatographic columns [205, 206].

The mass spectrometer has found very important application in recent years in the determination of the molecular weights and structure of complex organic compounds [187, 207].   It has proved to be of great value in the identification of a large number of compounds encountered in the chemistry of natural products [208–210].   The use of high-resolution techniques [71] in the accurate identification of substituent groups and other molecular fragments is of great significance for the analysis of organic compounds.

The study, by mass spectrometry, of the ions produced in glow discharges [211] has provided a substantial amount of new information on the processes occurring in such systems.

The variety of applications of mass spectrometry in research referred to in this chapter will serve to illustrate the extreme versatility of this powerful research tool. Detailed discussions of the main areas of research interest in mass spectrometry are to be found in the chapters that follow.

REFERENCES

1. J. J. Thomson: "Rays of Positive Electricity and Their Application to Chemical Analysis," Longmans, Green & Co., Ltd., London, 1913.
2. J. W. Aston: *Phil. Mag.*, **38**, 709 (1919).
3. A. J. Dempster: *Science*, Dec. 10, 1920.
4. K. T. Bainbridge: *J. Franklin Inst.*, **215**, 509 (1933).
5. A. O. Nier: *J. Am. Chem. Soc.*, **60**, 1571 (1938).
6. A. O. Nier: *Phys. Rev.*, **79**, 450 (1950).
7. A. O. Nier: *Science*, **121**, 737 (1955).
8. F. L. Mohler: *Science*, **122**, 334 (1955).
9. H. Ewald: Mass Spectroscopy in Physics Research, *Natl. Bur. Std. (U.S.) Circ.* 522, p. 37, 1953.
10. J. Mattauch: *ibid.*, p. 1.
11. J. T. Kerr, G. R. Bainbridge, J. W. Dewdney, and H. E. Duckworth: *Advan. Mass Spectrometry Proc.*, 1959.
12. H. Hintenberger: "Nuclear Masses and Their Determination," Pergamon Press, London, 1957.
13. C. P. Keim: *J. Appl. Phys.*, **24**, 1255 (1953).
14. J. Kistemaker and C. J. Zilverschoon: Mass Spectroscopy in Physics Research, *Natl. Bur. Std. (U.S.) Circ.* 522, p. 179, 1953.
15. J. Koch: *ibid.*, p. 165.
16. R. H. V. M. Dawton and M. L. Smith: *Quart. Rev. (London)*, **9**, 1 (1955).
17. M. L. Smith: "Electromagnetically Enriched Isotopes and Mass Spectrometry," Butterworth & Co. (Publishers), Ltd., London, 1956.
18. H. G. Thode: *Nucleonics*, **3**, 14 (1948).
19. M. G. Inghram: Mass Spectroscopy in Physics Research, *Natl. Bur. Std. (U.S.) Circ.* 522, p. 151, 1953.
20. H. E. Duckworth: "Mass Spectroscopy," Cambridge University Press, London, 1958.
21. A. O. Nier, E. T. Booth, J. R. Dunning, and A. V. Grosse: *Phys. Rev.*, **57**, 546, 748 (1940).
22. A. O. Nier: *Phys. Rev.*, **55**, 150 (1939).
23. K. H. Kingdon, H. C. Pollock, E. T. Booth, and J. R. Dunning: *Phys. Rev.*, **57**, 749 (1940).
24. M. Lounsbury: *Can. J. Chem.*, **34**, 354 (1956).
25. R. D. Russell and R. M. Farquhar: "Lead Isotopes in Geology," Interscience Publishers, Inc., New York, 1960.
26. W. R. Smythe and A. Hemmendinger: *Phys. Rev.*, **51**, 178 (1937).
27. M. A. Martell and W. F. Libby: *Phys. Rev.*, **80**, 977 (1950).
28. M. G. Inghram, R. J. Haydon, and D. C. Hess: *Phys. Rev.*, **72**, 349, 967 (1947).
29. H. G. Thode and R. L. Graham: *Can. J. Res.*, **A25**, 1 (1947).
30. J. Macnamara and H. G. Thode: *Phys. Rev.*, **80**, 471 (1950).
31. D. R. Wiles, B. W. Smith, R. Horsley, and H. G. Thode: *Can. J. Phys.*, **31**, 419 (1953).
32. W. Fleming, R. H. Tomlinson, and H. G. Thode: *Can. J. Phys.*, **32**, 522 (1954).
33. R. K. Wanless and H. G. Thode: *Can. J. Phys.*, **33**, 541 (1955).

34. J. H. Reynolds: *Phys. Rev.*, **79**, 789 (1950).
35. A. O. Nier: *Phys. Rev.*, **55**, 153 (1939).
36. R. K. Wanless and H. G. Thode: *Can. J. Phys.*, **31**, 517 (1953).
37. C. B. Collins, R. D. Russell, and R. M. Farquhar: *Can. J. Phys.*, **31**, 402 (1953).
38. C. Patterson, G. Tilton, and M. G. Inghram: *Science*, **121**, 69 (1955).
39. G. H. Curtis, J. Lipson, and J. F. Eversden: *Nature*, **178**, 1360 (1956).
40. A. K. Monsef: *Phys. Rev.*, **88**, 150 (1952).
41. K. F. Chackett: *Phys. Rev.*, **81**, 1057 (1951).
42. L. F. Herzog and W. H. Pinson, Jr.: *Am. J. Sci.*, **254**, 555 (1956).
43. H. C. Urey: *ASTM Comm. E-14 on Mass Spectrometry, Paper* 41, Cincinnati, 1956.
44. H. C. Urey and L. J. Grieff: *J. Am. Chem. Soc.*, **57**, 321 (1935).
45. J. Bigeleisen: *J. Chem. Phys.*, **17**, 675 (1949).
46. L. Friedman: *Geochim. Cosmochim. Acta*, **4**, 89 (1953).
47. H. Craig: *ibid.*, **3**, 53 (1953).
48. C. R. McKinney, J. M. McCrea, S. Epstein, H. A. Allen, and H. C. Urey: *Rev. Sci. Instr.*, **21**, 724 (1950).
49. Reference 20, p. 171.
50. J. Macnamara and H. G. Thode: *Phys. Rev.*, **78**, 307 (1950).
51. H. G. Thode, H. Kleerekoper, and D. McElcheran: *Research*, **4**, 581 (1951).
52. J. Macnamara and H. G. Thode: *Research*, **4**, 582 (1951).
53. H. G. Thode, J. Macnamara, and W. H. Fleming: *Geochim. Cosmochim. Acta*, **3**, 235 (1953).
54. H. W. Washburn: Physical Methods in Chemical Analysis, vol. I, p. 587, Academic Press Inc., New York, 1950.
55. H. Powell and G. N. Ross: "Applied Mass Spectrometry," p. 6, The Institute of Petroleum, London, 1953.
56. G. P. Barnard: *Analyst*, **79**, 594 (1954).
57. J. D. Waldron: *Research*, **9**, 306 (1956).
58. V. H. Dibeler, "Organic Analysis," vol. 3, Interscience Publishers, Inc., New York, 1956.
59. I. W. McLafferty, *Anal. Chem.*, **28**, 306 (1956).
60. M. J. O'Neal: "Applied Mass Spectrometry," p. 27, The Institute of Petroleum, London, 1954.
61. A. W. Tickner, W. A. Bryce, and F. P. Lossing: *J. Am. Chem. Soc.*, **73**, 5001 (1951).
62. F. H. Field and S. H. Hastings: *Anal. Chem.*, **28**, 1248 (1956).
63. H. E. Lumpkin: *Anal. Chem.*, **30**, 321 (1958).
64. A. G. Sharkey, Jr., C. F. Robinson, and R. A. Friedel: "Advances in Mass Spectrometry" (J. Waldron, ed.), p. 193, Pergamon Press, New York, 1959.
65. A. Hood and M. J. O'Neal: "Advances in Mass Spectrometry" (J. Waldron, ed.), p. 175, Pergamon Press, New York, 1959.
66. J. D. Morrison: *Roy. Australian Inst. J. and Proc.*, **17**, 339 (1950).
67. S. M. Rock: *Anal. Chem.*, **23**, 261 (1951).
68. J. M. McCrea: *Anal. Chem.*, **25**, 526 (1953).
69. G. H. Lane, H. S. Katzenstein, and S. S. Friedland: *Phys. Rev.*, **93**, 363 (1954).
70. R. D. Craig and G. A. Errock: "Advances in Mass Spectrometry" (J. Waldron, ed.), p. 66, Pergamon Press, New York, 1959.
71. J. H. Benyon: "Mass Spectrometry and Its Applications to Organic Chemistry," Elsevier Publishing Company, Amsterdam, 1960.
72. P. D. Zemany: *Anal. Chem.*, **24**, 1709 (1952).
73. S. L. Madorsky, S. Strauss, D. Thompson, and L. Williamson: *J. Res. Natl. Bur. Std.*, **42**, 499 (1949).
74. B. G. Achhammer, M. J. Reincy, L. A. Wall, and F. W. Reinhart: *J. Polymer Sci.*, **8**, 555 (1952).

75. C. A. Burkhard and F. J. Norton: *Anal. Chem.*, **21**, 304 (1949).
76. S. Strauss and S. L. Madorsky: *J. Res. Natl. Bur. Std.*, **50**, 165 (1953).
77. J. R. Sites and R. Baldock: *Phys. Rev.*, **87**, 171 (1952).
78. A. V. Grosse, S. G. Ikindin, and A. D. Kirschenbaum: *J. Am. Chem. Soc.*, **68**, 2119 (1946).
79. D. C. Hess, H. Brown, M. G. Inghram, C. C. Patterson, and G. R. Tilton: Mass Spectroscopy in Physics Research, *Natl. Bur. Std. (U.S.) Circ.* 522, p. 183, 1953.
80. R. F. Glascock: "Isotopic Gas Analysis for Biochemists," Academic Press Inc., New York, 1954.
81. R. K. Webster: "Advances in Mass Spectrometry" (J. Waldron, ed.), p. 103, Pergamon Press, New York, 1959.
82. G. H. Palmer: "Advances in Mass Spectrometry" (J. Waldron, ed.), p. 89, Pergamon Press, New York, 1959.
83. M. G. Inghram: Mass Spectroscopy in Physics Research, *Natl. Bur. Std. Circ.* 522, 1953.
84. L. T. Aldrich: *J. Appl. Phys.*, **22**, 1168 (1951).
85. R. D. Craig, G. A. Errock, and J. D. Waldron: "Advances in Mass Spectrometry" (J. Waldron, ed.), p. 136, Pergamon Press, New York, 1959.
86. J. G. Gorman, E. J. Jones, and J. A. Hipple: *Anal. Chem.*, **23**, 438 (1951).
87. W. M. Hickam: *Anal. Chem.*, **24**, 362 (1952).
88. R. E. Honig: *Anal. Chem.*, **25**, 523 (1953).
89. R. G. Johnson, D. E. Hudson, W. C. Caldwell, F. H. Spedding, and W. R. Savage: *J. Chem. Phys.*, **25**, 917 (1956).
90. J. Dowart and R. E. Honig: *J. Chem. Phys.*, **25**, 581 (1956); **61**, 980 (1959).
91. Y. V. Kernev and E. Z. Vintaikin: *Zh. Fiz. Khim.*, **30**, 1540 (1956).
92. W. Weiershausen: "Advances in Mass Spectrometry" (J. Waldron, ed.), p. 120, Pergamon Press, New York, 1959.
93. E. W. R. Steacie: "Atomic and Free Radical Reactions," 2d ed., Reinhold Publishing Corporation, New York, 1954.
94. W. A. Chupka and M. G. Inghram: *J. Chem. Phys.*, **22**, 1472 (1954); **59**, 100 (1955).
95. R. E. Honig: *J. Chem. Phys.*, **21**, 573 (1955); **22**, 127 (1954).
96. T. L. Cottrell: "The Strengths of Chemical Bonds," 2d ed., Butterworth & Co. (Publishers), Ltd., London, 1958.
97. P. Bradt, F. L. Mohler, and V. H. Dibeler: *J. Res. Natl. Bur. Std.*, **57**, 223 (1956).
98. J. S. Kanes and J. A. Reynolds: *J. Chem. Phys.*, **25**, 342 (1956).
99. L. T. Aldrich: *J. Appl. Phys.*, **22**, 1168 (1951).
100. M. G. Inghram, W. A. Chupka, and R. F. Porter: *J. Chem. Phys.*, **23**, 2159 (1955).
101. R. F. Porter, W. A. Chupka, and M. G. Inghram: *J. Chem. Phys.*, **23**, 339 (1955).
102. M. G. Inghram, R. F. Porter, and W. A. Chupka: *J. Chem. Phys.*, **25**, 498 (1956).
103. J. Berkowitz, W. A. Chupka, and M. G. Inghram: *J. Chem. Phys.*, **27**, 85 (1957).
104. J. Berkowitz, M. G. Inghram, and W. A. Chupka: *J. Chem. Phys.*, **26**, 842 (1957).
105. W. A. Chupka, J. Berkowitz, and M. G. Inghram: *J. Chem. Phys.*, **26**, 1207 (1957).
106. R. F. Porter and R. C. Schoonmaker: *J. Phys. Chem.*, **62**, 234 (1958); **31**, 830 (1959); **35**, 318 (1961).
107. P. Goldfinger and M. Jeunehomme: "Advances in Mass Spectrometry" (J. Waldron, ed.), p. 534, Pergamon Press, New York, 1959.
108. E. Leifer and H. C. Urey: *J. Am. Chem. Soc.*, **64**, 994 (1942).

109. N. D. Coggeshall and N. F. Kerr: *J. Chem. Phys.*, **17**, 1016 (1949).

110. P. D. Zemany and M. Burton: *J. Am. Chem. Soc.*, **13**, 499 (1951).

111. P. D. Zemany and M. Burton: *J. Phys. & Colloid Chem.*, **55**, 949 (1951).

112. W. A. Bryce and P. Kebarle: *Trans. Faraday Soc.*, **54**, 1660 (1958).

113. J. R. Nesby, C. M. Drew, and A. S. Gordon: *J. Chem. Phys.*, **24**, 1260 (1956).

114. J. R. Nesby and A. S. Gordon: *J. Am. Chem. Soc.*, **79**, 4593 (1957).

115. R. E. Mardaleishvili et al., *Otd. Khim. Nauk*, 516 (1956).

116. S. Toby and H. I. Schiff: *Can. J. Chem.*, **34**, 1061 (1956).

117. A. R. Olson and J. L. Hyde: *J. Am. Chem. Soc.*, **63**, 2459 (1941).

118. M. L. Bender: *J. Am. Chem. Soc.*, **73**, 1626 (1951).

119. J. P. Hunt and J. Taube: *J. Chem. Phys.*, **18**, 757 (1950).

120. K. W. Wiberg and R. Stewart: *J. Am. Chem. Soc.*, **77**, 4719 (1955).

121. J. W. Ladbury and C. F. Cullis: *Chem. Rev.*, **58**, 403 (1958).

122. G. C. Eltenton: *J. Chem. Phys.*, **10**, 403 (1942).

123. G. C. Eltenton: *J. Chem. Phys.*, **15**, 455 (1947).

124. F. P. Lossing and A. W. Tickner: *J. Chem. Phys.*, **20**, 907 (1952).

125. K. U. Ingold and F. P. Lossing: *J. Chem. Phys.*, **21**, 368 (1953).

126. F. P. Lossing, D. G. H. Marsden, and J. B. Farmer: *Can. J. Chem.*, **34**, 701 (1956).

127. A. J. B. Robertson: *Trans. Faraday Soc.*, **48**, 228 (1952).

128. S. N. Foner and R. L. Hudson: *J. Chem. Phys.*, **21**, 1374 (1953).

129. S. N. Foner and R. L. Hudson: *J. Chem. Phys.*, **23**, 1364 (1955); **36**, 2681 (1962).

130. K. U. Ingold and W. A. Bryce: *J. Chem. Phys.*, **24**, 360 (1956).

131. D. J. Fabian and W. A. Bryce: *Seventh Symp. Combustion*, 1959, p. 150.

132. T. Tsuchiya: *J. Chem. Phys.*, **22**, 1784 (1954).

133. D. S. Jackson and H. I. Schiff: *J. Chem. Phys.*, **23**, 2333 (1955).

134. G. B. Kistiakowsky and G. G. Volpi: *J. Chem. Phys.*, **27**, 1141 (1957).

135. E. G. Leger and C. Quellet: *J. Chem. Phys.*, **21**, 1310 (1953).

136. L. P. Blanchard, J. B. Farmer, and C. Quellet: *Can. J. Chem.*, **35**, 115 (1957).

137. G. B. Kistiakowsky and P. H. Kydd: *J. Am. Chem. Soc.*, **79**, 4825 (1957).

138. F. P. Lossing and I. Tanaka: *J. Chem. Phys.*, **25**, 1031 (1956).

139. R. F. Herzog and F. F. Marmo: *J. Chem. Phys.*, **27**, 1202 (1957).

140. P. F. Knewstubb and T. M. Sugden: *Nature*, **181**, 474 (1958).

141. P. F. Knewstubb and T. M. Sugden: *Seventh Symp. Combustion*, 1959, p. 247; *Proc. Roy. Soc. (London)*, **A255**, 520 (1960).

142. J. Deckers and A. Van Tiggelen: *Bull. Soc. Chim. Belges.*, **66**, 664 (1957).

143. J. Deckers and A. Van Tiggelen: *Eighth Symp. Combustion*, 1959, p. 254.

144. V. H. Dibeler and T. I. Taylor: *J. Chem. Phys.*, **16**, 1008 (1948).

145. O. Beeck, J. W. Otvos, D. P. Stevenson, and C. D. Wagner: *J. Chem. Phys.*, **16**, 255, 745, 993 (1948).

146. P. H. Emmett and J. T. Kummer: *J. Chem. Phys.*, **47**, 67 (1950).

147. J. Turkevich, F. Bonner, D. Schissler, and P. Irsa: *J. Phys. & Colloid Chem.*, **55**, 1078 (1951).

148. C. Kemball; *Proc. Roy. Soc. (London)*, **A207**, 539 (1951).

149. H. C. Mattraw, C. F. Pachuski, and L. M. Dorfmann: *J. Chem. Phys.*, **20**, 926 (1952).

150. J. E. Douglas and B. S. Rabinovitch: *J. Am. Chem. Soc.*, **74**, 2486 (1952).

151. K. Hannerz: *Acta Chem. Scand.*, **10**, 655 (1956).

152. C. Kemball: *Proc. Roy. Soc. (London)*, **A214**, 413 (1952).

153. R. L. Burwell and R. H. Tuxworth: *J. Phys. Chem.*, **60**, 1043 (1956).

154. P. LeGoff: *J. Chem. Phys.*, **50**, 423 (1953).

155. P. LeGoff and M. Letort: *J. Chim. Phys.*, **53**, 480 (1956).

156. P. LeGoff and M. Letort: *Compt. Rend.*, **239**, 970 (1954).

157. L. P. Blanchard and P. LeGoff: *Can. J. Chem.*, **34**, 233 (1956).
158. D. J. Fabian and A. J. B. Robertson: *Proc. Roy. Soc. (London)*, **A237**, 1 (1956).
159. D. J. Fabian and A. J. B. Robertson: *Trans. Faraday Soc.*, **53**, 363 (1957).
160. G. F. Crable and N. F Kerr: *Anal. Chem.*, **29**, 1281 (1957).
161. P. Wargo and W. G. Shepherd: *Phy., Rev.*, **106**, 694 (1957).
162. J. G. Lindsay, D. E. McElcheran, and H. G. Thode: *J. Chem. Phys.*, **17**, 589 (1949).
163. A. Fry and M. Calvin: *J. Phys. Chem.*, **56**, 897 (1952).
164. A. E. Cahill and H. Taube: *J. Am. Chem. Soc.*, **74**, 2442 (1952).
165. F. Stacey, J. G. Lindsay, and A. N. Bourns: *Can. J. Chem.*, **30**, 135 (1952).
166. K. B. Wiberg: *Chem. Rev.*, **55**, 713 (1955).
167. P. E. Yankwich and M. Calvin: *J. Chem. Phys.*, **17**, 109 (1949).
168. P. E. Yankwich, A. L. Promislow, and R. F. Nystrom: *J. Am. Chem. Soc.*, **76**, 5893 (1954).
169. G. P. Barnard: "Modern Mass Spectrometry," The Institute of Physics, London, 1953.
170. J. W. Otvos and D. P. Stevenson: *J. Am. Chem. Soc.*, **78**, 546 (1956).
171. J. A. Hipple, R. E. Fox, and E. U. Condon: *Phys. Rev.*, **69**, 347 (1946).
172. H. D. Hagstrum: *Phys. Rev.*, **104**, 309 (1956).
173. W. M. Hickam and D. Berg: "Advances in Mass Spectrometry" (J. Waldron, ed.), p. 458, Pergamon Press, New York, 1959.
174. D. C. Frost and C. A. McDowell: "Advances in Mass Spectrometry" (J. Waldron, ed.), p. 413, Pergamon Press, New York, 1959; *J. Chem. Phys.*, **29**, 503 (1958); *Can. J. Chem.*, **38**, 407 (1960).
175. F. H. Field and J. L. Franklin: "Electron Impact Phenomena," p. 217, Academic Press Inc., New York, 1958.
176. J. L. Franklin, F. H. Field, and F. W. Lampe: "Advances in Mass Spectrometry" (J. Waldron, ed.), p. 308, Pergamon Press, New York, 1959; *J. Am. Chem. Soc.*, **79**, 2419, 2665, 6132 (1957); **80**, 5583, 5587 (1958); **83**, 3555, 4509 (1961).
177. A. Langer and C. P. Johnson: *J. Phys. Chem.*, **61**, 891 (1957).
178. C. P. Johnson and A. Langer: *J. Phys. Chem.*, **61**, 1010 (1957).
179. P. N. Rylander and S. Meyerson: *J. Am. Chem. Soc.*, **78**, 5799 (1956); **79**, 1058 (1959); **81**, 2606 (1959); **83**, 1401 (1961).
180. P. N. Rylander, S. Meyerson, and H. M. Grubb: *J. Am. Chem. Soc.*, **79**, 842 (1957).
181. S. Meyerson and P. N. Rylander: *J. Phys. Chem.*, **68**, 2 (1958).
182. W. A. Bryce and P. Kebarle: *Can. J. Chem.*, **34**, 1949 (1956).
183. W. A. Bryce and E. W. C. Clarke: to be published.
184. D. P. Stevenson and C. D. Wagner: *J. Chem. Phys.*, **19**, 11 (1951).
185. W. H. McFadden and A. L. Wahrhaftig: *J. Am. Chem. Soc.*, **78**, 1572 (1956).
186. H. M. Rosenstock, M. Wallenstein, A. L. Wahraftig, and H. Eyring: *Proc. Natl. Acad. Sci. U.S.*, **28**, 667 (1952); *J. Chem. Phys.*, **23**, 2200 (1955).
187. F. W. McLafferty: "Advances in Mass Spectrometry" (J. Waldron, ed.), p. 355, Pergamon Press, New York, 1959.
188. K. P. Lanneau: "Applied Mass Spectrometry," p. 197, The Institute of Petroleum, London, 1954.
189. D. E. Charpentier: *ASTM Comm. E-14 on Mass Spectrometry, Paper 2*, Cincinnati, 1956.
190. P. F. Varadi and L. G. Sebestyen: *J. Sci. Instr.*, **33**, 392 (1956).
191. F. W. McLafferty, G. E. Clock, and R. S. Gohlke: *Anal. Chem.*, **25**, 526 (1953).
192. R. A. Friedel: *Anal. Chem.*, **28**, 1806 (1956).
193. G. P. Happ, D. W. Stewart, and H. F. Brockmyre: *Anal. Chem.*, **20**, 1224 (1950).

194. J. T. Gorman, E. J. Jones, and J. A. Hipple: *Instruments*, **24,** 294 (1951).
195. H. Landsberg and S. A. Dawkins: *Instr. Soc. Am. Proc.*, 9 (1954).
196. A. W. Pratt, B. E. Burr, M. Eden, and E. Lorens: *Rev. Sci. Instr.*, **22,** 694 (1951).
197. C. S. Jones, J. M. Saari, R. A. Devloo, A. Faulconer, and E. J. Baldes: *Anesthesiology*, **14,** 490 (1953).
198. P. M. Stier: *Phys. Rev.*, **83,** 877 (1951).
199. F. J. Norton: *J. Am. Ceram. Soc.*, **36,** 90 (1953).
200. A. W. Tickner and F. P. Lossing: *J. Phys. & Colloid Chem.*, **55,** 733 (1951).
201. J. C. Holmes and F. A. Morrell: *Appl. Spectry.*, **11,** 86 (1957).
202. C. M. Drew, J. R. McNesby, A. S. Gordon, and S. R. Smith: *Anal. Chem.*, **28,** 979 (1956).
203. J. Drowart, R. P. Burns, D. de Maria, and M. G. Inghram: *J. Chem. Phys.*, **31,** 1131 (1959).
204. H. Hurzler, M. G. Inghram, and J. D. Morrison: *J. Chem. Phys.*, **28,** 76 (1958).
205. R. S. Gohlke: *Ann. Chem.*, **31,** 535 (1959); **34,** 1332 (1962).
206. P. F. Varadi: *Proc. Instr. Soc. Am.*, 133 (1961).
207. J. H. Beynon: "Mass Spectrometry and Its Application to Organic Chemistry," Elsevier Publishing Company, Amsterdam, 1960.
208. K. Biemann: *Angew. Chem.*, **74,** 102 (1962).
209. K. Biemann: "Mass Spectrometry: Organic Chemical Applications," McGraw-Hill Book Company, Inc., New York, 1962.
210. R. I. Reed: *J. Chem. Soc.*, 3432 (1958).
211. P. F. Knewstubb and A. W. Tickner: *J. Chem. Phys.*, **36,** 674 (1962); **36,** 684 (1962); **37,** 2941 (1962).

4

Ion Sources

R. M. Elliott

Associated Electrical Industries Limited
Instrumentation Division
Trafford Park, Manchester, England

4.1. Introduction

Positive and negative ions may be formed in a number of different ways, and a variety of ion sources for mass spectrometry have been developed. No one of these ion sources is ideally suited to all applications, and the choice of source for a particular problem depends on the nature of the sample and the type of information desired. However, it is instructive to consider the features which an ideal source should possess.

1. The ion beam should be of sufficient intensity for accurate measurement of the separated beams; a total ion-beam intensity of 10^{-10} amp is desirable. The limit of detection with electronic amplifiers is about 10^{-15} amp, and with an electron multiplier, 10^{-19} amp.

2. The ion beam should be stable. In some cases, the nature of the ionization process may make this impossible. For example, the ion beam from a vacuum-spark source is both intermittent and erratic.

3. The energy spread of the ion beam should be as low as possible if the highest resolving power is to be obtained in a single-focusing instrument. If the energy spread is large, a double-focusing analyzer system may be necessary and usually only a narrow range of energies can be used.

4. The source should be designed to produce the minimum intensity of background ions. This may mean provision for outgassing the source at high temperatures, a high pumping speed in the source region, and the use of pure materials, etc.

5. There should be no cross contamination or "memory effects" between successive samples.

6. Mass discrimination should be negligible; i.e., the probability that an ion formed in the ionization region enters the analyzers should be independent of its mass.

The material used in the construction of the source should, of course, be nonmagnetic; resistance to corrosion, low catalytic activity, and little tendency to adsorb gases are also desirable. The most commonly used metals for source construction are Nichrome V, Monel, and stainless steel. Machining and assembly to fine limits and a high degree of surface finish are essential requirements for satisfactory and dependable performance.

In the majority of ion sources both positive and negative ions are formed, but for analytical purposes positive ions are almost always used, since they are produced in much greater numbers. Interest in negative ions [1] has been centered chiefly on their role in electrical discharges and in upper-atmosphere phenomena and on the ionization process itself. Accordingly, the various ion sources described below are considered with reference to positive ions, unless negative ions are specifically mentioned.

4.2. Electron-impact Sources

The electron-impact source, in which molecules in the gas phase are ionized by collision with energetic electrons, is the one most widely used in mass spectrometry. It was first used by Dempster, but was developed to its present form mainly by Bleakney and Nier. As this method produces ion beams with excellent stability, low energy spread, and intensities of the order of 10^{-9} amp, it has been extensively used for isotope abundance measurements, atomic-mass measurements, chemical analysis, and studies of the ionization and dissociation of molecules, and it is the type of source usually employed in mass-spectrometer leak detectors. It has the disadvantage that residual gases are ionized along with the sample, and the consequent background spectrum can often be troublesome.

A. Ionization and dissociation by electron impact

The numbers and types of positive ions formed on collision between electrons and molecules depend on the electron energy. At energies just above the ionization potential of the molecule, little fragmentation occurs, but at higher energies ions of all the possible fragments of the molecule are produced. The ionization efficiency curves for a number of positive ions from acetylene [2] are shown in Fig. 4.1. The shape of these curves and the extent of the differences between them are typical of all gases. The magnitude of the maximum ionization cross section varies considerably for different gases, being about 9×10^{-17} cm^2 for hydrogen and 6×10^{-16} cm^2 for C_3—C_5 hydrocarbons at 0°C. Since the maximum of the curve usually occurs between 50 and 90 volts, an electron accelerating voltage of about 70 is generally chosen for analytical work. For the study of molecular structure the portion of the curves just above the threshold is the region of greatest interest.

Negative ions [1] may be formed simultaneously in some of the processes leading to positive ions, by a mechanism such as $XY + e \rightarrow X^+ + Y^- + e$, or they may be formed by an electron-capture process of the

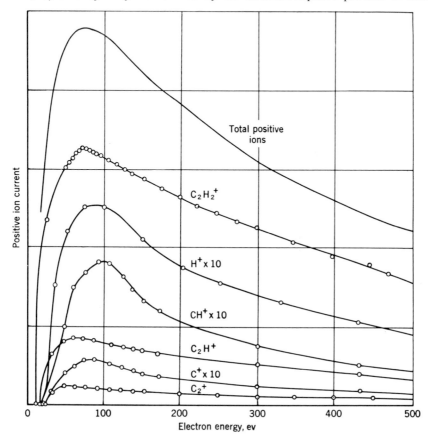

Fig. 4.1. Ionization efficiency curves for ions from acetylene (*From Tate, Smith, and Vaughan* [2].)

form $XY + e \rightarrow XY^-$, or $XY + e \rightarrow X + Y^-$; the latter are resonance processes since there is no electron to carry away excess energy, and they occur at electron energies which are usually below 10 ev and sometimes as low as 0.05 ev [3]. Such processes are not readily applicable to chemical analysis, and at higher energies, where processes of the first type take place, the ionization cross section for negative ions is much less than for positive ions, so that for analytical purposes negative ions have not so far been used.

B. Design and operation of electron-impact sources

The great majority of electron-impact sources used at present are based on those developed by Bleakney [4] and Nier [5, 6]. A typical example is shown diagrammatically in Fig. 4.2. The sample vapor enters the ionization chamber through a glass pipe fitting tightly into a hole in the top plate. The electron accelerating voltage is applied between the filament and plate A, which is connected to the ionization chamber. Electrons emitted by the heated filament are accelerated through the aperture in A (typically 3×0.4 mm) and pass through the ionization chamber to be collected by the trap, which is usually maintained at a potential some twenty volts higher than the ionization

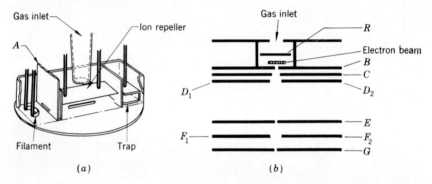

(a)   (b)

Fig. 4.2. Electron-impact source. (a) Ionization chamber; (b) schematic diagram of source, showing the ion repeller R, draw-out plate C, focusing half-plates D, collimating plates E and G, and beam-centering half-plates F.

chamber, in order to prevent the escape of secondary electrons. Divergence of the electron beam is prevented by a magnetic field directed along the beam, provided in 180° instruments by the main analyzer field and in sector instruments by an auxiliary permanent magnet with a field strength of a few hundred gauss. The electrons describe helical paths, the radius and pitch of which are given by Barnard [7] for various fields and transverse electron energies.

Ions formed along the path of the electron beam are accelerated through the exit slit in B (Fig. 4.2b) by a weak field determined partly by field penetration from plate C and partly by a potential applied to the ion repeller R. The main acceleration of the ions takes place between B and E, with the slit plate C and the half-plates D at intermediate potentials. E and G contain narrow slits limiting the width and angular spread of the ion beam, and F_1 and F_2 are a further pair of half-plates by means of which the beam can be deflected to correct for asymmetry or misalignment or cut off completely by application of a much higher deflecting voltage.

To achieve the maximum stability, ion intensity, energy homogeneity, and ion-beam collimation with such a source, attention must be paid to several aspects of design and operation, which may be conveniently divided into those affecting (1) the electron beam, (2) the ionization chamber, and (3) ion-beam formation.

The electron beam.   The number of ions produced depends on both the electron current and electron energy, and therefore these quantities must be carefully regulated.   Measurement of the electron current to the trap is complicated by the production of secondary electrons at various points in the ionization chamber, but nevertheless regulation of this quantity has been found in practice to give greater stability of operation than regulation of the filament emission.   The trap current is usually stabilized (see Chap. 6) to within 1 part in 10^4 at a value of 10 to 100 μa.

The energy spread in the conventional electron beam may amount to two or three electron volts [3], which is not important for analytical work but is a serious limitation in the measurement of appearance potentials and detection of fine struction in ionization efficiency curves.   To overcome this difficulty, Clarke [8] has used an electrostatic energy filter to provide a monoenergetic beam of electrons, and Fox et al. [9] have devised the *retarding-potential-difference* technique in which the electron energy distribution is modified in a five-plate electron gun, in such a way as to simulate a monoenergetic electron beam.   (These techniques are described more fully in Chap. 12.)

The most commonly used filament material in electron-impact sources is tungsten, being free of poisoning by sample gases, but it is far from ideal, because of the high operating temperatures and the dependence of the operation of the source on the degree of carbonization of the filament [10].   For the most stable operation, the surface layers of the filament should consist of W_2C, but exposure to excessive amounts of aromatic or oxygenated compounds disturbs this condition.   The behavior of the source then depends on the nature of the previous sample, and it is necessary to restore the stable condition by running an appropriate compound through the source for some time.   Rhenium has a great advantage as a filament material in this respect, as it does not form carbides [10].   Dependence on previous samples is greatly reduced, and temperature stability is improved, so that although the evaporation rate is greater than that of tungsten for a given emission, rhenium filaments are likely to be increasingly used in the future.

The ionization chamber.   In order to achieve the highest ratio of sample to residual gases in the ionization region and at the same time the highest sample pressure for a given flow rate through the mass spectrometer, the ionization chamber should be made as gastight as possible; that is, the only openings should be those for entry of the electron beam and for ion extraction. In studies of free radicals or appearance potentials, the probability of foreign radicals being formed on the filament and passing into the ionization chamber can be minimized by replacing the slit aperture through which the electrons enter the ionization chamber (Fig. 4.2a) by a long narrow canal. This impedes the flow of molecules and radicals, but, if properly aligned, does not affect the electron beam.

The mass spectrum of a compound is affected by the temperature of the gas at the time of ionization [11], the major effect being a decrease in the parent peak relative to the fragment peaks as the temperature is increased, so that for accurate analytical work it is necessary to control the temperature of the ionization chamber to within 1°C. The ion source normally reaches a temperature of 100 to 200°C due to radiation from the heated filament alone, and the temperature at which the ionization chamber is maintained is usually about 250°C.

If an ion beam of low energy spread is required, then since a weak electric field is present in the ionization chamber, the thickness of the electron beam must be restricted. This restriction is ensured by the presence of the source magnetic field. Accurate alignment of the filament and of the magnetic field ensures that the electron beam is parallel to the equipotentials, and provided the ion draw-out field is kept below a few volts per centimeter, the required low energy spread is obtained. Fragment ions often possess considerable initial kinetic energy, but for parent ions the energy spread in a carefully designed source can be as low as 0.05 ev. On the other hand, although a low energy spread is required for the highest resolving power, when the ion draw-out field is low, the stability and linearity of the source are poor because of space-charge and surface-charge effects, and in cases where some resolving power can be sacrificed, it is preferable to operating the source with a high-draw-out field (20 to 30 volts/cm). The effects of space charge and surface charges are then reduced, and the stability and linearity are greatly improved [12].

Adjustment of the potentials on the ion repeller R and the plates C and D in Fig. 4.2b provides great flexibility in the control of the ion trajectories. The methods of ion optics have not yet been applied very widely to this type of source, and optimum settings have usually been found empirically. However, calculation of ion trajectories has shown that in one case [13], where the potential of the plates D is close to the final accelerating potential, the slit system behaves as a strong immersion lens, and a demagnified image of the electron-beam region is formed in the vicinity of D. Thus, if the width of the ion beam is to be restricted in order to achieve high resolving power, this can be done with minimum loss of intensity if the defining slit E is placed at or near this image position.

Some mass discrimination has been found to occur in all sources of this type, whether operated by voltage or magnetic scanning, with or without a magnetic field. As Bainbridge [14] pointed out, mass discrimination must occur if magnetic scanning is used with a fixed source magnetic field, because the ion trajectories, and hence the effective solid angle for collection, must be different for each mass. However, Coggeshall's simplified analysis for a magnetic-field-free source [15] shows that mass discrimination must also occur for voltage scanning, and it is found in practice [16] that the increase in peak height with ion accelerating voltage is greater than predicted by Coggeshall. The increase is particularly large for fragment ions formed with large initial energies, since discrimination against these ions with initial kinetic energy [17] decreases with increasing accelerating voltage. Mass discrimination should be a minimum for magnetic scanning in a 180° instrument where the source is immersed in the main magnetic field, since H^2/M is constant during scanning and the ion trajectories for all masses should be identical, but some discrimination is observed even in this case. For isotopic analysis there is probably little to choose between voltage and magnetic scanning, and this applies also to scanning over a wide mass range in 180° instruments; but the variation of ion-beam intensity with accelerating voltage is much greater in sector instruments [6], and for these, magnetic scanning is essential if a wide mass range is to be covered.

C. Internal furnace sources

Any substance having a vapor pressure higher than 10^{-1} mm at room temperature may be analyzed by admitting the sample from an external reservoir through a leak. Some liquids and solids can be introduced by heating the entire inlet system provided that a temperature can be found at which the compounds have a vapor pressure of at least 10^{-1} mm but do not decompose. For organic compounds, 350 C appears to be the maximum permissible temperature. Compounds of even lower vapor pressure must be introduced into the vacuum enclosure and evaporated directly into the ionization chamber. The analysis of high-molecular-weight organic compounds in this way is a new and promising field; heating the sample in a tube attached to the gas-inlet line and evaporation from a heated filament near the ionization chamber are two possible ways of handling such compounds.

On the other hand, evaporation of inorganic solids into the ionization chamber from a heated crucible is a well-established technique, particularly for isotopic analysis. This arrangement has a number of serious disadvantages: (1) The vacuum must be broken for introduction of each sample, or a vacuum-lock system incorporated. (2) Considerable fractionation may occur during the evaporation because of the large differences in evaporation rates of compounds in a mixture. Even in isotopic analysis fractionation of isotopes may be serious, and the degree of fractionation depends on whether the sample is liquid or solid at the evaporation temperature and whether mixing of isotopes can occur if it is solid.   Fractionation can be reduced by choosing a compound of high molecular weight such as the iodide.   (3) To avoid memory effects the crucible, and possibly other adjacent source parts, must be carefully cleaned or replaced between successive runs.   (4) Because of outgassing of the source, the background spectrum is very large.

These points show clearly that, if it is at all possible, a solid sample for isotopic analysis should be converted to a volatile compound or analyzed using a thermal-ionization source.   However, for a number of elements, no convenient volatile compounds exist, while thermal ionization is too inefficient for many elements unless an electron multiplier detector is used.

A number of furnace arrangements have been described, notably by Nier [18], Palmer and Aitken [19], Shaw [20], and Cameron [21].   Tungsten, tantalum, and less commonly platinum, Nichrome, nickel, and iridium have been used as crucible materials.   The crucible is usually a tube, into which the sample is packed, and is suspended over the ionization chamber.   Resistance heating or heating by electron bombardment [20] is used, and the vapor issues through a small hole into the ionization chamber.   Background may be minimized by outgassing the assembly for some hours at a temperature slightly below that required for evaporation.   Palmer and Aitken [19] have described a simple nickel tube furnace for materials evaporating at temperatures below 500°C which can be thrown away at the end of each run.   In Cameron's source [21], the furnace and ionization chamber are machined from a single piece of tubing, so that the whole ionization chamber can easily be removed for cleaning.

This method is not generally suitable for chemical analysis, but Honig [22] has obtained an analysis for impurities in germanium at the level of 5 parts in 10^5 by evaporating the sample completely from a graphite-lined alumina crucible and integrating the ion currents.   He considers that the limit of detection for impurities might be lowered to 1 part in 10^8, but the time for the complete evaporation was several days, and the method cannot compare for convenience and sensitivity with analysis by the vacuum spark.

Sensitivity and accuracy.   The ion current passing into the analyzer from an electron-impact source can be expressed as

$$i = \beta n_0 Q_i s i_e \tag{4.1}$$

where i, i_e = ion and electron currents, respectively

s = effective electron path length

Q_i = ionization cross section

n_0 = density of molecules in ionization chamber

β = ion-extraction efficiency

The useful electron current is limited to about 10^{-4} amp because of space charge, and s is determined by the dimensions of the instrument. The effects of positive-ion space charge on the electron beam, the need to maintain a high vacuum in the analyzer, and the probability of secondary reactions between ions and molecules normally limit the sample pressure in the ionization chamber to about 10^{-4} mm, and n_0 therefore has a maximum value of about 2.5×10^{16} moles/cm³. β depends on a number of factors, including the energy homogeneity, beam width, and collimation required and the mass of the ion and may vary from 2% to something approaching 100% in a source designed specifically for high sensitivity. In the case of a typical source for precision analysis, one ion enters the analyzer for every hundred thousand molecules passing through the ionization chamber.

The size of sample normally used in routine analyses with the electron-impact source is about one standard cubic centimeter for gases and a few milligrams for solids evaporated in a furnace. No difficulty is experienced in obtaining stable beams of 10^{-10} amp for mass analysis. The minimum size of gas sample which can be analyzed is determined mainly by the background spectrum due to residual gases; these can be reduced to a very low level by modern high-vacuum techniques. As little as 2×10^{-9} cm³ (5×10^{10} atoms) of a xenon isotope has been detected by Kennett and Thode [23], and by suspending pumping after admission of the sample, Reynolds [24] was able to detect even smaller quantities, 5×10^5 atoms of xenon and 1.4×10^7 atoms of ^{36}Ar.

The limit of detection of impurities and rare isotopes in gases is about 1 part in 10^7, determined partly by the background spectrum and partly by the maximum permissible value of sample pressure in the ionization chamber.

Very high accuracy is possible with the electron-impact source. The electron current can be regulated to 1 part in 10^4, and the short-term stability of the ion beams is of this order, although changes in sensitivity of up to 2% may occur over a period of hours. Precisions of $\pm 0.01\%$ or better in isotope abundance measurements and of ± 0.3 mole % in analysis of complex mixtures can be achieved.

4.3. Vacuum-spark Sources

In 1935, Dempster developed the vacuum spark as a source of positive ions for his investigations of the isotopic constitution of the elements [25, 26]. For this purpose it had the advantage over the discharge source that it was easier to maintain a high vacuum and there was no hydride formation. Apart from this, little use was made of it until about 1946. Since then it has been used extensively for mass measurement, chiefly by Duckworth and his coworkers. Moreover, following work by Dempster [27] and Hannay [28, 29], it has been established as a powerful tool for the general chemical analysis of solid materials [30, 31]. It is in fact the only source suitable for this type of analysis because, in contrast to the thermal-ionization, furnace, and discharge sources, the sensitivities for different elements do not differ by more than an order of magnitude.

A. The vacuum spark

The spark is formed between two electrodes by application of a high voltage of the order of 50 kv, usually supplied in the form of pulsed radio frequency. Because of the intense local heating of the electrodes, solid material is evaporated and is ionized in the discharge. The use of short pulses ensures that the electrodes do not melt, and there is little, if any, mass fractionation such as would be expected if the material were evaporating as a liquid in equilibrium at high temperature.

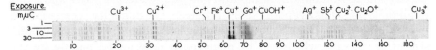

Fig. 4.3. Spark source mass spectrum of a copper sample, showing impurity lines due to Cr (200 ppm atomic), Fe (100 ppm), Ga (150 ppm), Ag (200 ppm), and Sb (200 ppm).

Because of the high energy available in the discharge, multiply charged ions, with up to six charges or more, are produced in considerable numbers. This is an advantage for mass-comparison work, but is something of a disadvantage for chemical analysis, since it complicates the otherwise simple mass spectrum of an element. However, the abundance of multiply charged ions decreases by approximately a factor of 5 for each additional charge. Complex singly charged ions are also produced, for example, Fe_2^+, Fe_3^+, $NiCr^+$, $NiAl^+$, FeO^+, $FeOH^+$, etc. The abundance of these ions is generally small and decreases with increasing complexity. The mass spectrum of a copper sample recorded on a photographic plate is shown in Fig. 4.3.

The chief advantages of this source for chemical analysis are the approximate equality of ionization efficiencies for all elements and the simplicity of the mass spectrum of an element, which is easily recognized by its isotopic constitution. The disadvantages are the very large energy spread of the ions ($\sim$1,000 ev), which makes a double-focusing analyzer necessary, and the difficulty of measuring the fluctuating ion currents.

B. Design of the vacuum-spark source

The electrodes. Two types of spark gap have been described. The rod-and-cylinder arrangement consists of a rod of sample material fitting into a hollow cylinder with a small clearance. The cylinder is mounted axially at the center of a plate, and ions formed in the gap between the rod and cylinder drift into the accelerating system of the ion gun. The second arrangement is simply a pair of rod electrodes mounted with their ends close together, either in line or at an angle. This is somewhat more difficult to align, but the preparation of the electrodes is much simpler. In either case the position of the electrodes must be adjustable from outside; this adjustment is usually made through a bellows.

The rods are usually of the order of 0.001 in.^{2} cross section and $\frac{1}{2}$ in. long. If the sample is electrically conducting, the rods should preferably be made of the sample material itself, but for nonconducting materials the sample may be packed into a thin-walled conducting tube [29], mixed in powder form with a conducting material such as graphite and compressed to form an electrode or used in conjunction with a conducting electrode lying alongside the sample [32]. If these techniques are used in impurity analysis, the foreign material that maintains the spark must be free from any impurities expected in the sample.

Formation of the ion beam. The ion gun in the spark source is relatively simple. Since the actual position of the spark may vary considerably during the running of the sample and since the ion energy spread is so large, there is little to be gained by incorporating lenses such as are used in other sources to form a demagnified image of the ionization region. The spark electrodes are usually placed just in front of a small hole in a plate to which one of the electrodes is connected and to which the ion accelerating voltage is applied; in the rod-and-cylinder source this plate contains the cylinder. This is followed by a narrow slit at earth potential, which limits the width of the beam. The angular spread

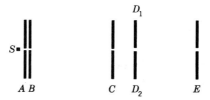

Fig. 4.4. Vacuum-spark source. S is the spark gap; B and E are collimating plates; C and D are plates for ion-beam suppression.

of the beam is limited by another slit at earth potential some distance beyond the first. A simple source of this type is shown schematically in Fig. 4.4. The additional plates C, D_1, and D_2 are used as an electrical shutter by applying a high voltage between them.

General design of the source. Although the spark source is basically simple, the high voltages used present considerable technical problems. The radio-frequency voltage supplied to the spark must be of the order of 50,000 volts, and preferably as high as 100,000 volts, and is introduced into the source through glass-to-metal seals of special design having long electrical-leakage paths. The insulators through which the motion of the electrodes is controlled must also be capable of withstanding the high voltage.

Tantalum electrodes are frequently used in the region of the spark source since this avoids any risk of the introduction into the ion beam, by ion bombardment or stray high-voltage breakdown, of any foreign metallic ions other than tantalum ions.

The pressure in the ion source may rise considerably during the passage of the discharge, and differential pumping of the source is essential if the analyzer pressure is to remain low, an essential requirement for the achievement of low background, and hence low detection limits, in impurity work.

C. Sensitivity, accuracy, and limitations of the method

The sensitivity obtainable with the spark source may be judged from the estimate of Hannay [29] that 10^{11} atoms can be detected either as a bulk or a surface impurity and from the limits of detection for impurities in silicon given by Craig et al. [31].   More recent work on silicon has shown that although this case is an unfavorable one, 9 out of 73 impurity elements considered had detection limits of 0.001 ppm or less, 50 between 0.002 and 0.01 ppm, 12 between 0.02 and 0.1 ppm, and only 2 above 0.1 ppm.   Only 10 mg or so of material is used up in an analysis of this kind, although 100 mg or more is needed to form electrodes of the size required.   Although only a fraction of the ion beam is transmitted through the analyzers because of the high-energy spread, the useful ion current is of the order of 10^{-9} amp.

The sensitivity for a particular element is governed partly by factors such as the mass of the ions, the extent of overlap with other ions in the spectrum, and the division of the element into a number of isotopes, and partly by the ionization efficiency.   The variations in the latter are not large, and taking iron as a standard, the ionization efficiencies for the majority of elements lie between 0.8 and 3 [31].

The accuracy of a quantitative analysis carried out with the spark source is limited by the difficulty of measuring the ion currents, first, because of interference with electrometer circuits by the r-f voltages in the source, and second, because of the large and erratic fluctuations in the ion current.   Both problems are solved to some extent by using an integrating device such as the photographic plate as an ion detector, but even with careful calibration and photometric examination of the plates, the error in an abundance measurement cannot at present be reduced below about $\pm 20\%$.   However, in impurity analysis the advantages in speed and reliability of detection are often more important than high accuracy.   More accurate results can be obtained with electrical detection of the ions if the recording system is arranged so that ratios of ion currents are measured, thereby greatly reducing the effect of the fluctuations in the ion currents, since any two ion beams may be expected to vary in the same manner.   By this means Gorman et al. [33]

measured the abundance of chromium and nickel in stainless steels with an accuracy of $\pm 2\frac{1}{2}\%$.

4.4. Thermal-ionization Sources

The phenomenon of surface or thermal ionization, in which positive ions are produced directly from a solid when it is heated on a metal filament, was discovered by Gehrke and Reichenheim and was used by Dempster in his original mass spectrograph [34]. Because of the enormous variations in ionization efficiency between different elements, it cannot be used directly for quantitative chemical analysis and its main application is to the measurement of isotopic abundances, but in this field it has become of increasing importance and is now probably the most widely used method of ionization. Its chief merits are relative freedom from background and the small quantities required for an analysis.

A. Mechanism of thermal ionization

The method is based upon the fact that when an atom or molecule evaporates from a surface, there is a probability that it will evaporate as a positive ion. For the case of a solid in thermal equilibrium with a hot surface, the relative probability of evaporation as a positive ion is given by [35]

$$\frac{n_+}{n_0} = \frac{1 - r_+}{1 - r_0}\frac{g_+}{g_0}\exp\left(\frac{\phi - I}{kT}\right) \tag{4.2}$$

where n_+ = number of positive ions

n_0 = number of neutral particles

r_+, r_0 = internal reflection coefficients for ions and neutral particles, respectively

g_+, g_0 = appropriate statistical weights

ϕ = work function of surface, ev

I = ionization potential of the element, ev

T = absolute temperature of surface

In a practical case, where a layer of a solid sample many atoms thick is evaporated from a heated filament, the quantities in the above equation are not well defined, but the formula

$$\frac{n_+}{n_0} \propto \exp\left(\frac{\phi - I}{kT}\right) \tag{4.3}$$

may still be taken as a rough guide to the variation of ionization efficiency with temperature, work function, etc.

It is clear from this equation that if the ionization potential of the molecule is lower than the work function of the surface and $\phi - I \gg kT$, then n_+/n_0 is very large and nearly every evaporating particle is ionized. Cases of 100% ionization have indeed been observed for the vapors of alkali metals in contact with heated tungsten and nickel surfaces [35, 36].

In contrast, if the ionization potential of the molecule is higher than the work function, n_+/n_0 may be extremely small. For instance, if $\phi = 4.6$ volts (tungsten) and $I = 7.7$ volts (copper), then for the maximum reasonable value of $T = 2700°K$, $n^+/n_0 = 1.6 \times 10^{-6}$.

Thus the ionization efficiency may vary over several orders of magnitude for different elements; furthermore, the variation of ionization efficiency with temperature is quite different in the two cases mentioned above. For if $I < \phi$, the temperature should be as low as possible consistent with efficient evaporation (or reevaporation). On the other hand, if $I > \phi$, the temperature of the ionizing filament must be as high as possible, so that it is not usually possible to study the ions of two elements at the same time. The thermal-ionization source is therefore quite unsuitable for the chemical analysis of mixtures of elements. However, it may reasonably be assumed that the ionization efficiency is the same for different isotopes of the same element, so that isotopic abundance measurements can be made using a thermal-ionization source, subject to the usual precautions regarding fractionation of isotopes in the evaporation process and mass discrimination in the ion gun.

Since the ions have only thermal energies on formation, the energy spread in the ion beam is small and mass analysis can be carried out with a single-focusing magnetic analyzer.

B. Thermally emitting surfaces

The arrangements used for the production of positive ions from a solid sample can be divided into three types: (1) single filament, (2) specially prepared surfaces, and (3) multiple filaments.

Single filament. This consists simply of a ribbon or wire filament on which the sample is placed. The ideal filament material should have a high work function and low vapor pressure (high permissible working temperature) and be free of impurities. Tungsten appears to be the most useful material, though tantalum is also commonly used. The work function of tungsten (4.6 volts) is lower than that of platinum (6.2 volts), but the temperature at which it may be operated is considerably higher; furthermore, at temperatures below 2100°K the surface is probably the oxide, which has a work function of 6.5 volts [37, 38]. Other filament materials which have been used are platinum-plated tungsten [39, 40] and iridium [41], while for some metals a filament of the metal itself can be used as the source of ions. Work by Stevens [42] has shown that rhenium may be an even better filament material than tungsten, giving greater sensitivity for tin and a number of other elements.

An element is often presented for analysis in the form of the oxide, in which case the sample is placed on the filament as a paste and heated to dryness. More intimate contact between sample and filament is achieved if a solution of a soluble salt such as the nitrate or sulfate is used.

Specially prepared surfaces. A variety of more complicated methods of preparing the ion-emitting surface have been described. In most cases the purpose was to obtain a more intense and more stable beam of ions; in others, to reduce or eliminate the fractionation of isotopes during evaporation [43]. Factors which may be expected to affect the intensity of the ion beam are the chemical form of the sample, the nature, and hence the work function, of the filament surface, and the adherence of the sample to the filament.

The two most important techniques of this kind are the use of cement binders and the so-called borax-bead technique. Sauereisen cement has been used by several workers, and this can result in an increase in ion intensity by a factor of 5 or more, but has the disadvantage that background peaks are greatly increased [19, 45], so that for small samples there is probably little overall advantage. In the borax-bead technique, a small quantity of powdered fused borax is added to the drop of solution on the filament. On heating to red heat, the borax melts and wets the filament, forming a glasslike bead. This preheating may be carried out in air. Once again, a considerable increase in ion current is achieved, sometimes by several orders of magnitude [46]. The mechanism is not well understood, but it seems certain that part of the increase is due to the improved adherence of the sample; the sample is presumably distributed uniformly throughout the borax bead, and there is therefore no tendency for the sample to fall off in lumps. However, it is possible that other factors contribute even more to the increase. Observation of MBO_2^+ ions suggests that the sample may be combined chemically with the borax. Furthermore, the wetting of the filament by the borax may ensure better electrical and thermal contact with the filament surface and may also increase the work function.

Gordon and Friedman [47] report the use of a "flux" of ammonium sulfate to increase the ionization efficiency for Cs on a tungsten filament.

Multiple filaments. An alternative and very elegant method of increasing the ionization efficiency is the use of the multifilament source first reported by Inghram and Chupka [48]. This is particularly useful for elements of high ionization potential, where, as shown earlier, the temperature of the ionizing filament should be as high as possible. With a single filament a high filament temperature cannot normally be used because the sample then evaporates before the temperature for efficient ionization is reached. If, however, the sample is placed on a separate filament, the rate of evaporation and the temperature of the ionizing filament can be varied independently. The ionizing filament is run at the highest temperature consistent with reasonable life.   Although only a fraction of the particles evaporated from the sample filament strike the ionizing filament, the overall ionization efficiency is greatly improved. As Palmer has shown [45], it is also possible to minimize isotope fractionation in lithium analysis by using lithium nitrate in a multifilament source.

The three most generally useful of the arrangements described above are the triple filament, the borax-bead single filament, and the plain single filament.   The single filament has the advantage of simplicity and is quite adequate for a number of elements.   The triple-filament method is generally the most sensitive, but the borax-bead technique approaches it and even exceeds it in some cases [49] and enables lower source temperatures to be used.   However, it remains true that, for the ultimate in sensitivity, each element presents a problem of its own, and if the greatest sensitivity is desired, some preliminary investigation of the best method and chemical form of the sample will be necessary in most cases.

C. Design of ion sources for thermal ionization

Filament assemblies. Some workers have used coiled, V-shaped, or kinked filaments in order to ensure that the sample remains in the center of the filament, but ordinary ribbon filaments are usually suitable and are easier to manufacture.

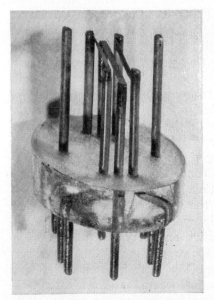

A photograph of a triple-filament assembly as used in the A.E.I. thermal-ionization sources is shown in Fig. 4.5.   The sample is placed on one or both of the side filaments, two of which are included for symmetry, and ionization takes place at the center filament.   The filament legs are of Nilo-K and are sealed into the glass bead, which is sintered around them.   Current supplies for the filaments are taken to a plug into which the filament legs fit, and the assembly is readily removed from the source for replacement of the sample. The assemblies can be refilamented once or twice.

Fig. 4.5. Thermal-ionization triple-filament assembly.

Formation of the ion beam. The majority of ion sources used for thermal ionization have been modifications of a Nier-type electron-impact source.   In many cases the filament has been placed in the ionization chamber in the position formerly occupied by the electron beam [19, 47,

48, 50]; in others, a furnace arrangement has been placed at the center of the top plate of the ionization chamber [51], so that the source could, in principle, be used for gas analysis, for electron bombardment of vapor from the furnace, or for thermal ionization. This arrangement is attractive at first sight, but such a source is not ideally suited to thermal ionization, and it is preferable to design the ion-optical system afresh, bearing in mind the special requirements of a thermal-ion source. Among these requirements are:

1. Alignment should not be critical, since the filament assembly must be frequently replaced, and with some designs of vacuum lock, one half of the source is entirely removed. This means that all slits should be as wide as possible.

2. In contrast with the electron-impact source, the pumping speed in the ionization region should be as high as possible, and the ionization chamber should therefore be of open construction.

3. Secondary ions and thermal ions produced elsewhere than at the ionizing filament should not be accelerated into the ion beam. Thus, in a triple-filament system, thermal ions of residual hydrocarbons may be formed at the side filaments or other parts of the source which become heated. Ions striking the first plate of the ion gun may produce secondary electrons, which are accelerated back toward the filament and produce tertiary ions. Inghram et al. [52] have observed ions of this kind at mass numbers 24, 25, 26, 27, 28, and 38.2, when the sample was a mixture of sodium and potassium, and show that they can be greatly reduced in number by a grid placed before the first slit and held at a potential of -300 volts with respect to it, thus preventing the secondary electrons from returning to the filament. Since secondaries may now be formed at the grid, the suppression ratio is approximately equal to the ratio of the open and closed areas of the grid.

Finally, it should be noted that, in the thermal-ionization source, a high electric field can be permitted in the region of the filament, since all ions are effectively formed on an equipotential surface.

A source designed along these lines specifically for thermal ionization has been described by Craig [53]. In this source, shown in Fig. 4.6, a demagnified image of the filament is formed by a strong immersion ion lens immediately below the filament. All the slits are wide (0.040 in.), and the only slit which limits the ion beam geometrically is the final one, which determines the angular spread. Secondary electrons formed at this plate are not accelerated back through the source, so that tertiary ions are not observed. Ions from the side filaments can be suppressed simply by making these filaments a few volts negative with respect to the center filament.

General remarks. The filament must, of course, be removed from the vacuum enclosure after each run for replacement or application of a fresh sample, and it is clearly desirable to incorporate a vacuum-lock system so that the filament can be removed without breaking the vacuum in the main analyzer.   Vacuum locks for this purpose have been described by Roberts and Walsh [54] and Stevens [55], by means of which the sample can be changed and operating pressures reached in as little as one minute [45].

A high pumping speed is desirable at the source, first, to reduce the partial pressure of residual hydrocarbons, which may be thermally ionized, and second, because considerable outgassing may occur when the filament is heated.

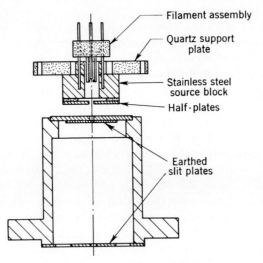

Fig. 4.6.   Thermal-ionization source (A.E.I. M.S.5 mass spectrometer).

Since the major portion of any sample will settle on surfaces near the filaments, the source should be easy to dismantle and all surfaces should be accessible for cleaning.

D. Operation

The technique of running a sample varies considerably, but in general the temperature of the sample filament is gradually raised, while the region of the spectrum which is of interest is continuously scanned. When the sample peaks appear they are usually varying with time, normally decreasing because of the gradual diminution of sample, but occasionally increasing.   This makes it necessary to record several sets of peaks equally spaced in time or, by means of a special collector, to record the abundance of each isotope as a fraction of the total [56].   Accurate control of the filament current is necessary since the rate of ion emission varies with temperature according to a power law where the exponent may be as high as 35 [57, 58]. Sufficiently close control can usually be achieved using a constant-voltage transformer and Variac, but several electronic current stabilizers have been developed [45, 58], and recently Kendall has reported the use of an ion current stabilizer [57], which, however, is limited in its application to certain special problems.

E. Sensitivity, accuracy, and limitations

Despite the wide variations in ionization efficiency, it is nevertheless true that, for the majority of elements, an isotopic analysis can be made on an extremely small sample, that is, a microgram or less. This is of great advantage in cases where the amount of material available is very small or where it is desirable to use only small amounts, as in the case of highly radioactive samples. An example of the extreme sensitivity which is attainable in favorable cases is the work of Gordon and Friedman [47], in which they were able to determine, to within 5%, 10^8 atoms of cesium, that is, 2×10^{-14} g.

Background ion currents from the thermal-ionization source are relatively low, but for many elements the maximum obtainable ion currents are very small, and though they can be amplified greatly by an electron multiplier, the background then becomes important and may limit the usable sensitivity. There are two main sources of the background ions: (1) impurities in the filament and (2) hydrocarbons, possibly due to pump-oil vapor condensed on the walls of the source or introduced on the filament. The most commonly encountered ions of the first type are sodium and potassium. Enormous beams of these ions can be obtained from a tungsten filament, but fortunately, these impurities can be removed by preheating the filaments, if possible monitoring the ion currents until they disappear. Other ions often obtained from a tungsten filament are W^+, Mo^+, and the other alkali metals. The hydrocarbon background, which gives a peak at almost every mass number from 12 up to several hundred [49], can also be reduced by preheating, and if the ions are originating at the side filaments, they can be largely suppressed by biasing these filaments negative.

The accuracy of isotopic analysis is limited by fractionation in the evaporation process, variation of the ion beams with time, the necessity for removing part of the source in order to charge samples, and memory effects.

Fractionation is most serious for elements of low atomic weight; in the case of lithium, which has been extensively studied, the preferential evaporation of the Li^6 isotope results in a measured Li^7/Li^6 ratio, which is not only too low initially, but which rises with time as the sample becomes impoverished in the light isotope. Even in a comparison between two samples, there may be an error if the measurements are not taken at the same stage of the evaporation.   Various solutions have been proposed, including (1) integration of the ion beams over the life of the sample [59], a tedious and uncertain method, (2) correction of the ratio measured at the start of the evaporation according to a square root of the mass law [19, 60], and (3) use of a compound of high molecular weight such as the nitrate with a triple filament [45].

Apart from the variation of ion intensity due to diminution of the sample, which is approximately inversely proportional to the sample size and the sensitivity of the instrument, rapid fluctuations occur which are small, but larger than in the electron-impact source.   Careful loading of the sample onto the filament minimizes these fluctuations.

Mass discrimination in a thermal-ionization source can be made small because of the absence of a magnetic field, and provided alignment is not critical, changes in discrimination between one sample and the next should be negligible.

Memory effects can usually be avoided by using a fresh filament assembly for each sample.   In the case of relatively volatile samples, it may be necessary to clean the parts of the source adjacent to the filaments since material deposited on a previous run may be evaporated later because of general heating of the source.   This type of memory effect is often less serious for the borax-bead technique than for the triple filament, as the source as a whole does not get so hot.

If care is taken to reduce the above sources of error to a minimum, isotope ratios can be measured with a reproducibility of $\pm 0.1\%$ and an accuracy of $\pm 1\%$.

Apart from the sources of error mentioned above, the only serious limitation of the method as a general one for isotopic analysis is that it is difficult, and occasionally impossible, to use it for elements of high ionization potential, but the number of cases is quite small.   Thermal-ionization mass spectra have been observed for all but 16 of the solid elements up to atomic number 92.   The 16 elements are beryllium, carbon, silicon, phosphorus, sulfur, germanium, arsenic, selenium, antimony, tellurium, rhenium, osmium, gold, polonium, astatine, and palladium, and it will be noted that all these elements have ionization potentials higher than 8 volts and are mostly members of Groups IV, V, and VI of the periodic table.

4.5. Gas-discharge Sources

The discharge in a gas at low pressure is among the most prolific sources of ions, and it was therefore natural that it should have been the source most commonly used in early studies of positive ions, including mass spectrometry.   More recently, discharge sources have been developed to provide the intense ion beams needed for large-scale electromagnetic separation of isotopes. However, all discharge sources suffer from certain disadvantages:

1. The discharge is unsteady, resulting in a fluctuating ion-beam intensity.
2. The rate of consumption of the sample is high.
3. The high pressures used in the source make it difficult to maintain a high vacuum in the analyzers.
4. The energy of the ionizing electrons is not easily controlled.

The principal advantage of discharge sources is the intensity of the ion beams produced, in many cases as high as several milliamperes. However, in the majority of applications considered in this book, such large ion currents are not necessary and, indeed, may be undesirable because of the difficulties of focusing due to the effects of space charge. The role of discharge sources in analytical mass spectrometry is therefore a minor one, and only a brief description of the various types will be given here. A detailed account has been given by Ewald and Hintenberger [61].

The main division of the discharge sources is between high- and low-voltage types, and a further classification can be made according to the presence or absence of a magnetic field. The high-voltage source gives ions having a very large range of energies, but is much simpler in construction than low-voltage sources and requires no heated filament.

A. High-voltage glow-discharge source

This source consists in essence of two electrodes placed some tens of centimeters apart in an envelope containing vapor at a pressure of 10^{-3} to 10^{-1} mm. A glow discharge is maintained by a voltage of 10,000 to 60,000 volts between the electrodes; the ions formed are accelerated through a hole in the center of the cathode and pass into the analyzers. The discharge current is of the order of 10 ma, and ion currents up to 1 ma have been obtained. Operation of the discharge is unpredictable, and the energies of the ions vary from zero to the full discharge voltage.

Solid materials may be ionized by placing them, in the form of a paste, in wells drilled in the end of the anode [62] or surrounding the cathode exit canal [63] and introducing a noble gas to support the discharge. The sample is evaporated by electron bombardment in the former case and by ion bombardment in the latter, and the vapor is then ionized in the discharge.

A source of this kind was used by Aston [64] in his pioneer work on isotopes, and with it he was able to obtain isotopic analyses for almost all the elements. In isotopic abundance measurements it has now been largely superseded by the electron-impact and thermal-ionization sources, although it is still used in one or two laboratories for mass measurements [61].

B. Low-voltage arc source

The arc source depends for its operation on an independent supply of electrons from a heated filament. The arc current is much higher (100 ma to 4 amp) than in the glow discharge, but since the operating voltage is of the order of 100 volts, the ion-energy spread is much lower. In order to obtain high arc and ion-current densities, the arc is usually confined either by a capillary or by a magnetic field.

The capillary type of arc source has been used by Koch [65] for small-scale isotope separation. This source, shown in Fig. 4.7, is similar to those developed by Lamar et al. [66], Tuve [67], and others for the production of intense proton beams for particle accelerators. Ions from the

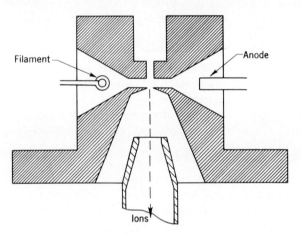

Fig. 4.7. Capillary-arc discharge source.

positive-ion sheath which forms the outermost part of the discharge emerge through a small aperture in the wall of the capillary and are then accelerated. With an arc current of 0.2 to 0.8 amp and pressures of 2×10^{-3} to 6×10^{-3} mm, ion currents of several hundred microamperes having an energy spread of only 0.2 ev have been obtained [68].

Sources in which the arc is confined by a magnetic field have been described by Bernas and Nier [69], Normand et al. [70], and Dawton [71], among others. The operating conditions are similar to those for the capillary-arc source, though the minimum working pressure is somewhat lower. Ion extraction is usually in a direction at right angles to the arc as before; this arrangement gives a lower ion energy spread than extraction through the anode, but nevertheless the energy spread is rather greater than in the capillary-arc source. These sources are, however, more suited to the ionization of solids than the capillary source. Solid materials are vaporized in an oven adjacent to the arc [69], or in the case of refractory elements, the charge material can be vaporized by bombardment with electrons from an auxiliary filament [70].

C. Oscillating discharge sources

These sources operate on the principle of the PIG (Philips ionization gauge) source first proposed by Penning for use as a cold-cathode ionization gauge. A hollow cylindrical anode is placed between two plane cathodes, and a magnetic field is applied parallel to the line joining the cathodes. Any electrons formed are accelerated toward the center of the anode, but cannot strike it because their lateral motion is confined to tight helices by the magnetic field. They pass through the anode and are slowed down and reflected before the other cathode, and thus execute an oscillatory motion. Under these conditions, the probability that an electron will make an ionizing collision is so high that a self-supporting discharge can be maintained at pressures of 10^{-4} to 10^{-3} mm and voltages of 1,000 to 2,000 without the use of a heated filament. Backus [72] has found the ion energy spread in these circumstances to be about 25 ev. The cold-cathode source has the advantage that there is no heated filament to burn out, although the life of the cathodes is nevertheless limited because of sputtering. However, although cold-cathode sources have been used in mass spectroscopy [72], for isotope separation it is more common for one or both cathodes to be a thermionic emitter; the discharge voltage is then of the order of 100 volts, and the system is thus a variation of the arc discharge in a magnetic field, employing the principle of oscillatory electron motion. Sources of this type have been reported by Heil [73], Kistemaker [68], and Almen and Nielsen [74], who obtained ion currents up to 120 μa and an overall efficiency in their isotope separator (ions collected to sample atoms used up) of up to 13%.

4.6. Photoionization Sources

Mass analysis of ions produced by photon impact was first reported by Terenin and Popov in 1932 [75]. Interest in the technique was revived in 1956, when Lossing and Tanaka [76] described some preliminary experiments on the performance and characteristics of a photoionization source for mass spectrometry. The potentialities of photoionization as an analytical tool have been further investigated by Herzog and Marmo [77], and Hurzeler, Inghram, and Morrison [78] have employed photoionization for the study of ionization potentials and vibrational states of molecules.

Photoionization differs from ionization by electron impact chiefly in the fact that the cross section for ionization to a particular ionic state is finite at the threshold energy and remains constant for energies above the threshold. Photoionization cross sections are one or two orders of magnitude less than those for impact by 70-volt electrons, but the ion currents obtained are smaller by two or three orders of magnitude, because the flux density in terms of ionizing particles is much less for any presently available light source than in the case of electron beams.

Since the ionization potentials of the majority of gases are greater than 8 ev, ultraviolet light of wavelengths below 1,500 A is required. The light is produced in a discharge tube containing hydrogen or krypton at a pressure of 10^{-1} to 10 mm Hg and passes into the ionization chamber through a lithium fluoride window about 1 mm thick.   An arrangement of this kind is shown in Fig. 4.8.

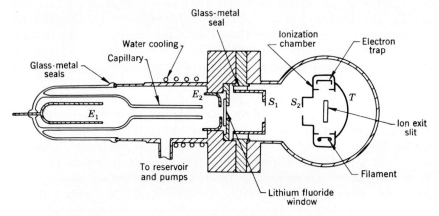

Fig. 4.8. Photoionization source: diagram of the arrangement of light source and ionization chamber. E_1 and E_2 are the discharge electrodes; S_1 and S_2 are collimating slits; a magnetic field prevents the escape of photoelectrons from the "photon trap" T.   The auxiliary filament and electron trap may be used for electron impact.

The design of the source as a whole is similar to that of the electron-impact source, with the light beam replacing the electron beam.   It is important to avoid ionization by photoelectrons by ensuring that no light strikes any surfaces at a potential lower than that of the ionization chamber.

Unfortunately, light of wavelengths below 1,050 A is completely absorbed by the window, and the compounds which can be ionized are limited to those with ionization potentials below 11.8 ev.   This limitation can be removed [96] by eliminating the window, but in view of the pressures used in the lamp, a very high degree of differential pumping is required in order to maintain the analyzer vacuum.   This involves a rather severe collimation of the light beam and consequent reduction in photon flux through the ionization chamber (see Chaps. 11 and 12).

Photoionization mass spectra are much simpler than those obtained with 70-volt electrons, in many cases consisting of parent peaks only.

Fragment peaks, when they occur, are always in lower abundance than in the corresponding 70-volt electron-impact spectrum, although for a few compounds, such as isopropyl alcohol and neopentane, fragment peaks predominate.

Given adequate sensitivity, the simplicity of photoionization mass spectra would be an advantage for some types of mixture analysis. However, it has been found that electron-impact spectra for 10-volt electrons are very similar, and the ion intensities are higher. Nevertheless, photoionization retains some advantages over low-voltage electron

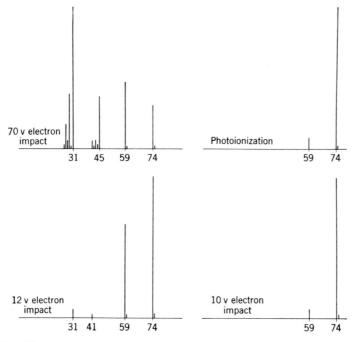

Fig. 4.9. Mass spectra of diethyl ether obtained by photoionization and by electron impact.

impact because of the nature of the threshold ionization law. Operation of the electron-impact source at low voltages is rather variable because small changes in effective electron accelerating voltage, due to contact potential variations, etc., give rise to large changes in absolute and relative ion intensities. Analogous variations in the photoionization source, that is, in the spectral composition of the light, do not lead to such serious changes in ion intensities. Thus both these techniques for the production of simple mass spectra have disadvantages in respect to either sensitivity or stability, and low-voltage electron impact is probably to be preferred only because the same source can be used without modification for conventional work with 70-volt electrons. The mass spectra of diethyl ether for 70-, 12-, and 10-volt electron impact and for photoionization are shown for comparison in Fig. 4.9.

For the study of appearance potentials and the upper energy states of molecules, photoionization possesses two great merits compared with electron-impact methods. First, the electron-impact ionization efficiency curve falls to zero at the threshold, so that the more closely the appearance potential is approached, the more difficult it becomes to measure the ion current. In contrast, the photoionization efficiency is finite at the threshold, and fine structure is more easily observed. Second, the energy of the photons can be accurately controlled and measured, and the energy spread reduced to an arbitrarily small value, by means of a vacuum monochromator. The latter, however, is a large and rather costly item, and the advantages are gained only at the expense of a considerable loss of ion intensity. The maximum ion current attained by Hurzeler et al. [78] was 10^{-15} amp. Nevertheless, they were able to measure the vertical and adiabatic ionization potentials of a number of gases with a precision of ± 0.1 ev and to detect and measure the spacings of several vibrational levels in nitric oxide, n-propyl alcohol, and acetaldehyde. These topics are discussed further in Chaps. 11 and 12.

4.7. Field-ionization Sources

In the field-ionization source, ions are formed in the intense electrostatic field set up at the tip of a sharpened tungsten wire when a high voltage is applied between the wire and a coaxial ring. In Muller's field-ion microscope [79, 80] the ions are accelerated through the ring to a screen. By allowing some of the ions to pass through a hole in the screen into a magnetic analyzer, Gomer and Inghram [81] have shown that the ions formed from molecules on and around the tip are almost always parent ions. For instance, no fragment peaks greater than 0.1% of the parent peak are observed for acetone, which, under electron bombardment by 70-volt electrons, gives 12 peaks of over 2% relative abundance [82]. No measurable fragment peaks are observed for ethylene, oxygen, nitrogen, or methane. Methyl alcohol and hydrogen are exceptions, the CH_3O^+ peak in methyl alcohol and the H^+ peak in hydrogen being comparable with the parent peaks.

Although the ion current density at the tip may be as high as 100 amp/ cm² [80], the effective emitting area is only of the order of 10^{-13} cm², and the ion currents available for mass analysis are very small, requiring an electron multiplier for their detection. The molecules which are ionized may be chemically or physically adsorbed on the surface, but at high fields autoionization takes place at some distance from the tip, and the energy spread in the ion beam increases with the applied voltage. The minimum energy spread is about 20 ev at fields of 5 volts/A. Certain applications of the field emission source are discussed in Sec. 11.3E.

The absence of fragmentation in the ionization process means that transient dissociation products from other processes, such as photo-decomposition, are more easily studied. Information on catalytic frag-mentation at the surface may be obtained [81], and the study of these processes is aided by the fact that perfectly clean surfaces can be prepared by applying very high fields at the tip, when the surface layers of the tungsten itself are completely stripped.

4.8. Positive-ion-bombardment Sources

A. Ion bombardment of solids

Sputtering, or the vaporization of surface particles of a solid under ion bombardment, has been well known for many years. Some of the sputtered atoms are ionized, and several attempts have been made [37, 83] to utilize these ions for the study of surfaces. A careful study of the sputtering process by Wehner [84] has shown that, for reliable results, the surface must be atomically clean and free of insulating deposits in particular. This requires that the bombarding-ion current density must be more than sufficient to remove all foreign atoms impinging on the surface, and Honig [85] estimates that, in a practical case, the ion current density required is 5×10^{-5} amp/cm^2.

Honig [85] used argon ions produced in a Heil-type source to bombard silver and silicon-germanium targets and performed mass analysis on the sputtered ions, but was not able to produce a focused-argon-ion current density greater than 10^{-5} amp/cm^2. Neutral atoms could be ionized in an auxiliary electron beam. At bombarding energies above 50 ev, a variety of positive and negative sputtered ions were observed, repre-senting the solid, the gaseous and solid impurities in the lattice, and surface-adsorbed gases. The majority of these ions had initial kinetic energies less than 5 ev. The results of an attempt to determine the composition of the bulk material by means of the sputtered ions were disappointing, the Ge$^+$/Si$^+$ ratio for a sample of 11% Si in Ge varying from 5 at 100 ev bombarding energy to 18 at 400 ev. The reasons for this are not clear, but improved results may be expected with higher bombarding-ion-current densities.

B. Ion bombardment of gases

The ionization and dissociation of molecules by bombardment with positive ions has been studied by Lindholm [86, 87], using a beam of ions of about 500 ev energy produced in an auxiliary-gas-source mass spec-trometer. Ions of the sample molecules are formed by charge exchange, and this process has a high probability only if the appearance potential of the ion is close to the recombination energy of the bombarding ion. By proper choice of the bombarding ion it is therefore possible to produce ions in excited states without forming ions in the states of low energy, a situation which is not possible for electron impact.

In view of the low ion currents produced, this method is unlikely to have any analytical applications, but is a useful technique, complementary to electron impact and photoionization, for the study of molecular-energy states.

4.9. Other Types of Ion Source

A. Electron bombardment of solids

Dempster [34] observed that ions were formed when aluminum phosphate was bombarded by 128-volt electrons, and the technique was investigated by Plumlee and Smith [88] and Dart [89] as a method for studying the composition of surface layers of solid materials. They found that, apart from a few exceptional cases, no ionization took place which could not be accounted for by thermal ionization due to electron-bombardment heating of the sample.

B. Thermal-ion equilibrium

The operation of this source [37] depends on the emission of molecular ions such as Rb_2Cl^+ by thermal ionization and a subsequent equilibrium reaction in the gas phase of the type $Rb_2Cl^+ \rightleftarrows Rb^+ + RbCl$. Its applications are mainly in the field of physical chemistry.

4.10. Ion Optics

The ion-optical problem in ion sources is simply to accelerate as many as possible of the ions formed into a narrow beam of width w_1 and divergence angle 2α, where the quantities of w_1 and α are predetermined by other considerations. The obvious and most common method of ensuring that the design parameters w_1 and α are not exceeded is to insert two slits in the final stages of the ion gun, the first a *source slit* of width w_1, and the second an *angle-limiting slit* of width w_2 at a distance d such that $(w_1 + w_2)/d = 2\alpha$. For instance, in Fig. 4.2b, plate E contains the source slit and plate G the angle-limiting slit. The accelerating system (plates B, C, and D in Fig. 4.2b) is then designed, and the potentials adjusted so that the maximum emergent ion-beam intensity is obtained. This approach works quite well in practice, but finding the optimum conditions is very much a process of trial and error, and in the end the ion gun is likely to be more complex than is really necessary. Furthermore, it may not be possible to achieve the maximum efficiency for the stipulated w_1 and α values since, in practice, it is likely that the values of either or both parameters will be less than those computed, because of the formation of a virtual image of the ionization region or of some slit.

The alternative approach is to set out to use a demagnified image of the ionization region in place of the source slit. This can lead to a very simple and versatile ion gun with an efficiency equal to the theoretical maximum determined by applying the Lagrange-Helmholtz law:

$$\sqrt{V_o}\, y_o \tan \alpha_0 = \sqrt{V_i}\, y_i \tan \alpha_i \tag{4.4}$$

where V = potential relative to point where the ions are at rest

y = beam width

α = semiangular spread of beam

and the subscripts o and i refer to the object and image positions, respectively.

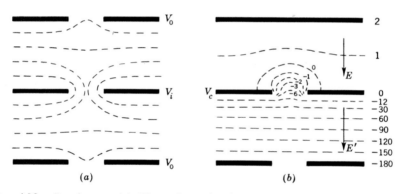

Fig. 4.10. Ion lenses. (*a*) Three-electrode einzel lens. (*b*) An aperture lens; equipotential levels referred to the potential V_c of the aperture.

The corresponding problems in electron optics have received a great deal of attention both theoretically and experimentally, and the methods can be readily applied to ion sources, since ion trajectories are identical with electron trajectories in fields of opposite sign. Unfortunately, the majority of electron optical systems studied have been rotationally symmetrical, but line-focus systems have recently been investigated more fully [90], and in many cases a useful qualitative picture of the lens action of systems of long slits can be gained from the results for circular apertures.

Two types of ion lens are shown in Fig. 4.10, the *einzel* lens in Fig. 4.10*a* and an *aperture* lens in Fig. 4.10*b*. The potentials V_o, V_i, and V_c are expressed relative to the ionization region, where the ions are at rest. The einzel lens is so called because of its similarity to an isolated glass lens, the potential, and hence the effective refractive index (see below), being the same on either side of the lens. The potential of the central slit may be above or below that of the outer slits; in either case the lens is convergent. Archard [90] has investigated the properties of such lenses theoretically and tabulated focal lengths for various slit widths and separations. He showed that the focal length can be expressed as a function of $\overline{V_i}/\overline{V_o}$, where $\overline{V_i}$ and $\overline{V_o}$ are the potentials at the center of the inner and outer slits, respectively.

Since no overall acceleration takes place in einzel lenses, their incorporation in an ion source involves additional electrodes, and they have been little used. On the other hand, focusing and acceleration take place simultaneously in aperture lenses. They depend for their focusing action on field penetration through the slit when the field strengths on either side of the slit are unequal. For weak lenses of this type, the focal length is found theoretically [91] to be $f = -2V/(E' - E)$, where V_c is the potential of the aperture and E and E' are the field strengths before and after the aperture. Since V_c is negative, it is evident that if E' is less than E, the lens is divergent. In the initial acceleration region, for instance, where the ions are being drawn out through the exit slit in the ionization chamber in Fig. 4.2, the situation is similar to that shown in Fig. 4.10b, E' being much greater than E. The lens is then very strongly convergent, the focal point often being within the lens field. The simple formula for focal length no longer holds, and the characteristics of the lens can be determined only by plotting the field and tracing a number of trajectories through the lens.

The techniques available for field plotting and ray tracing have been reviewed by Liebmann [92]. Determination of the potential distribution by means of an analytical solution of Laplace's equation is possible only with simple configurations, and graphical sketching of the field is inaccurate; numerical solution using relaxation methods is accurate and requires no special apparatus, but is very laborious. Of the more rapid methods employing electrical analogs, the resistor network [93] is capable of higher accuracy than the electrolytic tank, particularly where the field is changing rapidly.

Having found the potential distribution, the simplest and quickest method of ray tracing is based on an analogy with Snell's law in optics, the refractive index being identified with the ion velocity and hence with $\sqrt{V}$. The field is considered to be divided by the equipotentials into regions of potential V_1, V_2, etc., and the refraction at the boundaries of these regions is calculated from Snell's law $\sin i/\sin r = \sqrt{V_{n+1}}/\sqrt{V_n}$. (Various automatic ray-tracing devices, based on mechanical or electrical analogs, have been described [92].) However, these methods are rather inaccurate, and for accurate results it is necessary to find the derivatives of the potential from the field plot and to employ these in an iterative

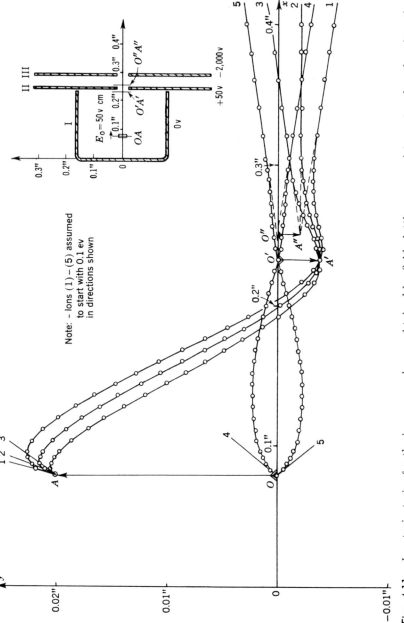

Fig. 4.11. Ion trajectories for the ion source shown, obtained by field plotting on a resistor network and an iterative numerical method of ray tracing. (*Craig* [13].)

numerical solution of the general ray equation. A detailed description of such a calculation has been given by Liebmann [94], and ion trajectories obtained in this way for a mass-spectrometer ion source are shown in Fig. 4.11. The work is greatly simplified by the fact that only the potential along the axis is required for the calculations.

Some of the special properties of ion lenses are illustrated by consideration of the analogy between refractive index and ion velocity mentioned above. In particular, very short focal lengths are possible because of the high ratios of velocities at different points, and large demagnifications can be achieved because the image lies in a region of high ion velocity, and hence of high effective refractive index. However, it is important to note that, as shown by Eq. (4.4), the product of the angular and linear magnifications is constant for a given ratio of initial and final velocities, so that, for fixed potentials, an increased linear demagnification is achieved only at the expense of a larger angular spread of the beam.

Surprisingly few accounts have been published of the application of these methods to mass-spectrometer ion sources. The work of Vauthier [95] may be mentioned, and two applications are briefly described in Secs. 5.4 and 5.17. There is little doubt that wider use of these techniques would result in great improvements in the design of ion sources, but it should perhaps be emphasized that overelaboration and too great a reliance on the computations should be avoided, since misalignments, space charge, and stray magnetic fields will probably cause the characteristics of a source to differ somewhat from those predicted. Considerable success can in fact be achieved without detailed calculations, and if they are employed, the aim should be a simple design, with flexibility provided by control of electrode potentials.

It will be seen that a great variety of ion sources exist, but many of these are specialized and only three, the electron-impact, thermal-ionization, and vacuum-spark sources, are of general importance. Of these, the electron-impact source, with its high stability, sensitivity, low energy spread, and ease of sample handling, is to be preferred whenever it can conveniently be used. For the study of gases and the vapors of organic compounds, the electron-impact source has no rival, and indeed there are only two situations commonly encountered where any other type of source is necessary. These are (1) the isotopic analysis of solid elements, where the problems associated with the evaporation of inorganic solids make the electron-impact source undesirable and the thermal-ionization source is to be preferred, and (2) the chemical analysis of inorganic solids, where the vacuum-spark source is the automatic choice because it is the only source for which the sensitivity is approximately equal for all elements. One might add that in the future the photo-ionization source may prove to be the most useful for the accurate determination of appearance potentials and other molecular properties.

REFERENCES

1. H. W. S. Massey: "Negative Ions," Cambridge University Press, London, 1950.
2. J. T. Tate, P. T. Smith, and A. L. Vaughan: *Phys. Rev.*, **48**, 525 (1935).
3. W. M. Hickam and D. Berg: "Advances in Mass Spectrometry" (J. Waldron, ed.), p. 458, Pergamon Press, New York, 1959.
4. W. Bleakney: *Phys. Rev.*, **34**, 157 (1929).
5. A. O. Nier: *Rev. Sci. Instr.*, **11**, 212 (1940).
6. A. O. Nier: *Rev. Sci. Instr.*, **18**, 398 (1947).
7. G. P. Barnard: "Modern Mass Spectrometry," p. 60, The Institute of Physics, London, 1953.
8. E. M. Clarke: *Can. J. Phys.*, **32**, 764 (1954).
9. R. E. Fox, W. M. Hickam, D. J. Grove, and T. Kjeldaas, Jr.: *Rev. Sci. Instr.*, **26**, 1101 (1955).
10. A. G. Sharkey, Jr., C. F. Robinson, and R. A. Friedel: "Advances in Mass Spectrometry" (J. Waldron, ed.), p. 193, Pergamon Press, New York, 1959.
11. F. H. Field and J. L. Franklin: "Electron Impact Phenomena," p. 202, Academic Press Inc., New York, 1957.
12. G. P. Barnard: "Modern Mass Spectrometry," p. 209, The Institute of Physics, London, 1953.
13. R. D. Craig: unpublished.
14. K. T. Bainbridge: Solvay Report, *Seventh Congr. Chem.*, R. Stoops, Brussels, 1947.
15. N. D. Coggeshall: *J. Chem. Phys.*, **12**, 19 (1944).
16. G. Careri and G. Nencini: *J. Chem. Phys.*, **18**, 897 (1950).
17. H. W. Washburn: "Physical Methods in Chemical Analysis," vol. 1, p. 600, Academic Press Inc., New York, 1950.
18. A. O. Nier: *Phys. Rev.*, **53**, 282 (1938).
19. G. H. Palmer and K. L. Aitken: "Applied Mass Spectrometry," p. 47, The Institute of Petroleum, London, 1954.
20. A. E. Shaw: *Phys. Rev.*, **75**, 1011 (1949).
21. A. E. Cameron: *Rev. Sci. Instr.*, **25**, 1154 (1954).
22. R. E. Honig: *Anal. Chem.*, **25**, 1530 (1953).
23. T. J. Kennett and H. G. Thode: *Phys. Rev.*, **103**, 323 (1956).
24. J. H. Reynolds: *Rev. Sci. Instr.*, **27**, 928 (1956).
25. A. J. Dempster: *Proc. Am. Phil. Soc.*, **75**, 755 (1935).
26. A. J. Dempster: *Rev. Sci. Instr.*, **7**, 46 (1936).
27. A. J. Dempster: *Manhattan District Declassification Comm. Rept.* 370, 1946.
28. N. B. Hannay: *Rev. Sci. Instr.*, **25**, 644 (1954).
29. N. B. Hannay and A. J. Ahearn: *Anal. Chem.*, **26**, 1056 (1954).
30. J. D. Waldron: *Research*, **9**, 306, 1956.
31. R. D. Craig, G. A. Errock, and J. D. Waldron: "Advances in Mass Spectrometry" (J. Waldron, ed.), p. 136, Pergamon Press, New York, 1959.
32. J. A. James and J. L. Williams: "Advances in Mass Spectrometry" (J. Waldron, ed.), p. 157, Pergamon Press, New York, 1959.
33. J. G. Gorman, E. J. Jones, and J. A. Hipple: *Anal. Chem.*, **23**, 438 (1951).
34. A. J. Dempster: *Phys. Rev.*, **11**, 316 (1918).
35. S. Datz and E. H. Taylor: *J. Chem. Phys.*, **25**, 289 (1956).
36. P. B. Moon and M. L. Oliphant: *Proc. Roy. Soc. (London)*, **A137**, 463 (1932).
37. M. G. Inghram and R. J. Hayden: A Handbook on Mass Spectroscopy, *Natl. Acad. Sci.-Natl. Res. Council, Nucl. Sci. Ser., Rept.* 14, 1954.
38. W. Weisershausen: "Advances in Mass Spectrometry" (J. Waldron, ed.), p. 120, Pergamon Press, New York, 1959.

39. E. A. Melaika, M. J. Parker, J. A. Petruska, and R. H. Tomlinson: *Can. J. Chem.*, **33**, 830 (1955).
40. J. P. Blewett and E. J. Jones: *Phys. Rev.*, **50**, 464 (1936).
41. C. P. Keim and C. R. Baldock: "Electromagnetically Enriched Isotopes and Mass Spectrometry," p. 145, Butterworth & Co. (Publishers), Ltd., 1956.
42. C. M. Stevens: "Advances in Mass Spectrometry" (J. Waldron, ed.), p. 101, Pergamon Press, New York, 1959.
43. K. Ordzhonikidze and V. Shiuttse: *Soviet Phys. JETP*, **2**, 396 (1956).
44. J. E. Hand: *Rev. Sci. Instr.*, **24**, 181 (1953).
45. G. H. Palmer: "Electromagnetically Enriched Isotopes and Mass Spectrometry," p. 156, Butterworth & Co. (Publishers), Ltd., 1956.
46. G. Nief: "Applied Mass Spectrometry," p. 253, The Institute of Petroleum, London, 1954.
47. B. M. Gordon and L. Friedman: *Phys. Rev.*, **108**, 1053 (1957).
48. M. G. Inghram and W. A. Chupka: *Rev. Sci. Instr.*, **24**, 518 (1953).
49. G. H. Palmer: "Advances in Mass Spectrometry" (J. Waldron, ed.), p. 89, Pergamon Press, New York, 1959.
50. W. T. Leland: *Phys. Rev.*, **77**, 634 (1950).
51. G. H. Palmer and K. L. Aitken: *J. Sci. Instr.*, **30**, 314 (1953).
52. M. G. Inghram, G. Wetherill, and D. C. Hess: *Rev. Sci. Instr.*, **22**, 838 (1951).
53. R. D. Craig: *J. Sci. Instr.*, **36**, 38 (1959).
54. R. H. Roberts and J. V. Walsh: *Rev. Sci. Instr.*, **26**, 890 (1955).
55. C. M. Stevens: *Rev. Sci. Instr.*, **24**, 148 (1953).
56. M. G. Inghram and C. M. Stevens: *Rev. Sci. Instr.*, **24**, 987 (1953).
57. B. R. F. Kendall: *Rev. Sci. Instr.*, **29**, 851 (1958).
58. C. Reuterswärd: *J. Sci. Instr.*, **29**, 184 (1952).
59. Lu Hoff: *Phys. Rev.*, **53**, 845 (1938).
60. A. K. Brewer: *J. Chem. Phys.*, **4**, 350 (1936).
61. H. Ewald and H. Hintenberger: "Methoden und Anwendungen der Massenspektroskopie," Verlag Chemie GmbH, Weinheim, Germany, 1953.
62. F. W. Aston: *Phil. Mag.*, **47**, 385 (1924).
63. K. T. Bainbridge and E. B. Jordan: *Phys. Rev.*, **50**, 282 (1936).
64. F. W. Aston: "Mass Spectra and Isotopes," Edward Arnold (Publishers) Ltd., 1942.
65. J. Koch: Mass Spectroscopy in Physics Research, p. 165, *Natl. Bur. Std. (U.S.) Circ.* 522, 1953.
66. E. S. Lamar, E. W. Samson, and K. T. Compton: *Phys. Rev.*, **48**, 886 (1935).
67. M. A. Tuve, O. Dahl, and L. R. Hafstad: *Phys. Rev.*, **48**, 241 (1935).
68. J. Kistemaker: Mass Spectroscopy in Physics Research, p. 179, *Natl. Bur. Std. (U.S.) Circ.* 522, 1953.
69. R. Bernas and A. O. Nier: *Rev. Sci. Instr.*, **19**, 895 (1948).
70. C. E. Normand, L. O. Love, W. A. Bell, and W. K. Prater: "Electromagnetically Enriched Isotopes and Mass Spectrometry," p. 1, Butterworth & Co. (Publishers), Ltd., 1956.
71. R. H. Dawton: "Electromagnetically Enriched Isotopes and Mass Spectrometry," p. 37, Butterworth & Co. (Publishers), Ltd., 1956.
72. J. Backus: in Buthrie and Wakerling, (eds.), "Characteristics of Electrical Discharges in Magnetic Fields," p. 345, McGraw-Hill Book Company, Inc., New York, 1949.
73. H. Heil: *Z. Phys.*, **120**, 212 (1943).
74. O. Almen and K. O. Nielsen: "Electromagnetically Enriched Isotopes and Mass Spectrometry," p. 23, Butterworth & Co. (Publishers), Ltd., 1956.
75. A. Terenin and B. Popov: *Physik. Z. Sowjetunion*, **2**, 299 (1932).

76. F. P. Lossing and I. Tanaka: *J. Chem. Phys.*, **25**, 1031 (1956).
77. R. F. Herzog and F. F. Marmo: *J. Chem. Phys.*, **27**, 1202 (1957).
78. H. Hurzeler, M. G. Inghram, and J. D. Morrison: *J. Chem. Phys.*, **27**, 313 (1957).
79. E. W. Muller: *Phys. Rev.*, **102**, 618 (1956).
80. E. W. Muller and K. Bahadur: *Phys. Rev.*, **102**, 624 (1956).
81. R. Gomer and M. G. Inghram: *J. Am. Chem. Soc.*, **77**, 500 (1955).
82. Catalog of Mass Spectral Data, *Am. Petrol. Inst. Res. Project* 44, National Bureau of Standards, Washington, D.C.
83. R. F. Herzog and F. P. Viehbock: *Phys. Rev.*, **76**, 855 (1949).
84. G. K. Wehner: *Advan. Electron. Electron Phys.*, **7**, 239 (1955).
85. R. E. Honig: "Advances in Mass Spectrometry" (J. Waldron, ed.), p. 162, Pergamon Press, New York, 1959.
86. E. Lindholm: "Applied Mass Spectrometry," p. 191, The Institute of Petroleum, London, 1954.
87. E. Lindholm: *Arkiv. Fysik*, **8**, 433 (1954); **18**, 219 (1960).
88. R. H. Plumlee and L. P. Smith: *J. Appl. Phys.*, **21**, 811 (1950).
89. F. E. Dart: *Phys. Rev.*, **78**, 761 (1950).
90. G. D. Archard: *Brit. J. Appl. Phys.*, **5**, 395 (1954).
91. O. Klemperer: "Electron Optics," p. 291, Cambridge University Press, London, 1953.
92. G. Liebmann: *Advan. Electron. Electron Phys.*, **2**, 102 (1950).
93. G. Liebmann: *Nature*, **146**, 149 (1949).
94. G. Liebmann: *Proc. Phys. Soc.*, **B62**, 753 (1949).
95. R. Vauthier: *Compt. Rend.*, **231**, 764 (1950).
96. G. L. Weissler, J. A. R. Sampson, M. Ogawa, and G. R. Cook: *J. Opt. Soc. Am.*, **49**, 338 (1959).

5

Ion Optics

L. Kerwin

Department of Physics, Laval University
Quebec, P.Q.

5.1. Introduction

The fundamental problem confronting the mass spectrometrist is this: given a sample of mixed components at A, he wishes somehow to gather together all those of one mass at X and all those of another mass at Y. In the fifty years since he has been doing this successfully, he has utilized as his separating mechanism the action of electric and magnetic fields on the charged particles into which he converts his sample.

Various methods of applying electric and magnetic fields to the problem have been devised, but the great majority of mass spectrometers have utilized a simple electric field to accelerate the charged particles to a common energy and a homogeneous magnetic field to effect a momentum analysis which may be immediately interpreted as a mass analysis. This chapter will be confined to the study of the ion optics of this almost universal system.

In the more specialized field of mass spectrography, radial electric fields are generally used in conjunction with the basic system mentioned in the last paragraph. A discussion of several variations of such combinations will be found in Chaps. 2 and 7. Recently, several types of nonmagnetic mass spectrometers have been developed, using high-frequency pulsed and radio-frequency electric fields to effect mass separation. The ion optics of these instruments is just beginning to develop and will borrow heavily from that of the electron microscope and the cathode-ray tube (see Sec. 2.3).

As the techniques of mass spectrometry have developed, the requirements of obtaining ever greater resolution and ever higher intensity have provoked contributions to the study of mass-spectrometer *ion optics* from a great number of workers. Any specialist in the field could easily name

From these expressions we see that ions of equal charge and energy will follow trajectories of different radii according to their mass, as indicated in Fig. 5.1.   In this way the magnetic field can be used to sort out or analyze masses.   More strictly speaking, it is a momentum filter, which becomes a mass filter when the boundary conditions of equal ion energy and charge are introduced.   Table 5.1 gives a few typical values

TABLE 5.1

Magnetic field		50 gauss		200 gauss		500 gauss		1,000 gauss		5,000 gauss	
Ion	Typical mass	Energy, ev										
		500	2,000	500	2,000	500	2,000	500	2,000	500	2,000
H^+	1	64.4	130	16.1	32.2	6.4	12.8	3.2	6.4	0.64	1.28
H_2O^+	18	273	546	68	137	27	54	13.7	27	2.7	5.4
CO_2^+	44	427	854	107	214	43	86	21.4	43	4.3	8.6
Mo^+	100	645	1,290	161	322	64	128	32	64	6.4	12.8
U^+	235	990	1,980	247	494	99	198	49	99	9.9	19.8
$U_3N_4^+$	770	1,790	3,580	447	895	179	358	89	179	17.9	35.8

Radii are in centimeters for circular trajectories of ions of 500 or 2,000 ev, in uniform magnetic fields of from 50 to 5,000 gauss.

of mass-spectrometer parameters, while Fig. 5.2a and b gives the variation of R, M, B, and V for singly charged ions.   Doubly charged ions behave as singly charged ions of half the mass.

5.3. The Ideal Focusing Field

Figure 5.1 indicates how the action of a magnetic field effects a separation of the various mass components of an ion beam.   In this figure the ion beam is shown as a line, which in practice would correspond to a beam of infinitely small intensity.   Measurable beams require breadth, and in Fig. 5.3a we see how a practical divergent beam is deflected by the magnetic field.   In general, the various beam components overlap and no effective separation is obtained.   To remedy this defect, a magnetic field must be devised which will act as a lens to *focus* a divergent ion beam, as well as a prism to *disperse* it.   Such a field is indicated in Fig. 5.3b.   In this section we shall consider how such magnetic focusing is obtained.

One approach to the problem of magnetic focusing is illustrated in Fig. 5.4.   Here we consider a beam of ions diverging from a point source S and consider what shape of magnetic field boundary will result in perfect focusing of the beam at a point D.   To begin with, the symmetrical problem is considered, so that the centers of curvature of the ion trajec-

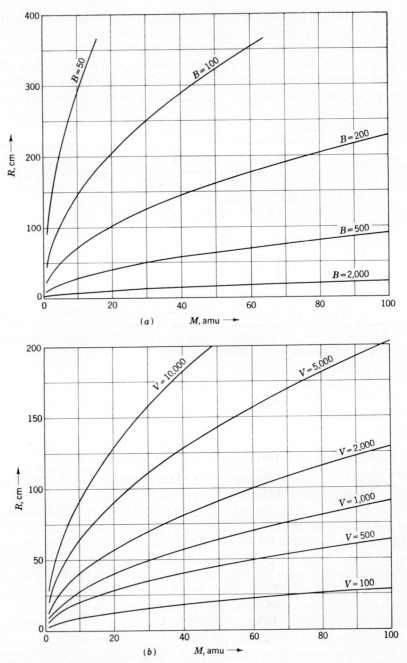

Fig. 5.2. (*a*) Radii of curvature of 1,000-ev ion trajectories as a function of mass and magnetic field intensity; (*b*) radii of curvature of ion trajectories in 500-gauss magnetic field as a function of mass and energy.

tories in the magnetic field lie on Y, which normally bisects the line SD. The two ion paths 1 and 2 will become circular when they enter the magnetic field region bounded by M and become parallel when crossing Y, since their centers of curvature lie on the latter.   From the symmetry of the problem, the two paths will converge on D.   We require the nature of M.

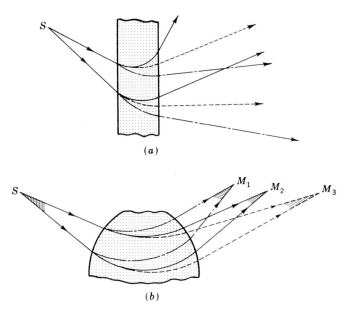

(a)

(b)

Fig. 5.3.   (a) Nonfocusing magnetic prism; (b) focusing magnetic prism.

Consider beam 1, which, forming an angle θ with SD, enters the magnetic field at $P(x, y)$, to follow a circular trajectory of radius R.   Placing $SD, = 2a$, we have, from the trigonometry of Fig. 5.4,

$$x = R \sin \theta \tag{5.8}$$
$$y = (a - x) \tan \theta \tag{5.9}$$
$$\frac{x/R}{y/(a - x)} = \frac{\sin \theta}{\tan \theta} = \cos \theta$$
$$= \frac{(R^2 - x^2)^{1/2}}{R}$$
$$y = \frac{x(a - x)}{(R^2 - x^2)^{1/2}} = f_m(x) \tag{5.10}$$

From symmetry, the boundaries of the magnetic field M on either side of Y and above and below SD will be similar.   A typical field shape is shown in Fig. 5.5 for arbitrary values of R and a.   Such a field will be referred to as an *ideal focusing field*.   It must be of opposite sign on either side of SD.   It will focus to a point on D, an ion beam of very wide divergence issuing from S.

The choice of parameters a and R will affect the shape of the ideal field, as is seen in Fig. 5.6, giving various curves M as R is varied from very small to very large fractions of a.   In general, for a given distance SD

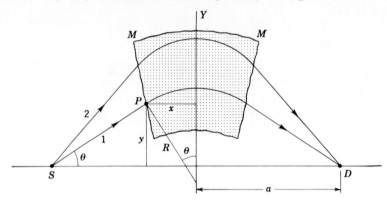

Fig. 5.4

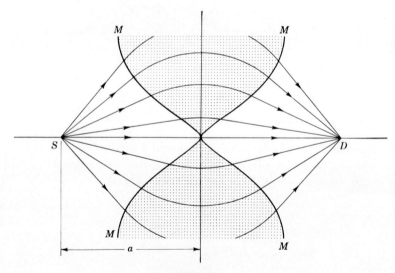

Fig. 5.5

(between source and detector), the size of the magnetic field varies roughly with R.   The intensity of the field required is given by Eq. (5.6) and is inversely proportional to R.   Thus the various magnetic fields shown in Fig. 5.6 require roughly the same total magnetic flux.

One case is shown for R greater than a.   Such a field does not lend itself readily to the focusing of widely divergent beams, since in many cases the circular and straight portions of different rays intersect.

In practice, magnetic fields of the ideal shapes illustrated in Fig. 5.6 may not be achieved, for various reasons, including difficulty of machining the pole pieces, field inhomogeneities, and fringing-field effects. However, practical fields may be used which are close approximations to the

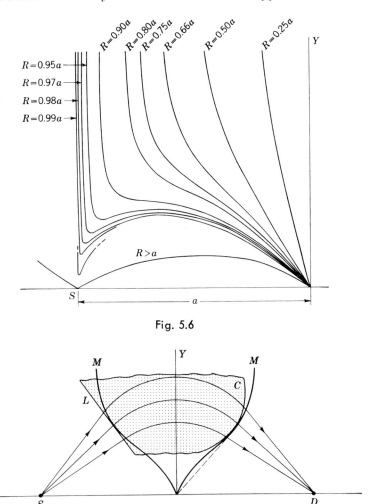

Fig. 5.6

Fig. 5.7

ideal fields. Thus, in Fig. 5.7, one side of an ideal field is approximated by a straight line L, and the other side by a curved line C.

If a practical magnetic field shape as indicated by the shading is used, then it will produce a certain amount of focusing at D. The precise amount of defocusing will depend on how good the approximations at L and C are. It is therefore necessary to develop an expression giving the amount of defocusing at D as a function of the nature and position of L and C.

5.4. Image Displacement—General

To calculate the ion optics of a practical magnetic field considered as an approximation to an ideal field, certain fundamental equations may be developed [5–9]. We begin by tracing a ray of ions which leaves the source S (Fig. 5.8) at an angle θ with SD, to enter the magnetic field bounded by A_1 at x_1y_1, follow a curved path of radius R [Eq. (5.6)], and

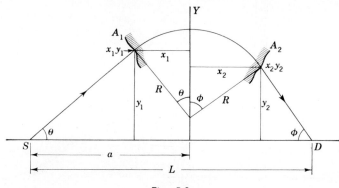

Fig. 5.8

emerge through boundary A_2 at x_2y_2 to cross the base line at D. First, we determine the distance $SD = L$ in terms of the ray and field parameters θ, R, x_1, x_2, and a.

From Fig. 5.8 we have

$$L = (a - x_1) + R \sin \theta + R \sin \phi + y_2 \cot \phi \qquad (5.11)$$

Since at the moment we are considering symmetrical cases and reasonable approximations to the ideal boundaries M, we may place $\phi = \theta + \epsilon$, where ϵ is small. When this is done, and various trigonometric relationships obtainable from Fig. 5.8 are introduced, Eq. (5.11) may be shown to be

$$L = 2a - \frac{(a - x_1)(x_2 - x_1)}{R \sin \theta \cos^2 \theta} \qquad (5.12)$$

For a symmetrical ideal field, $L = 2a$. The second term in (5.12) represents the variation in the focusing point D from the ideal symmetrical distance $2a$ caused by the boundaries A_1 and A_2 being various approximations to M and at slightly unequal distances from Y. We shall now express this displacement of the focusing point D as a function of the nature of A_1 and A_2.

In Fig. 5.9, a beam from source S which passes through the magnetic field B bounded by the ideal boundaries M will cross the base line at D, whose distance from S is given by Eq. (5.12). Since, with ideal boundaries M, $x_1 = x_2$, Eq. (5.12) gives the distance as simply $2a$. However, if approximate boundaries as at A are used, then in general the point P_1 (where the beam enters the field) is not shifted from x_1y_1 by the same amount that the point P_2 (where the beam leaves the field) is shifted from x_2y_2. Thus $x_1 \neq x_2$, and the second term of Eq. (5.12) does not equal zero. The beam crosses SD at a point removed dL from D. dL is the sum of the shift caused by the point of initial deviation moving to P_1 (a distance dx_1, which, were the shift of the exit point to P_2 the same,

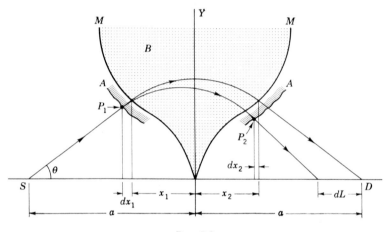

Fig. 5.9

would move the image point a total of $2dx_1$) plus the additional shift caused by the fact that these quantities are not equal and which is given as the second term in Eq. (5.12). Now we may write

$$(a - x_{p1}) = (a - x_1)$$

since the relative variation in these terms is of an order smaller than the relative variation in $(x_2 - x_1)$. We may also write exactly

$$(x_{p2} - x_{p1}) = (dx_1 - dx_2)$$

From all these considerations we have

$$dL = 2\,dx_1 + \frac{(a - x_1)(dx_1 - dx_2)}{R \sin\theta \cos^2\theta} \tag{5.13}$$

There remains to express dx_2 and dx_1 as functions of M. In the case of dx_1, Fig. 5.10 shows a beam from S entering the magnetic field through the ideal boundary M at the point x_1y_1.

If an approximate boundary A is used, the entry point x_1 is displaced an amount dx_1. From the figure we see that

$$dx_1 \tan \theta = t_1$$
$$= n_1 - dy_1$$

where n is the vertical distance between the curves M and A in the vicinity of x_1 (variations in n_1 over the region dx_1 are of an order smaller than n_1). Thus

$$dx_1 \tan \theta = n_1 - dx_1 f'_m(x)$$

where $f_m(x)$ is given by Eq. (5.10), and finally

$$dx_1 = \frac{n_1}{\tan \theta + f'_m(x)} \tag{5.14}$$

In the case of dx_2 we refer to Fig. 5.11, which is a completion of Fig. 5.10. The beam emerges through A at a point P_2 removed dx_2 from x_2.

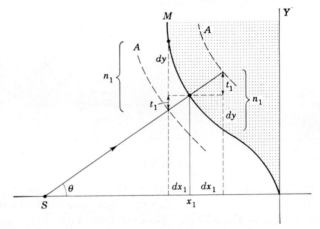

Fig. 5.10

From the figure we can obtain

$$dx_2 \tan \theta = t_2$$
$$= 2t_1 - n_2 - dy_2$$
$$= 2 dx_1 \tan \theta - n_2 - dx_2 f'_m(x_2)$$
$$dx_2 = \frac{2 dx_1 \tan \theta - n_2}{\tan \theta + f'_m(x_2)} \tag{5.15}$$

The same general remarks applied to the derivation of Eq. (5.14) are pertinent here.

The expressions (5.14) and (5.15) for dx_1 and dx_2 may now be inserted in Eq. (5.13), together with the following relations which obtain for the symmetrical case: $n_1 = n_2 = n$, and $f_m(x_1) = f_m(x_2) = f_m(x)$. After simplifying, we get

$$dL = \frac{2n}{[\tan \theta + f'_m(x)]^2} \left[\tan \theta + f'_m(x) + \frac{(a - x)f'(x)}{R \sin \theta \cos^2 \theta} \right] \quad (5.16)$$

This basic expression gives the displacement of the focal point D as a function of n, the ordinate difference between the ideal focusing curve

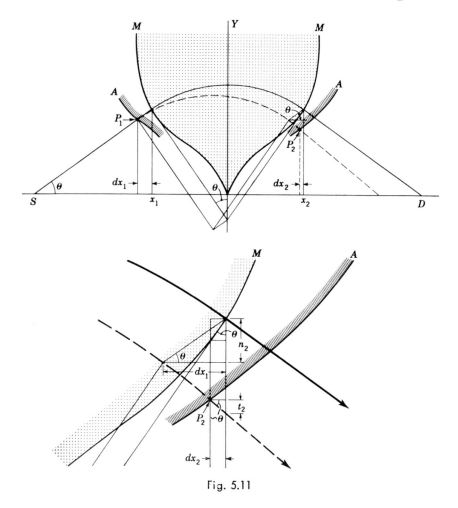

Fig. 5.11

and the approximation to it causing the displacement. It may be used for calculating aberrations, dispersions, image widths, and other parameters of magnetic lenses.

To complete the derivation of the expression for dL in its most useful form for calculating α aberrations, we may express n as a function of beam divergence α, as shown in Fig. 5.12.

Here a beam of ions of divergence α is seen leaving S at an angle θ with the base line, to be perfectly focused at D by the ideal magnetic field M.   If an approximation to the ideal field such as A is used, then, if A coincides with M at a point such as x_1y_1, the ray from S through this point will also be focused by the field A to the point D.   However, the ray through P_1 forming an angle α with the first will cross the base line at a point distant dL from D, which is a function given by Eq. (5.16) of n, the ordinate distance between M and A.   Here dL is the aberration from

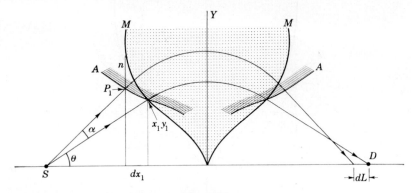

Fig. 5.12

a point image caused by the use of an ion beam of divergence α and an approximation A to the ideal curve M.

If A is a continuous function, then

$$n = f_n(dx) \tag{5.17}$$

Now
$$y = f_m(x)$$
$$= (a - x) \tan \theta$$

so
$$dy = -\tan \theta \, d\theta + \frac{(a - x) \, d\theta}{\cos^2 \theta}$$

Now putting $\alpha = d\theta$ for any curve A, which is a first or better approximation to M, $dy = f'_m(dx)$, and substituting with (5.17) in (5.16), we obtain

$$dL = \frac{2}{[\tan \theta + f'_m(x)]^2} f_n \left[\frac{(a - x)\alpha}{[\tan \theta + f'_m(x)] \cos^2 \theta} \right]$$
$$\left[\tan \theta + f'_m(x) + \frac{(a - x)f'_m(x)}{R \cos^2 \theta \sin \theta} \right] \tag{5.18}$$

α FOCUSING

In most mass spectrometers, the bulk of the ion beam may be considered as following trajectories in planes parallel to those of the pole faces. For convenience, the aberrations produced by the focusing action of magnetic fields on beams in such planes will be referred to as α *aberrations*, and the focusing action as α *focusing*. The motion of ion beams in planes at angles to those considered here gives rise to other aberrations and focusing, which will be referred to subsequently as β *focusing*.

5.5. Straight-line Approximations—General

In Sec. 5.4 it was shown how an approximation A to the ideal curve M will cause an ion beam divergent from a point source S to refocus to an

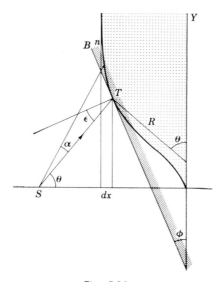

Fig. 5.13

aberrated image at D. The aberration dL is proportional to n, which is a function of the beam divergence α.

We must now consider approximations to M, which are interesting and/ or practical. The simplest approximation, as shown in Fig. 5.13, is a straight line B tangent to M at some point T.

Now the ideal curve M may be expressed as $f_m(x)$, as in Eq. (5.10). If this function is expanded in a Taylor series about the point T, we have

$$M(x_T + dx_T) = f_m(x_T) + dx_T f'_m(x_T) + \frac{dx_T^2}{2} f''_m(x_T) + \frac{dx_T^3}{6} f'''_m(x_T) + \cdots$$

The straight line B may also be expressed as a function of x, which is $f_B(x)$. Expanded in a Taylor series about the point T, this becomes

$$B(x_T + dx) = f_B(x_T) + dx_T f'_B(x_T) + 0 + 0 + \cdots$$

Second-order and subsequent terms are zero because B is a straight line. Now, since the point T is common to M and B, we have $f_M(x_T) = f_B(x_T)$. Similarly, because the lines are tangent at T, they have the same slope at this point, so that $f'_M(x_T) = f'_B(x_T)$. Accordingly, the two lines M and B differ in their Taylor expansions about the point T only in the third and higher terms. We may thus write

$$n_B = M(x_T + dx_T) - B(x_T + dx_T) = \frac{dx_T^2}{2} f''_m(x_T) + \frac{dx_T^3}{6} f'''_m(x_T) + \cdots$$

(5.19)

Placing this value for n in Eq. (5.16), we obtain, by neglecting orders higher than the second,

$$dL = \frac{(a - x)^2 \alpha^2 f''_m(x_T)}{[\tan \theta + f'_m(x_T)]^4 \cos^2 \theta} \left[\tan \theta + f'_m(x_T) + \frac{(a - x) f'_m(x_T)}{R \sin \theta \cos^2 \theta} \right] \quad (5.20)$$

This expression gives the aberration produced by any linear approximation to the ideal focusing field. Since a linear approximation is a second-order approximation, a second-order term appears in the aberration (that is, $dL = k\alpha^2$). This amount of focusing is called *first-order direction focusing*. There are an infinite number of cases possible, since an infinite number of lines may be tangent to M.

A very important expression may be derived from Fig. 5.13. Considering the distance from the source S to the point of entry T as an *object distance*, it may be shown, from Fig. 5.13, that

$$ST = l' = R \cos^2 \theta (\cot \phi + \tan \theta) \quad (5.20a)$$

and further that

$$\frac{R \cos^2 (\phi - \theta)}{\sin^2 \phi} = R \cos^2 \theta (\cot \phi + \tan \theta) - \frac{R \cos (\theta + \phi) \cos (\theta - \phi)}{\sin^2 \phi}$$

Now placing

$\epsilon = \theta - \phi$ = angle of beam entry

$\Omega = 2\phi$ = angle of magnet sector

$f = \dfrac{R \cos^2 \epsilon}{\sin \Omega}$ = focal length

$l' = R \cos^2 \theta (\cot \phi + \tan \theta)$ = object distance from field boundary

$g = \dfrac{R \cos (\Omega + \epsilon) \cos \epsilon}{\sin \Omega}$ = focal-point distance from field boundary

we have
$$f = l' - g \quad (5.21)$$

As indicated in Fig. 5.14, this important relation, first derived by Herzog, shows the focusing action of a magnetic lens as being similar to that of a Newtonian optical system. For the symmetrical case being considered, the image and object distances are the same.

To summarize, simple symmetrical approximations to the ideal focusing field may be made by straight lines, tangent to the ideal field at any point. The geometry may be determined from that of the ideal field or from consideration of the field as an optical thick lens. The aberration is calculated by means of Eq. (5.20).

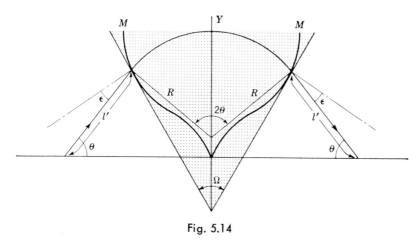

Fig. 5.14

We shall now see that some straight-line approximations are of particular interest.

5.6. Straight-line Approximations—Special

Equation (5.20) gives the aberration produced by any linear approximation to the ideal focusing curve. Some linear approximations are of particular interest

A. The normal case

An examination of Fig. 5.6 will show that each ideal focusing boundary has a point at minimum distance from S. Thus, in Fig. 5.15, the point P_n on M is such that SP_n is minimum. Therefore, since the derivative of the curve M with respect to SP_n at this point is zero, the tangent to M at this point will be at right angles to SP_n. In other words, a beam from S entering at P_n will do so at right angles to M.

To find P_n, we may express SP_n as a function of θ:

$$SP_n = SB - P_nB$$
$$= a \sec \theta - R \tan \theta$$

Then
$$\frac{d(SP)}{d\theta} = a \tan \theta \sec \theta - R \sec^2 \theta$$

Placing $d(SP)/d\theta$ equal to zero, we have as a condition that SP be minimum

$$R = a \sin \theta \qquad (5.22)$$

From Fig. 5.15, we see that P_nO also equals $a \sin \theta$. Therefore, for this case, $P_nO = R$, and the center of curvature coincides with O. Therefore, if as an approximation to M we use the straight line N tangent to M at the point P_n of minimum distance from S, the center of curvature of the ion beam will pass through O, as will the apex of the wedge formed by both sides of N, as seen in Fig. 5.15.

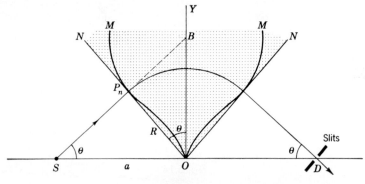

Fig. 5.15

This particular case of first-order focusing was developed by Barber [12] and Stephens [13] and is used in the familiar sector instruments, in which the source, magnet apex, and detector are in line and the ion beam enters and leaves the magnetic field at *right angles*. We have therefore called it the *normal* focusing field.

To calculate the aberration produced by this field, we use Eq. (5.20), introducing the condition (5.22), and the corresponding fact $f'_m(x) = \cot \theta$. When this is done, the expression (5.18) collapses to the following:

$$dL_n = a\alpha^2 \qquad (5.23)$$

It must be remembered that dL is measured along the line SD. In most mass spectrometers, the ion-beam detecting slits are placed normal to the beam, at an angle $(90° - \theta)$ with SD, as in Fig. 5.22. Therefore the aberration in this case, which we shall call A_n, becomes

$$A_n = dL \sin \theta = a\alpha^2 \sin \theta$$
$$= R\alpha^2 \qquad (5.24)$$

This is the familiar expression first derived by Barber and Stephens.

Equations (5.23) and (5.24) are independent of θ. This may be visualized qualitatively from Fig. 5.6 by noting that, as θ increases, the worsening approximation of a straight line to the ideal curve is compensated by the smaller spread of the ion beam as it enters the field for a given divergence α, since the distance SP decreases as θ increases. In Fig. 5.16 we see several possible cases of normal focusing fields. Those on the left have in common the source-detector distance $2a$ and the aberration $a\alpha^2$ measured along SD. Those on the right have in common the radius of curvature R and the aberration $R\alpha^2$ measured normally to the ion beam.

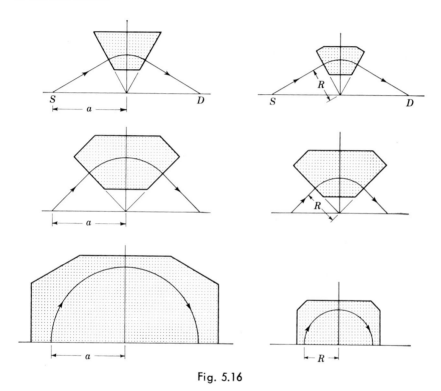

Fig. 5.16

The various cases differ in the angle θ, the size of the magnet required, and the extra-magnetic-field trajectory length, which is of consequence in evaluating fringing-field effects. Figure 5.16 depicts cases for $\theta = 30°$, 45°, and 90°, which have proved popular with designers. Of particular interest is the 90° case, for which the two sides of the magnetic field become a single straight line and the ion deviation becomes 180°. This particular case may be derived very simply from the consideration of circular trajectories as indicated in Fig. 5.17. It was known to Classen [10] and used by Dempster [11] in his first mass spectroscope.

Examination of Fig. 5.17 shows that perfect direction focusing occurs after the beam has turned through 360°. However, this feature is not useful in itself, because of the necessity of placing the source and detector at the same point and because *all* masses are perfectly focused at the same point (i.e., no net dispersion after 360° deviation). These difficulties may be avoided by combining the magnetic field with an electric field, as has been done by Richards, Hays, Goudsmit, and Smith. One of Smith's [14] arrangements is shown in Fig. 5.18. The ion beam from the source after a preliminary 180° (orbit 1) turn is decelerated by pulser slits P. It then makes a number of circular turns along orbit 2, missing the source and increasing the time dispersion between masses at every turn. After n turns, it receives a second decelerating pulse, which causes it to follow orbit 3 into the detector at the perfect focusing 360° deflection position.

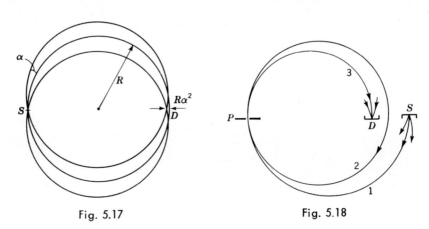

Fig. 5.17   Fig. 5.18

The group of normal fields of constant a on the left of Fig. 5.16 are such that, as θ increases, R and the size of the magnet also increase. However, the total magnetic flux required is about the same. The group of normal fields of constant R also requires larger magnets as θ is increased, and so the total flux required increases. Referring to Fig. 5.6, it may be noted that the various points P of entry in the normal cases lie on a semicircle of radius $a/2$.

B. The inflection case [5–9, 15–18]

An examination of the ideal-focusing-field equation (5.10) and the field shapes of Fig. 5.6 shows that the ideal curve has an inflection point. This may be found by equating the second derivative of Eq. (5.10) to zero, when we find

$$x_f = \frac{3a + (9a^2 - 8R^2)^{1/2}}{2} \tag{5.25}$$

In Fig. 5.19 such an inflection point is indicated at P_f.

Obviously, a straight-line approximation to the curve M will have an optimum fit when tangent to M at the inflection point.   Referring to Eq. (5.19), we see that, in fact, the difference between the ideal curve M and a straight line at this point is equal to the third-order term in the

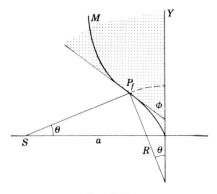

Fig. 5.19

Taylor expansion of the function for M.   Proceeding as in Sec. 5.5, we thus find

$$n_f = \frac{dx_T{}^3}{6} f_m'''(x_T) \text{ and higher terms} \tag{5.26}$$

The aberration produced by this approximation to the ideal field may be calculated using Eq. (5.18) and condition (5.26).   We then obtain

$$dL_f = \tfrac{2}{3}\alpha^3 \frac{(3a - x)(3a - 2x)}{[6x(a - x)]^{\frac{1}{2}}} \tag{5.27}$$

If the detector slit is placed normally to the ion beam, then the aberration becomes

$$A_f = \tfrac{2}{3}\alpha^3 \frac{(3a - x)(3a - 2x)}{[6x(a - x)]^{\frac{1}{2}}} \sin\theta \tag{5.28}$$

Since the inflection tangent is a third-order approximation to M_3, the aberration produced contains the third-order term α^3.   This is of an order smaller than that produced by other straight-line approximations, and this type of second-order magnetic focusing field is called an *inflection field*.

In Fig. 5.20 are seen several examples of inflection fields.   As before, the group on the left have common a and aberration dL_f.   Those on the right have common R and aberration A_f.

Comparison with Fig. 5.16, drawn to the same scale for a, shows that the inflection field requires a smaller magnet than the normal field for a given a or for a given R, in addition to providing improved focusing. However, the inflection field involves nonnormal entry of the ion beam and provides less dispersion (because of the smaller magnetic flux traversed) than the normal case.   A comparison of the effectiveness of the two types of field will vary somewhat, depending on whether a, R, magnet size, or other parameter is used as the common factor.

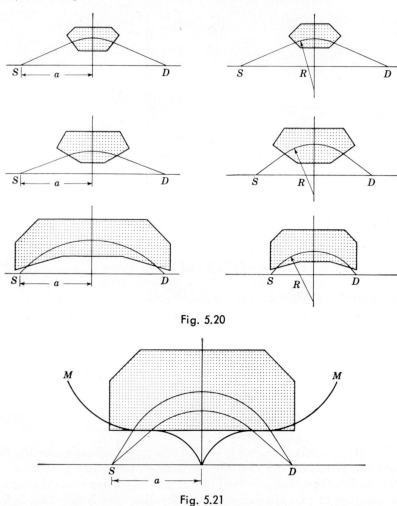

Fig. 5.20

Fig. 5.21

An inflection case of particular interest is shown in Fig. 5.21.   Here the ratio of R to a is such that the inflection tangent is parallel to SD, so that the magnet profile becomes a straight line, in somewhat analogous fashion to the 180° normal case.   However, the magnet required is smaller, and of course second-order focusing is produced.   The parameters of this case are included in Table 5.2, which gives various useful design data for normal and inflection cases.   For this case $\theta = 54.7°$, and, as will be shown later, it is the optimum inflection case (Fig. 5.47).

TABLE 5.2. Design Parameters for Various Normal and Inflection Fields

θ	R_n	R_f	x_n	x_f	ϕ (Fig. 5.19)
0°	0.000a	0.000a	0.000a	0.000a	0.0°
10°	0.174a	0.257a	0.030a	0.045a	15.0°
20°	0.342a	0.485a	0.117a	0.166a	30.2°
30°	0.500a	0.666a	0.250a	0.333a	46.0°
40°	0.647a	0.799a	0.413a	0.513a	62.5°
45°	0.707a	0.848a	0.500a	0.600a	71.5°
50°	0.767a	0.888a	0.587a	0.680a	80.7°
54.7°	0.816a	0.918a	0.666a	0.750a	90.0°
60°	0.866a	0.945a	0.750a	0.818a	101°
70°	0.940a	0.978a	0.884a	0.919a	125°
80°	0.985a	0.995a	0.970a	0.980a	150°
90°	1.000a	1.000a	1.000a	1.000a	180°

C. Other linear approximations

In Fig. 5.22, magnetic field boundary M represents any given ideal focusing field. In part A of this section we discussed the normal approximation N to this curve, and in part B, the inflection approximation I. Linear approximations remote from the center such as at B will give first-order focusing, and because of the long trajectory in the magnetic

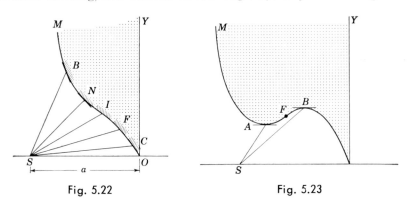

Fig. 5.22   Fig. 5.23

field, the dispersion will be greater. This advantage may be reduced by the problems raised by the long ion-beam path. Approximations closer to Y than are N and I such as F will give first-order focusing, with a small aberration approaching second-order focusing as F approaches I. Linear approximations nearer to the center such as at C will give first-order focusing and the advantage of a very small magnet and short ion path. However, the small magnetic flux traversed will result in correspondingly small dispersion. In all these cases the aberration may be calculated by means of Eq. (5.18).

One more linear approximation to the ideal curve M which is of a certain interest is shown in Fig. 5.23. When R is greater than $0.92a$, a portion of the ideal curve M has negative slope. This means that there are two points, A and B in Fig. 5.23, which have slope zero, i.e., are parallel to line SD. Accordingly, straight lines tangent to M at these points would give magnetic field boundaries which are continuous straight lines, similar to the inflection case illustrated in Fig. 5.21. The first-order focusing produced by these approximations would be about the same as those of the corresponding normal cases. The positions of points A and B may be calculated by placing the derivative of Eq. (5.10) equal to zero and solving the third-order equation which results. For R smaller than $0.92a$, this equation has no roots corresponding to real points on the useful curve M. For R greater than $0.92a$, there are two real roots, corresponding to the cases A and B, and a third which is not on M. The aberrations produced by these two cases are of opposite sign, i.e., lie on the opposite sides of the central ray. As R approaches $0.92a$, these aberrations therefore become smaller, vanishing when $R = 0.92a$, for which value the equation has two roots that are equal. At this point, of course, points A and B have converged to the inflection point, and this provides an independent derivation of the particular inflection case illustrated in Fig. 5.21, where $R = 0.92a$.

5.7. Circular Approximations—General [7–9, 19, 20]

In Fig. 5.24 is shown how the ideal focusing curve M may be closely approximated at any point P by a tangent circle C. For best fit, the

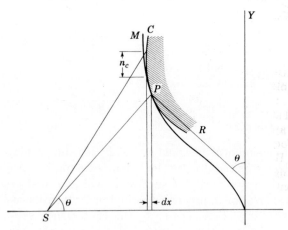

Fig. 5.24

radius of C should equal the radius of curvature of M at the point P. Referring to Eq. (5.17), we see that the line M and such a circle tangent to it differ in the third-order term of their Taylor expansions.   Proceeding as in Sec. 5.5, we thus find

$$n_c = \frac{dx_T^3}{6} f_m'''(x_T) \text{ and higher terms} \qquad (5.28a)$$

This expression is the same as (5.26), because the inflection case may be considered as a circular approximation of radius infinity at the inflection point.

Upon substituting expression (5.28a) in the general equation (5.19), the aberration produced by any circular approximation to the ideal curve M may be obtained.   It is proportional to $R\alpha^3$ and thus of an order smaller than that of the general linear approximation.

The radius of curvature of the circle C which must be used may be derived from the general expression for the curvature of a function,

$$R_c = \frac{[1 + f_m'(x)^2]^{3/2}}{f_m''(x)} \qquad (5.29)$$

where $f_m(x)$ is given by Eq. (5.10).

The focal length and other optical parameters of the circular approximations are the same as for the linear approximations, the only difference being in the reduction of aberration.   Accordingly, Herzog's expression (5.21) may be used for these cases.

5.8. Circular Approximations—Special [7–21]

A. Normal-circle case

In this case the point P_{nc} is such that the ion beam enters the field normally.   As shown in Fig. 5.25, the center of curvature of C then lies on the extension of the central ray. If the conditions for normal entry given in Sec. 5.6A are then introduced in Eq. (5.29), we obtain, for the radius of curvature for this case,

$$R_{nc} = R \cot^3 \theta \qquad (5.30)$$

This second-order-focusing magnetic field will be referred to as the *normal-circle* field.   Of the various values of θ which may be used, the 45° case is interesting in that R_{nc} is then equal to R, the radius of curva-

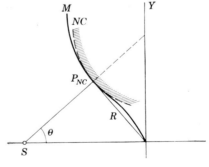

Fig. 5.25

ture of the ion-beam trajectory in the magnetic field.   The center of curvature thus lies on Y, and the complete magnet profile becomes a circle, which is easy to machine.   This case is shown with some typical others in Fig. 5.26.   The normal-circle case was first described by Bainbridge [19].

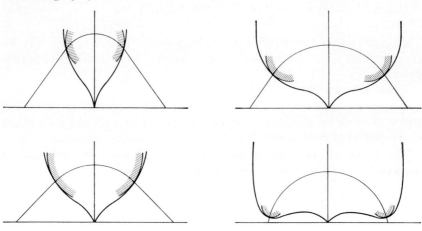

Fig. 5.26

The aberration produced by the normal-circle case is obtained by substituting condition (5.28a) and the normal-entry conditions into general

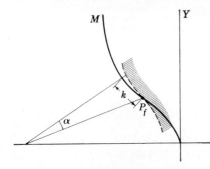

Fig. 5.27

equation (5.18).   The latter then simplifies greatly to

$$dL_{nc} = R\alpha^3 \sec \theta \quad (5.31)$$

Measured normally to the ion-beam path, we have

$$A_{nc} = R\alpha^3 \tan \theta \quad (5.32)$$

B. Inflection-circle case

As shown in Fig. 5.27, the ideal curve M may be very closely approximated at the inflection point by two arcs of circles tangent at P_f. The radius of curvature of M at this point is of course infinite, but the radii of the approximating circles may be chosen so that the *average* of the residual third-order aberration term is zero over a region k on either side of the central ray.   Calculations which become somewhat lengthy show that the radii of curvature of the two circles are equal and are given by

$$R_{fc} = \frac{4}{f_m'''(x)k} \quad (5.33)$$

In practice, such a field may not be obtained because of fringing-field effects, and so it is of academic interest only.

C. Third-order-circle case

The ideal curve M has a point of maximum curvature, and at this point a circular approximation is a third-order approximation. Persson has calculated the parameters for this high-order-focusing field and finds that it generally involves large deflection, as may be appreciated from the curves in Fig. 5.6. Because the acute angle of entry favors β focusing, and in view of the considerable dispersion, this case is interesting.

5.9. Asymmetric Cases

A. Slightly asymmetric cases

Some slightly asymmetric cases may be treated by the following method. As indicated in Fig. 5.28, M is the ideal field boundary and

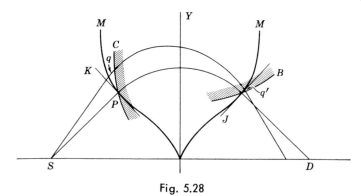

Fig. 5.28

C is any curve coincident with M at P. Then any ray from S entering the field defined by C at any point other than P will pass through an interval of magnetic field q, which will cause it to be aberrated from the ideal focusing point D. If, using a magnetic field boundary C on the left, we still wish to focus perfectly an ion beam at D, then the point where an extra-P beam leaves the field boundary on the right must not be on M, but at a point distant q' from it, such that the aberration produced by q' is equal to that produced by q, but of the opposite sign. Although not immediately obvious, it may be shown that, to the third degree of approximation, q' must equal q. By finding q' for each ray, we establish a curve B which differs from M by q' and corrects the defocusing effect of C. Placing an axis of reference K tangent to M at P, we have

$$C = C(K) \tag{5.34}$$

Placing axis J tangent to M at Q, we have

$$B = B(J) \tag{5.35}$$

Then, since $q = C(K) - f_m(K)$ and $q' = B(J) - f_m(J)$ and since $q = q'$ and $f_m(K) = f_m(J)$, we have

$$C(K) + B(J) = 2f_m(K) \tag{5.36}$$

Expanding these functions in Taylor series at the points P and Q, neglecting fourth-order terms, and remembering that to this degree of approximation $\Delta J = \Delta K$, we have

$$\Delta K \left[C'(K) + B'(J) \right] + \frac{\Delta K^2}{2} \left[C''(K) + B''(J) \right] + \frac{\Delta K^3}{6} \left[C'''(K) + B'''(J) \right]$$

$$= 2 \left[\Delta K f_m'(K) + \frac{\Delta K^2}{2} f_m''(K) + \frac{\Delta K^3}{6} f'''(K) \right] \tag{5.37}$$

Equation (5.37) enables us to determine the exit boundary shape B, which will correct aberrations produced by an entrance boundary shape C.

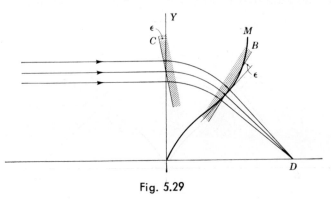

Fig. 5.29

Many cases could be considered. For example, if C is the straight line of the normal case, then it may be shown, by Eq. (5.37), that B must be a circle of radius

$$R_B = \frac{R(\cot^3 \theta)}{2} \tag{5.38}$$

if second-order focusing is to be achieved. If C is the straight line of the inflection case, then B must also be a straight line for second-order focusing (i.e., the ordinary inflection case) or tangent circles for third-order focusing (i.e., the inflection-circle case). If C is the normal-circle approximation, then B is given by Eq. (5.37) as a circle of the same radius as C for second-order focusing (i.e., the normal-circle case).

As shown in Fig. 5.29, this method may be applied to the case of "half" an ideal focusing field, which is used to focus a parallel beam of ions to a point at D. If the boundary of entry Y is approximated by a line C, then perfect focusing will be accomplished by a boundary of exit B which is a function of C. In particular, if C is a straight line at an angle ϵ with Y, then B must be a straight line at an angle ϵ with M at any point for first-order focusing, if the beam deflection is not too large.

B. Asymmetric cases

All of the ideal magnetic-focusing lenses considered so far have been symmetric. We may extend the theory to the more general case by considering the ideal focusing field M as being composed of two halves as shown in Fig. 5.30a.

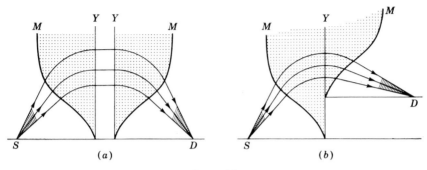

Fig. 5.30

Here we see that the first half of the field converts a diverging beam to a parallel one, which the second half focuses to a point image at D.

Since all that is necessary for refocusing is that the second half receive a parallel beam, any section of the first half may be used with any section of the second to produce perfect focusing, as indicated in Fig. 5.30b. If the two halves are united in a single magnet, then it is necessary that the radius of curvature of the ion beam be the same in each half. However, a need not be the same for each half.

C. Linear approximations—general case

In the general asymmetric case, any section of the first half of the field is joined with any section of the second half, and both boundaries are approximated by straight lines as shown in Fig. 5.31.

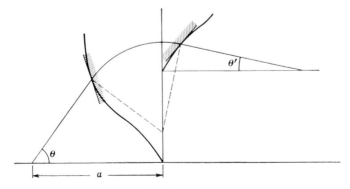

Fig. 5.31

Since the field boundaries in this case are linear, in general such asymmetric cases produce first-order direction focusing. This construction may be adapted to any sector magnetic field, for any angle of beam

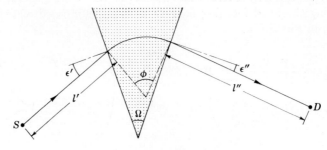

Fig. 5.32

entry. When the methods of Sec. 5.5 are extended to this case, the general Herzog equation [8] may be obtained:

$$(l' - g')(l'' - g'') = f^2 \qquad (5.39)$$

where $f = R\,\dfrac{\cos \epsilon \cos \epsilon'}{\sin \Omega} = $ focal length

$l' = $ object distance (from field boundary)

$l'' = $ image distance (from field boundary)

$g' = R\,\dfrac{\cos \epsilon' \cos (\phi - \epsilon'')}{\sin \Omega}$

$g'' = R\,\dfrac{\cos \epsilon'' \cos (\phi - \epsilon')}{\sin \Omega}$

The quantities involved in this equation are shown in Fig. 5.32.

D. Linear approximations—special cases

There are obviously a very great number of asymmetric lenses with linear boundaries which might be considered. In Fig. 5.33 several are given as examples. Part (*a*) shows a rather arbitrary type, giving first-order focusing. Part (*b*) is an example of a normal-entry asymmetric type. Part (*c*) shows an asymmetric inflection lens, giving second-order focusing.

In Fig. 5.34 is shown another asymmetric case of special interest because it produces second-order focusing in spite of the use of second-order approximations to the ideal curve M. This case is based on the observation that the aberration is positive or negative (i.e., on one or the other side of the central ion ray), depending on whether n is positive or negative, i.e., whether the concave or convex part of the ideal curve is approximated. We may therefore obtain a better correction if we combine the convex part of the boundary P at the entrance with the

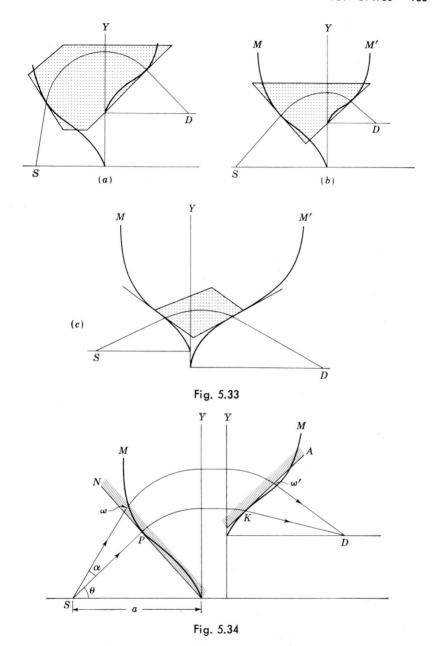

Fig. 5.33

Fig. 5.34

concave section K at the exit.   The result is an interesting combination in which the defocusing produced by one boundary is compensated by an opposite defocusing produced at the other boundary.   In Table 5.3 are given the parameters of Fig. 5.34 for several values of θ.   These were initially calculated by Kerwin and Geoffrion [69], but the published values contained small errors which were found by Persson [15], whose values are given in the table. Figure 5.35 indicates the appearance of these cases.

TABLE 5.3

θ	R	θ'	l''
30°	0.500a	11.5°	1.86R
45°	0.707a	19.3°	1.20R
90°	1.000a	47.0°	0.50R

It is not necessary that a be the same on both sides of Y, nor is it necessary to have normal entry of the ion beam as given in the above example. Thus there are many second-order-focusing combinations possible, some of which involve *acute* entry of the ion beam into the field boundary, which is of significance in considering β aberrations, as we shall see.

This method of correcting aberrations is quite powerful, and in general it may be said that a magnetic field boundary of any shape may be

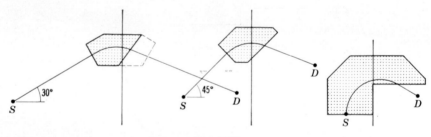

Fig. 5.35

matched by one designed to produce an aberration of any order and of any required focal length.

Among other special asymmetric magnetic lenses is that of Smythe, Rumbaugh, and West [21], which they designed following experiments with a field identical to half an ideal focusing field (Fig. 5.29). The particular field which they finally used is shown in Fig. 5.36. C is the center of a circle of any radius r. C'' is located R from C (R being the radius of curvature of the ion trajectory in the magnetic field) and is the center of a circle of radius $(R^2 + r^2)^{1/2}$. It is obvious from the drawing that the points C, C', M, N form a parallelogram, so that the final ion-beam direction is parallel to the x axis regardless of the value of θ. This field focuses a divergent beam to a parallel one, in the manner of Fig. 5.29.

Persson [15] has used Cartan's [9] graphical methods (Sec. 5.17) to investigate many types of fields and, among others, has calculated an asymmetrical one somewhat similar to the symmetrical inflection field.

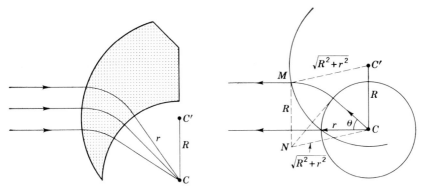

Fig. 5.36

5.10. Minimizing Aberrations

The various approximations to the ideal-magnetic-field boundary discussed thus far have this in common: the central ion beam passes through the points of coincidence between the ideal boundary and the approximation, and hence is not aberrated, while rays to either side of the central ray are aberrated by an amount which is a function of their distance from the central ray. For all practical purposes, the aberration

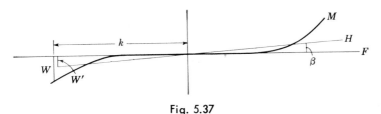

Fig. 5.37

of an instrument is its maximum aberration, which is provided by the outer rays of the beam. It is possible to reduce the instrument's effective aberration by reducing this maximum aberration, at the expense of aberrating the central rays. In other words, it is possible to minimize the maximum aberration by simple methods, which will be illustrated for the inflection case.

Referring to Fig. 5.37, curve M represents the ideal-magnetic-field boundary, and line F, the inflection approximation. Then, over a beam width k, the maximum aberration is proportional to w. If, instead of F, the line H at an angle β with F were used as the effective field boundary, then the maximum aberration would be proportional to w'. The condition for which w' will be a minimum is best found by averaging the squares of the deviations and finding the condition for which this is a minimum. For this case it turns out that, for minimum aberration,

$$\tan \beta = \frac{f'''(x_m)k^2}{10}$$

In a typical case, if the inflection field boundaries are tilted so as to coincide with H, the aberration will be reduced by a factor of about 2.

β FOCUSING

As opposed to α *focusing*, which concerns ion optics of rays in planes parallel to the pole face, the term β *focusing* concerns the ion optics of rays which move at an angle with these planes.

5.11. Homogeneous Fields

Viewed in the plane of the magnetic-pole face, most mass spectrometers have source slits which are very narrow; however, the ion beam divergent from the slit by $\pm \alpha$ results in a beam width at the center of the trajectory that can be considerable.

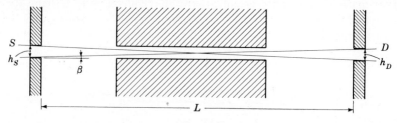

Fig. 5.38

Viewed at right angles to this, the beam is usually much narrower at the center of the trajectory, being confined to the pole-gap width at most. It is usual for the source slit to extend right across this width, a situation such as indicated in Fig. 5.38.

From this figure, we see that the maximum angle which an ion ray can make with the pole face is given by the following expression:

$$\beta = \frac{h_S + h_D}{2L} \tag{5.40}$$

where h_S, h_D = source and detector slit lengths

L = length of central ion trajectory

In practice, h_S, h_D, and the effective gap are often equal.

Ions following a path at an angle with the pole face will traverse a homogeneous field (neglecting fringing-field effects) that is a component of the main field H equal to $H \cos \beta$. Consequently, these ions will be deflected less than those following paths parallel to the pole faces and arrive at the detector separated from these by an amount which is here called β *aberration*.

The variation in the radius of curvature may be obtained as follows [22]. From Eq. (5.6), we see that

$$\Delta R = R \frac{\Delta H}{H} \tag{5.41}$$

Here ΔH is given by $(H - H \cos \beta)$, whence, by expanding $\cos \beta$ in series and neglecting high-order terms, we have

$$\Delta R = R \frac{H(\beta^2/2)}{H} = \frac{R\beta^2}{2} \tag{5.42}$$

The effect of a variation in R on the image displacement may be calculated by the methods developed in Secs. 5.13 and 5.14, for calculating dispersion. For example, when applied to the normal case, the dispersion is given as $2 \Delta R$ (for a given variation ΔR, which, in Secs. 5.13 and 5.14, is coupled with M and ΔM). As a result the β aberration for the normal case is

$$A_{\beta n} = R\beta^2 \tag{5.43}$$

This expression also applies closely to the normal-circle case.

For the inflection and inflection-circle cases, the β aberration is

$$A_{\beta f} = \frac{2R\beta^2}{3} \tag{5.44}$$

β aberration must be added to the α aberration and other components in calculating total image width, as discussed in Sec. 5.16.

5.12. Fringing Fields

Figure 5.39a represents the ion beam of a mass spectrometer passing through the magnetic field created by two pole pieces. The view is perpendicular to the plane of these pole faces.

The central ion beam always crosses the field lines at right angles. However, the beam following a trajectory in a plane parallel to the central plane crosses the field lines at an angle, one component of the field being in the same direction as the main field H_m, the other being normal to the pole boundary H_n. If the angle of entry of the beam into the field with the normal to the field boundary is ϵ, as shown in Fig. 5.39b, H_n will also have two components, one along the beam path H_{nb} and one normal to the beam path H_{nn}. This latter component will accordingly bend the ion beam in the direction of the main field lines, so that the beam will either spread out or converge in the β direction. Focusing occurs when ϵ is positive, i.e., increasing toward the center of curvature. Defocusing occurs when ϵ is negative.

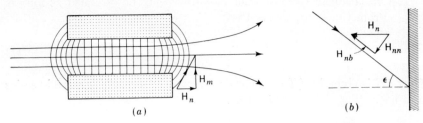

(a)   (b)

Fig. 5.39

If ϵ is positive, Cotte [23] has shown that the β-focal point is situated at a distance l'' from the field boundary given by the following expression, which is due to Cross (placing $R = 1$):

$$\frac{1}{l''} = \tan \epsilon_2 - \frac{1}{\phi - \cot n} \tag{5.45}$$

where n is given by

$$\tan n = \tan \epsilon_1 - \frac{1}{l'}$$

and l' = distance of source from pole boundary.

Herzog has shown that the β focusing takes place in such a way that the fringing field may be considered as a cylindrical lens of focal length

$$f = \frac{R}{\tan \epsilon} \tag{5.46}$$

if a parallel beam enters the field at 90° as in Fig. 5.29. Consideration of this figure will show that expressions (5.45) and (5.46) are equivalent.

Figure 5.40 given by Herzog [24] shows the variation of the β-focal point with angle ϵ, as well as the corresponding α focusing for a given half deviation of beam ϕ. It is seen that, for the particular angle $\phi = 50°$ illustrated, both occur at the same point when $\epsilon = 31°$. In this way two-directional focusing of the beam occurs, which may be very advantageous in concentrating weak beams on a photographic plate.

Cross [25] has worked out the two-directional-focusing conditions in a general way and has calculated them for various values of beam deviation. His values for $\phi = 45°$ are given in Fig. 5.41.

Cross's general condition is given as follows:

$$\frac{1}{l_d''} = \frac{1}{2}\left[\tan\left(\phi - \gamma\right) - \frac{1}{\phi - \cot n}\right] \qquad (5.47)$$

where l_d'' is the distance of the two-directional-focusing image from the pole boundary, n is as defined for (5.45), $R = 1$, and $\tan\gamma = \tan\epsilon_1 + 1/l'$.

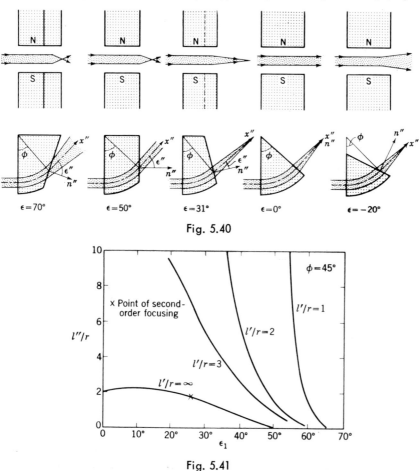

Fig. 5.40

Fig. 5.41

In the case of normal focusing fields and normal-circle fields, it is obvious that that β focusing cannot be achieved. For these cases $\epsilon = 0$, and thus the focal length of the β-focusing field is infinite, from (5.46). This is the case illustrated in Fig. 5.40 for $\phi = 0°$.

In the case of inflection fields and inflection-circle fields, there occurs β defocusing.

However, as discussed in Sec. 5.9, there are asymmetric cases where second-order focusing can be achieved with acute entry of the ion beam, so that two-directional focusing can be achieved. Such cases have been calculated by Cross, and one is marked x in Fig. 5.41.

Camac [26] has worked out β-focusing conditions applied to successive magnetic analyzers and derived some particular solutions of Herzog's and Cotte's more general theories.

These theories assume that the pole gap is small and measurements are difficult. However, Herzog has achieved two-directional focusing with the magnetic field of a double-focusing spectrograph. It was necessary to shield out much of the fringing field, i.e., reduce it to conditions approximating the assumptions of the theory. When unshielded, the fringing field of his magnet ruined the β-focusing action, although the gap width was relatively modest.

β focusing should receive more consideration in future design.

DISPERSION

5.13. Dispersion—General Equation

Figure 5.1 illustrates how ion beams of differing masses are deviated by magnetic fields by various amounts. The total deviation produced in a

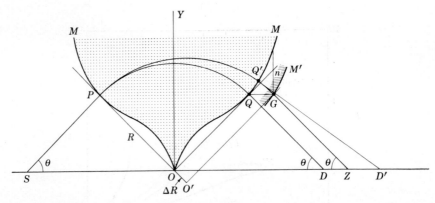

Fig. 5.42

given case depends on the mass, charge, and energy of the ion, on the magnetic field strength, and on the magnetic field extent [5, 6].

For a given magnetic field, ions of a given charge and energy will be variously deviated according to their mass. The variation of this deviation *per percentage mass difference* is called the dispersion of the magnetic prism. In general, the dispersion of any prism is a function of its total magnetic flux. The theory developed in this chapter permits us to calculate the dispersion produced by any type of plane homogeneous magnetic field, as follows.

In Fig. 5.42, the ideal field M causes a beam from S to be refocused at D after being curved in the magnetic field in an arc of radius R. If, upon entering the field at P, the particle is curved through a radius $(R + \Delta R)$ instead of R, then it will leave the field at Q' instead of at Q and will cross the base line at D' instead of at D. For a given magnetic field and ion energy, the change in radius will be caused by a change in mass ΔM of the particle. For ΔM equal to a unit per cent mass, the distance DD' is the dispersion produced by the field.

This dispersion is seen to consist of two parts. If the ray leaving the field at Q' had remained in the field until its direction was parallel to QD, then it would have focused at Z. Under these conditions it may be shown that QG is parallel to DZ, which is then given by

$$DZ = 2 \Delta R \sin \theta \tag{5.48}$$

The point G may then be considered as lying on a curve M', which would focus the beam perfectly at Z. Actually, the beam leaves the field at Q', on a boundary M, which may be considered as being displaced from the "ideal" boundary M'. Equation (5.16) applies for calculating the distance ZD' as a function of n. We must remember to divide the displacement by 2, as the ideal field is displaced on one side only. We thus obtain as the general equation for the dispersion produced by any symmetric approximation to the ideal field

$$D = 2 \Delta R \sin \theta + \frac{n}{[\tan \theta + f'(x)]^2} \left[\tan \theta + f'(x) + \frac{(a - x)f'(x)}{R \cos^2 \theta \sin \theta} \right] \tag{5.49}$$

5.14. Dispersion—Special Cases [5,6,21]

A. Normal case

If the geometric conditions for the normal case as discussed in Sec. 5.6A are substituted in Eq. (5.49), we obtain, for the dispersion of the normal field,

$$D_n = \frac{2a \, \Delta R}{R} \frac{M}{\Delta M} = \frac{2 \, \Delta R}{\sin \theta} \frac{M}{\Delta M} \tag{5.50}$$

It must be remembered that this is measured along the line SD. Measured normally to the central ion path, as is the case with many spectrometers, this expression becomes

$$D_n = 2 \, \Delta R \frac{M}{\Delta M} \tag{5.51}$$

It is often convenient to use the following expressions for ΔR, obtained from Eq. (5.6):

$$\Delta R = R \frac{\Delta M}{2M} \tag{5.52}$$

$$\Delta R = R \frac{\Delta V}{2V} \tag{5.53}$$

B. Inflection case

The geometric conditions corresponding to the inflection case were discussed in Sec. 5.6B. Inserting these in Eq. (5.49), we obtain

$$D_f = \frac{4\,\Delta R}{3\,\sin\theta}\frac{M}{\Delta M} \tag{5.54}$$

Measured normally to the line SD, this becomes

$$D_f = \tfrac{4}{3}\,\Delta R\,\frac{M}{\Delta M} \tag{5.55}$$

If we compare two magnetic lenses of given R/a, then, measured normally to the central ion beam, the dispersion of the inflection field is 2/3 that of the normal field, which is due to the lesser magnetic flux traversed. Measured along the line SD, as is done in some spectrographs, the ratio of the two dispersions is a function of θ. The ratio may then be conveniently simplified to the following expression:

$$\frac{D_f}{D_n} = \frac{a - x/3}{a} \tag{5.56}$$

Depending on the ratio R/a used (Fig. 5.7), the ratio varies from 1 to $\tfrac{2}{3}$.

C. Normal-circle case

Because the normal-circle case uses essentially the same magnetic flux as the normal case, we should expect the dispersion to be the same. The insertion of the geometric conditions corresponding to this case in Eq. (5.49) gives for the dispersion, measured normally to the ion beam,

$$D_{nc} = 2\,\Delta R\left(1 + \frac{\Delta R}{R}\tan^2\theta\right)\frac{M}{\Delta M} \tag{5.57}$$

The correction term is of the second order in ΔR and may be neglected in most cases.

D. Inflection-circle case

When the geometry pertinent to the inflection-circle case is applied to Eq. (5.49), we obtain for the dispersion

$$D_{fc} = \frac{4\,\Delta R}{3}\frac{M}{\Delta M} + (2\,\Delta R\,\sin\theta)^3 f(x)f'''(x)\frac{M}{\Delta M} \tag{5.58}$$

Since this case uses essentially the same magnetic flux as the inflection case, the dispersion is the same except for a negligible correction factor.

E. Asymmetric cases

As discussed in Sec. 5.9, asymmetric cases may be considered to be composed of two sections, each of which is a different approximation to half of an ideal focusing field. Since the dispersion is proportional to the total magnetic flux traversed, it will be given closely by the mean dispersion of the two approximations considered.

5.15. Wide Sources

A. General

The discussion thus far has been confined to point sources. A wide source may be considered as a series of points, each giving rise to an aberrated image which overlaps the others to produce an extended total image. In the 180° normal case, it is easily seen, from Fig. 5.43, that the image due to the source (neglecting aberrations) is equal to the source width, but is inverted in the sense that the extreme rays have crossed each other. The image

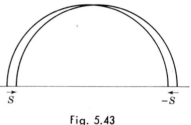

Fig. 5.43

width produced by other cases may be calculated [1, 6, 8, 9] by a method illustrated for the normal case.

In Fig. 5.44a, SS' represents a source of width s, measured normally to the central-beam trajectory. Considering the central ray and that

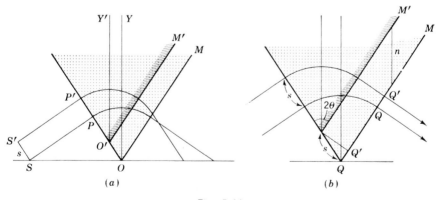

Fig. 5.44

from the upper edge of the source, we see, from Fig. 5.44b, that if each were deflected through the same angle 2θ, they would emerge still parallel with each other, but with the distance separating them changed. From the figure we see that this distance QQ' is given by

$$QQ' = s \cos 2\theta \qquad (5.59)$$

If, indeed, this deflection took place (i.e., if the edge beam left the field at M' instead of at M), then the image width would be given by $s \cos 2\theta$. For example, in the 180° case this applies, and placing $\theta = 90°$ in this formula gives us an image width of $-s$, the negative sign indicating that it is inverted.

The line M' may therefore be considered as the "ideal" boundary for the edge ray. This does not leave the field at M', however, but at M, which may be considered as being displaced from the "ideal" by a quantity whose ordinate is n. This ray will therefore cross SD at a point displaced from its ideal "focus," and this displacement may be calculated by Eq. (5.16). Upon subtracting the displacement so calculated from Eq. (5.42) as required by the geometry, the total image width is obtained.

B. Special cases: symmetrical

(1) **Normal case.** In this case it may be shown that n is given by

$$n = 2s \cos \theta \qquad (5.60)$$

When this value is inserted in Eq. (5.16), we obtain, for the corresponding image contribution,

$$I_n = 2s \cos^2 \theta \qquad (5.61)$$

Subtracting this from Eq. (5.59), we obtain for the image width produced by a normal magnet

$$I = -s \qquad (5.62)$$

(2) **Inflection case.** Here n is given by

$$n = \frac{2s \sin \phi}{\sin (\phi + \theta)} \qquad (5.63)$$

and this contribution to image width is

$$I_f = \frac{2s \sin \phi \cos \theta}{\sin (\phi + \theta)} \qquad (5.64)$$

Unlike the normal case, the expression corresponding to Eq. (5.59) is here

$$I' = \frac{s \sin (\phi - \theta)}{\sin (\phi + \theta)} \qquad (5.65)$$

When these are subtracted to find the total image width, we obtain

$$I_f = -s \qquad (5.66)$$

The image width produced by an inflection field is thus the same as that produced by the normal field.

(3) **Normal-circle and inflection-circle cases.** The expressions for these cases are more complicated and indicate the formation of an image slightly smaller than $-s$. However, the correcting terms are usually negligible, being of the order of $s^2/2a$ and s^3/a, respectively.

For all practical purposes, these symmetrical approximations may be considered as having unit magnification, producing an image of the same width as the source, both being measured normally to the central ion beam.

C. Asymmetric cases

Asymmetric fields do not in general have unit magnification. Figure 5.45 illustrates this for the normal case. The image size may be cal-

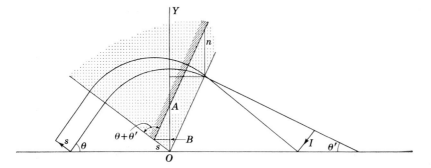

Fig. 5.45

culated by means of Eq. (5.16), as for the symmetrical case. However, Eq. (5.59) becomes

$$QQ' = s \cos (\theta + \theta') \tag{5.67}$$

and Eq. (5.60) must be written

$$n = AB + BO$$
$$= s(\sin \theta \cot \theta' + \cos \theta) \tag{5.68}$$

From (5.68) and (5.16) we obtain

$$I_n = 2s' \cos^2 \theta \tag{5.69}$$

and finally
$$I = s \cos (\theta + \theta') - 2s' \cos^2 \theta'$$
$$= -s \frac{\sin \theta}{\sin \theta'} \tag{5.70}$$

The equivalent of this equation is given by Bainbridge. For the symmetrical case, $\theta = \theta'$, and the magnification is unity. Magnifications less or greater than unity may be obtained by judicious choice of angles.   In mass spectrometry it is useful to have as small an image as possible, and in Fig. 5.46 we see a normal asymmetric magnet of magnification one-half. θ' is 70° (arbitrary), and thus θ equals 47°.   The smallest image may be obtained for $\theta' = 90°$, and $\theta = 0°$, as shown in Fig. 5.46b (α aberration neglected).

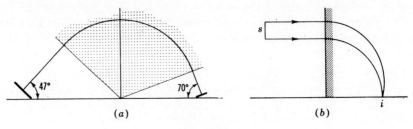

(a) (b)

Fig. 5.46

For the inflection case, the calculations are more complicated.   For Eq. (5.65) we must use

$$I' = \frac{s \sin (\phi - \theta')}{\sin (\phi + \theta)} \tag{5.71}$$

and for Eq. (5.63)

$$n = \frac{s(\sin \phi + \sin \phi')}{\sin (\phi + \theta)} \tag{5.72}$$

when we obtain

$$I = -s \frac{\sin \theta' \cos \phi + \sin \phi' \cos \theta'}{\sin (\phi + \theta)} \tag{5.73}$$

Here again, the judicious choice of angles will provide a magnification of less than or greater than unity.   For the symmetrical case $\theta = \theta'$ and $\phi = \phi'$ and the magnification is unity, as previously derived.

The normal-circle and inflection-circle asymmetric cases give magnifications corresponding, respectively, to the normal and inflection asymmetric cases, except for small correction terms.

The general expression for the magnification produced by a sector lens as derived by Herzog's theory [8] is

$$m = \frac{f}{l' - g'} = \frac{l'' - g''}{f} \tag{5.74}$$

where the various symbols are defined as in Sec. 5.5.   The resemblance of this formula to optical analogy is obvious.

5.16. Resolving Power

The purpose of a mass spectrometer is to separate ions of one mass from those of another, and in general this purpose will be more effectively accomplished if the components are simultaneously well separated in space and well focused to small images. Thus a measure of the resolving power of an instrument is the ratio of its dispersion to its beam width. We shall now develop this idea more exactly [28–35].

The dispersion of a magnetic prism was defined as the difference in the deviation of two ion beams per percentage difference in mass (assuming that they have the same charge). Thus, if a spectrometer focuses two ion beams differing by ΔM per cent at two points differing by Δx, the dispersion is

$$D = \frac{\Delta x}{\Delta M} \tag{5.75}$$

If the spectrometer just resolves an ion beam of mass M from another differing in mass by ΔM, then the resolving power of the instrument is defined as

$$RP = \frac{M}{\Delta M} \tag{5.76}$$

From (5.75) and (5.76) we have

$$RP = \frac{DM}{\Delta x} \tag{5.77}$$

Now, two beams are just resolved when the separation between them approaches zero, i.e., when the distance between their centers becomes their mean width W. Thus

$$RP = \frac{DM}{W} \tag{5.78}$$

where D = dispersion
      W = beam width
      M = mass measured at limit of resolution
In other words, the resolution of a spectrometer is 100 if it focuses mass 101 one beam width away from mass 100.

A fundamental property of a magnetic lens prism is therefore its ratio of dispersion to α aberration, although in practice other factors enter into the image width than α aberration. This ratio may be called the *reduced dispersion* (RD) and taken as a measure of the inherent resolving power of a given magnetic field.

For example, combining expressions for the dispersion and aberration of a normal field from Secs. 5.6 and 5.14, we obtain, for the reduced dispersion of this field,

$$RD_n = \frac{2\,\Delta RM/\Delta M}{R\alpha^2} \tag{5.77}$$

Remembering from Eq. (5.52) that $\Delta R/R = \Delta M/2M$, we then have

$$RD_n = \frac{R}{R\alpha^2} \tag{5.78}$$

This formula applies whether the reduced dispersion is measured normally to the ion-beam path or along the base line from source to detector SD.

The reduced dispersion of the inflection field, measured normally to the ion-beam path or along the base line, is obtained, in the same way, as

$$RD_f = \frac{2R}{3a_f} \qquad (5.79)$$

For the normal-circle case the reduced dispersion is

$$RD_{nc} = \frac{R}{R\alpha^3 \tan\theta} \qquad (5.80)$$

A comparison of these field types based on R as the common parameter shows that the normal-circle type has a larger reduced dispersion by a factor of $1/\alpha$ than the normal field. For $\alpha = 2°$, this is a theoretical factor of about 27 times.

The ratio of the reduced dispersions of the inflection and normal fields is

$$\frac{RD_f}{RD_n} = \frac{0.66a_n}{a_f}$$

$$= \frac{R}{(3a - 2x)\alpha \sec\theta} \qquad (5.81)$$

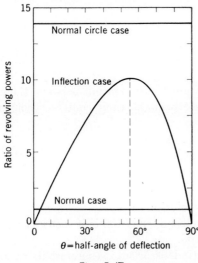

Fig. 5.47

This ratio varies with θ, that is, with the particular ideal focusing curve of Fig. 5.6 chosen. The variation of the ratio with θ is plotted in Fig. 5.47. There is seen to be a maximum at $\theta = 54.7°$, where the inflection field has a reduced dispersion about ten times that of the normal field for $\alpha = 2°$. As mentioned in Sec. 5.6, this corresponds to the field with a single straight-line boundary (Fig. 5.21), which is thus the *optimum inflection field*.

The inherent resolving power of the magnetic field is modified by several factors [30]. The total image width contains other components than the α aberration. Thus we must consider:

1. A: the α aberration (Sec. 5.6).
2. B: the β aberration (Sec. 5.11).
3. I: the image of the source slit (Sec. 5.15).
4. E: the energy dispersion. Generally speaking, the ions in a mass-spectrometer beam are assumed to possess the same energy. However,

a small variation caused by the source is usually present. The ratio $\Delta V/V$ gives rise to a corresponding $\Delta R/R$ [Eq. (5.6)], and so to an image-width component. This is

For the normal and normal-circle case: $\quad E = \dfrac{R \, \Delta V}{V}$

For the inflection case: $\quad\quad\quad\quad\quad\quad\quad E = \dfrac{2R \, \Delta V}{3V}$

5. *SC*: the space-charge effect, as investigated by Becker and Walcher. This occurs when the ion beam is intense and finally focused, so that the ions become closely bunched at the exit slit and the beam is broadened. The effect is thus directly proportional to beam intensity and inversely to beam width. It may be reduced by operating at relatively high pressure, so that the electrons produced by ionization of the residual gas

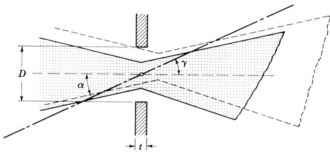

Fig. 5.48

help to neutralize the space charge. The effect is usually negligible for currents smaller than 10^{-9} amp. For heavier currents the image distance is affected, depending on the type of focusing used.

6. *F*: the fringing-field aberration [Eq. (5.118)].

7. *D*: the detector slit width, which for a scanned beam is effectively added to the beam width.

8. γ: the effect of slit width and line of focus. This effect is illustrated in Fig. 5.48 and is discussed by Persson [15]. It is seen here that the line of focus of the ion beams does not lie in the plane of the slit, but at an angle γ to it. Therefore the ion beam is focused only at the center of the slit, and as it sweeps to one side, the portion of it that grazes the slit edge is wider than the focused beam width. The effect is complicated by the slit thickness t. As shown in the figure, we define D and t as the slit width and thickness, α as the half angle of beam divergence, γ as the angle between the line of focus and the central ion beam, and w as the width of the ion beam at perfect focus. Under these conditions the *effective* beam width is:

Normal case:

$$\frac{(D + w)(1 - \tan \alpha/2 \tan \gamma) - (t \tan \alpha)/2}{1 - \tan \alpha/\tan \gamma} \quad (5.82)$$

Inflection and normal-circle cases:

$$\frac{(D + w) - t \tan \alpha}{1 - \tan \alpha/\tan \gamma} \quad (5.83)$$

If slits of negligible thickness (beveled) are used and the edges placed so that the plane of the slit lies along the line of focus, then these expressions simplify to $(D + w)$. The effective widening effect is greater for the higher-order-focusing instruments, as in their case the focused beam diverges by $\pm\alpha$ at the image, while in the case of the first-order instruments, the focused beam diverges on one side only.

9. T: If the mass spectrometer employs automatic scanning, the response time of the detector may be considered as adding a component T to the total width.

When all these effects are considered, the expressions for the resolving power of the instrument become:

Normal case:

$$\frac{R(1 - \tan \alpha/\tan \gamma)}{(D + I + A + B + E + SC + F)(1 - \tan \alpha/2 \tan \gamma) - (t \tan \alpha)/2 + T(1 - \tan \alpha/\tan \gamma)} \quad (5.84)$$

Inflection case:

$$\frac{2R(1 - \tan \alpha/\tan \gamma)}{3(D + I + A + B + E + SC + F) - 3(t \tan \alpha) + 3T(1 - \tan \alpha/\tan \gamma)} \quad (5.85)$$

Normal-circle case:

$$\frac{R(1 - \tan \alpha/\tan \gamma)}{(D + I + A + B + E + SC + F) - t \tan \alpha + T(1 - \tan \alpha/\tan \gamma)} \quad (5.86)$$

Most contemporary spectrometers should consider almost all these factors in calculating resolving power. However, proper design can eliminate some of them and improve the resolution. Thus, if beveled slits are used, the exit-slit plane is placed approximately along the focus line, the sweep time is sufficiently slow, and space-charge effects are reduced, then the equations become

Normal and normal-circle cases:

$$\frac{R}{D + I + A + B + E + F} \quad (5.87)$$

Inflection case:

$$\frac{2R}{3(D + I + A + B + E + F)} \tag{5.88}$$

To reduce any of the remaining image contributions involves reducing the beam intensity, and this should not be done arbitrarily, as is often the case when α and β are made negligible by the use of very small slits. An optimum choice of parameters exists, as will be discussed in Sec. 5.17.

From Eq. (5.87) we obtain the approximate rule of thumb that the resolving power of a normal spectrometer is equal to the radius R divided by the sum of the entrance and exit slit widths $(D + I)$.

Determination of the actual resolving power of a spectrometer usually involves making measurements on a spectrogram record or photograph. For this purpose, certain formulas are useful:

1. The base of a *mass peak* is, of course, the total effective width given by Eqs. (5.82) and (5.83). Regardless of the height of any particular peak, the base widths are all the same width if the sweep is linear. This may not always appear to be the case, because the tails of small peaks disappear into the base line. Another exception is the case of very intense beams, where the space charge SC and the detector T are functions of beam intensity.

2. The width of the mass peak at half intensity is about equal to D when D is greater than w, regardless of intensity distribution in the beam. The only exception to this is the improbable case of the intensity being zero for a finite distance within the beam, i.e., there being effectively two beams.

When D is smaller than w, the peak width at half intensity is equal to w, when the ion-beam intensity is homogeneous.

3. When the exit slit D is wider than w and the sweep time is sufficiently slow, the peaks will have flat tops. The width of the flat top is a function of the difference between D and w. The equations for flat-top width are:

Normal case:

$$\frac{(D - w)(1 + \tan \alpha/2 \tan \gamma) - (t \tan \alpha)/2}{1 + \tan \alpha/\tan \gamma} \tag{5.89}$$

Inflection and normal-circle cases:

$$\frac{(D - w) - t \tan \alpha}{1 + \tan \alpha/\tan \gamma} \tag{5.90}$$

Here the effect of the tails may be to cause smaller peaks to appear to have wider tops than the intense peaks.

A. Slit location

It was mentioned in previous sections that the effect of the line of focus not being normal to the central ion beam could be avoided by placing the plane of the exit slit approximately along the line of focus.   More precisely, the slit edges should be staggered so that their remote faces are separated by a normal distance equal to (for the inflection and normal-circle cases)

$$d_{f,nc} = \frac{D + w}{\tan \gamma} \qquad (5.91)$$

The two edges are located symmetrically on each side of the point of central focus.   In the normal case, only the slit edge nearest the source is moved, so that the final separation is given by

$$d_n = \frac{D + w}{2 \tan \gamma} + \frac{t}{2} \qquad (5.92)$$

If this is done and the other conditions mentioned are met, then the resolving power will be given by Eqs. (5.87) and (5.88).

Fig. 5.49

However, this staggering of the slit edges also affects another measurement, that of the flat-top width, which is reduced.   Under conditions described in this section, the flat-top width will be given by

Normal case:

$$(D - w) - \frac{w}{(\tan \gamma / \tan \alpha) - 1} \qquad (5.93)$$

Inflection and normal-circle cases:

$$(D - w) - \frac{2w}{(\tan \gamma / \tan \alpha) - 1} \qquad (5.94)$$

A mechanical exit slit, whose effective width is the geometric width, is commonly used.   However, Berry has discussed the use of a virtual exit slit effectively obtained by applying a high voltage to a wide exit slit.   As shown in Fig. 5.49, a high voltage placed on an insulated slit produces a saddle-shaped potential distribution near it.   If the center of this potential distribution is slightly less than the ion energy, the ion beam will be able to penetrate the barrier. However, if simultaneously the edges of the potential distribution are higher than the ion energy, they will not be able to penetrate at the edges, and thus the effective width through which they may pass is narrowed. This technique has the advantage of providing an electrically controlled slit width.

The geometric and chromatic aberrations of the spectrometer (factors 1 to 3, p. 148) have been treated in detail by Boerboom, [31], Tasman [32], and by Tasman, Boerboom, and Wachsmuth [33–35].

Inhomogeneous fields have been extensively studied by Tasman and his associates. They have calculated the characteristics of fields produced by various means and bounded by both straight and circular lines [32–35]. Boerboom and his colleagues [31, 31a, 35a] have made important contributions to the knowledge of slit systems, and as a result of their work the designer is now able to proceed much more confidently in estimating the positions of the virtual objects formed by the ion beams.

5.17. Optimum Geometrical Conditions

When the parameters of a mass spectrometer are known, the methods of the previous section may be used to calculate the resolving power. However, various combinations of parameters will give the same resolving power, and in designing an instrument, some arbitrary decisions have to be made.

Equations (5.87) and (5.88) show that the resolving power of an instrument may be improved by reducing any of the components of the total image width. However, to do so also reduces the beam intensity. Geoffrion [36] has shown that an optimum choice may be made which will ensure maximum beam intensity for a given resolving power. This is possible because both beam intensity and resolving power are functions of the same parameters. Figure 5.50 shows the typical geometry of a mass spectrometer. We see that the beam divergence α is determined by the widths and separation of diaphragm $S2$ and object $S1$ (slit $S1$ may be virtual, as will be seen in Sec. 5.20); beam divergence β is determined by slits $S1$ and $S3$ and their separation which is a function of R and a (Sec. 5.12); the object width is determined by $S1$. Then, from Eqs. (5.87) and (5.88), we see that

$$RP = f_1(S1, S2, S3, a, R) \tag{5.95}$$

which function f_1 has been derived by Geoffrion [36].

Assuming that an ion source delivers a uniform isotropic beam of ions to the analyzer, the transmission of the instrument (which determines the fraction of the ion beam which emerges to the detector) is also a function of the various slit widths and separations, since these determine the solid angle of the beam and the fraction of it detected. The beam intensity I may then be written

$$I = f_2(S1, S2, S3, a, R) \tag{5.96}$$

where f_2 has also been determined by Geoffrion [36].

In designing a spectrometer, a practical factor which must often be considered, in addition to beam intensity and resolving power, is the size of the magnet which may be constructed. This depends on the

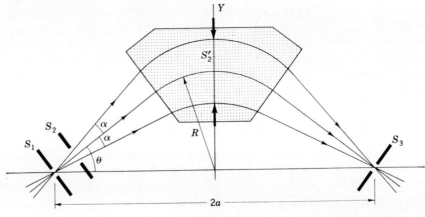

Fig. 5.50

dimensions of the ion beam which will traverse its pole gap and therefore may be expressed as a function of the beam geometry. We may thus write, for the pole surface S required,

$$S = f_3(S1, S2, S3, a, R) \tag{5.97}$$

Lagrange's method of undetermined coefficients is used to define the optimum conditions. A function is defined which is

$$\psi = RP + kI \tag{5.98}$$

The derivatives of this function with respect to all parameters are then obtained and equated to zero. The solution of the resulting set of equations gives the optimum parameters, which are, for the *unrestricted case:*

1. R must be as great as possible.
2. The width of the object slit should be

$$nDR(1 - T)\,\frac{1}{4n + 2}\,RP \tag{5.99}$$

3. The diaphragm $S2$ (determining α) should be placed at mid-trajectory (shown at $S2'$ in Fig. 5.50) and be of width

$$2a \left[\frac{(RP)(1 - T)D}{A(4n + 2)} \right]^{1/n} \tag{5.100}$$

4. All slits in the spectrometer should be of the same length, given by

$$2C \left[\frac{n(1 - T)RP}{4n + 2} \right]^{\frac{1}{2}} \tag{5.101}$$

5. The exit slit width should be

$$\frac{(1 + T)DR(RP)}{2} \tag{5.102}$$

where, in expressions (5.99) to (5.102):

n = power of α aberration (2 for first-order focusing, 3 for second-order, etc.)

D = dispersion coefficient (normal case, 1; inflection case, $\frac{2}{3}$; etc.)

R = radius of curvature

T = fraction of recorded ion-beam peak which it is desired to have "flat-topped"

RP = desired resolving power

a = half the source-detector distance

A = α-aberration coefficient, given by Eqs. (5.24), (5.26), (5.32), etc.; for example, $A_{nc} = R$

C = half the length of ion-beam trajectory

As an example of the application of these conditions, Geoffrion has calculated the optimum parameters for normal, inflection, and normal-circle spectrometers, all being designed to the following arbitrary specifications: $RP = 100$, $T = \frac{1}{4}$, θ (ion-beam angle with source-detector line) $= 45°$, $R = 15$ cm. The optimum parameters are given in Table 5.4.

The consideration of optimum dimensions provides a further basis of comparison between various spectrometers. Comparison with contemporary spectrometers of nonoptimum design indicates that *an improvement of from 5 to 15 times in the luminosity could be effected.* This order of magnitude would be appreciated by all operators. The optimum-normal-circle instrument has about twice the efficiency of the others (ratio: luminosity/magnet surface). This could be translated as requiring a magnet $2^{\frac{3}{2}}$ times smaller, a considerable saving. Given an optimum-parameter spectrometer, the luminosity can be improved only by making it larger, since the ratio L/S is approximately independent of R.

TABLE 5.4.   Optimum Parameters—Unrestricted Case

	Normal	Inflection	Normal-circle
Luminosity L	$0.8K$	K	$2.7K$
a, cm	21.2	17.7	21.2
C	$1.78R$	$1.45R$	$1.78R$
n	2	3	3
D	1	0.66	1
A	1	2	1
Slit length, cm	2.08	1.75	2.14
Diaphragm, cm	1.16	2.0	3.45
Object slit, mm	0.225	0.16	0.24
Exit slit, mm	0.94	0.62	0.94
Pole surface S, cm^2	125	130	182
L/S	$1.0K'$	$1.22K'$	$2.4K'$

K and K' = constants.

In many cases it is not convenient to increase the slit length to its optimum value, either because the magnet-pole gap is fixed or the required gap width is undesirable because of particular fringing conditions, etc. The fixing of the parameter h changes the other optimum conditions, which become:

1. The object slit width should be

$$\frac{Rn(1 - T)D}{2n + 2} \tag{5.103}$$

2. The width of diaphragm $S2$ (placed as before) should be

$$2a\left[\frac{(1 - T)(D/A)e}{2n + 2}\right]^{1/n} \tag{5.104}$$

3. The exit slit width should be

$$(1 + T)\frac{DR}{2}RP \tag{5.105}$$

where the various symbols are defined as before and

$$e = RP - \frac{2h^2}{(1 - T)4C^2} \tag{5.106}$$

As an example, Table 5.5 gives the various optimum parameters for three types of spectrometer based on the same arbitrary values of $RP = 100$, $T = 1/4$, $R = 15$ cm, and $\theta = 45°$ as were chosen for Table 5.4, and in addition h is set at 1 cm.

TABLE 5.5.   Optimum Parameters—Fixed h Case

	Normal	Inflection	Normal-circle
Luminosity L	$3.5K$	$5.6K$	$11K$
Relative luminosity	1	1.6	3.5
a, cm	21.2	17.7	21.2
n	2	3	3
D	1	0.66	1
A	1	2	1
C	$1.78R$	$1.45R$	$1.78R$
Object slit, mm	0.33	0.24	0.38
Diaphragm width, cm	1.42	2.28	4.02
Exit slit, mm	0.93	0.62	0.93
Pole surface S, cm^2	81	101	142
L/S	$1.0K'$	$1.3K'$	$1.8K'$

K and K' = constants.

Contrary to the unrestricted case, there is no marked improvement in luminosity if R is increased, once h is fixed and the resulting optimum conditions applied. This is because, as R is increased, the aberrations are increased also, and the slits must be reduced to maintain the required resolving power. Consequently, for a given gap width, the radius R could be made as small as is convenient, without introducing unusual fringing fields, etc. With unrestricted optimum conditions, on the contrary, h will be increased as R increases, and a gain in luminosity is effected.

Another case of particular interest is the 180° (semicircular) normal case, which is widely used in β-ray spectrometry as well as in mass spectrometry. For a required resolving power RP and radius of curvature R, the optimum parameters for this case are:

1. Source and detector slit widths should be

$$h = \frac{1.33R}{RP} \tag{5.107}$$

2. Diaphragm width S_2 should be

$$1.63R(RP)^{\frac{1}{2}} \tag{5.108}$$

3. Slit lengths should be

$$\frac{3.63R}{(RP)^{\frac{1}{2}}} \tag{5.109}$$

A survey of a number of spectrometers described in the literature indicated that most could improve their beam intensity by about 50% by simply applying restricted optimum dimensions, which in practice consists in readjusting the object, exit, and diaphragm slit widths.

Adjusting the slit lengths added a factor 5, while changing to normal-circle focusing gave a factor 15.

In considering optimum parameters, the effects of fringing field must be taken into consideration, and the total effective field as discussed in Sec. 5.18 should be used. This will modify in some respects the numerical values in Tables 5.4 and 5.5. However, a valuable rule of thumb emerges from these considerations: the various contributions to the final ion beam width which also contribute to beam intensity should be about equal. To reduce one arbitrarily so as to gain resolution involves a needless sacrifice of beam intensity, and resolution should be improved by reducing *all* such components somewhat, in accord with the principles discussed here.

MISCELLANY

5.18. Fringing-field Effects

One difficulty which presents itself in mass-spectrometer design is that the magnetic field extends beyond the limits of the pole-piece boundaries, dropping gradually to zero as the distance increases. The trajectory of an ion beam originating outside the pole pieces will therefore curve a certain amount before entering the region between the pole pieces. Thus the focusing discussed in previous sections will be disturbed. Second, the fringing field is not composed of straight lines of force, as is the case with the main field between the poles; the lines are curved, and this introduces added complications. The practical effects of the fringing field may be considered as three: (1) a focusing action in the β plane; (2) a displacement of the focal point in the α plane; (3) an additional contribution to the image width.

A. Focusing in the β plane

This has been discussed in Sec. 5.12.

B. Displacement of the focal point in the α plane

The first step in considering this effect of the fringing field is to plot its intensity. This has been done theoretically by Coggeshall and Reuterswärd, who obtain essentially the same curve, shown in Fig. 5.51.

This figure gives the magnetic field intensity $h(x)$ (as a fraction of the maximum uniform field measured well within the pole boundaries) as a function of the normal distance from the pole boundary. The function is for the field in the central plane, which the fringing-field lines all cross normally. The curve is largely independent of the thickness of the pole pieces. Careful measurements made by Barnard [2], König and Hintenberger [40–42], and others, of the fringing field of their magnets, confirm this theoretical curve, and it is to be noted that Barnard uses a relatively wide pole gap.

The next step is to consider the precise trajectory of the ion beam in this varying fringe field. Coggeshall's [37, 38] approach, which is the basis for several final kitchen recipes, is indicated in Fig. 5.52.

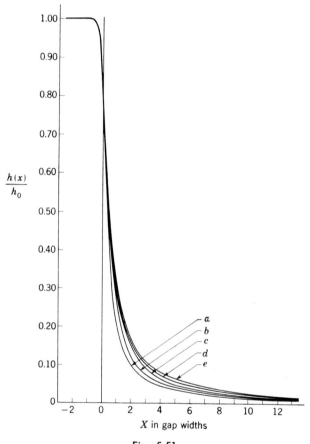

Fig. 5.51

An ion source at x_0 sends an ion beam into the magnetic field, and in the absence of fringing field the ion trajectory would be as indicated by the dotted line. We note that the center of curvature C of this trajectory in the magnetic field is situated on the pole boundary.

Now, in the presence of a fringing field, the ion beam follows the trajectory indicated by the continuous line, which curves increasingly as it approaches the pole boundary. Once within the region of constant field, this trajectory becomes circular, with the same radius of curvature

as the dotted trajectory.   However, the center of curvature C' is situated
outside the pole boundary, a distance d from it.

To calculate d, one may proceed as follows.   Coggeshall has shown
that the differential equation of the ion trajectory is given by

$$\frac{dy}{dx} = \frac{f}{(R^2 - f^2)^{1/2}} \tag{5.110}$$

where R is the radius of curvature mentioned and f is given by

$$f = \int_{x_0}^{x} h(x)\, dx \tag{5.111}$$

where the integration is performed over the distance from the source
x_0 to the point x at which dx/dy is measured.

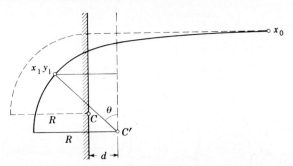

Fig. 5.52

In Fig. 5.52 d is given by

$$d = R \sin \theta - x_1 \tag{5.112}$$

From the geometry of the figure we have

$$\frac{dy}{dx} = \tan \theta = \frac{R \sin \theta}{(R^2 - R^2 \sin^2 \theta)^{1/2}} \tag{5.113}$$

From (5.100) and (5.113) we have

$$R \sin \theta = f = \int_{x_0}^{x} h(x)\, dx$$

If x_1 in Eq. (5.112) is conveniently chosen one gap width inside the pole
boundary (where the field has become constant for all practical purposes), then the value of d becomes (in gap widths)

$$d = \int_{x_0}^{x_1} h(x)\, dx - 1 \tag{5.114}$$

To calculate d, a numerical integration of the fringing field as given
by Fig. 5.51 must be made between the limits x_0 and x_1.   For this and
other purposes (such as the calculation of y to be mentioned), Table 5.6
is useful.   It gives the value $\int h(x)\, dx$ between various limits of x.

TABLE 5.6

(Both columns in gap widths, G)

Δx	$\int h(x)\, dx =$
-2 to -1	1.00
-1 to 0	0.98
0 to 1	0.55
1 to 2	0.21
2 to 3	0.12
3 to 4	0.08
4 to 5	0.06
5 to 6	0.05
6 to 7	0.04
7 to 8	0.04
8 to 12	0.12
12 to 20	0.08

Bainbridge [1] gives the result as shown in Fig. 5.53.

It is important to note that d depends *not only on the magnet, but also on the source or detector distance from the magnet.*

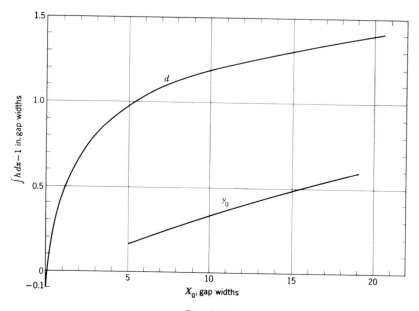

Fig. 5.53

Once d (the distance of the center of curvature C' from the pole boundary) has been calculated, the third step is to decide how to apply it, and unfortunately not enough measurements have been published giving a clear indication of what experimental arrangements correspond to an optimum allowance for this fringing field. Several procedures have been proposed as indicated in Fig. 5.54.

Procedure (*a*) has been followed by Nier and many others. The value *d* is calculated from Eq. (5.114), read from Fig. 5.53 or arbitrarily chosen as one gap width (Nier [43] and others). "Effective" boundaries are drawn parallel to the real boundaries at a distance *d* from them, and the effective field apex, source, and detector are placed in a straight line (for the normal case, here being considered as an example). Analogous procedures are used for inflection or asymmetric magnets. Usually

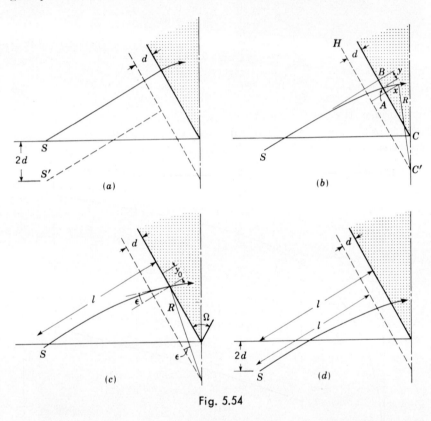

Fig. 5.54

fine adjustment is made for optimum focusing by moving the spectrometer tube relative to the magnet. The precise amount of this final adjustment is never published and may convert this procedure into one of the others.

Procedure (*b*) has been suggested by Coggeshall [37, 38]. Source *S* and detector *D* are placed relative to the pole faces as though no fringe field existed. The ray trajectory from *S* is then traced into the magnet as previously discussed, and *d* and the center of curvature *C'* established. Line *C'C* at $\theta/2$ with *C'H* is then drawn, establishing the point *C*, and distance *CB* may then be calculated from a knowledge of point *S*. The advantage of this method is that, as shown by Coggeshall, the angular and spatial separation of the rays is the same in the effective field as in the ideal field, and symmetrical conditions are maintained. The procedure is criticized by Bainbridge [1] and Reuterswärd [45] on the grounds that, since the ion beam does not enter the effective magnetic field normally, the conditions for normal focusing do not apply, and thus the image will not be formed symmetrically at D, but somewhat further, at D'. There is no doubt on this score, and anyone applying Coggeshall's procedure would have to make some adjustments.

Procedure (c) is suggested by Bainbridge [1]. Having calculated d by Eq. (5.114) or otherwise, it is used to calculate ϵ, the angle of entry of the ion beam into the pole boundaries, as indicated in Fig. 5.54c. ϵ is given by

$$\tan \epsilon = \frac{d}{(R^2 - d^2)^{1/2}} \tag{5.115}$$

From this value of ϵ and the design angle Ω of the magnet apex, Eq. (5.20a) of the section on focusing may be used to calculate the image distance l. There remains to be calculated y_0, the normal displacement from the ideal ion trajectory of the source. y_0 is the sideways displacement of the ion beam as it travels from the source to the pole boundary, and it is given from Eq. (5.110) as

$$y_0 = \int_{x_0}^{x_b} \frac{\int h(x) \, dx}{\{R^2 - [\int h(x) \, dx]^2\}^{1/2}} \, dx \tag{5.116}$$

and is obtained by numerical integration over the limits x_0 to x_b at the pole boundary. y_0 has been calculated by Bainbridge for the special normal case of $R = 10$ gap widths and is shown in Fig. 5.53. To summarize, d and Ω being given, ϵ, l, and y_0 are calculated, which situates the source and detector symmetrically about the magnet.

Procedure (d) has been evolved by Reuterswärd [45]. After deriving graph (5.110) and an expression for d which is similar to Eq. (5.114) by somewhat different methods, he places the source and detector at distances from the *effective* field boundary, which are the same as those which would be used were no fringing field present. Except for the determination of d, this method produces a result similar to procedure (a). Reuterswärd adopts it because, for the normal case, he has shown that the total deviation produced by the total field is the same as would be produced for the fringeless field. His correction is shown in Fig. 5.54d.

Bloch and Walcher [44] have also considered the fact that the effective field is not only displaced laterally from the real field boundary, but also angularly. They find, for the normal case, that there would be an effective angular shift $\epsilon = d/R$, and thus the deflection would be changed by a similar amount.

Herzog [39] has extended the consideration of fringing field to the general case of oblique incidence. He sets up, first, a sharply delineated substitute field which will produce the same deflection of the ion beam as the real field (including its fringe). The initial and final directions of the real beam are thus parallel to the initial and final directions of the substitute ion beam (which coincides with the real one inside the pole gap) drawn for the substitute field. The substitute beam is used to calculate the positions of substitute source and image, and the displacement of the real source and image from these is given.

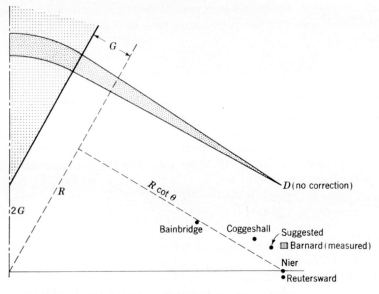

Fig. 5.55

Some of these corrections are now calculated for the precision mass spectrometer described by Barnard [2]. In this instrument the magnet gap width is 1.20 in. = 1 G, the radius of curvature of the ion trajectory in the gap is 4.00 in. = 3.33 G, and the angle of the normal sector magnet is 60°. These data are indicated in Fig. 5.55. We see that, were no fringing field present, the source would be placed at a distance $R \cot \theta = 6.93$ in., or 5.77 G, from the pole boundary.

(a) **Nier's method** [43]. It may be calculated from Eq. (5.12), or read from Fig. 5.53, that the effective-field-boundary distance from the pole boundary is 1.02 G. (Note that this is calculated for the source as shown.) Nier has suggested that d be arbitrarily taken as equal to G, and in the case of Barnard's magnet, this is an excellent approximation. These effective boundaries lower the apex of the effective field a distance $2d = 2\,G = 2.4$ in. R and θ are the same, so that the source position according to Nier should be placed at the point so indicated in Fig. 5.55.

(b) **Coggeshall's method** [37, 38].   Here the source is placed at the distance $5.8\ G = 6.93$ in. from the pole boundary as though no fringing field existed.   Integrating the fringing-field effect to one gap width inside the real boundary [Eq. (5.111)] gives $f = 2.04\ G$, which is equivalent to d calculated in part (a).   From Eq. (5.110) we next obtain $dx/dy = 0.775$; whence the slope of the ion beam at this point is $37.8°$ and y (the lateral displacement of the ion beam at this point from the initial beam direction) $= 1.14\ G$, or 1.37 in.   Added to R cos $37.8°$, this gives us the lateral displacement of the effective center of curvature C' from the initial ion-beam direction.   This is

$$3.77\ G = 4.52 \text{ in.}$$

The point C, where the real magnet apex must be placed, is given as being $1.04\ G$ cosec $30° = 2.50$ in. from C', in the vertical direction of Fig. 5.55.   From these values and the geometry of Coggeshall's method, we finally obtain that S must be 7.17 in. from the center of the field and 1.37 in. below the vertex of the real magnet.   This point is shown in Fig. 5.55.

(c) **Bainbridge's method** [1].   A preliminary estimation of d $(= 1\ G)$ applied to Eq. (5.113) gives an angle of entry $\epsilon = 17.5°$. Image distance l may then be calculated from Eq. (5.20a) as being 4.30 G.   This gives a more precise limit over which to integrate the fringing-field effect and a better value of d $(= 0.96\ G)$, giving in turn a better $\epsilon = 16.6°$ and a better $l = 4.37\ G = 5.25$ in.

From Eq. (5.116), y_0 is determined by numerical integration and is equal to 0.47 in.   From the geometry of the figure it may then be established that the source should be placed at the point indicated in Fig. 5.55.

(d) **Reuterswärd's method** [45].   The source is placed at the same distance from the effective field as it would be from the real field were no fringing field present.   This is R cot $30° = 6.93$ in.   The source is thus farther from the real field than the previous cases, and thus d is calculated by successive approximations as being $1.10\ G = 1.3$ in. His source, effective apex, and detector are in line, and thus the source is placed $2.2\ G = 2.64$ in. below the real magnet apex, as shown in Fig. 5.55.

Experimental focus.   Barnard [2] has made very careful measurements of the position of best focus for his instrument, and the experimentally determined optimum source location is as indicated in Fig. 5.55, 8 in. from the central axis and 1.69 in. below the magnet apex. Although the source-detector separation seems to have been fixed during the experiment, rotation of this line relative to the magnet produced no improvement, as might have been expected if this distance were not indeed optimum.

It is to be noted that, of the various corrections considered thus far, those of Nier and Reuterswärd correspond most closely to the measured position, and so, for this particular instrument at least, the rule of thumb that the effective field extends a gap width from the real magnet is as good as any.

Perusal of Fig. 5.55 permits the following criticisms of the various proposed corrections. Nier's [43] method does not take into account the oblique entry of the beam into his effective field, which results in his calculated point being laterally displaced from the measured focal point. Coggeshall's method, as indicated by Bainbridge and Reuterswärd, calculates the curvature faithfully; however, the source being displaced toward the magnet, the image will not be located symmetrically, but at a greater distance from the magnet than the source distance. This is indicated by his calculated point of best focus being displaced toward the magnet. Bainbridge's [1] method correctly takes into account the oblique entry of the ion beam into the real pole-face boundary because of the fringing-field effect, but neglects the effective enlargement of the real field by the magnetic field. He thus calculates for a smaller field than is actually present, as indicated by the displacement of his calculated point toward the magnet. Reuterswärd [45] has taken into account both the oblique entry of the beam and the effective extension of the real field by the fringing field, but in his rule for the normal-case correction omits the lateral-displacement effect of ϵ and so obtains a point close to Nier's.

(e) **Suggested method.** Consideration of the points raised in the last paragraph leads inevitably to a combination of the arguments advanced by the various authors in the following method. Physically, the ion beam will have a center of curvature which, if symmetrically located below the apex of the real magnet, locates two effective boundaries that may be considered as containing the bulk of the effective field. These boundaries may therefore be used to calculate the object and image positions of a beam entering them obliquely, at an angle determined by the bending of the ion beam in the fringing field *outside* of the *effective* boundaries. The calculation of the source position by this method follows most closely the procedure outlined by Bainbridge.

Applied to Barnard's magnet this method gives the following results:

1. d is selected to a first approximation as $= 1\ G$.
2. l is selected to a first approximation as being equal to

$$R \cot 30° = 5.77\ G$$

from the effective boundary (that is, $6.77\ G$ from the real pole boundaries).

3. From Eq. (5.110), the slope of the ion beam as it enters the effective field is calculated, using Eq. (5.111), with limits 6.75 G and 1 G, and Fig. 5.53 or Table 5.6. This gives $dy/dx = \tan \epsilon = 0.150$; whence $\epsilon = 8.5°$.

4. Using the design angle of the effective field of 60° and $\epsilon = 8.5°$, l may be calculated by Herzog's method, and is found equal to 5.16 G.

5. Result 4 indicates that the inner limit of calculation 3 is more precisely 1.01 G [from Fig. 5.53 or Eq. (5.114)—this new limit affects the calculation but little], while of course the more precise outer limit is 6.17 G. A more accurate ϵ is thus calculated equal to 9.15°.

6. From result 5 a more precise value for l is calculated as in result 4 and is equal to 5.12 G from the effective field, or 6.13 $G = 7.4$ in. from the real pole boundary.

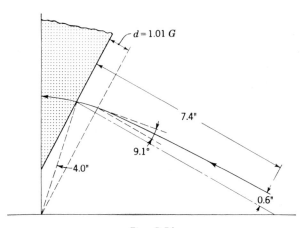

Fig. 5.56

7. Finally, y_0 is calculated by numerical integration from Eq. (5.110), with limits 6.13 G and the real boundary. This calculation gives the lateral displacement of the source from the line normal to the real field at the point of entry to it and is 0.60 in.

The construction for this method of correcting for fringing field is indicated in Fig. 5.56. It has the advantage of taking into account the effective extension of the real field by the fringe field (Nier), the oblique entry of the ion beam into the magnetic field (Bainbridge), and the lateral displacement of the ion beam caused by the exterior fringing field (Cogge-shall) and provides the same total deflection and initial and final ion-beam directions as the ideal field (Reuterswärd). This would seem to approximate closely the actual behavior of the real ion beam. In Fig. 5.55 the source position as calculated by this suggested method is shown. It is seen to coincide very closely with that measured by Barnard. More important, this calculated point is at the correct vertical height, and the slight sideway displacement might not have been detected by Barnard's tube-rocking experiment.

C. Image broadening

The general conditions for the focusing action of magnetic lenses developed in the last section have assumed that the ion beam traveled in the median plane parallel to the pole faces. Depending on the type of lens used, the image of a point source may be located. Now rays traveling through the spectrometer in a plane parallel to the median one will not traverse exactly the same total magnetic flux, because, while in the fringing field, they cross only a component (though a major one) of the field lines, which, being curved, are normal to the ion trajectory only in the median plane.

By introducing certain simplifications, Berry has derived a useful approximate expression for the change in image position due to this effect. This expression must be added to the image width produced by other effects and may be considered as an extra aberration.

Berry's [29] simplified case postulates a long source-magnet distance, so that the fringing field is negligible at S, a narrow magnet gap compared with other pole-piece dimensions, negligible β focusing, and relatively small contribution of the fringing field to the total deviation. The strength of the magnetic fringing field in a plane off the median is then given by

$$h_z(x) = h_0(x) - \frac{z^2}{2} h_z''(x) \tag{5.117}$$

which is obtained by applying Maxwell's equations

$$\text{curl } h = 0 \qquad \text{and} \qquad \text{div } h = 0$$

to a series expansion of $h_z(x)$. Applying the assumptions mentioned, the fringing-field aberration is shown to be

$$A_z = \frac{z^2}{R} \tag{5.118}$$

where z is the half gap width.

Within wide limits, this is independent of the shape of the fringing field and of the angle of beam deflection.

Because of the simplifications, Berry's formula cannot be expected to give the complete effect, but he has confirmed, by careful examination of mass-spectrograph image plates, that it is correct to within a factor of 2.

The fringing-field aberration may easily attain the dimensions of other image components. An instrument using 10 cm radius of curvature and 1 cm gap width will, by Berry's formula, have a fringing-field aberration equal to 0.25 mm.

5.19. Inhomogeneous Fields

Consideration of Fig. 5.57, which shows the ordinary 180° focusing action of a homogeneous magnetic field, indicates that the inner and outer rays of the beam would focus more perfectly with the central ray if they were deviated less. This implies that excellent focusing would be obtained with a magnetic field that was stronger along the central ray and weaker to either side of it. The gap profile of such a field as compared with a homogeneous field is indicated in the same figure.

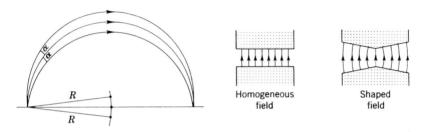

Fig. 5.57

Some attention has been paid to the precise way in which such a field should vary. Bock [46] has proposed such a field, and, as modified by the work of Beiduk and Konopinski [47], it is given by

$$H = H_0(1 - \tfrac{3}{4}d^2 + \tfrac{7}{8}d^3 - \tfrac{9}{16}d^4 + \cdots) \qquad (5.119)$$

where $d = (R - R_0)/R_0$
  R = distance from center of curvature at which H is measured
  H_0 = maximum field at position of central ion beam
  R_0 = radius of curvature of central beam

Several magnets producing such "shaped fields" have been built, including one by Kistemaker and Zilverschoon [49] and one by Langer and Cook [48] for a β-ray spectrometer. The profile of the latter magnet is shown in Fig. 5.58. The pole face was designed with rectangular cuts for ease in machining. The curves in the lower part of the figure permit comparison of the theoretical curve (dotted line), proposed by Bock [46], the corrected curve (solid line) according to Beiduk and Konopinski [47], and the experimentally determined points of the Cook-Langer magnet.

Another approach has been that of McMillan [50] and Svartholm and Siegbahn [51], whose work has been further developed by Shull and Dennison [52, 53]. Their magnetic field is proportional to the radial distance in the following manner:

$$H = H_0 - (r - a)\alpha \frac{H_0}{a} + (r - a)^2 \beta \frac{H_0}{a^2} \qquad (5.120)$$

and focusing takes places after a deflection of $\pi\sqrt{2}$, or 255°.   Magnets for β-ray spectrometers patterned after this theory have been built by Siegbahn and Svartholm [55] and Kurie, Osoba, and Slack [54].   Svartholm [55] has extended the theory to cover sector fields, where the source and/or detector are removed beyond the magnet gap.

Although applied with great success to β-ray spectrometers, the design of shaped fields must be approached with caution in the case of mass

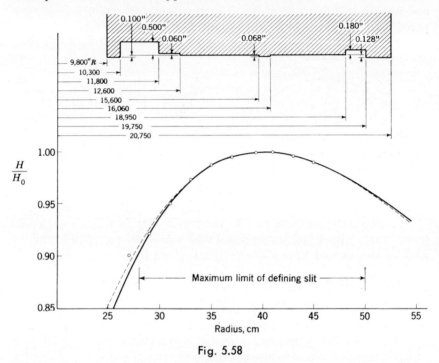

Fig. 5.58

spectrometers, where the higher fields used introduce problems of iron inhomogeneity and uneven hysteresis.   The Kistemaker and Zilverschoon [49] instrument has worked well, but is of considerable size, so that shimming and adjustments are somewhat easier than in a machine of ordinary size.

5.20. Graphical Methods

The ion-optical parameters of homogeneous magnetic prism lenses as developed analytically in previous sections may be verified to the first order by a graphical method developed by Cartan [9].   This technique is extremely useful for quick verification of calculations or for preliminary first-order investigation of a proposed design.   The method is indicated in Fig. 5.59.

It is given that a ray from the source S is to enter the magnetic field bounded by M and M' at the point P, making an angle ϵ with the normal N to the tangent T to M at P. The radius of curvature of the ion beam in the magnetic field is R, so that the beam curves about a center of curvature at O and emerges from the magnetic field at P', in a direction $P'D''$, making an angle ϵ' with the normal N' to the tangent T' to the field boundary M' at P'.

The problem is to find the first-order focusing point D for rays making a small angle $\pm\alpha$ with the central ray SP.

The construction is made quickly, as follows. The normal to the line SP through S is drawn to cut the line N at S'. The line $S'O$ is then drawn and extended to cut the normal N' at D'. From D' a line is drawn normally to the ray $P'D''$, cutting it at D, which is the image point required.

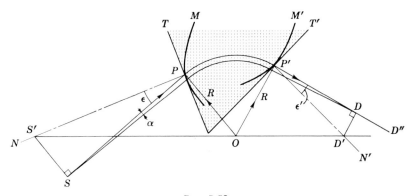

Fig. 5.59

It may be deduced from the figure that if ϵ and ϵ' are zero (normal case), then S and S' coincide, D and D' coincide, S, O, and D are in a straight line, and T and T' intersect at O.

This technique was greatly elaborated by Cartan, who investigated many of the first-order-focusing schemes following from Herzog's [8] work. It is routinely used by many designers and was the object of special attention by Persson [15], who used it to assist in the investigation of some second-order-focusing fields.

A machine based on this graphical construction and capable of drawing variously shaped focusing fields automatically has been constructed and described by Walton [56]. It is illustrated schematically in Fig. 5.60.

The instrument consists of two pieces of flat plastic material joined together by a pivot at E. Equidistant from E are pens P and P', the distance being adjustable and corresponding to various radii of curvature of the ion beam in the magnetic field. At right angles to the lines E–pen, a slot is cut into each member of the instrument, and it may then be slipped over two pins O and I fixed on a drawing board.

O then represents the source of an ion beam which it is desired to deflect and focus to an image *I*, in a magnetic field in which the radius of curvature *R* is fixed by the adjustable distance *E*–pen.

It is clear from the illustration that if the pen *P* is made to follow some arbitrary curve *HGF*, the other pen *P'* will trace out a corresponding curve *H'G'F'*. At each position of the instrument, while tracing out the curves, the "ion beam" follows a straight line from *O* to *P*, a circular trajectory of radius *E*–pen from *P* to *P'*, and a straight line from *P'* to *I*. Thus the region bounded by the two curves represents a magnetic field which will accomplish the necessary focusing. If *HGF* is arbitrarily

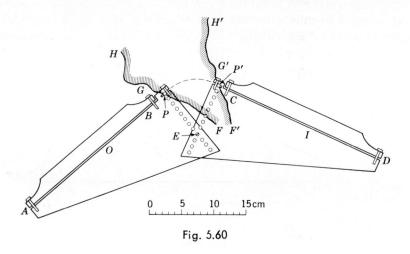

Fig. 5.60

made a straight line, for example, then *H'G'F'* will—to a second approximation—be traced out as a circle of radius $R \cot^3 \theta/2$ as determined in Sec. 5.9.

5.21. Source Optics

From the designer's point of view, the ion optics of spectrometer sources (Chap. 4) is in an unsatisfactory state. The various theories giving the focusing action of the slits through which the ions pass are based on conditions rarely found in mass spectrometers. Most sources use auxiliary magnetic fields to increase the ion production, and these complicate the optics considerably. Inevitably, source parameters are adjusted arbitrarily to give optimum performance.

Several efficient types of source have been developed, with the help of much cutting and trying. Examination of the operating characteristics of some of the optimum designs is beginning to throw some light on the precise way in which the ion beams are created. Barnard [2] has done extensive experimenting with very precisely machined sources of various designs. His extensive discussion of the problem should be consulted.

In so far as this chapter is concerned, the most important aspect of source optics concerns the manner in which the effective source's width and position are affected. In the sections on ion optics, ion beams were considered as emerging from a point or wide source S coincident with the object slit. In Fig. 5.61 it is seen that the divergence of the ion beam may

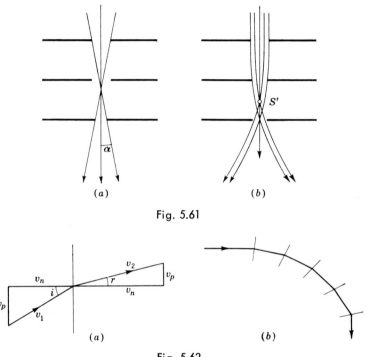

Fig. 5.61

Fig. 5.62

not be determined by the source geometry as in 5.61a, but rather by the focusing action of the slits as in 5.61b. A knowledge of the position of the virtual source S' is necessary for calculating α and the expected aberrations.

Barnard, citing several standard references, describes how such knowledge may be acquired. In Fig. 5.62a, we see an ion beam crossing a potential discontinuity. Its velocity component parallel to the equipotential surface remains constant, while its normal velocity component will be changed by the change in potential. We thus have the ordinary law of refraction

$$v_1 \sin i = v_2 \sin r \tag{5.121}$$

where

$$\frac{v_1}{v_2} = \left(\frac{V_1}{V_2}\right)^{1/2}$$

since the velocity of the ions is proportional to the square root of their energy. In Fig. 5.62b, we see how the beam may thus be traced across a region of varying potential.

In practice, the potential usually varies continuously and the integration of the refraction must be made judiciously, since any errors due to approximations are likely to be cumulative.

The slits of a typical ion source, such as indicated in Fig. 5.63a, produce such curved equipotentials, since the stronger field will always penetrate through a slit into the weaker. A focusing action is thus produced, and in the case illustrated, the ion beam is focused on the lower plate B.

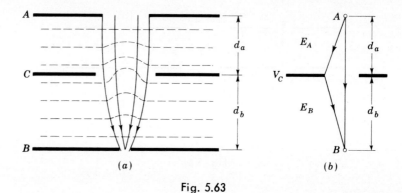

Fig. 5.63

For such focusing to take place, the lens parameters must be properly designed. In Fig. 5.63b, an ion beam is seen diverging from A and converging on B. The d's in the figure represent distances, the E's field strengths. Then it may be shown that

$$\frac{1}{d_b} - \frac{1}{d_a} = \frac{E_B - E_A}{2V_C} \qquad (5.122)$$

where V_c is the potential of the slit plate C with respect to A. From its resemblance to the thin-lens formula, we may write as the focal length of this slit lens

$$\frac{1}{f} = \frac{E_B - E_A}{2V_C} \qquad (5.123)$$

Although potentially useful to the designer, the derivation of this formula is based on many assumptions and simplifications, namely, small slit width compared with slit separation, negligible slit end effects (very long slits), negligibly thin slit plates, ions of zero energy at A, small field penetration, and absence of magnetic field. This last assumption is particularly harmful, since even sector instruments with no auxiliary magnetic field for the source would be disturbed by the action of the main magnet's fringing field on ions, leaving A with initially small velocity.

In the usual electron-bombardment source there are at least two slit lenses, and variation of the applied potentials permits changing the position of the final image and its magnification.   In many cases this is done by trial and error.

The design of an ion source benefits hugely from the use of an electrolytic tank to trace out the equipotentials of the slit lenses, of which large-scale models are placed in the tank.   The auxiliary magnetic field's effects may also be more easily visualized.   This technique was applied successfully to the design of the inflection-spectrometer source built by Leger and Ouellet [57], who desired negligible electric fringing field in their ionizing region.

Although not affecting the ion optics of a given ion beam, the various electric and magnetic fields present in a mass-spectrometer source do not act in identical fashion on all ion masses, and discrimination results [58]. This effect is discussed at some length by Barnard and in the papers cited by him.

5.22. Performance

It may be of some interest to the designer who has considered the various aspects of the ion optics of mass spectrometers in this chapter to read how some of the principles fare in practice, and the references at the end of this section concern instruments whose ion optics are of special interest at the moment.

The first-order instruments have been operated successfully for years, and the focusing of ion beams to well-defined images is a matter of many records.   Among those where careful measurements have been made on the image widths are Barnard's [2] 60° instrument (radius 4 in.) and Duckworth's [59–62] 180° machine (radius 108 in.).

The effects of the fringing field on sector instruments were long held to be such as to mask second-order focusing, but recent advances in the calculation of such effects have largely allayed these fears, and a number of second-order instruments are now in operation.   Continued study may be expected to provide most information on image components, since the α aberration may be made very small.   The author [68] has operated an inflection instrument satisfactorily for several years, and his colleagues in the Laval Chemistry Department have operated another equally successfully.   Each of these instruments has attained the theoretical resolving power to within the experimental error expected.   Persson [15] has designed and built a normal-circle instrument, as have Paul [63] and Bainbridge and Moreland [64].   Inflection spectrometers have also been built by Rickard [64a], Ordzhonikidze and Shutse [65], and Yokota, Nagatani, and Nonaka [66], all of whom report satisfactory performance. Prowse and Gibson [67] have built the optimum inflection lens of Sec. 5.6 (Fig. 5.21).

Geoffrion has built a β-ray spectrometer based on optimum parameters (Sec. 5.17) and finds that it behaves according to the theory. In Sec. 5.20, reference is made to several shaped-field instruments that have been constructed. In Sec. 5.19 it is shown how fringing-field effects as calculated seem to provide reasonable agreement with the measured effects. A double-focusing zero dispersion magnetic spectrometer has been built by Alvarez, Brown, Panofsky, and Rockhold [79] which incorporates corrections for second-order aberrations. Workable magnetic shims to correct second-order aberrations have been described by Balestrini and White [80] and Dietz [81, 82]. Source optics are still largely empirical, and performance ill related to theory, but Barnard has indicated that much may be done on this difficult problem. The use of quadrupole lenses has received great impetus from the work of Paul and his associates [70–73]. (An account of this work is given in Chap. 2, p. 34.) Such lenses may be used as source elements in mass spectrometers to provide intense beams or in combination with high-frequency fields to provide effective mass selection [74–78].

The theory of spectrometer ion optics and the development of instrumental techniques have progressed in closely parallel manner to the point where the elegant formulation of the one is well matched by the performance of the other.

REFERENCES

1. K. T. Bainbridge: *Exptl. Nucl. Phys.*, **1**, 559 (1953).
2. G. P. Barnard: "Modern Mass Spectrometry," The Institute of Physics, London, 1953.
3. H. Ewald and H. Hintenberger: "Methoden und Anwendungen der Massenspectroskopie," Verlag Chemie GmbH, Weinheim, Germany, 1958.
4. H. E. Duckworth: "Mass Spectrometry," Cambridge University Press, London, 1938.
5. H. Hintenberger: *Z. Naturforsch.*, **3a**, 125 (1948).
6. L. Kerwin: *Rev. Sci. Instr.*, **20**, 36 (1949).
7. W. Henneberg: *Ann. Phys.*, **19**, 335 (1934).
8. R. Herzog: *Z. Physik*, **89**, 447 (1934).
9. L. Cartan: *J. Phys. Radium*, **8**, 453 (1937).
10. J. Classen: *Z. Physik.*, **9**, 762 (1908).
11. A. J. Dempster: *Phys. Rev.*, **11**, 316 (1918).
12. N. F. Barber: *Proc. Leeds Phil. Lit. Soc. Sci. Sec.*, **2**, 427 (1933).
13. W. E. Stephens: *Phys. Rev.*, **45**, 513 (1934).
14. L. G. Smith: *Natl. Bur. Std. (U.S.) Circ.* 522, 1953, see also *Proc. Intern. Conf. Nuclidic Masses*, p. 418, University of Toronto Press, Toronto, Canada, 1960.
15. R. Persson: *Arkiv Fysik*, **3**, 455 (1951).
16. J. Geerk and C. Heinz: *Z. Physik*, **133**, 513 (1952).

17. L. Musermeci: *Nuovo Cimento*, **9**, 429 (1952).
18. D. F. Dempsey: *Rev. Sci. Instr.*, **26**, 1141 (1955).
19. K. T. Bainbridge: Solvay Report, *Seventh Congr. Chem.*, p. 5, R. Stoops, Brussels, 1947.
20. M. Spighel: *J. Phys. Radium*, **10**, 207 (1949).
21. W. R. Smythe, L. H. Rumbaugh, and S. S. West: *Phys. Rev.*, **45**, 724 (1934).
22. C. Geoffrion: *Rev. Sci. Instr.*, **23**, 224 (1952).
23. M. Cotte: *Ann. Phys.*, **10**, 333 (1938).
24. R. Herzog: *Natl. Bur. Std. (U.S.) Circ.* 522, 1953.
25. W. Cross: *Rev. Sci. Instr.*, **22**, 717 (1951).
26. M. Camac: *Rev. Sci. Instr.*, **22**, 197 (1951).
27. J. Mattauch: *Natl. Bur. Std. (U.S.) Circ.* 522, 1953.
28. E. W. Becker and W. Walcher: *Natl. Bur. Std. (U.S.) Circ.* 522, 1953.
29. C. E. Berry: *Natl. Bur. Std. (U.S.) Circ.* 522, 1953.
30. L. Kerwin: *Can. J. Phys.*, **30**, 503 (1952).
31. A. J. H. Boerboom: De Ionenoptiek van de Massaspectrometer, Thesis, Amsterdam, 1957.
31a. A. J. H. Boerboom: *Z. Naturforsch.*, **14a**, 809 (1959); **15a**, 350, 824 (1960).
32. H. A. Tasman: "Advances in Mass Spectrometry" (J. Waldron, ed.), p. 63, Pergamon Press, New York, 1959.
33. H. A. Tasman and A. J. H. Boerboom: *Z. Naturforsch.*, **14a**, 121 (1959); **15a**, 78, 736 (1960).
34. H. Wachsmuth, A. J. H. Boerboom, and H. A. Tasman: *Z. Naturforsch.*, **14a**, 818 (1959).
35. H. A. Tasman, A. J. H. Boerboom, and H. Wachsmuth: *Z. Naturforsch.*, **14a**, 822 (1959).
35a. A. J. H. Boerboom, H. A. Tasman, and H. Wachsmuth: *Z. Naturforsch.*, **14a**, 816 (1959).
36. C. Geoffrion: *Rev. Sci. Insts.*, **20**, 638 (1949).
37. N. D. Coggeshall and M. Muskat: *Phys. Rev.*, **66**, 187 (1944).
38. N. D. Coggeshall: *J. Appl. Phys.*, **18**, 855 (1955).
39. R. Herzog: *Z. Naturforsch.*, **10a**, 887 (1955).
40. L. A. König and H. Hintenberger: *Z. Naturforsch.*, **10a**, 877 (1958).
41. H. Hintenberger and L. A. König: "Advances in Mass Spectrometry" (J. Waldron, ed.), p. 16, Pergamon Press, New York, 1959.
42. L. A. König: *Proc. Intern. Conf. Nuclidic Masses*, p. 498, University of Toronto Press, Toronto, Canada, 1960.
43. A. O. C. Nier: *Rev. Sci. Instr.*, **11**, 212 (1940).
44. W. Bloch and W. Walcher: *Z. Physik*, **127**, 274 (1950).
45. C. Reuterswärd: *Arkiv Fysik*, **3**, 53 (1952).
46. C. D. Bock: *Rev. Sci. Instr.*, **4**, 575 (1933).
47. F. M. Beiduk and E. J. Konopinski: *Rev. Sci. Instr.*, **19**, 594 (1948).
48. L. M. Langer and C. S. Cook: *Rev. Sci. Instr.*, **19**, 257 (1948).
49. J. Kistemaker and C. J. Zilverschoon: *Natl. Bur. Std. (U.S.) Circ.* 522, p. 179, 1953.
50. E. M. McMillan: see Ref. 52.
51. N. Svartholm and K. Siegbahn: *Arkiv Mat. Astron. Fysik*, **A33**, no. 21 (1946).
52. F. B. Shull and D. M. Dennison: *Phys. Rev.*, **71**, 681 (1947).
53. F. B. Shull and D. M. Dennison: *Phys. Rev.*, **72**, 256 (1947).
54. F. D. Kurie, J. S. Osoba, and L. S. Slack: *Rev. Sci. Instr.*, **19**, 771 (1948).
55. N. Svartholm: *Arkiv Fysik*, **2**, 115 (1950).
56. E. T. S. Walton: *Proc. Roy. Irish Acad.*, **A57**, 1 (1954).
57. E. Leger and C. Ouellet: *J. Chem. Phys.*, **21**, 1310 (1953).

58. J. R. Pierce: "Theory and Design of Electron Beams," D. Van Nostrand Company, Inc., Princeton, N.J., 1949.
59. H. E. Duckworth: *Advan. Electron. Electron Phys.*, **3**, 229 (1956).
60. H. E. Duckworth: "Mass Spectrometry," Cambridge University Press, London, 1958.
61. J. T. Kerr, G. R. Bainbridge, J. W. Dewdney, and H. E. Duckworth: "Advances in Mass Spectrometry" (J. Waldron, ed.), p. 1, Pergamon Press, New York, 1959.
62. N. R. Isenor, R. C. Barber, and H. E. Duckworth: *Proc. Intern. Conf. Nuclidic Masses*, 1960, p. 438. University of Toronto Press, Toronto, Canada, 1960.
63. M. Paul: *Z. Naturforsch.*, **12a**, 634 (1957); **13a**, 745 (1958).
64. K. T. Bainbridge and P. E. Moreland, Jr.: *Proc. Intern. Conf. Nuclidic Masses*, p. 460, University of Toronto Press, Toronto, Canada, 1960.
64a. J. A. Rickard: Internal Report, Humble Oil and Refining Co., Houston, Texas.
65. K. Ordzhonikidze and V. Shutse: *J. Exptl. Theoret. Phys. U.S.S.R.*, **29**, 479 (1955).
66. M. Yokota, M. Nagatani, and I. Nonaka: *Mem. Fac. Sci. Kyushu Univ.*, **2B**, 22 (1956).
67. D. J. Prowse and W. M. Gibson: *J. Sci. Instr.*, **33**, 129 (1956).
68. L. Kerwin: *Rev. Sci. Instr.*, **21**, 96 (1950); see also Ref. 59.
69. L. Kerwin and C. Geoffrion: *Rev. Sci. Instr.*, **20**, 381 (1949).
70. W. Paul and M. Raether: *Z. Physik*, **140**, 262 (1955).
71. W. Paul, H. P. Reinhard, and U. von Zahn: *Z. Physik*, **152**, 143 (1958).
72. R. Köhler, W. Paul, K. Schmidt, and U. von Zahn: *Proc. Intern. Conf. Nuclidic Masses*, p. 507. University of Toronto Press, Toronto, Canada, 1960.
73. F. von Busch and W. Paul: *Z. Physik*, **164**, 581, 588 (1961).
74. E. D. Courant and L. Marshall: *Rev. Sci. Instr.*, **31**, 193 (1960).
75. H. A. Enge: *Rev. Sci. Instr.*, **29**, 885 (1958).
76. H. A. Enge: *Rev. Sci. Instr.*, **30**, 248 (1959).
77. H. A. Enge: *Rev. Sci. Instr.*, **32**, 662 (1961).
78. C. F. Giese: *Rev. Sci. Instr.*, **30**, 260 (1959).
79. R. A. Alvarez, K. L. Brown, W. K. H. Panofsky, and C. T. Rockhold: *Rev. Sci. Instr.*, **31**, 556 (1960).
80. S. J. Balestrini and F. A. White: *Rev. Sci. Instr.*, **31**, 633 (1960).
81. L. A. Dietz: *Rev. Sci. Instr.*, **30**, 225 (1959).
82. L. A. Dietz: *Rev. Sci. Instr.*, **32**, 859 (1961).

6

Electronic Techniques

D. C. Frost

Department of Chemistry
The University of British Columbia
Vancouver, B.C.

6.1. Introduction

The number of electronic circuits needed for a mass spectrometer are very few compared with the number of acceptable varieties in published design. In general, the needs are electric and magnetic fields with stability in accordance with the resolution requirements, and ion production and detection arrangements of sufficient efficiency and sensitivity to cope with the particular problem on hand. Usually, some form of regulation of ion-source potentials and electric and magnetic fields employed in ion acceleration and dispersion is essential.

In the conventional ion source, ions are formed by electron impact and then accelerated through a series of electrodes. They are then separated according to their mass-charge ratio in a magnetic field before collection and measurement. It will be convenient, then, to divide the discussion into sections dealing with ion-beam production, dispersion, and measurement.

6.2. Emission Regulation

In electron-bombardment ion sources it is necessary, for many applications, to maintain a constant ionizing current despite fluctuations in line voltage or variations in filament emissivity with different sample vapors in the ion chamber. Very often the ion-source filament transformer is connected in series with a variable impedance. The electron current, amplified, controls the loading of a second transformer whose primary constitutes the variable impedance. Thus any change of emission causes an opposing change to occur in the filament supply current. This type

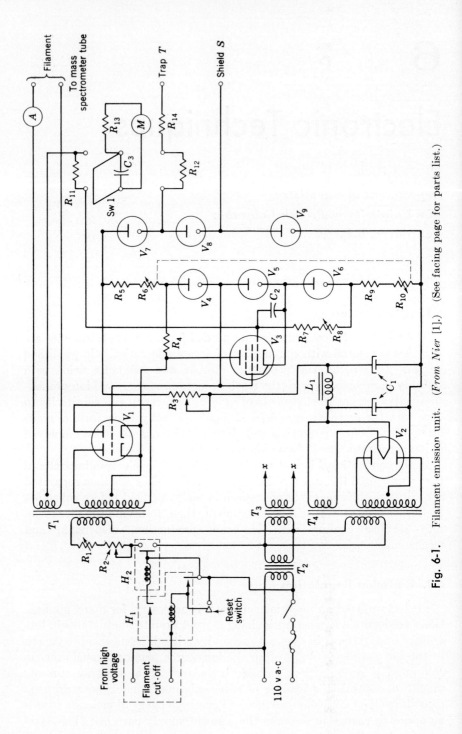

Fig. 6-1. Filament emission unit. (*From Nier* [1].) (See facing page for parts list.)

of servo system can be highly sensitive, although the stabilization range is quite small in such a series-impedance circuit.

Various emission-regulation circuits have been described, such as the one proposed by Nier [1] in Fig. 6.1.

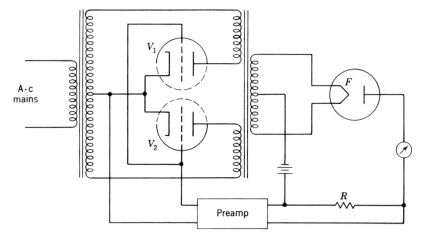

Fig. 6.2. Filament emission unit.

An arrangement providing a much wider range of stabilization has been proposed by Steckelmacher and Van der Meer [2]. This is shown in Fig. 6.2.

The power required for the source filament is controlled by V_1 and V_2, which can operate from cutoff to full load.

Feedback in the preamplifier may be used to prevent any oscillatory currents encouraged by the use of physically large filaments. The electron current develops a voltage across R whose difference from a reference

Parts List for Fig. 6.1

R_1	150-ohm 50-watt potentiometer		lent; mass-spectrometer filament supplied by 6.3-volt winding
R_2	250-ohm 50-watt adjustable		
R_3	5,000-ohm 25-watt adjustable	T_2	1:1 isolation transformer, 115-volt 60-cycle to 115-volt, secondary insulated to 5,000-volt 200-va capacity
R_4	0.5-megohm ½-watt		
R_5	500-ohm 1-watt		
R_6	5,000-ohm potentiometer, General Radio 214 A		
R_7	150,000-ohm, 1-watt	T_3	T19 F97 Thordarson or equivalent
R_8	0.5-megohm carbon potentiometer	T_4	T13 R14 Thordarson or equivalent
R_9	500-ohm 1-watt	L_1	15-henry 60-ma
R_{10}	5,000-ohm potentiometer, General Radio 214 A	V_1	6AS7
		V_2	5U4G
R_{11}, R_{12}	Wire-wound shunts for meter M of such size as to convert M to 500 and 200 μa, respectively	V_3	6SJ7
		V_4–V_8	VR-75
		V_9	VR-150
		A	Meter, 0–10 amp, a-c
R_{13}	2,000-ohm precision-wire-wound	M	Meter, 0–100 μa, d-c
R_{14}	20,000-ohm ½-watt	$Sw\,1$	DPDT toggle switch
C_1	8-8 μf 450 volts	H_1	Relay, DPST 110-volt a-c
C_2, C_3	0.1 μf 400 volts	H_2	Relay, SPST 110-volt a-c, 2,500-volt insulation
T_1	T13 R14 Thordarson or equiva-		

voltage is applied to the preamplifier grid. The preamplifier output adjusts the grids of V_1 and V_2 in such a sense as to cause positive stabilization of the emission from F. A 20% change in line voltage or 0.5- to 2.5-amp change in filament current changes the emission by only 0.5%; the stabilization for a mass-spectrometer filament emission should be similar.

A rather simple type of emission regulator has been proposed by Winn and Nier [3]. It operates by controlling the space charge instead of the filament heater current and in performance compares favorably with electronic regulators. For instance, a tenfold change in oxygen pressure, which caused a 1.5% emission change in the electronically regulated filament [1], caused a 2% change in the space-charge-regulated one. A control plate is situated between the filament and ion chamber, and the control voltage is obtained by passing the emission current through a biasing battery and resistor in series so that a change in emission causes a change in control-plate voltage such as to correct the emission change.

If current leakage is liable to occur in the ion chamber, it is often preferable to regulate the trap current. This can be done if the trap current is used to control the potential of an electron-gun electrode. Figure 6.3 gives an example [4]; the low output impedance means that leakage of insulators on the first slit causes little trouble. The filament power is supplied from a Sorensen 150S voltage regulator, and line-voltage effects on the trap current are further reduced by an electronic regulator to the point where they become negligible. A 40% emission change produces a 0.1% trap-current change.

Regulating units often dissipate quantities of energy out of proportion to the power they supply. The use of transistors makes for higher efficiency, lower cost and physical size, and less maintenance. Russell and Kollar [5] have described a transistorized emission regulator which, drawing only 30 watts from the mains supply, controls the rms variation of electron current to better than 1 part in 10,000. Their circuit is shown in Fig. 6.4.

The series transistor can easily regulate the 5-amp filament current at the expense of a voltage drop less than 2.5 volts. Control is effected by two 2N456 transistors in the Darlington compound arrangement. They provide a large current gain and a low base current (typically 0.7 ma). The driving stages are a 2N369 voltage amplifier and a 2N366 emitter follower. The error voltage is the difference between the voltage developed across the 100-kilohm helipot with series resistor and the 27-volt reference across the Zener diode and is almost zero at balance since the forward voltage drops across the emitter-base diodes of the 2N369 and the 2N366 nearly cancel.

Supplies are required for the electron energy, trap bias, and emitter follower. The first two are fed from a small filament transformer con-

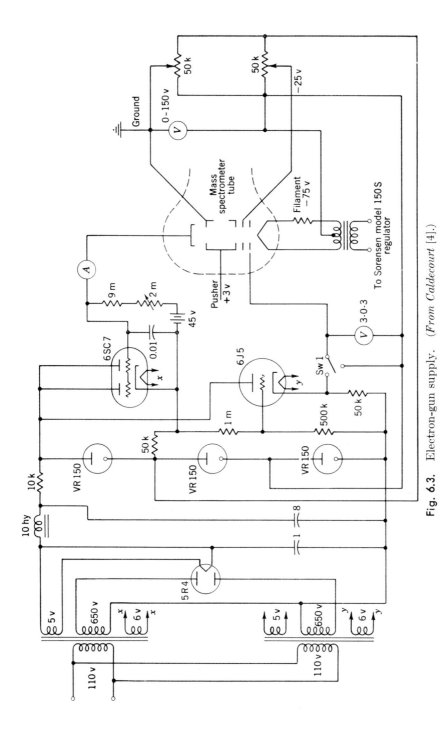

Fig. 6.3. Electron-gun supply. *(From Caldecourt [4].)*

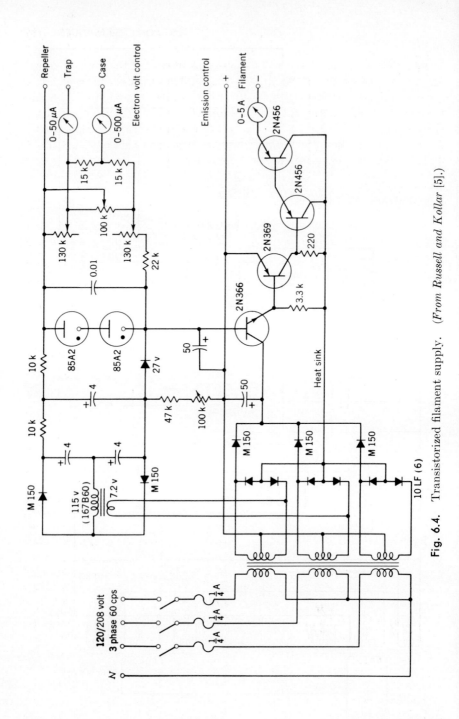

Fig. 6.4. Transistorized filament supply. *(From Russell and Kollar [5].)*

nected backwards so that the 7.2 volts rms output of the 3-phase transformer is applied to the 6-volt winding. This obviates the need for a second transformer insulated for the high voltages associated with the ion source. The emitter follower is supplied with current from a 3-phase silicon rectifier circuit. The 3-phase system can easily be replaced by a full-wave single-phase rectifier with a large condenser or LC filter. This emission unit has proved most reliable.

6.3. Voltage Stabilizers

A high-voltage supply stabilized by gas-discharge tubes alone is not stable enough for ion-acceleration purposes in mass spectrometry. A

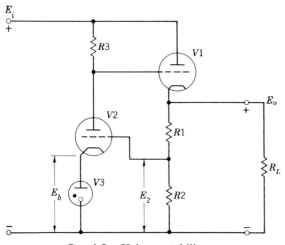

Fig. 6.5. Voltage stabilizer.

voltage which is more nearly constant can be obtained from a stabilizer using vacuum tubes with negative feedback. The tubes generally act in two ways: as class A amplifiers and as variable resistances to direct current. A tube, to be employed in the latter function, is usually inserted in series with the d-c lead from the power supply, for example, V_1 in Fig. 6.5.

The signal input of the amplifier V_2 is a fraction of the regulator output voltage, and its output controls the bias and resistance of V_1. The diode V_3 and the voltage drop across R_2 provide the proper bias for V_2, and the value of R_3 is determined jointly by the gain requirements of the amplifier and the bias requirements of V_1.

The load current passes through V_1 and can therefore be controlled by its grid potential. The output voltage appears across the cathode load of V_1, and a part of it, E_2, is impressed on the grid of V_2 in series with a bias voltage E_b. E_b is developed across the discharge tube V_3

and is therefore very nearly constant, and so any changes in the grid or anode voltages of V_2 are the same whether referred to ground or the V_2 cathode.

Now, if the supply voltage E_i or the load resistance R_L changes, the negative feedback provided by V_2 between the input and output circuits of V_1 will tend to keep the output voltage of V_1 constant.   For example, suppose an increase in the load resistance increases the output voltage by ΔE_0 with V_2 in, and $\Delta E_0'$ with V_2 out of, the circuit.   Then we may write

$$\Delta E_0 = \Delta E_0' - \Delta E_0''$$

where $\Delta E_0''$ is the reduction in E_0 due to the stabilizing action of V_2. The grid of V_2 will rise in potential by

$$\Delta E_{g2} = \Delta E_2 = \frac{R_2}{R_1 + R_2} \Delta E_0 \tag{6.1}$$

and if A is the amplification afforded by V_2, its anode voltage will fall by

$$\Delta E_{a2} = A \; \Delta E_{g2} = \frac{A R_2}{R_1 + R_2} \Delta E_0 \tag{6.2}$$

This fall in voltage, applied between the grid of V_1 and earth, will result in a very nearly corresponding voltage drop on the cathode due to the cathode-follower action of this tube.   This fall in output voltage due to the stabilizing action is therefore

$$\Delta E_0'' = \Delta E_{a2} = \frac{A R_2}{R_1 + R_2} \Delta E_0$$

Thus we have

$$\Delta E_0 = \Delta E_0' - \Delta E_0''$$
$$= \Delta E_0' - \frac{A R_2}{R_1 + R_2} \Delta E_0$$

whence
$$\frac{\Delta E_0}{\Delta E_0'} = \frac{1}{1 + A[R_2/(R_1 + R_2)]} \tag{6.3}$$

and this equation gives the ratio by which the stabilizer reduces changes in the output voltage E_0.   Clearly, for the best stabilization, both A and $R_2/(R_1 + R_2)$ should be as large as possible.   The value of $R_2/(R_1 + R_2)$ determines the positive voltage E_2, which, with the bias developed across V_3, fixes the grid potential of V_2.   If E_2 is too high for the tube when $R_2/(R_1 + R_2)$ is made large, it is a simple matter to insert the correct number of discharge tubes in series with V_3 so that the tube is able to operate.   It may therefore be assumed that the fraction $R_2/(R_1 + R_2)$ is large and A about 50, if V_2 is a triode, and many times larger if a pentode. We may then write, as a fair approximation,

$$\frac{\Delta E_0}{\Delta E'_0} = \frac{1}{A}\left(1 + \frac{R_1}{R_2}\right)$$ (6.4)

The output voltage E_0 can be varied by changing the grid potential of V_2. The lower the V_2 grid potential, the higher the V_1 output voltage. However, since this operation means increasing the ratio R_1/R_2, the stability of E_0 will be adversely affected.

If R_3 is connected between the anode of V_2 and the cathode, not the anode, of V_1, the output voltage is more stable with regard to changes in the supply voltage. This is because the supply voltage for V_2 is now

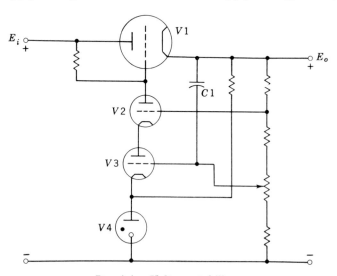

Fig. 6.6. Voltage stabilizer.

the stabilized output voltage E_0, and not the unstabilized supply, and because V_1 acts as an amplifier, not a cathode follower, thereby increasing the effective value of A. However, the voltage drop across R_3, which can be appreciable, now appears between the grid and cathode of V_1, with the former the more negative. As a consequence, if a large current is to be drawn from the regulator, there will be a large voltage drop across V_1. To avoid this the grid of V_1 may be connected to a tapping on R_3, but since the effectiveness of the tube is thereby reduced in high-current applications, R_3 is better connected to the positive voltage output.

The amplifier tube V_2 may be replaced by a pentode, with a marked increase in amplification and therefore stability, or another triode may be added in series to produce the same effect (Fig. 6.6). The current rating may be increased by adding power tubes in parallel with V_1. The capacitor C_1 effectively applies the entire ripple or rapid-fluctuation

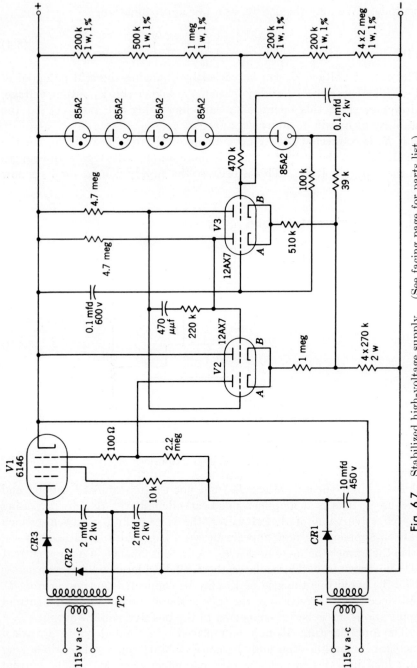

Fig. 6.7. Stabilized high-voltage supply. (See facing page for parts list.)

voltage to the grid of V_3, so that, with this type of circuit, the efficiency of the primary-power-supply filter need not be high—a simple resistance-capacitor filter is often quite adequate.

The circuit of Fig. 6.7 will provide 2,500 volts with less than 1 mv ripple. Opposite phases of the ripple signal are coupled to opposite grids of V_3, one of the regulation amplifiers. The other is V_2. The amplified d-c error signal and the amplified ripple signal are led from the V_2A anode and coupled directly to the V_1 grid. The output is stable to 1 in 30,000 under normal working conditions and will deliver up to 25 ma. Other useful circuits have been described by Donner [6] and Muraca [7].

The magnetic field stability required depends on the type of instrument and the uses to which it is to be put. For a given spectrometer the resolution depends on the ion beam width at the focal point, the dispersion, the resolving slit width, beam aberrations, and electronic instabilities. The last factor is the one with which we shall be primarily concerned here.

If the magnetic field varies fractionally by an amount $\Delta H/H$, then the incremental beam width is given by $\Delta W = 2r_m(\Delta H/H)$. Similarly, a fractional deviation in the electric field, $\Delta E/E$, produces a

$$\Delta W = r_e(\Delta E/E)$$

For a cycloidal instrument the radius of curvature r is replaced by the focal distance.

One can calculate how electric and magnetic field fluctuations affect the resolution, since two adjacent beams will be completely distinguished only when the dispersion is greater than the collector slit width plus the total image spread. Typically, a ripple of 0.1% in the electric field could well double the theoretical image width. Corresponding variations in the magnetic field will, of course, be more serious.

The theoretical field regulation necessary for a particular application may be determined from the geometry of the instrument, but one should bear in mind that ion-beam aberrations, never entirely absent, will affect the resolution and therefore increase the field-stability requirements.

In the constant-current regulator, the controlling voltage is usually the difference between the voltage drop across a portion of the load and a reference voltage. The series resistor must carry the full regulated output current, and unless it is of low value, it must dissipate a considerable quantity of heat. A low-value, low-temperature-coefficient resistor is therefore required, and in this application Minett [8] has described

---

Parts List for Fig. 6.7: $CR1$ Rectifier, audio devices 40-A1
    $CR2$ 9 rectifiers, general instrument PA430
    $CR3$ 9 rectifiers, general instrument PA430
    $T1$ Transformer, Thordarson 26R60
    $T2$ Transformer, Triad 67071

the use of an incandescent-lamp filament.   Equilibrium is reached very
soon because of the low heat capacity of the filament, and it is possible
to regulate a 250-ma current to 2 parts in 10,000 over several hours.

Several practical designs for current regulators are given in the literature [1, 9, 10], and various methods have been employed to keep a particular ion beam in focus.   Nier [11] has used an O1A tube as a magne-
tron in the magnetic field to compensate for changes therein by variations
in the accelerating voltage.

The conventional use of vacuum tubes in magnet current regulators
has called for high-impedance coil windings—quite a considerable waste-
ful dissipation of power in the regulator.   However, power transistors
operating with a potential drop of as low as a volt can change this state
of affairs, with a great improvement in efficiency.   Transistors able to
pass 10 amp at a voltage drop of a volt are easily obtainable, and Garwin
et al. [12] have described a magnet current regulator employing several
of these.   Fine regulation is accomplished by amplifying in a chopper
amplifier the difference between a reference voltage and the voltage drop
produced by the magnet current in a shunt, and their regulator maintained a stability of three parts per million for several hours.

Russell and Kollar [5] have designed a magnet current supply embodying transistor, vacuum tube, and magnetic amplifiers.   Their 800-cm²-
pole-area magnet can produce a 5,000-gauss field in a 1.9-cm gap with a
regulator power dissipation of only 50 watts.   The power transformer
primary, in series with a saturable reactor, is fed from an autotransformer.
The filtered output is connected to the magnet coils via the 2N457
control transistor and a series resistor.   Two 2N369 transistors constitute
a difference amplifier to compare the series-resistor voltage drop with a
reference and to control the 2N457.   The sum of the series-resistor and
control-transistor voltage drops serves to regulate the bias of a 6AQ5
vacuum tube controlling the saturable reactor current.   The two control
circuits combine to provide a magnet current stable to 1 part in 25,000
over several minutes.   The circuit is shown in Fig. 6.8.

Ions formed in radio-frequency mass spectrometers pass through a
system of coaxial disks or cylinders.   The frequency of an r-f voltage
fed in antiphase to alternate disks determines the mass-charge ratio of
the ions leaving the analyzer with sufficient energy to be collected.
This type of spectrometer requires a constant-amplitude r-f generator,
and one is shown in Fig. 6.9.   Boyd and Morris [13] found that, to cover
the 6- to 100-Mc/sec range without switching or coil changing, the most
satisfactory form of tuned circuit was one in which the values of L and
C could be varied simultaneously.

The electrical length of a resonant line is varied in synchronism with a
variable condenser connected across its open end.   The line, of 16 SWG
silver wire, is stretched around a pair of 24-in.-diameter disks spaced

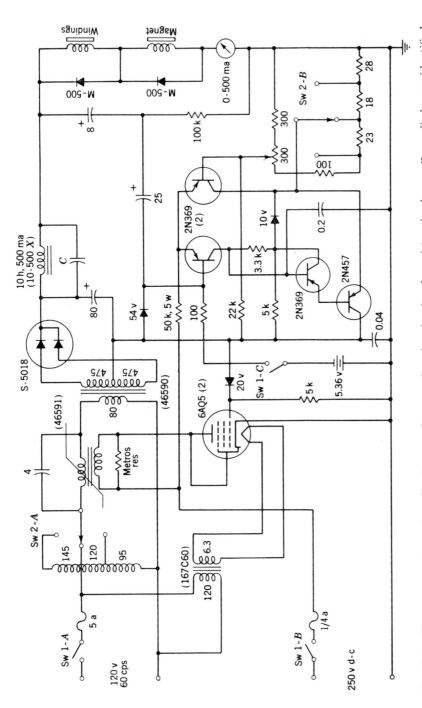

Fig. 6.8. Magnet current supply. Condenser values are given in microfarads, resistors in ohms. Zener diodes are identified by their breakdown voltages. Transformer and reactor numbers in parentheses are Hammond-type numbers. The condenser C is chosen for minimum ripple ($0.2\mu f$). (*From Russell and Kollar* [5].)

2 in. apart.   Its effective length is varied by a pair of silver contacts carried on an arm which rotates about the disks' centers.

Constancy of r-f amplitude is achieved through automatic control of the oscillator supply voltage.   A portion of the r-f output is rectified by a crystal diode, amplified by V_2, and used to control the voltage drop

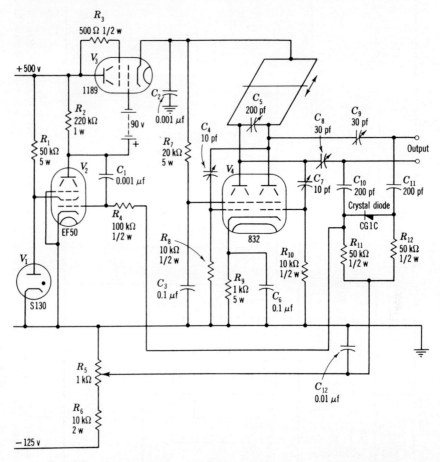

Fig. 6.9.   Radio-frequency oscillator.   (*From Boyd and Morris* [13].)

across V_3.   Any variations in oscillator output will therefore result in compensating changes in the V_4 supply voltage.   The r-f amplitude may be adjusted by altering R_5, the regulator-tube bias control.   This supply is suitable for a 12-stage spectrometer with a mass range of 1 to 300.

6.4. Ion-current Amplification

In order to measure the intensity of mass-spectrometer ion beams, it is necessary to have some form of current amplification.   The ion currents to be measured range from about 10^{-10} amp to a minimum which may be set by limitations of the measuring technique.

In most cases the ion beam provides a steady current in the first instance, so an electrometer tube can often be used either alone or in conjunction with subsequent stages of amplification with or without negative feedback. An electrometer tube is generally operated with potential differences of only a few volts between the electrodes to minimize secondary electron emission and residual-gas ionization. Exclusion of light and the use of a high-work-function grid help to prevent photoelectric emission. The grid potential is kept as near as possible to the value where the grid current is zero.

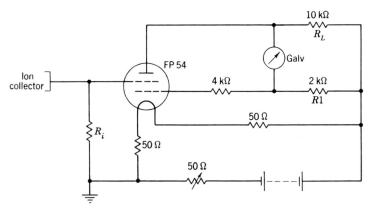

Fig. 6.10. Simple electrometer ion detector.

Electrometer circuits in which the balance point is independent of small changes in supply voltage have been developed, and most of them use tetrodes, with the galvanometer connected as in Fig. 6.10. When the galvanometer is undeflected, the voltage drop across the resistor R_L due to the anode current equals the drop across R_1 due to the screen current.

If the battery voltage drifts, the anode and screen currents change in the same proportion, so the galvanometer remains undeflected. The deflection varies with the control-grid potential, and ion currents of as low as 10^{-15} amp can be measured in this way.

The addition of directly coupled amplification stages allows the galvanometer to be replaced by a more robust current meter. The basic circuit of such an amplifier is shown in Fig. 6.11.

Suitable choice of the tube electrode voltages can result in a meter deflection proportional to the primary ion current, provided the dynamic characteristics of the tubes are straight within the operating regions.

Frequently, negative feedback is employed to reduce circuit noise and overcome nonlinearity of the amplifier due to variation of the preamplifier control-grid potential with input current. Also, the amplifier gain is made reasonably independent of fluctuations in the supply voltage.

The circuit of Fig. 6.12 employs 100% negative feedback. The output of the amplifier, of voltage gain A, appears across R_0, and the voltage amplification A' is

$$\frac{\Delta E_0}{\Delta E_i} = \frac{A}{1 + A} \tag{6.5}$$

ΔE_i can be replaced by E_i, the voltage developed across the input resistor R_i, since with no input this voltage is zero.

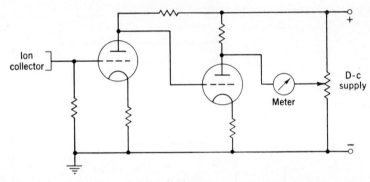

Fig. 6.11. Electrometer detection with subsequent amplification stage.

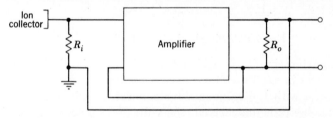

Fig. 6.12. Block diagram of d-c amplifier with 100% negative feedback.

A can easily be made several thousand, and so the voltage amplification is very close to unity and independent of small fluctuations in A. If E_i is due to an ion current I_i and causes a change ΔI_0 in the current through the load R_0, then

$$R_0 \, \Delta I_0 = R_i I_i$$

and the current amplification is therefore R_i/R_0.

With suitable choice of tubes, R_i can be as high as 10^{11} or 10^{12} ohms, and with R_0 a few hundred ohms, a current amplification of 10^8 is easily obtained. The amplified signal may be indicated by metering the output current, or on a recorder after further amplification of the voltage output. Figure 6.13 shows a circuit that has proved satisfactory. The preamplifier and RCA 954 acorn pentode can easily be mounted close to the spec-

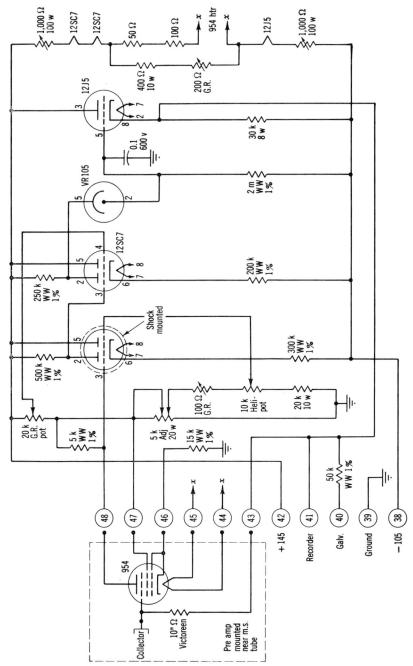

Fig. 6.13. Direct-current amplifier with facilities for recorder and galvanometer ion-current display.

trometer collector. The collector lead to the grid should of course be kept as short as possible.

Circuits of a similar type have been described in the literature [1, 14].

There are some precautions to be observed when using d-c amplifiers, apart from those already mentioned. For best results the electrometer tube should be carefully shock-mounted in an evacuated, magnetically screened box. The filament power used should be as low as possible to avoid thermionic grid emission, and great care must be taken to avoid contamination of the tube external surfaces.

Vibrating-reed electrometers manufactured in recent years, although expensive, provide a much more rugged and reliable means for measuring small currents [15]. The principle of operation is simple: a metal reed is made to vibrate close to a static anvil, and any d-c signal applied across them is reflected in an alternating voltage generated by virtue of their varying capacity. The signal is then fed to a narrow-bandwidth a-c amplifier before being rectified and applied to the input as negative feedback. For most purposes the zero drift is negligible.

Electron multiplier tubes, now coming into more general use, have been constructed for the purpose of detecting positive ions. The ions are caused to impinge on a cathode surface and to liberate electrons. These electrons are in turn caused to strike a succession of dynodes, each giving a secondary electron yield greater than unity. In conventional electron multipliers the electrons are directed from dynode to dynode by the potential difference across them. Often a magnetic field is used in addition, to increase the gain. The result is a form of low-noise, wide-bandwidth amplification.

Although multiplier response is dependent on time, vacuum, surface conditions, and ion type [16], it is possible, in conjunction with a d-c or vibrating-reed amplifier, to measure ion currents down to 10^{-20} amp in applications where the disadvantages mentioned are not serious. Multipliers using the secondary-electron-emitting properties of such materials as 2% beryllium copper are easily constructed [17, 18], and the gain, although deteriorating with time, may be soon restored through a simple activating process [19].

A high-resistance strip magnetic electron multiplier has recently been developed [20] in which the multielement dynode structure of the conventional multiplier has been replaced by two strips of semiconductor material—the dynode and field strips of Fig. 6.14. The potential differences applied across the strip ends are adjusted to produce equipotentials angled as in the figure.

The whole instrument, which can be 3 or 4 in. long, is placed in a magnetic field of about 400 gauss. Electrons, produced through ion collisions with the cathode while en route to the anode, suffer enough collisions with the dynode strip to produce a current gain of 10^7. This figure is unaffected by periodic exposure to air, unlike the case of the conventional multiplier. The dark current can correspond to a primary ion current of as low as 3×10^{-21} amp.

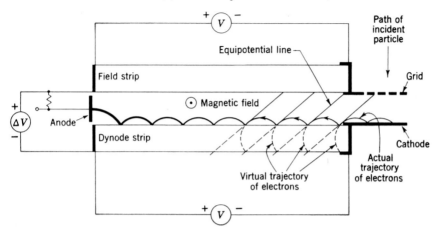

Fig. 6.14. Resistance strip magnetic multiplier. (*From Goodrich and Wiley* [20].)

6.5. Ionization-gauge Circuits

Ionization gauges provide a convenient means of measuring the high vacuum desired in most mass spectrometers. The gauge usually contains a plate, a grid, and a filament. The grid is positive and the plate

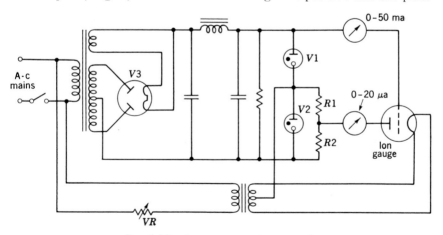

Fig. 6.15. Ion-gauge-control circuit.

negative with respect to the filament, and the electrode voltages are so arranged that electrons from the filament bombard and ionize some of the gas molecules present in the tube. The resulting positive ions are attracted to a negatively charged plate, usually cylindrical and enclosing the filament and grid.

The rate of flow of electrons from the plate circuit is used to indicate the gas pressure. In the conventional design the ratio of plate to grid current is proportional to the gas pressure. In the gauge-control circuit

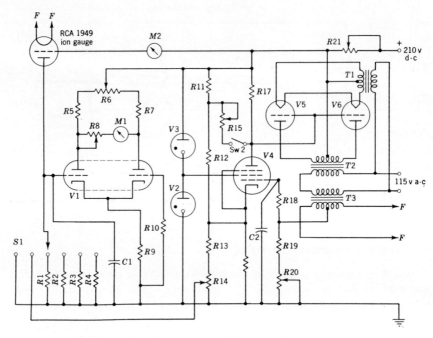

Fig. 6.16. Stabilized-emission ion-gauge-control circuit.

R_1	1k 1w Aerovox Carbofilm	R_{21}	2k 2w WW variable
R_2	10k 1w Aerovox Carbofilm	C_1, C_2	0.02 mfd 600-volt paper
R_3	100k 1w Aerovox Carbofilm	S_1	1P.6 pos. rotary switch
R_4	1 meg 1w Aerovox Carbofilm	Sw_2	1P.2 pos. on/off switch
$R_5, R_7, R_{11},$	10k 1w Aerovox Carbofilm	T_1	Transformer, Hammond 165X60
R_{12}, R_{13}		T_2	Transformer, Hammond 273X60
R_6	5k 1w Variable	T_3	Transformer, Hammond 1144X60
R_8, R_{14}	2k 1w Variable	V_1	6SC7 tube
R_{10}, R_{17}, R_{18}	100k 1w	V_2, V_3	VR105 tube
R_9	270Ω 1w	V_4	6SJ7 tube
R_{15}	10k 1w variable	V_5, V_6	6AS7 tube
R_{16}	3.9k 1w	M_1	0–1 ma d-c meter
R_{19}	5.6k 1w	M_2	0–10-ma d-c meter
R_{20}	3k 1w variable		

shown in Fig. 6.15, the stabilized voltages for the gauge are obtained from gas-discharge tubes V_1 and V_2. Their characteristics and the values of R_1 and R_2 are chosen to suit the voltage requirements of the gauge. V_3 and the associated components provide the necessary power supply. Since the gas pressure is related to the ion current for a certain fixed electron emission, and since the latter quantity will probably change from gas to gas, it is necessary to adjust VR to obtain a filament emission of the correct value before reading the ion current.

The control circuit just described is useful for pressures above 10^{-5} mm Hg. At this pressure the ion current may be about 1 μamp. To indicate lower pressures with any degree of accuracy, the control circuit must be provided with an amplifier. A circuit of this type, which also incorporates a regulated filament emission, is shown in Fig. 6.16. It is suitable for use with an RCA 1949 gauge. Tubes V_1 and V_2 constitute a differential amplifier for the gauge ion current. The grid of V_2 is fixed in potential; electrons flowing to the gauge plate through any one of the grid resistors result in a reading on the "gas pressure" meter M_1. The amplifier is calibrated by determining from the gauge characteristics what ion current corresponds to, say, 10^{-5} mm Hg pressure. The voltage this would develop across a resistor (1 megohm would probably be most convenient for this pressure) can easily be calculated. This voltage can be applied directly to the grid of V_1 from R_{14}, and the sensitivity control R_8 adjusted until M_1 reads full-scale. Rotation of S_1 causes the V_1 grid resistor to be changed by one or more factors of 10, and therefore the full-scale deflection of pressure meter M_1 is changed by the same factor but in the opposite sense. A separate gauge calibration must be performed for each gas or mixture of gases to be used, since the ionization efficiencies will be different in each case.

The filament emission is stabilized by means of a servo circuit. The primaries of T_2 and T_3 are in series with the main supply. The secondary of T_3 feeds the filament, and the secondary of T_2 a variable impedance, the tubes V_5 and V_6. The filament emission controls the grid of V_4 and therefore the grids of V_5 and V_6, and the reaction of tubes V_5 and V_6 to a change of filament emission is in such a way as to correct the change. R_{20} sets the emission required, and S_2 and R_{15} enable a fixed voltage to be applied to the grids of V_5 and V_6 for initial degassing of the gauge.

REFERENCES

1. A. O. Nier: *Rev. Sci. Instr.*, **18**, 398 (1947).
2. W. Steckelmacher and S. Van der Meer: *J. Sci. Instr.*, **27**, 189 (1956).
3. E. B. Winn and A. O. Nier: *Rev. Sci. Instr.*, **20**, 773 (1949).
4. V. J. Caldecourt: *Rev. Sci. Instr.*, **22**, 58 (1951).
5. R. D. Russell and F. Kollar: *Can. J. Phys.*, **38**, 616 (1960).
6. W. Donner: *Electronics*, **28**, 137 (1955).
7. R. F. Muraca: *Rev. Sci. Instr.*, **24**, 1152 (1953).
8. E. E. Minett, D. A. MacRae, and J. Townsend: *Rev. Sci. Instr.*, **20**, 136 (1949).
9. J. L. Lawson and A. W. Tyler: *Rev. Sci. Instr.*, **10**, 304 (1939).
10. H. S. Sommers, P. R. Weiss, and W. Halpern: *Rev. Sci. Instr.*, **22**, 612 (1951).
11. A. O. Nier: *Rev. Sci. Instr.*, **6**, 254 (1935).
12. R. L. Garwin, D. Hutchinson, S. Penman, and G. Shapiro: *Rev. Sci. Instr.*, **30**, 105 (1959).

13. R. L. F. Boyd and D. Morris: *Electron. Eng.*, **25**, 389 (1953).
14. R. M. Dowben: *Rev. Sci. Instr.*, **23**, 506 (1952).
15. H. Palevsky, R. K. Swank, and R. Grenchik: *Rev. Sci. Instr.*, **18**, 298 (1947).
16. H. E. Stanton, W. A. Chupka, and M. G. Inghram: *Rev. Sci. Instr.*, **27**, 109 (1956).
17. J. S. Allen: *Phys. Rev.*, **55**, 966 (1939).
18. C. F. Barnett, G. E. Evans, and P. M. Stier: *Rev. Sci. Instr.*, **25**, 1112 (1954).
19. F. F. Fitz-Osborne: *Can. J. Phys.*, **30**, 658 (1952).
20. G. W. Goodrich and W. C. Wiley: *Rev. Sci. Instr.*, **32**, 846 (1961).

7

High-resolution Mass Spectroscopes

H. E. Duckworth

Department of Physics
McMaster University
Hamilton, Ontario

S. N. Ghoshal*

Department of Physics
Presidency College
Calcutta, India

7.1. Early High-resolution Mass Spectroscopes

A. Introduction

Mass spectroscopy has come a long way since Sir J. J. Thomson's [1] positive-ray parabola apparatus, with which he obtained the first evidence for the existence of stable isotopes. In the intervening years mass spectroscopes have been utilized in increasing number in many diverse fields of scientific research, each of which imposes its own peculiar instrumental requirements. Thus, for example, high-resolution mass spectroscopes are needed in the precision determination of atomic masses, in certain isotopic-abundance studies, in the identification of organic compounds, and in the mass-spectroscopic analysis of solids. In this chapter, a systematic treatment of high-resolution mass spectroscopes will be undertaken. In doing this, however, we shall omit many technical details, such as those concerning ion sources, detector systems, problems of field stabilization, etc. These are of primary importance in the actual operation of an instrument, but are discussed elsewhere in the book.

It is difficult to obtain high resolution with a mass spectroscope that consists of a magnetic analyzer alone. This type of instrument, which had its beginning with Dempster's [2] semicircular magnetic analyzer,

---

* Most of this chapter was written while S. N. Ghoshal was on sabbatical leave at McMaster University.

possesses the property of direction focusing.† In it, monoenergetic ions of a given mass, from a narrow slit with small angular divergence, are brought to a line focus after passage through the magnetic field. However, a velocity (or energy) spread among the ions will lead to a broadening of the line image so formed. The resolving power of these instruments is therefore limited by the energy spread of the ions entering the analyzer [Eq. (7.9)]. A brief survey of the characteristics of the direction-focusing mass spectroscopes that have actually been constructed (see, for instance, the list compiled by Inghram and Hayden [3]) reveals that their resolving power is seldom greater than 200 to 400 and is rarely sufficient for the types of experiment mentioned above. It is possible to monochromatize the initial ion beam, as Bainbridge [4] did, by passing it through a Wien velocity filter. This results in a substantial improvement in resolving power, but it is gained at the cost of intensity.

Aston's [5] first and subsequent mass spectrographs represented a different approach to the problem. Here the ions were initially collimated to a very high degree by a suitable combination of slits. This reduced the angular divergence of the ions to such a point that the need for direction focusing was greatly diminished. He was consequently free to design the instrument so as to obtain velocity focusing. This was done by passing the ions through a linear combination of two fields, one electric and one magnetic, with the result that the velocity dispersion produced in the first field was counterbalanced by that produced in the second. By this strategem Aston was able to achieve a much higher resolving power than had been available with the parabola method. But it was still not very high, principally because of the geometrical spreading of the ion beam (i.e., no direction focusing).

Mass spectroscopes incorporating *both* direction *and* velocity focusing were wanted long before it was known how to construct them. During the early 1930s, however, much knowledge was gained concerning the focusing properties of deflecting fields. This made it possible to construct mass spectroscopes possessing both types of focusing, the so-called double-focusing instruments, which always employ a combination of electric and magnetic fields. In most of these instruments velocity focusing is obtained, as it was by Aston, by counteracting in a magnetic field the velocity dispersion produced in a preceding electric field. At the same time, direction focusing is maintained through the pair. The magnetic field alone is responsible for the mass dispersion.

Before discussing these double-focusing instruments, which, incidentally, provided a gain in resolving power of at least a factor of 10, we shall briefly describe the instruments of Aston and Bainbridge, which were the historical forerunners of modern high-resolving power devices.

---

† For a discussion of the theory of direction focusing by a magnetic field, see Chap. 5.

B. Aston's mass spectrograph

As mentioned in the last section, Aston built his first mass spectrograph in 1919. This was followed by two improved versions, which were completed in 1925 and 1937, respectively [6]. The original apparatus is shown schematically in Fig. 7-1. In this a discharge tube serves as the ion source, and consequently, the ions possess a large energy spread. The positive ions are collimated by two narrow slits S_1 and S_2, spaced some distance apart. They are then deflected through an angle θ by a uniform electric field between P_1 and P_2. Subsequently, they enter a

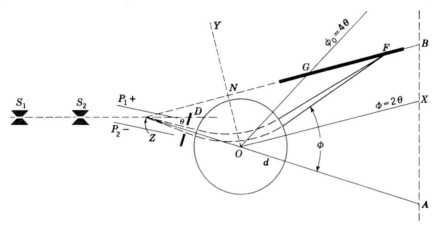

Fig. 7.1. The scheme of Aston's mass spectrograph. S_1 and S_2 are collimating slits; P_1 and P_2 are condenser plates. D is a diaphragm for selecting a portion of the beam emerging from the condenser; Z is a virtual source for the rays emerging from the condenser; O is the center of the uniform magnetic field; GF is the photographic plate; θ and ϕ are the deflections in the electric and magnetic fields, respectively. (*From Aston* [6].)

circular uniform magnetic field centered at 0, in which they are deflected in the opposite direction through more than twice the angle θ. The foci lie along the line ZB, which determines the position of the photographic plate GF.

As already stated, this arrangement possesses the velocity-focusing property. Thus, in spite of the considerable velocity spread among the ions emerging from the source, those having the same value of e/M are focused at one point. There is no direction focusing. The broadening of the image due to angular divergence of the initial ion beam is avoided to a great extent by use of the highly selective collimating system.

In his first apparatus, Aston achieved a resolution of about 1 part in 130. In later versions, instead of using a parallel-plate condenser to produce the electrostatic deflecting field, Aston employed cylindrical

condensers.   The resolving power achieved in his third apparatus was
about 2,000.

An essentially similar apparatus was also built in Paris by Costa [7],
who obtained a resolution of 1 part in 600.

Aston used his first apparatus to confirm Thomson's discovery of
isotopy among the stable elements.   He also observed small departures
from the whole-number rule of isotopic weights, which were subsequently
investigated by Costa with greater precision.   Later, Aston, with his
improved apparatus, undertook a systematic investigation of the isotopic
composition of a large number of elements in the periodic table.   He was
able to obtain fairly accurate values for the percentage abundances of the
different isotopes from the intensities of the mass-spectral lines recorded
on the photographic plate.   His painstaking determination, by the
doublet method, of the exact divergences of isotopic weights from the
whole-number rule led to his celebrated packing-fraction curve.   This
curve, together with the information which his work provided concerning
the distribution of isotopes, represents a fundamental step in our understanding of the conditions for, and variations in, nuclear stability.

C. Bainbridge's apparatus

While Aston was perfecting his combined electric- and magnetic-field
mass spectrographs, possessing velocity-focusing properties, no noticeable
improvement was made in the design of the Dempster-type mass spectrometers employing magnetic field alone until about 1930.   In the
early 1930s Bainbridge first used a Dempster-type semicircular instrument, which had been built under the direction of Swann [8].   To
improve the resolution of the apparatus, Bainbridge [4] added a Wien
velocity filter [9], which consists of crossed electric and magnetic fields
through which ions with a very small spread in velocity can pass before
they enter the analyzing magnetic field.   An account of the focusing
properties of the Wien filter will be given in Sec. 7.3E: these were not
known at the time Bainbridge built his apparatus.   The resolution
obtained with the modified apparatus was about 1 part in 500.   A schematic diagram of Bainbridge's apparatus is shown in Fig. 7-2.   Bainbridge used this instrument to determine the masses of many of the light
atoms.   This work provided the first experimental proof of the Einstein
mass-energy relationship.

7.2.   First-order Focusing in Magnetic and Electric Fields

A. Review and consideration of results of magnetic analysis

Since the high-resolution instruments to be discussed in the following
section involve double focusing of the ions by the use of a combination of

magnetic and electrostatic fields, it is necessary to understand the focusing action of these fields on a beam of ions traveling through them.   We shall first summarize the results of focusing in a magnetic field which have been developed in Chap. 5.

The 180° magnetic field employed by Dempster [2] is actually a special case of the general sector magnetic field whose focusing properties have been investigated by Barber [10], Stephens [11], Henneberg [12], Herzog [13], and others.   Let us consider a beam of monoenergetic ions with

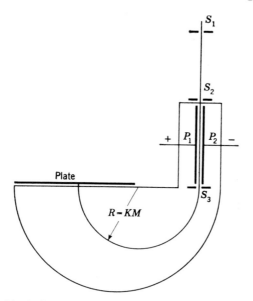

Fig. 7.2.   Bainbridge's first mass spectrograph. S_1, S_2, and S_3 are slits; P_1 and P_2 are condenser plates providing the electric field of the Wien velocity filter.   As indicated, the radius of curvature of the magnetic analyzer is linear with ion mass for constant degree of ionization.   (*From Aston* [6].)

different masses diverging from a point source with a small angular divergence and entering a magnetic sector field with plane boundaries. The magnetic field will separate these ions into different mass groups and focus them at different points on the other side of the field along a curve called the *image curve*.   The positions of these foci on the image curve can be most simply determined by a geometrical method due to Cartan [14], which has been described in Chap. 5 on Ion Optics (see Sec. 5.4).

The radius of curvature a_m of the orbit of ions of mass M within the magnetic field is given by

$$a_m = \frac{1}{B} \sqrt{\frac{2MV}{e}} \tag{7.1}$$

where all quantities, namely, the magnetic induction B, the ion's charge

e, the radius of its orbit a_m, its mass M, and the accelerating voltage V, are expressed in mks units.

It has been shown in Sec. 5.9C that if l'_m and l''_m denote the object and image distances from the entrance- and exit-field boundaries, respectively, the following relationships hold for any particular ion group:

$$(l'_m - g'_m)(l''_m - g''_m) = f_m{}^2 \tag{7.2}$$

$$g'_m = \frac{a_m \cos \epsilon'_m \cos (\phi_m - \epsilon''_m)}{\sin (\phi_m - \epsilon'_m - \epsilon''_m)}$$

$$g''_m = \frac{a_m \cos \epsilon''_m \cos (\phi_m - \epsilon'_m)}{\sin (\phi_m - \epsilon'_m - \epsilon''_m)} \tag{7.3}$$

$$f_m = \frac{a_m \cos \epsilon'_m \cos \epsilon''_m}{\sin (\phi_m - \epsilon'_m - \epsilon''_m)}$$

In the above equations, as shown in Fig. 7.3, ϕ_m is the angle of deflection for the ions following the mean direction, a_m is the radius of curvature of

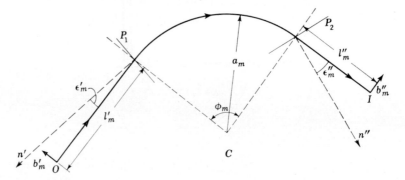

Fig. 7.3.   Focusing properties of a homogeneous magnetic field. O, I, and C are the positions of the object, image, and center of curvature, respectively. P_1 and P_2 are the boundaries of the pole pieces of the magnet; n' and n'' are normals to these boundaries. ϵ'_m and ϵ''_m are the angles made by the mean-incident and emergent ion beams with n' and n'', respectively. l'_m and l''_m are the distances of the object and image, respectively, from the boundaries of the magnetic field; b'_m and b''_m are used to describe the object or image widths, respectively, or to measure the displacement of object or image in a direction normal to the central path; a_m is the radius of curvature in the magnetic field; and ϕ_m is the angle of deflection.

the mean orbit in the field [given by Eq. (7.1)], ϵ'_m and ϵ''_m are the angles made by the central entrance and exit rays with the normals n' and n'' to the respective field boundaries P_1 and P_2.   These angles are considered positive if the outward-drawn normals from the field boundaries lie on the opposite side of the optic axis† from the center of curvature C of the

† We shall call the central ray corresponding to the mean mass M_0 and mean velocity v_0 of the ions emerging normally from the central point of the slit, the optic axis.

orbit in the field. f_m is the focal length of the magnetic lens so formed, while g'_m and g''_m are the distances of the principal foci from the entrance- and exit-field boundaries, respectively.

Equation (7.2) is analogous to the similar equation in geometrical optics for a thick lens and shows the lens action of a homogeneous magnetic field. The magnetic field also has a prism action in so far as it produces a dispersion of ions with different masses and energy.

In an actual mass spectroscope, the ions do not diverge from a single point, but from different points in a slit of finite width S'. This results in the broadening of the image. If we consider ions of a given mass M_0 and velocity v_0 emerging from a point on the slit at a distance b'_m from the optic axis (Fig. 7.3), then the point at which they will be focused after emerging from the field will be displaced from the optic axis by an amount b''_m given by

$$b''_m = -b'_m \frac{f_m}{l'_m - g'_m} \tag{7.4}$$

where the negative sign represents the inversion of the image due to the focusing property of the magnetic field [10]. The magnification of the image is then given by

$$G_m = \frac{b''_m}{b'_m} = -\frac{f_m}{l'_m - g'_m} = -\frac{1}{\cos \epsilon'_m} \left[\frac{l''_m}{a_m} \frac{\sin (\phi_m - \epsilon'_m - \epsilon''_m)}{\cos \epsilon''_m} \right.$$
$$\left. - \cos (\theta_m - \epsilon'_m) \right] \tag{7.5}$$

If now we consider, instead, an ion of mass $M = M_0(1 + \gamma)$ with a velocity $v = v_0(1 + \beta)$ emerging from the same point of the object slit, there will be an additional displacement of the image from the optic axis, normal to it, given by

$$D = \frac{K''_m \Delta a_m}{a_m} \tag{7.6}$$

a_m being the radius of curvature of the mean ray following the optic axis, while $(a_m + \Delta a_m)$ is the radius of curvature of the orbit of the ions under consideration. K''_m, which is known as the *coefficient of velocity dispersion*, is given by

$$K''_m = a_m(1 - \cos \phi_m) + l''_m[\sin \phi_m + (1 - \cos \phi_m) \tan \epsilon''_m] \tag{7.7}$$

From Eq. (7.1) it follows that $\Delta a_m / a_m = (\beta + \gamma)$, and hence

$$D = K''_m(\beta + \gamma) \tag{7.6a}$$

The total image displacement normal to the optic axis is then given by

$$b'' = b''_m + D \tag{7.8}$$

For normal incidence and emergence ($\epsilon'_m = \epsilon''_m = 0$), we get

$$g'_m = g''_m = g_m$$

and

$$K''_m = a_m \left(1 + \frac{f_m}{l'_m - g_m}\right) \tag{7.7a}$$

Hence, in this case, the total normal displacement of the image reduces to

$$b'' = a_m(\beta + \gamma)\left(1 + \frac{f_m}{l'_m - g_m}\right) - b'_m \frac{f_m}{l'_m - g_m} \tag{7.8a}$$

For ions of a given type ($M = M_0$, $\gamma = 0$), emerging from a given point with a spread in velocity amounting to $\pm v_0\beta$ about the mean velocity v_0, there will thus be an image broadening given by [Eq. (7.6a)]

$$D_0(\beta) = K''_m\beta \tag{7.6b}$$

on either side of the optic axis. This is called the *velocity dispersion* of the ions.

The ions of different mass and velocity are focused along the image curve. If the normal to the image curve makes an angle ω with the mean ray at its point of intersection with the former, the velocity dispersion along the image curve will be

$$D(\beta) = \frac{D_0(\beta)}{\cos \omega} \tag{7.6c}$$

On the other hand, if we consider ions of different masses having the same energy ($2\beta + \gamma = 0$), there will be a dispersion of the ions along the image curve due to the fact that the different masses have different velocities. This *mass dispersion* (or simply *dispersion*) normal to the central ray is given by

$$D_0(\gamma) = \frac{K''_m\gamma}{2} \tag{7.6d}$$

The dispersion along the image curve is then

$$D(\gamma) = \frac{D_0(\gamma)}{\cos \omega} \tag{7.6e}$$

The standard dispersion is usually referred to 1% mass difference; that is, $\gamma = 0.01$. This would give, from Eq. (7.6e), the value of standard dispersion as

$$D = \frac{1}{200} \frac{K''_m}{\cos \omega} \tag{7.6f}$$

for 1% mass difference.

In order that the lines due to two neighboring masses M_0 and $M_0(1 + \gamma)$ may appear clearly separated in spite of the finite entrance slit width and the finite energy spread of the ions, the distance between the lines corresponding to these masses given by Eq. (7.6d) must be larger than the line width $2b''$ for any one mass, which is obtained from Eqs. (7.8) and (7.6b) ($2b'_m = S' =$ slit width),

$$2b'' = 2b''_m + 2D_0(\beta) = \left| G_m \middle| S' + \frac{1}{2} \middle| \frac{\Delta V}{V} K''_m \right|$$

where $eV =$ ion energy

$V =$ ion accelerating potential [corresponding to $v = v_0(1 - \beta)$]

$\Delta V =$ spread in V [$V + \Delta V$ corresponds to $v = v_0(1 + \beta)$]

The condition for the resolution between two neighboring masses is then

$$D_0(\gamma) > 2b''$$

which gives

$$\frac{\Delta M}{M} > \frac{2S'}{K''_m} \left| G_m \right| + \left| \frac{\Delta V}{V} \right|$$

For two masses just separated, the above inequality should be replaced by the equation

$$\frac{M}{\Delta M} = \frac{1}{\dfrac{2S'}{K''_m} |G_m| + \left| \dfrac{\Delta V}{V} \right|} \tag{7.9}$$

This gives the resolving power of a mass spectrograph, where the ions are detected photographically. In mass spectrometers, where electrical methods of detection are employed, there is a collector slit of width S''. The plane of this slit is perpendicular to the direction of the ions entering the collector. In this case the resolving power is given by

$$\frac{M}{\Delta M} = \frac{1}{\dfrac{2S'|G_m| + 2S''}{K''_m} + \left| \dfrac{\Delta V}{V} \right|} \tag{7.9a}$$

For the special case of the symmetrical arrangement, in which the object and image distances are equal ($l'_m = l''_m$) and both the incident and emergent rays are normal to the field boundaries ($\epsilon'_m = \epsilon''_m = 0$), we have

$$G_m = -1 \qquad K''_m = 2a_m$$

$$f_m = \frac{a_m}{\sin \phi_m} \qquad g'_m = g''_m = a_m \cot \phi_m$$

so that

$$\frac{M}{\Delta M} = \frac{1}{\dfrac{S' + S''}{a_m} + \left| \dfrac{\Delta V}{V} \right|} \tag{7.9b}$$

It should be noted that the direction focusing considered here is valid only if the angular divergence $2\alpha \ll 1$, i.e., only to the first order of approximation.

In much of the foregoing, and also in much of the following section, we have followed the presentation of Ewald and Hintenberger [23].

B. Focusing by radial electrostatic fields

In this and in the next sections we shall consider the focusing action of electrostatic fields on beams of charged particles passing through them. The most commonly employed electrostatic field is the radial field between two coaxial sector-shaped cylindrical electrodes. Such a field possesses first-order direction focusing for ions emerging from a source with a small angular divergence in the radial direction. In recent years

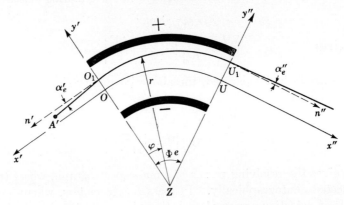

Fig. 7.4. Path of ions in a radial electric field between two sector-shaped cylindrical electrodes whose common axis is through the point Z. The central ray is along $x'OUx''$, OU being the mean circular trajectory of radius a_e in the field. The ions starting from the point A' along $A'O_1$, making an angle α_e' with the normal n' to the entrance field boundary, follow the trajectory O_1U_1 and emerge from the field, making an angle α_e'' with the normal n'' to the exit field boundary. (*From Ewald and Hintenberger* [23].)

spherical and toroidal electrodes have been employed to introduce, in addition, axial focusing of the ions. In this section, we shall consider the focusing action of a radial electric field employing cylindrical electrodes [13], which is the most commonly employed field in mass spectroscopy. The effect of the fringing fields will be neglected. In the next section, the more general toroidal electric field will be considered.

In Fig. 7.4 the paths of the ions in such a radial field are shown. The distance between the two electrodes is usually small compared with the mean radius a_e of the two electrodes. If an ion enters the electric field normally, it will describe a circular path throughout the field only if it has the correct energy to make the centrifugal force balance the electrostatic force acting upon it. Another ion entering the same field at the same point but with a different energy, or making a small angle to the normal, will usually describe a rather complicated orbit which cannot be easily calculated. However, this orbit will still be a nearly circular one, in the close vicinity of the exactly circular orbit of the ion mentioned above, if its energy differs only slightly from that of the first ion, or if its angular divergence from the normal is small. We shall confine our attention to ions describing such approximately circular orbits and investigate the focusing action of the radial electric field upon them.

The equations of motion of the ions in the cylindrical polar coordinates (r, ϕ, z) can be written as

$$M(\ddot{r} - r\dot{\phi}^2) = eE_r$$

$$M\frac{d}{dt}(r^2\dot{\phi}) = 0 \qquad (7.10)$$

and $$M\ddot{z} = 0$$

The electric field E_r can be written as

$$E_r = -\frac{E_0 a_e}{r} \qquad (7.11)$$

where we assume the field to have the value $-E_0$ at the mean radius $r = a_e$. We also assume that the cylindrical surface $r = a_e$ is the zero-potential surface.† The other equipotentials are cylindrical surfaces coaxial with the electrodes. The potential at a radius $r = a_e(1 + \rho)$ is given by

$$V = E_0 a_e \ln(1 + \rho) = E_0 a_e \rho \qquad (7.12)$$

to a first order of approximation.

Outside the field boundaries, we assume the field to drop abruptly to zero and also the potential to be zero. Hence ions entering the field at any radius except at $r = a_e$ will lose an amount of energy equal to eV, where V is given by Eq. (7.12). Along $r = a_e$, of course, there will be no change in the energy of the ion on entering the field, since $V = 0$ at this radius.

Let v_0 be such a velocity that ions of mass M_0 entering the field with this velocity normal to the field boundary at $r = a_e$ describe a circular path within the field. Hence $\rho = 0$ for such ions, and we have the condition

$$\frac{M_0 v_0^2}{a_e} = -eE_0 \qquad (7.13)$$

Let us now consider an ion of mass $M = M_0(1 + \gamma)$ and of velocity $v' = v_0(1 + \beta)$ entering the field at a radius r. The velocity v that the ion will have on entering the field can be deduced from the energy equation

$$\tfrac{1}{2}Mv^2 = \tfrac{1}{2}Mv'^2 - eV$$

† Actually the surface of zero potential is given by $\sqrt{(a_e - d/2)(a_e + d/2)}$, where d is the electrode spacing, but this reduces to a_e when $a_e \gg d$.

which, with the help of Eqs. (7.12) and (7.13), gives

$$v = v_0(1 + \beta - \rho) \qquad (7.14)$$

to a first approximation.

The solutions of the equations of motion are obtained by remembering that we neglect motion along the z axis. Thus the third of Eqs. (7.10) gives a constant value of z. The second of Eqs. (7.10) gives

$$r^2\dot{\phi} = \text{constant} \qquad (7.15)$$

We introduce the coordinate systems x', y' and x'', y'' at the entrance and exit boundaries of the electric field, respectively (Fig. 7.4). Let us consider ions emerging from the point A' to the left of the entrance field boundary. The coordinates of A' are $x' = l'_e$ and $y' = a_e\rho_0$. If these ions enter the field at $x' = 0$ and $y' = a_e\rho_1$ [that is, $r = r_1 = a_e(1 + \rho_1)$ and $\phi = 0$], then, from Eq. (7.15), we have, for any point on the orbit of an ion inside the field,

$$r^2\dot{\phi} = r_1{}^2\dot{\phi}_1 \qquad (7.15a)$$

Here $\dot{\phi}_1$ is the angular velocity of the ion at $\phi = 0$.

If α'_e is the angle which the ray makes with the normal as it enters the field, then, to a first approximation,

$$-\alpha'_e = \frac{1}{a_e}\left[\frac{d(a_e\rho)}{d\phi}\right]_{\phi=0} = a_e\,\frac{\rho_1 - \rho_0}{l'_e} \qquad (7.16)$$

The angle α'_e is taken as positive if the outward drawn normal to the entrance field boundary lies on the same side of the ion path as the center of curvature of the beam.

The initial angular velocity $\dot{\phi}_1$ will be given by

$$\dot{\phi}_1 = \frac{v \cos \alpha'_e}{r_1} = \frac{v_0}{a_e}(1 + \beta - 2\rho_1) \qquad (7.17)$$

in which we have substituted for v from Eq. (7.14) and neglected squares and higher powers of α'_e.

Then, from Eq. (7.15a),

$$\dot{\phi} = \frac{r_1{}^2\dot{\phi}_1}{r^2} = \frac{v_0}{a_e}(1 + \beta - 2\rho) \qquad (7.18)$$

to the first order of approximation. Then, to a zero-order approximation,

$$\phi = \frac{v_0 t}{a_e} \qquad (7.18a)$$

We now seek the solution of the radial equation [first of Eqs. (7.10)], which, with the help of Eqs. (7.11) and (7.18), can be transformed as follows:

$$\ddot{\rho} = 2 \left(\frac{v_0}{a_e}\right)^2 (\delta - \rho) \tag{7.19}$$

where
$$\delta = \beta + \frac{\gamma}{2} \tag{7.19a}$$

In Eq. (7.19), only first-order terms have been retained.

The solution of this equation with the initial conditions $\rho = \rho_1$ and $\dot{\rho} = (v_0/l'_e)(\rho_1 - \rho_0)$ at $\phi = 0$ is

$$\rho = \delta + \frac{a_e(\rho_1 - \rho_0)}{\sqrt{2}\, l'_e} \sin \sqrt{2}\, \phi + (\rho_1 - \delta) \cos \sqrt{2}\, \phi \tag{7.20}$$

If we call $y = r - a_e = a_e\rho$, we have, at the exit boundary,

$$(y)_{\phi_e} = a_e \left[\delta + \frac{a_e(\rho_1 - \rho_0)}{\sqrt{2}\, l'_e} \sin \sqrt{2}\, \phi_e + (\rho_1 - \delta) \cos \sqrt{2}\, \phi_e \right] \tag{7.21}$$

and the angle α''_e which the beam makes with the normal to the exit boundary at emergence is

$$\alpha''_e = \frac{1}{a_e} \left[\frac{d(a_e\rho)}{d\phi} \right]_{\phi=\phi_e} = \frac{a_e(\rho_1 - \rho_0)}{l'_e} \cos \sqrt{2}\, \phi_e - \sqrt{2}\, (\rho_1 - \delta) \sin \sqrt{2}\, \phi_e \tag{7.22}$$

The lateral displacement of the beam at a distance x'' after emerging from the field is then

$$\begin{aligned}
y'' &= (y)_{\phi_e} + x'' \alpha''_e \\
&= a_e \left[\delta + \frac{a_e(\rho_1 - \rho_0)}{\sqrt{2}\, l'_e} \sin \sqrt{2}\, \phi_e + (\rho_1 - \delta) \cos \sqrt{2}\, \phi_e \right] \\
&\quad + x'' \left[\frac{a_e(\rho_1 - \rho_0)}{l'_e} \cos \sqrt{2}\, \phi_e - \sqrt{2}\, (\rho_1 - \delta) \sin \sqrt{2}\, \phi_e \right] \tag{7.23}
\end{aligned}$$

Let us now consider a beam of monoenergetic ions emerging from the source point with a small angular divergence. For such a group of ions, $\delta = 0$ (that is, $M_0 v_0^2/2 = M_0(1 + \gamma)v_0^2(1 + \beta)^2/2$, whence $\gamma + 2\beta = 0$ to first order). Thus, ions with the same energy, same point of origin, and same initial direction will follow the same path. They will then be brought to a point focus at a distance l''_e on the other side of the field, if the lateral displacement y'' of all particles in the beam at l''_e is independent of the radial distance r_1 at which these particles enter the field at the entrance boundary; that is, y'' should be independent of ρ_1 when $x'' = l''_e$. Thus, by setting the coefficient of ρ_1 in Eq. (7.23) equal to zero, we can find the required distance l_e'' from the exit boundary of the field to the point of focus:

$$l_e'' = \frac{a_e^2/2 + (l_e' a_e/\sqrt{2}) \cot \sqrt{2}\,\phi_e}{l_e' - (a_e/\sqrt{2}) \cot \sqrt{2}\,\phi_e} \tag{7.24}$$

For a parallel beam of ions incident at the entrance boundary of the field, $l_e' \to \infty$ and $\alpha_e' \to 0$ for all values of ρ_1. Hence the focus point in this case is beyond the exit boundary a distance g_e'', given by

$$g_e'' = \frac{a_e}{\sqrt{2}} \cot \sqrt{2}\,\phi_e$$

Also, if the particles emerging from the field are rendered into a parallel beam, $l_e'' \to \infty$ and $\alpha_e'' \to 0$. In this case the distance g_e' of the point of divergence of the ions from the entrance boundary is given by

$$g_e' = \frac{a_e}{\sqrt{2}} \cot \sqrt{2}\,\phi_e$$

Thus the distances of the object and image focal points from the corresponding field boundaries are equal and can be written as

$$g_e = \frac{a_e}{\sqrt{2}} \cot \sqrt{2}\,\phi_e \tag{7.25}$$

Substituting this in Eq. (7.24), we get

$$(l_e' - g_e)(l_e'' - g_e) = f_e^2 \tag{7.25a}$$

where
$$f_e = \sqrt{g_e^2 + \frac{a_e^2}{2}} = \frac{a_e}{\sqrt{2}\,\sin\sqrt{2}\,\phi_e} \tag{7.25b}$$

f_e is called the focal length of the cylindrical electrostatic lens.

The lateral displacement of the image at the focal point according to Eqs. (7.23), (7.25), (7.25a), and (7.25b) becomes

$$y_2 = (y'')_{l_e''} = a_e \left[\delta\left(1 + \frac{f_e}{l_e' - g_e}\right) - \rho_0 \frac{f_e}{l_e' - g_e} \right] \tag{7.26}$$

This lateral displacement thus depends upon ρ_0 and δ. For monoenergetic ions of slightly different masses, $\delta = \beta + \gamma/2 = 0$ (see above). Hence monoenergetic ions are all focused at the same point, independently of their masses, after emerging from the radial electrostatic field. This means that the resolving power of the radial electrostatic field is zero as far as monoenergetic ions are concerned. This result is a special case of the general result that the mass dispersion of an electric field is

zero for ions of constant energy.   It is apparent that, since there is no
mass sensitivity, this field arrangement by itself has little practical value
in mass spectroscopy.   Its importance arises from its combination with
other fields in double-focusing instruments.

On the other hand, if we consider a group of ions of the same mass
($\gamma = 0$) but with a small energy spread, δ has to be replaced by β in Eq.
(7.26).   The position of focus is now linearly dependent upon β, and
hence particles of the same mass but with a small spread in velocity
display a velocity dispersion as they emerge from the field.   The lateral
displacement y_2 of ions of a given mass is given by

$$ y_2 = a_e \left[\beta \left(1 + \frac{f_e}{l'_e - g_e} \right) - \rho_0 \frac{f_e}{l'_e - g_e} \right] \qquad (7.26a) $$

In actual electrostatic analyzer arrangement, the ions are allowed to
emerge from an object slit of width $S'_e = 2a_e\rho_0$, placed symmetrically with
respect to the mean path of the ions.   Then, for ions emerging with a
given energy from different points of the object slit, the image formed at
the focal plane has a width S''_e, which can be easily obtained from Eq.
(7.26a):

$$ S''_e = y_2(+\rho_0) - y_2(-\rho_0) = -2a_e\rho_0 \frac{f_e}{l'_e - g_e} $$

or

$$ S''_e = -S'_e \frac{f_e}{l'_e - g_e} = S'_e G_e \qquad (7.27) $$

where

$$ G_e = - \frac{f_e}{l'_e - g_e} \qquad (7.27a) $$

G_e being the magnification of the analyzer.

In a double-focusing apparatus, a slit is usually placed in the image
plane of the electrostatic field, prior to the magnetic field.   The width
of this slit B (Fig. 7.7) should be larger than the width of the image S''_e,
so that ions belonging to different energy groups may pass through it.
The slit width should, however, be two or three orders of magnitude
smaller than a_e.   The separation between the cylindrical electrodes
should also not exceed a few per cent of a_e.   On the other hand, the object
slit width S'_e and the image width S''_e should be smaller than the width of
B by a factor of about 10^2.

The velocity dispersion of ions of a given mass ($\gamma = 0$) can be obtained
from Eq. (7.26a) by considering the lateral displacement in the image
plane of ions entering the field along the central ray ($\rho_1 = \rho_0 = 0$) with
a velocity $v = v_0(1 + \beta)$, namely,

$$ y_2(\beta) = \beta a_e \left(1 + \frac{f_e}{l'_e - g_e} \right) = K''_e \beta \qquad (7.28) $$

where K''_e, the coefficient of velocity dispersion in a radial electrostatic field, is given by

$$K_e'' = a_e \left(1 + \frac{f_e}{l_e' - g_e} \right) = a_e(1 - \cos \sqrt{2}\, \phi_e) + \sqrt{2}\, l_e'' \sin \sqrt{2}\, \phi_e$$

(7.29)

C. Toroidal electrostatic field

Double-focusing mass spectrographs, using a radial electrostatic field in conjunction with a homogeneous magnetic field, have focusing only in the central plane of deflection and not perpendicular to it, i.e., not in the direction of the mass-spectral lines. There is thus a great loss in the available ion intensities at the collector or at the photographic-plate detector. Since, for accuracy, only very short entrance slits are employed, the mass lines may be 10 to 100 times longer than the slits [15]. To make the mass-spectral lines shorter in length, and thus increase the available ion intensity at the detector, it is desirable to have focusing, either total or partial, in the direction of the mass lines as well.

The situation is analogous to the case of stigmatic-image formation in optics. Neither a radial electrostatic field nor an abruptly terminating homogeneous magnetic field can achieve this stigmatic focusing of the ions. The first successful attempt in producing stigmatic focusing of an ion beam was carried out by Herzog [16]. He employed a spherical condenser, in conjunction with a homogeneous sector magnetic field, and also utilized the axial-focusing property of the fringing field of the magnet on ions entering it at an angle $\epsilon_m' > 0$.

Ewald and his coworkers [17, 19] have shown that it is possible to obtain a stigmatic image in a double-focusing apparatus by the use of a toroidal condenser in conjunction with a homogeneous sector magnetic field, without the necessity of utilizing the additional axial focusing of the fringing magnetic field at the boundary. One has to choose suitable axial radii of curvature for the toroidal surfaces, which, in general, are different from the radial ones.

Ewald and Liebl [17] have investigated the ion optics in a toroidal condenser and have shown that the cylindrical and spherical condensers in mass spectroscopy are special cases of the former. We shall now briefly outline the method of obtaining the focusing properties of a toroidal electrostatic field: it is analogous to that used in discussing the radial electrostatic field developed in the previous subsection.

We assume the sector electrodes of a toroidal condenser to have a common rotational axis (the z axis in the r, ϕ, z cylindrical coordinate system) and a common plane of symmetry ($z = 0$). The axial and radial radii of curvature of the outer and inner electrodes are R_a and r_a and R_b and r_b, respectively (Fig. 7.5). r_a and r_b are always positive, while R_a and R_b can be positive or negative, depending on whether the circles of intersection of the toroid sectors with the meridian plane, $\phi = $ constant, have curvatures concave or convex with respect to the z axis.

The electrostatic deflecting field is purely radial in the $z = 0$ plane. If we call E_0 the radial field at the mean circle of radius $r = a_e$ in this

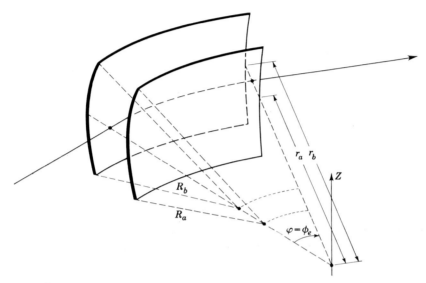

Fig. 7.5. Diagram of a toroidal condenser. (*From Ewald and Liebl* [17].)

plane, then the radial field in the same plane for any other value of r can be given by a series expansion

$$E(r, 0) = E_0(1 + \beta_1\rho + \beta_2\rho^2 + \cdots) \tag{7.30}$$

where $r = a_e(1 + \rho)$. We assume the potential to be zero for the $r = a_e$. Then the potential $V(r, z)$ and the field components $E_r(r, z)$ and $E_z(r, z)$ can be written as [20, 21]

$$V(r, z) = -E_0 a_e[\rho + \tfrac{1}{2}\beta_1\rho^2 + \tfrac{1}{3}\beta_3\rho^3 - \tfrac{1}{2}(1 + \beta_1)\zeta^2$$
$$+ \tfrac{1}{2}(1 - \beta_1 - 2\beta_2)\rho\zeta^2 + \cdots]$$
$$E_r(r, z) = E_0[1 + \beta_1\rho + \beta_2\rho^2 + \tfrac{1}{2}(1 - \beta_1 - 2\beta_2)\zeta^2 + \cdots] \tag{7.31}$$

and

$$E_z(r, z) = E_0[-(1 + \beta_1)\zeta + (1 - \beta_1 - 2\beta_2)\rho\zeta + \cdots]$$

where $z = a_e\zeta$, $\zeta \ll 1$.

The parameters β_1 and β_2 can be expressed as functions of the radial and axial radii of curvature a_e and R_e of the zero potential surface:

$$\beta_1 = -\left(1 + \frac{a_e}{R_e}\right)$$

and
$$\beta_2 = 1 + \frac{a_e}{R_e} + \frac{a_e{}^2(1 + R_e')}{2R_e}$$

(7.31a)

where $R_e' = (dR/dr)_{r=a_e, z=0}$, R being the axial radius of curvature of an equipotential surface adjacent to the zero potential surface in the $z = 0$ plane. Then, for an ion of mass $M = M_0(1 + \gamma)$ and charge e, which moves in the field along a path near the orbit of the central ray, the equations of motion can be written as

$$M\ddot{r} - Mr\dot{\phi}^2 = eE_r$$

$$M\frac{d}{dt}(r^2\dot{\phi}) = 0$$

$$M\ddot{z} = eE_z$$

(7.32)

in place of Eqs. (7.10) for a radial field.

We define a velocity v_0 such that a particle of mass M_0 follows the mean orbit in accordance with the equation of motion

$$\frac{M_0 v_0{}^2}{a_e} = -eE_0$$

If we consider an ion of mass M and velocity $v_0(1 + \beta)$, $\beta \ll 1$, outside the field, entering the deflecting field at a radius r, then the change in velocity suffered by it on entering the field can easily be calculated in the same way as in Eq. (7.14) to give its total velocity within the field as

$$v = v_0\left[1 + \beta - \rho + \frac{1}{2}\frac{a_e}{R_e}(\rho^2 - \zeta^2) + \rho(\beta + \gamma)\right]$$

(7.33)

to a second order of approximation.

If $r_1 = a_e(1 + \rho_1)$, $\phi = \phi_1$, and $z_1 = a_e\zeta_1$ denote the coordinates of a definite point of the orbit at which the tangent to the orbit makes a small angle σ_1 with the tangent to the circle $r_1 = a_e(1 + \rho_1)$ at $z_1 = a_e\zeta_1$, then, from Eq. (7.33), we have

$$\phi = \frac{v_0}{a_e}\left[1 + \beta - 2\rho - 2\rho\beta + 3\rho^2 - \rho_1{}^2 + 2\rho_1\beta + \rho_1\gamma - \frac{\sigma_1{}^2}{2}\right.$$
$$\left. + \frac{1}{2}\frac{a_e}{R_e}(\rho_1{}^2 - \zeta_1{}^2)\right]$$

(7.34)

To a first approximation, we can write

$$E_r = E_0 \left[1 - \left(1 + \frac{a_e}{R_e} \right) \rho \right]$$

$$E_z = \frac{E_0 \zeta a_e}{R_e} \tag{7.35}$$

and

$$\phi = \frac{v_0}{a_e} (1 + \beta - 2\rho)$$

The radial equation of motion [the first of Eqs. (7.32)] can, to a first approximation, be transformed to

$$\ddot{\rho} = \chi^2 \left(\frac{v_0}{a_e} \right)^2 (\delta - \rho) \tag{7.36}$$

where

$$\chi^2 = 2 - \frac{a_e}{R_e} \qquad \delta = \frac{\gamma + 2\beta}{\chi^2} \tag{7.36a}$$

The axial equation of motion [the third of Eqs. (7.32)] can also be transformed to

$$\ddot{\zeta} = - \left(\frac{v_0^2}{a_e R_e} \right) \zeta \tag{7.37}$$

to a first approximation.

It should be noted that, for a radial condenser ($R_e = \infty$), Eq. (7.36) degenerates into Eq. (7.19).

Equations (7.36) and (7.37) can easily be solved with the initial conditions that the ions enter the field at $r_1 = a_e(1 + \rho_1)$, $\zeta = \zeta_1$, and $\phi_1 = 0$, making small angles α'_r and α'_z with the plane tangential to the cylinder of radius r_1 and the plane $z_1 = a_e \zeta_1$, respectively.

$$\alpha'_r = - \left(\frac{1}{a_e} \right) \left(\frac{dr}{d\phi} \right)_{\phi=0} = \frac{a_e(\rho_0 - \rho_1)}{l'}$$

$$\alpha'_z = - \left(\frac{1}{a_e} \right) \left(\frac{dz}{d\phi} \right)_{\phi=0} = \frac{a_e(\zeta_0 - \zeta_1)}{l'} \tag{7.38}$$

for ions emerging from a point with the coordinates $r = r_0(1 + \rho_0)$ and $z = z_0 = a_e \zeta_0$, at a distance l' from the field boundary. The ions are focused behind the condenser at distances l''_r and l''_z, but in general making two astigmatic lines normal to the central ray, parallel and perpendicular, respectively, to the z axis (Fig. 7.6).

The solutions of Eqs. (7.36) and (7.37) are given by

$$\rho = - \left(\frac{\alpha'_r}{\chi} \right) \sin(\chi\phi) + \delta[1 - \cos(\chi\phi)] + \rho_1 \cos(\chi\phi) \tag{7.39}$$

and

$$\zeta = \zeta_1 \cos \left(\sqrt{\frac{a_e}{R_e}} \, \phi \right) - \alpha'_z \sqrt{\frac{R_e}{a_e}} \sin \left(\sqrt{\frac{a_e}{R_e}} \, \phi \right) \tag{7.40}$$

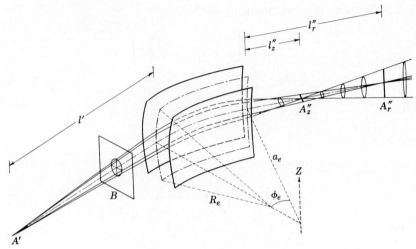

Fig. 7.6. Astigmatic-image formation by a toroidal condenser. The zero-potential surface and the mean ray are shown by the dashed lines with dots. (*From Ewald and Liebl* [17].)

There are two lens equations in this case, corresponding to radial and axial focusing, given by

$$(l' - g_r)(l''_r - g_r) = f_r^2 \tag{7.41}$$
$$(l' - g_z)(l''_z - g_z) = f_z^2 \tag{7.42}$$

where the subscripts r and z stand for radial and axial, respectively. The focal distances g_r, g_z and the focal lengths f_r, f_z are given by

$$g_r = \frac{a_e}{\chi} \cot \chi\phi_e \qquad\qquad f_r = \frac{a_e}{\chi \sin \chi\phi_e} \tag{7.41a}$$
$$g_z = \sqrt{a_e R_e} \cot \left(\sqrt{\frac{a_e}{R_e}}\, \phi_e \right) \qquad\qquad f_z = \frac{\sqrt{a_e R_e}}{\sin \sqrt{a_e/R_e}\, \phi_e} \tag{7.42a}$$

The velocity dispersion is $D(\beta) = K''\beta$, where

$$K''_r = \frac{2a_e}{\chi^2} \left(1 + \frac{l''_r - g_r}{f_r} \right) \tag{7.43}$$

The displacement of the image in the lateral direction for a displacement of the entrance slit by an amount b'_e is given by

$$b''_r = -b'_e \frac{l''_r - g_r}{f_r} \tag{7.44}$$

while the displacement in the axial direction, for slit displacement b'_z, is

$$b''_z = -b'_z \frac{l''_z - g_z}{f_z} \tag{7.45}$$

It should be noted that, for a spherical condenser, $R_e = a_e$, $g_r = g_z$, and $f_r = f_z$, so that the two astigmatic focal lines degenerate into a single stigmatic point at the image position.

The practical application of the above considerations will be discussed in Sec. 7.3C.

7.3. Double-focusing Mass Spectrographs

A. The double-focusing principle

In order to understand the principle underlying double focusing by the use of consecutive electrostatic and magnetic fields, we shall first discuss the principle of *velocity focusing*, which was employed by Aston in his mass spectrographs.

If a very narrow beam of ions, of identical specific charge e/M but heterogeneous in energy, enters a radial electric field with negligible angular divergence, it undergoes velocity dispersion. The lateral displacement of the ions of different velocities, after emerging from the electric field, is given by Eq. (7.28). The beam now appears to diverge from a point E as shown in Fig. 7.7. This divergent beam is subsequently allowed to enter a homogeneous magnetic field in which ions of different velocities again suffer different deflections. Hence ions of similar e/M but differing in energy recombine at a single point after emergence from the magnetic field. This is the point of velocity focusing. If the initial beam contains ions of different masses, the ions undergo mass analysis by the magnetic field, and ions of different masses converge to different velocity focusing points after emerging from the magnetic field. This is illustrated in Fig. 7.7. The locus of the velocity focusing points is called the velocity focusing curve (the line g in Fig. 7.7).

If, on the other hand, an initially monoenergetic beam of ions, again with identical specific charge but with small angular divergence, enters a radial electrostatic field, it is brought to a point focus (direction focusing) after emergence from the electric field [Eq. (7.25)]. Subsequently, these ions, which appear to diverge from the point of direction focus, are allowed to enter a homogeneous magnetic field, after emergence from which they are again brought to a direction focus. If the beam consists of ions differing in mass, they are subjected to a mass analysis by the magnetic field, and there are different direction focusing points, corresponding to the several mass components. The locus of these points is called the direction focusing curve.

Finally, if we consider a beam of identical ions, with a small angular divergence and a small but finite spread in energy, entering a radial electrostatic field, then the ions of different velocities are brought to a direction focus at the same distance from the exit boundary of the field but laterally displaced from one another. The velocity spread suffered by the ions is compensated for by a subsequent homogeneous magnetic field, after emerging from which the *central* rays corresponding to the different velocities are brought together at the velocity focusing point M_2 on the velocity focusing curve g (Fig. 7.8). However, the ions of different velocity groups that enter the magnetic field with small angular

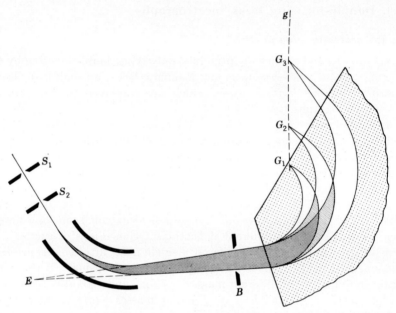

Fig. 7.7.　Velocity focusing of ions by a combination of electrostatic and magnetic sector fields.　Rays emitted with a very small angular divergence from the slit S_2 suffer velocity dispersion by the electric field, and ions of different velocities follow different paths after emerging from the field.　They are therefore spread out, as represented by the shaded region between the two fields.　After passing through the intermediate slit B between the two fields, the ions enter the magnetic field, after emerging from which ions of the same mass but different velocities are again brought to focus on the velocity focusing curve g.　The magnetic field also analyzes the ions into different mass groups which are focused at different points on g.　(*From Ewald and Hintenberger* [23].)

divergences are brought to a direction focus at the point A_2'', which is somewhat displaced from the point M_2 on the velocity focusing curve, as shown in Fig. 7.8. Ions of different specific charges are focused at different points, both on the velocity focusing curve g and the direction focusing curve r. In general, these two curves are different from one another. With a suitable combination of the fields, however, the two curves may be made to intersect, or even to overlap over extended regions. Those ions which are focused at such points of intersection of

g and r are said to have undergone double focusing by the combined electric and magnetic fields. In Fig. 7.8 this is illustrated by the point A_1''. It is obvious from Fig. 7.8 that, for those ions for which the velocity and direction focusing points are different, neither of the focusing points represents a sharp point focus. There is a spread of the ions, and consequent broadening of the mass-spectral lines, over a small finite region around both the focusing points. At the velocity focusing point, this is due to the "directional spread" of the ions of different velocities. At the

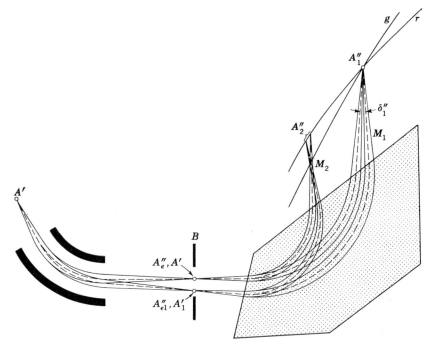

Fig. 7.8. Schematic diagram of a double-focusing apparatus. g is the velocity focusing curve; r is the direction focusing curve; and A_1'' is the double-focusing point. (*From Ewald* [82].)

direction focusing point the broadening is due to the velocity spread. Only at the point of double focusing are ions sharply focused, and both velocity broadening and directional broadening disappear, as illustrated by the point A_1'' in Fig. 7.8.

The condition of double focusing therefore requires that the velocity dispersion suffered by the ions in the electrostatic field must be exactly compensated by the magnetic field, direction focusing conditions (7.2) and (7.25) of the two fields being maintained at the same time. This would mean that if the central rays of two different velocity groups were made to proceed from the double-focusing point A_1'' in Fig. 7.8 in the opposite direction, then, with the magnetic field reversed, they would be brought to direction focusing after emerging from the magnetic analyzer, at the points A' and A_1', which are coincident with the points A_e'' and A_{e_1}'', respectively, at which these two groups are actually focused by the electrostatic field, when they are following their normal course. This is true because of the reversibility of the rays. Actually, since the angular divergence of the beams of each velocity group proceeding from A_1'' in the reverse direction is small, and also since the inclination of the central ray of each velocity group (dashed lines in Fig. 7.8) to the mean direction between them (not shown in Fig. 7.8) is small, we can think of a beam of these two groups of ions proceeding along the mean direction up to the field boundary, and being then subjected to velocity dispersion by the magnetic field, and finally being brought to the same direction focusing points A' and A_1' after emergence from the field. It is now easy to calculate the lateral separation D between A' and A_1' from the velocity-dispersion relation (7.6b) in the magnetic field. This is given by

$$D_0(\beta) = K_m'\beta$$
where $K_m' = a_m(1 - \cos \phi_m) + l_m' [\sin \phi_m + (1 - \cos \phi_m) \tan \epsilon_m']$ \qquad (7.46)

the symbols having their usual significance. The above equations have been obtained by replacing K_m'', l_m'', and ϵ_m'' in Eq. (7.7) by K_m', l_m', and ϵ_m', respectively, because we are considering the rays proceeding backward in the present case. The separation $D_0(\beta)$ between A' and A_1' is also equal to $K_e''\beta$, where K_e'' is given by the dispersion relation (7.29) in the radial electrostatic field with the auxiliary condition

$$l_e'' = \Delta - l_m' \qquad (7.47)$$

where Δ is the distance between the two fields.

We therefore obtain the following condition for velocity focusing:

$$K_m' = K_e''$$
which gives

(7.48)

$$a_m(1 - \cos \phi_m) + l_m'[\sin \phi_m + (1 - \cos \phi_m) \tan \epsilon_m']$$
$$= a_e(1 - \cos \sqrt{2}\, \phi_e) + l_e''\sqrt{2} \sin \sqrt{2}\, \phi_e$$

The above equations hold when the deflections in both fields are in the same sense. For *deflections in opposite senses* in the two fields, K_e'' should be replaced by $-K_e''$ in Eq. (7.48),† so that the velocity focusing

† The sign convention for the various parameters involved in the equation for velocity focusing when the two deflections are in opposite senses followed here is different from that followed by Herzog and Hauk [22] and agrees with that used by Ewald and Hintenberger [23]. Herzog and Hauk use Eq. (7.48) with negative values of a_m and ϕ_m for the case of opposite deflections. The sign of ϵ_m' is, however, taken the same as that for the case of deflections in the same sense in the two fields, since the coordinate system chosen by them at the entrance boundary of the magnetic field is

condition becomes

$$K'_m = -K''_e$$

$$a_m(1 - \cos \phi_m) + l'_m[\sin \phi_m + (1 - \cos \phi_m) \tan \epsilon'_m]$$
$$= -[a_e(1 - \cos \sqrt{2} \, \phi_e) + l''_e \sqrt{2} \sin \sqrt{2} \, \phi_e] \quad (7.49)$$

Equations (7.48) and (7.49) define the shape of the exit-pole boundary of the magnetic field. Usually this is a complicated curve, as has been shown by Herzog and Hauk [22]. However, for special arrangement of the fields, the curve degenerates into a straight line. This happens when the image of the object slit produced by the electric field approaches infinity ($l''_e \to \infty$). As we shall see later, under this condition, both the magnetic-pole boundary and the image curve are straight lines. This arrangement is therefore especially suitable for construction of mass spectrographs with double focusing at all masses, using photographic plates as detectors. The photographic plate is to be placed along the linear image curve.

In general, with a linear pole boundary, double focusing is achieved for only one radius of curvature in the magnetic analyzer. Consequently, only one species of ion can be brought to the position of double focus at a time. The ion so favored can, of course, be replaced by another species by simply changing the strength of the magnetic field. Thus, in practice, by use of an electrical detector located at the double-focusing point, it is possible to achieve double focusing for different masses by varying the magnetic field. Even with photographic detection, the above condition of double focusing at one mass is not a very serious limitation, since the image does not rapidly broaden when it is moved to one side or the other of the double-focusing position, provided the velocity band that is transmitted by the instrument is a narrow one.

It should be noted that the velocity focusing condition [Eqs. (7.48) and (7.49)] is necessary but not sufficient for the design of a double-focusing apparatus. This condition may be satisfied at points where direction focusing does not occur, i.e., at points where the position of the particle is not independent of the initial angular divergence of the beam. Under this condition, the beam suffers velocity focusing only, because all particles leaving the source *in a given direction* but with different velocities are brought back together (Fig. 7.8).

For linear magnetic-pole boundary and normal entry into the magnetic field, Eq. (7.48) may be transformed as follows:

$$a_m \left(\frac{l'_m - g_m}{f_m} + 1 \right) = a_e \left(1 + \frac{f_e}{l''_e - g_e} \right) \quad (7.50)$$

the same for both types of deflections. The resulting equation for velocity focusing with deflections in opposite senses is, however, identical with Eq. (7.49) above, obtained by changing the sign of K''_e, taking a_m and ϕ_m to be positive, and following the usual sign convention for the entrance angle ϵ'_m to the magnetic field.

The resolving power of the double-focusing mass spectrometer can be obtained from Eq. (7.9), which gives the resolving power of a direction-focusing magnetic sector field, by putting the term $\Delta V/V$ in it equal to zero; i.e.,

$$\frac{M}{\Delta M} = \frac{K_m''}{2S_m' G_m}$$

Here S_m' is to be put equal to the width S_e'' of the image of the entrance slit produced by the radial electric field. If this is substituted from Eq. (7.27), we obtain, with the help of Eqs. (7.5) and (7.7),

$$\frac{M}{\Delta M} = \frac{K_m''}{2S_e' G_e G_m} = \frac{a_m(1 - \cos \phi_m) + l_m''[\sin \phi_m + (1 - \cos \phi_m) \tan \epsilon_m'']}{2S_e'[f_e/(l_e' - g_e)][f_m/(l_m' - g_m')]}$$

ignoring the negative sign in Eq. (7.5) for G_m. Then, with the help of Eqs. (7.2) and (7.3), we obtain

$$\frac{M}{\Delta M} = \frac{a_m(1 - \cos \phi_m) + l_m'[\sin \phi_m + (1 - \cos \phi_m) \tan \epsilon_m']}{2S_e'[f_e/(l_e' - g_e)]}$$

From Eq. (7.46), the numerator of the above expression is seen to be equal to K_m'. Since, under double-focusing conditions, this equals K_e'', we get

$$\frac{M}{\Delta M} = \frac{K_e''}{2S_e'[f_e/(l_e' - g_e)]} \tag{7.51}$$

Finally, with the help of Eq. (7.29), we obtain the resolving power of a double-focusing apparatus as

$$\frac{M}{\Delta M} = \frac{a_e}{2S_e'}\left(1 + \frac{l_e' - g_e}{f_e}\right) \tag{7.52}$$

It should be noted from Eq. (7.52) that the resolving power of a double-focusing apparatus depends only on the constants of the electrostatic analyzer and on the entrance slit width S_e'. It increases linearly with increasing value of a_e.

The dispersion is given by Eq. (7.6e) and depends only on the constants of the magnetic analyzer.

B. Double-focusing instruments for one mass only

The first single-mass double-focusing apparatus was completed by Dempster [24] at the University of Chicago. A second type was independently designed and constructed by Bainbridge [25] at Harvard University. Later, a number of other instruments were built in different parts of the world. Mattauch and Herzog [26] had meanwhile, in Vienna, developed the theory for, and had constructed, a mass spectrograph which was double-focusing at *all* masses. A number of other instruments of this type have also since been constructed. These will be discussed in the next section: in this section we shall discuss the single-mass type only.

For the simplicity of construction, normal entry and exit for both the electric and magnetic fields have been employed by all workers; that is, $\epsilon_e = 0$, $\epsilon'_m = \epsilon''_m = 0$. Also, in all but one case, symmetrical image formation has been chosen in one or both the fields. Three distinct types of instrument have been constructed.

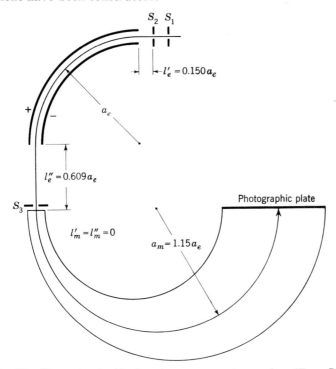

Fig. 7.9. The Dempster double-focusing mass spectrograph. (*From Duckworth* [27].)

Type 1—Dempster Double-focusing Mass Spectrograph. This type of instrument [24, 27] employs an electrostatic analyzer with $\phi_e = \pi/2$ radians, followed by a magnetic analyzer with $\phi_m = \pi$ radians (Fig. 7.9). It follows from Eqs. (7.2) and (7.3) that

$$l'_m = l''_m = 0$$

Hence the image of the entrance slit formed by the electrostatic field is located at the entrance boundary of the semicircular magnetic field. The final image is formed at the exit boundary of the magnetic field, where either the photographic-plate detector or the collector slit of some suitable electrical detector is located.

In the original apparatus of Dempster, the principal entrance slit S_2 is located at a distance $l'_e = 0.15a_e$ in front of the electrostatic field, so that the image is formed at a distance $l''_e = 0.609a_e$ behind it. The distance between the two fields is $\Delta = l''_e$. According to Eq. (7.48), velocity focusing occurs only for the magnetic-field orbit radius $a_m = 1.15a_e$. From Eq. (7.52), the resolving power then becomes

$$\frac{M}{\Delta M} = \frac{0.88a_e}{S'_e}$$

The intermediate slit S_3 allows only those ions to enter the magnetic field whose energies lie in the narrow range limited by Herzog's theory.

The ions are detected photographically. The plate surface coincides with the exit boundary of the magnetic field, which is also the position of the direction focusing curve. The velocity focusing curve intersects the former at an angle of 41° [23]. Hence sharply focused mass lines are observed only in a limited region of the plate near the intersection of the two image curves. In practice, the region of acceptable focus extends over a mass range of about 10%.

In Dempster's original instrument, $a_e = 8.5$ cm and $a_m = 9.8$ cm, which gives a resolving power of 3,000, for an entrance slit width of 0.0025 cm. In a larger apparatus of the same type, built by Duckworth [27], $a_e = 20$ cm, $a_m = 23$ cm, giving a resolving power of 7,000. The dispersion of the Dempster-type instruments is [Eq. (7.6e)] $D = a_m/100$ cm for 1% mass difference, which, in Dempster's original instrument, was 0.098 cm per 1% mass difference.

A much larger instrument of this type has been constructed at McMaster University [28, 29]. In this case the electrostatic analyzer is employed symmetrically, with $l'_e = l''_e = 0.351a_e$. This gives $f_z = l'_e - g_e$. Also, since $l'_m = l''_m = 0$, we have $g_m = -f_m$. The double-focusing condition reduces to $a_e = a_m$. With a_e and a_m both equal to 274 cm, this instrument provides a high resolving power, with relatively large slit widths.

It employs electrical detection (electron multiplier), and the mass spectra are presented on an oscilloscope screen in order to utilize the peak-matching technique (Sec. 7.5D) for determining atomic mass differences. The initial operating resolving power of 100,000 (at the *base* of the peaks) agrees with the anticipated value.

This and a few other recently conceived large instruments have the obvious advantage over the smaller high-resolution instruments in that relatively larger slit widths can be employed, and hence one can dispense with the exacting mechanical requirements associated with the construction and alignment of very fine slits (with widths of the order of a few microns; see Table 7.4).

Type 2—Bainbridge Double-focusing Mass Spectrograph. This instrument, the first model of which was constructed by Bainbridge and Jordan

[25], and which is shown in Fig. 7.10, employs an electrostatic field with $\phi = \pi/\sqrt{2}$ radians whose focusing properties correspond to those of a semicircular magnetic field in that image and object distances are zero;

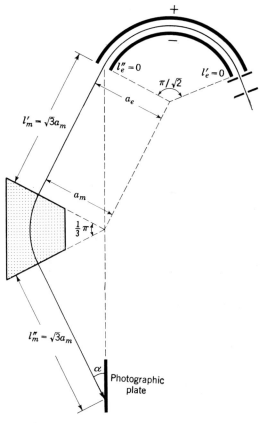

Fig. 7.10. The Bainbridge-Jordan double-focusing mass spectrograph. (*From Duckworth* [79].)

that is, $l'_e = l''_e = 0$. The entrance slit is located at the entrance boundary of the electric field. The magnetic field is 60° sector type and is employed symmetrically so that [from Eqs. (7.2) and (7.3)]

$$l'_m = l''_m = a_m \sqrt{3}$$

Since, for symmetrical fields, $K''_e = 2a_e$ and $K'_m = 2a_m$, the velocity focusing condition (7.48) reduces to $a_e = a_m$.

In the first instrument of Bainbridge and Jordan, $a_e = a_m = 25$ cm. From Eq. (7.52) the resolving power is

$$\frac{M}{\Delta M} = \frac{a_e}{S'_e}$$

Thus, for a principal slit of width 0.0025 cm, the resolving power is 10,000. From Eq. (7.6e), the dispersion is

$$D = \frac{a_m}{50}$$

for 1% mass difference, with the normal to the photographic plate making an angle of 60° with the plate.   In Bainbridge's first apparatus the dispersion was 0.50 cm per 1% mass difference.

A main advantage of the arrangement used by Bainbridge and Jordan is that the mass scale is linear.   Further, it has been shown by Herzog and Hauk [22] that the velocity and direction focusing curves make very small angles to one another and to the photographic plate (which makes an angle of 30° to the mean ion beam).   Hence the lines are sharply focused over extended regions of the plate.

A similar instrument was built by Asada, Okuda, Ogata, and Yoshimoto [30] in Japan.   Later, Ogata [31, 32] made a number of technical improvements in this apparatus and obtained a resolving power of 40,000 to 60,000.   The photographic plate in this instrument was placed along the direction focusing curve, which makes an angle of 24° to the mean ion beam.

Another instrument, which is a variant of the Bainbridge-Jordan type, has been constructed by a group of Russian workers [33].   In this instrument, the electrostatic deflection angle used was $\pi/2 \sqrt{2}$ radians (63°38'), instead of the usual $\pi/\sqrt{2}$ radians used in the Bainbridge type of instrument.   The magnetic deflection ϕ_m was 60°, the same as in the Bainbridge-type instrument.   With $a_e = a_m = 30$ cm and the principal slit width between 5 and 10 μ, a resolving power of 70,000 has been achieved, with the photographic plate placed at 30° to the central beam.   The instrument has been used to measure the masses of many atoms.

Type 3—Ogata and Matsuda Instrument.   Ogata and Matsuda [31, 32] have constructed a mass spectrograph of very high resolution, using asymmetric arrangements both for the electric and the magnetic analyzers. The asymmetric arrangement in the electric field has been used with a view to improving the resolution.   From Eq. (7.52), it can be seen that the resolving power depends on the quantity $(l'_e - g_e)/f_e$; hence, for given values of a_e, ϕ_e, and S'_e, the resolving power increases linearly with object distance l'_e.   Again from Eq. (7.6d), the normal dispersion for 1% mass difference for normal entry and exit is given by

$$D = \frac{a_m}{200} \left(1 + \frac{l''_m - g_m}{f_m} \right)$$

This increases with increasing l''_m, the image distance in the magnetic field.   In designing their apparatus, Ogata and Matsuda selected the following values of the parameters, which can be chosen arbitrarily:

$$a_m = 120 \text{ cm} \qquad l_e'' = 0 \qquad l_m' = \Delta = 120 \text{ cm} \qquad \phi_m = 60°$$

They then calculated the value of the function $(l_e' - g_e)/f_e$ for different values of the electrostatic deflection angle ϕ_e and chose

$$\phi_e = \pi \sqrt{2}/3 = 84°50'$$

to obtain a resolving power 50% higher than that associated with a symmetric arrangement.   The resolving power is given by

$$\frac{M}{\Delta M} = \frac{3a_e}{2S_e'}$$

From Eqs. (7.2), (7.25), and (7.48) we also obtain

$$a_e = \frac{a_m(1 + \sqrt{3})}{3} = 109.3 \text{ cm}$$
$$l_e' = \sqrt{1.5}\, a_e = 134.0 \text{ cm}$$
$$l_m'' = 448.0 \text{ cm}$$

The theoretically expected value of the resolving power is 700,000, with $S_e' = 0.0023$ mm. The dispersion is $D = 7.6$ cm per 1% mass difference.

In preliminary tests, an actual resolving power of 500,000 was achieved before an unfortunate flooding of the laboratory destroyed the instrument.

C. Double focusing at all masses

(1) **Theoretical considerations.** For a given arrangement of the electric field, and for given positions of the entrance slit and the separating diaphragms between the two fields, the right-hand sides of Eqs. (7.48) and (7.49) assume definite values.   The mass dependence of the velocity focusing relations (7.48) and (7.49) comes through the terms containing a_m, the radius of curvature in the magnetic field.   In order to make these equations independent of the mass of the ion, and thus obtain the condition of velocity focusing for all masses, it is therefore necessary to eliminate the terms containing a_m.   This can be done by making l_e'', the image distance in the electric field, approach infinity [26].   From Eqs. (7.47) and (7.49), we obtain

$$\frac{l_m'}{l_e''} = \frac{\Delta}{l_e''} - 1$$

$$\frac{a_m}{l_e''}(1 - \cos \phi_m) + \frac{l_m'}{l_e''}[\sin \phi_m + (1 - \cos \phi_m)\tan \epsilon_m'] \qquad (7.53)$$

$$= -\frac{a_e}{l_e''}(1 - \cos \sqrt{2}\,\phi_e) - \sqrt{2}\sin \sqrt{2}\,\phi_e$$

In the limit of $l_e'' \to \infty$, these equations reduce to

$$\frac{l_m'}{l_e''} = -1$$

$$\sin \phi_m + (1 - \cos \phi_m) \tan \epsilon_m' = \sqrt{2} \sin \sqrt{2} \, \phi_e$$

(7.54)

Equation (7.54), which is valid for deflections in *opposite* senses in the two fields, is independent of the mass of the ions.   For deflections in the same sense in the two fields, the right-hand side of the second equation in (7.54) should be negative.

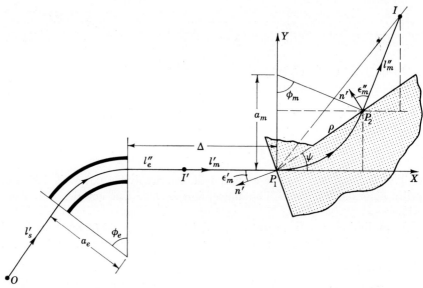

Fig. 7.11. Schematic diagram of a mass spectrograph which is double-focusing at all masses.   The line P_1P_2 represents the linear exit boundary of the magnetic field, while P_1I is the linear image curve. ρ and ψ are the polar coordinates of the point P_2 at which the ion crosses the exit field boundary. O, I', and I are the object, intermediate image, and the final image, respectively.

Since ϕ_e and ϵ_m' are the same for all masses, the same has also to hold for ϕ_m, the angle of deflection in the magnetic field.   Referring to Fig. 7.11, one can easily see that if $\rho(= P_1P_2)$ and ψ denote the polar coordinates of the points (for example, P_2) where an ion crosses the exit field boundary, we have

$$\rho = 2a_m \sin \psi \qquad \psi = \frac{\phi_m}{2} = \text{constant}$$

(7.55)

Then the cartesian components of ρ are

$$x = \rho \cos \psi = a_m \sin \phi_m$$
$$y = \rho \sin \psi = a_m(1 - \cos \phi_m)$$

(7.56)

$$\frac{y}{x} = \tan \psi = \text{constant}$$

(7.57)

This shows that the locus of the points at which the ions cross the exit field boundary is a straight line, passing through the mean entrance point of the ions into the magnetic field. Since ions of different masses suffer the same deflection, they all emerge parallel to one another. From Fig. 7.11, it is easily seen that the exit angle ϵ_m'' (which is negative) is given by

$$-\epsilon_m'' + \psi = \frac{\pi}{2}$$

$$\epsilon_m'' = \psi - \frac{\pi}{2} = \text{constant}$$

(7.58)

It should be noted that since, according to Eq. (7.54), $l_e'' = l_m' = \infty$, the ions emerge from the electrostatic field as a parallel beam and enter the magnetic field as a parallel beam. This is the essential condition for double focusing for all masses. It means that the entrance slit of the electrostatic analyzer should be located at the principal focus of that field, while the detector after the magnetic field is to be located at the principal focus of the latter. Since $l_m' = -\infty$, we must have

$$l_m'' = g_m'' = a_m \frac{\cos \epsilon_m'' \cos (\phi_m - \epsilon_m')}{\sin (\phi_m - \epsilon_m' - \epsilon_m'')}$$

If x_1 and y_1 denote the cartesian coordinates of the image point I, we have (Fig. 7.11)

$$x_1 = a_m \sin \phi_m + l_m'' \cos \phi_m$$
$$y_1 = a_m(1 - \cos \phi_m) + l_m'' \sin \phi_m$$

from which we obtain

$$\frac{y_1}{x_1} = \frac{(1 - \cos \phi_m) \sin (\phi_m - \epsilon_m' - \epsilon_m'') + \sin \phi_m \cos \epsilon_m'' \cos (\phi_m - \epsilon_m')}{\sin \phi_m \sin (\phi_m - \epsilon_m' - \epsilon_m'') + \cos \phi_m \cos \epsilon_m'' \cos (\phi_m - \epsilon_m')}$$

(7.59)

Since the right-hand side of the above equation is a constant, this represents a straight line through the mean point of ion entry into the magnetic field. Thus the locus of the final-image points is a straight line with a slope given by Eq. (7.59). Hence, in case of double focusing for all masses, not only can a linear exit-pole boundary be used, but a linear (plane) photographic plate can be used as the detector. Ions focused on this plate all fulfill the double-focusing condition.

It should be noted that the parallelism of the ion beams between the fields is a necessary but not sufficient condition for double focusing at all masses. Only appropriate combinations of the electric and magnetic deflection angles determined by Eq. (7.54) can give velocity focusing.

(2) **Actual instruments.** *Type 1—Mattauch Double-focusing Mass Spectrograph.* The above principles were first embodied in an instrument constructed by Mattauch and Herzog in Vienna in 1935 [26]. Figure 7.12 illustrates the principle of this instrument. They chose normal

entry into the magnetic field, $\epsilon'_m = 0$. Also, ϕ_m was so chosen as to cause the ions to focus at the exit boundary of the magnetic field. This eliminates the necessity of the ions passing through the stray exit field. Obviously, in this case, $g''_m = 0$. This happens if $\phi_m = \pi/2$, which gives $\epsilon''_m = \pi/4$. The corresponding electrostatic deflection angle ϕ_e is found

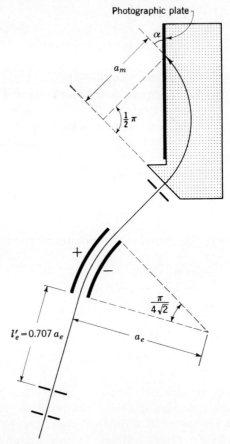

Fig. 7.12. The Mattauch-Herzog double-focusing mass spectrograph. (*From Duckworth* [79].)

from Eq. (7.54) to be $\pi/4 \sqrt{2}$ radians, or 31.8°, in consequence of which $l'_e = a_e/\sqrt{2}$. The photographic plate is placed along a line passing through the entrance point to the magnetic field, making an angle of 45° to the entrance beam and coinciding with the exit field boundary.

The resolving power and dispersion are $a_e/2S'_e$ and $0.00707a_m$, respectively. The magnification, which is equal to the product of the individual magnifications G_e and G_m of the two fields, given by the Eqs. (7.5) and

(7.27), is a_m/a_e. In the first instrument, $a_e = 28$ cm. Thus, for $S'_e = 0.0025$ cm the resolving power is 5,600. The dispersion and magnification change with the position of the line on the photographic plate, since a_m is different for different lines. The maximum values corresponding to $a_m = 24$ cm are 0.17 cm for 1% mass difference and 0.86, respectively.

The principal advantage of this type of mass spectrograph of course lies in the fact that a large portion of mass spectrum can be photographed simultaneously, since sharply focused lines are obtained throughout the entire length of the plate.

Other original scale versions of this instrument have been constructed by Mattauch and Lichtblau [34], Ewald [35], Mattauch and Bieri [36], and Everling et al. [37]. After meticulous adjustment these latter two instruments have provided resolving powers of 100,000. A much larger version of the apparatus (approximately eight times the original) has been built at Harvard University by Bainbridge [38]. Using electrical detection, a resolving power in excess of 100,000 (at the *base* of the peaks) has been achieved with this instrument, and it will undoubtedly play an important role in the determination of atomic masses. Also, for analytical work, Mattauch-type instruments have been built by Shaw and Rall [39] and Hannay [40]. Mass spectrometers of this type can now be purchased from A.E.I. Ltd, Trafford Park, Manchester, England, and C.E.C., Pasadena, Calif.

Reuterswärd [41] has built a modification of the Mattauch instrument in which the ions enter the magnetic analyzer obliquely. The angle of incidence ϵ'_m to the magnetic field was chosen by Reuterswärd to be equal to ψ. It can be shown that, in this case, if ω is the angle between the ion beam and the normal to the photographic plate, we get [42],

$$\tan \psi = \tfrac{3}{2} \cot \omega$$

ω was chosen to be 65°, which gives, from the above equation, $\psi = 35°$, and hence the magnetic deflection $\phi_m = 70°$. The double-focusing condition (7.54) reduces to

$$\sqrt{2} \sin \sqrt{2}\, \phi_e = 2 \tan \epsilon'_m \qquad (7.60)$$

which gives $\phi_e = 58°$. Also, from Eq. (7.58), $\epsilon''_m = -55°$.

The resolving power and dispersion are given by $a_e/2S'_e$ and $0.00776 a_m$ cm for 1% mass difference, respectively. With $a_e = 30$ cm and an entrance slit width of 34 μ, a resolving power of 4,400 has been reported.

Type 2—Ewald's Toroidal Condenser Instrument. Ewald and his coworkers [17–19] have used a toroidal condenser instead of a radial condenser to produce axial focusing of the rays in a mass spectrograph which is otherwise similar to the Mattauch apparatus. The theory of simultaneous axial and radial focusing by a toroidal electric field was developed in Sec. 7.2C. The schematic representation of the radial and axial focusing in their stigmatic-image mass spectrograph is shown in Fig. 7.13. By analogy with the Mattauch-Herzog type of mass spectrograph, Ewald has chosen $l'_e = g_r$ and $l''_r = -l'_m = \infty$, where the symbols have the same meaning as in Sec. 7.2C.

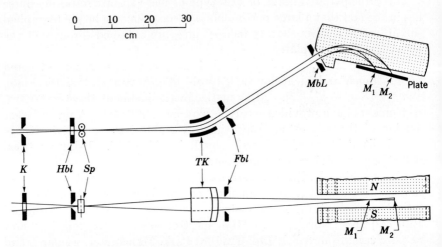

Fig. 7.13. Schematic representation of the radial (upper) and axial (lower) focusing in the stigmatic-image mass spectrograph. K is the entrance channel; Hbl, the horizontal diaphragm; S, the narrow entrance slit; TK, the toroidal condenser; Fbl, the interfield diaphragm; and MbL, the magnetic field diaphragm. The positions of focus for two different masses M_1 and M_2 are shown. (*From Ewald et al.* [19].)

With the help of Eqs. (7.7) and (7.43), and substituting K''_r of Eq. (7.43) on the right-hand side of Eq. (7.49), the double-focusing condition for deflections in opposite senses in the two fields then reduces to

$$\frac{2 \sin (\sqrt{2 - a_e/R_e}\ \phi_e)}{\sqrt{2 - a_e/R_e}} = \sin \phi_m + (1 - \cos \phi_m) \tan \epsilon'_m \qquad (7.61)$$

The following values were chosen for the various parameters $a_e = 12$ cm, $R_e = 9.6$ cm, $\phi_e = 29.7°$, $\phi_m = 87.5°$, and $\epsilon'_m = 0$. The axial image distance l''_z of the toroidal condenser is equal to the length of the ion path from the exit boundary of the electric field to the middle of the photographic plate. The resolving power of the apparatus can easily be found from Eq. (7.51) by replacing $K''_r, f_e, l'_e,$ and g_e by the corresponding radial parameters of the toroidal condenser [Eqs. (7.41), (7.41a), (7.43), and (7.44)],

$$\frac{M}{\Delta M} = \frac{a_e}{S'_e(2 - a_e/R_e)} \left(1 + \frac{l'_r - g_r}{f_r}\right) \qquad (7.62)$$

where S'_e is the object slit width. Since, in the present case, $l'_r = g_r$, we have

$$\frac{M}{\Delta M} = \frac{a_e}{S'_e(2 - a_e/R_e)} \qquad (7.63)$$

which is greater by a factor $2/(2 - a_e/R_e)$ than when a cylindrical condenser is used with the same value of a_e [Eq. (7.62)]. In the apparatus described, this factor is 2.67. A resolving power of 20,000 has been obtained.

In the apparatus described, the lengths of the photographed mass lines were approximately $b''_z = 0.32$ mm for an entrance slit length

$$b'_z = 0.2 \text{ mm}$$

This is in agreement with the value of 1.57 for the axial magnification, calculated from Eq. (7.45).

As was mentioned in Sec. 7.2C, Herzog [16] obtained axial focusing by utilizing the axial focusing property of a spherical condenser and that of the fringing field at the entrance boundary of the magnetic field, the ions entering the latter at an oblique angle of $\epsilon'_m = 31°$. The axial and radial image distances l''_z and l''_r were both infinite in this case.

D. Double-focusing mass spectrometers with higher-order focusing

In the double-focusing instruments described so far, the second-order image aberration due to the finite initial angular divergence and velocity dispersion of the ion beam has been neglected. In the derivation of the theory of focusing of charged particles by electric or magnetic fields, it is generally assumed that the rays from a point object, after passing through the field, are brought to a point focus. This is true only to a first order of approximation. In the expression for the lateral displacement of the ions emerging from the point object making a small angle α with the mean ray and having a velocity $v = v_0(1 + \beta)$ slightly different from the mean velocity v_0, only terms up to the first order in α and β are retained. Terms involving higher powers of α and β are small. However, it is sometimes advantageous to make α and β relatively large in order to improve the intensity of the ion beam at the collector. In the instruments described so far, this would mean an increase in the image aberration, with consequent loss in the resolving power. It is, however, possible, by suitable field combinations, to reduce the magnitude of the image aberrations in double-focusing instruments. In particular, the aberration dependent upon the second-order terms α^2, β^2, and $\alpha\beta$ can be reduced or completely eliminated by a suitable choice of the deflecting fields [43, 46, 53].

The first double-focusing mass spectrometer with second-order correction was built by Nier and Roberts [44], using a two-field combination in accordance with the theory of Johnson and Nier [45], in which the second-order angular aberration of the final image was eliminated, thus achieving second-order direction focusing.

(1) **Double focusing of first order with direction focusing of second order.** *Nier's Instrument.* As mentioned above, Johnson and Nier [45] have developed the theory needed to secure second-order direction focusing in instruments with first-order double-focusing properties. We shall briefly summarize their findings.

Consider an ion of a given mass and velocity $v = v_0(1 + \beta)$ entering a radial electric or sector-shaped homogeneous magnetic field, making an angle α with the mean ray composed of similar ions entering the field normally with mean velocity v_0. The lateral displacement of the former from the mean ray, after emerging from the field at a distance from the exit boundary, is given by

$$y = A_1\alpha + A_2\beta + A_{11}\alpha^2 \qquad (7.64)$$

where the coefficients A_1, A_2, and A_{11} are functions of the mean radius of curvature of the orbit of the ion beam in the field, the mean deflection of the beam, the object distance, and the distance x at which the displacement is considered. The coefficients are rather lengthy and will not be written down here. The above expression is obtained on the assumption that second- and higher-order terms in β and cross terms between α and β can be neglected. The only second-order term which is retained is the angular-aberration term in α^2.

In an arrangement using a radial electrostatic field and a homogeneous sector-shaped magnetic field in tandem, the first-order velocity focusing condition can easily be derived from the above equation. This is

$$G_m K_e'' + K_m'' = 0 \qquad (7.65)$$

where the different quantities have their usual significance. Equation (7.65) is the same as Eq. (7.48) obtained in Sec. 7.3A, since K_m''/G_m is equal to K_m' [see discussion preceding Eq. (7.51)].

The second-order angular aberration L_e [the last term on the right-hand side of Eq. (7.64) with appropriate value of the coefficient $A_{11}^{(e)}$] in the electrostatic field is magnified to the value $G_m L_e$ after going through the subsequent magnetic field. For second-order direction focusing at $x_m = l_m''$, this must be counterbalanced by the second-order angular aberration L_m [again equal to the last term on the right-hand side of Eq. (7.64) with appropriate value of the coefficient $A_{11}^{(m)}$] in the magnetic field. Hence the condition of second-order direction focusing is

$$G_m L_e + L_m = 0 \qquad (7.66)$$

To solve Eqs. (7.65) and (7.66), Johnson and Nier [45] have introduced an asymmetry parameter for the two fields,

$$z = \frac{\cos[\tan^{-1}(kl''/a)]}{\cos[\tan^{-1}(kl'/a)]} \qquad (7.67)$$

where a, l', and l'' are the mean radius of curvature and the object and image distances in the two fields. For the electrostatic field, $k = \sqrt{2}$, while for the magnetic field, $k = 1$. For a symmetric arrangement in the field considered, $l' = l''$ and $z = 1$.

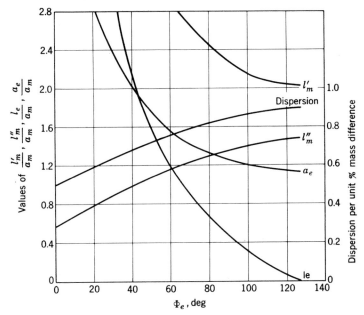

Fig. 7.14. Variations of the ratios of l'_m, l''_m, l_e, and a_e to a_m as a function of ϕ_e for instruments which employ electric and magnetic field in tandem and for which there is first-order velocity and second-order direction focusing. The figure also shows the corresponding dispersion per a_m per percentage mass difference. (*From Johnson and Nier* [45].)

It has been shown by Johnson and Nier [45] that Eqs. (7.65) and (7.66) can be simultaneously satisfied if

$$z_m = \tfrac{1}{2} \pm (A - \tfrac{3}{4})^{\frac{1}{2}} \tag{7.68}$$

$$\frac{a_c}{a_m} = \frac{1 + z_m}{1 + 1/z_c} \tag{7.69}$$

where $A = -2A_{11}{}^{(e)}/a_e(z_e{}^2 + z_e)$, $A_{11}{}^{(e)}$ being the second-order angular-aberration coefficient for the electric field which is a function of ϕ_e and z_e. Hence, from Eq. (7.68), z_m can be evaluated as a function of ϕ_e and z_e. Then, from Eq. (7.69), the ratio a_e/a_m can be obtained as a function of ϕ_e and z_e. Finally, from Eq. (7.67), for the magnetic-field case and Eq. (7.18), the ratios l'_m/a_m and l''_m/a_m can be obtained as functions of ϕ_e and ϕ_m.

In Fig. 7.14, the variation of the parameters l_e/a_m, l'_m/a_m, l''_m/a_m, and a_e/a_m as functions of the electric deflection angle ϕ_e are shown. The curves are drawn for $z_e = 1$, which means a symmetric arrangement in the electric field ($l'_e = l''_e = l_e$). The magnetic deflection angle ϕ_m has been set equal to 60°. To obtain large mass dispersion, small values of z_m are desirable, which means an asymmetrical arrangement in the magnetic field. It can be seen from Fig. 7.14 that, for values of $\phi_e > 60°$, the parameters l_e, l'_m, and a_e become prohibitively large. Hence, from practical considerations, the range of ϕ_e between 60° and 120° seems most desirable.

The theory developed was used by Nier and Roberts [44] to build a mass spectrometer for precision mass determinations. The various parameters chosen were $\phi_e = 90°$, $\phi_m = 60°$, $a_e = 18.87$ cm, and $a_m = 15.24$ cm. The values of l'_e, l''_e, l'_m, and l''_m were 6.61, 6.61, 34.77, and 20.73 cm, respectively. The spectrometer is shown schematically in Fig. 7.15. The resolving power attained was 14,000 at half maximum.

Besides introducing second-order direction focusing, Nier and Roberts [44] also employed, for the first time in the precision determination of atomic masses, an electrical detection system. They used an electron multiplier as the ion detector, the amplified output from which was recorded on a two-channel Brush high-speed recorder. The different masses were brought into focus by varying the electrostatic deflection voltage. Since the two peaks of a doublet were recorded at different instants, the fields in the instruments were required to be highly stable. The necessary stability was achieved by placing an auxiliary mass-spectrometer tube in the field of the magnetic analyzer. Ions of nearly the same mass as those under study in the main instrument were detected in this auxiliary mass spectrometer by a two-electrode collector. A differential amplifier measured and amplified the difference between the signals from the two halves of the collector. This amplified difference was then fed, as an error signal, into the 3,000-volt power supply, which provided the ion accelerating voltage for the auxiliary tube and the deflecting voltage for the electrostatic analyzer. Any tendency for these voltages to change was corrected by the signal. It also controlled, in an indirect way, the ion accelerating voltage in the main tube. The electrostatic deflection voltage was swept by means of a scanning motor which changed a resistance that controlled the value of the ion accelerating voltage in the auxiliary tube. The difference in mass ΔM between two ions of nearly equal specific charge was given by $\Delta M/M = \Delta R/R$ to a high degree of accuracy, where ΔR was the difference between the resistances required to bring the two peaks into focus.

The fact that the ions travel identical paths in an instrument employing electrical detection makes the achievement of a uniform dispersion law easier than in an instrument utilizing photographic detection.

An enlarged version of this instrument has been built by Nier and his coworkers [47], with $a_e = 50.31$ cm and $a_m = 40.64$ cm. With the slit

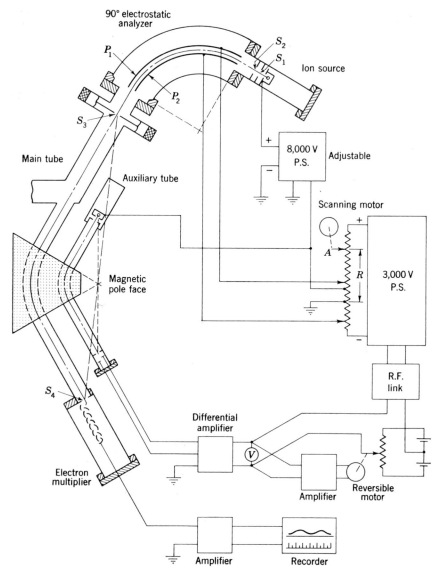

Fig. 7.15. Nier's double-focusing mass spectrometer. (*From Nier* [75].)

widths $S_1 = S_4 = 0.0001$ cm, a resolving power of 75,000 has been obtained, the working figure generally lying between 30,000 and 60,000 (all half-width values). The technique of recording the peaks in this instrument is different from that described above [76]. A variation of the peak-matching technique, first introduced by Smith [48, 49] is employed. As we shall see in Sec. 7.5D, the use of electrical detector systems, coupled with the peak-matching technique, has added a powerful new precision method, which has almost supplanted the photographic method in the determination of atomic masses.

The Mattauch-Herzog Instrument. Hintenberger, Wende, and König [50, 51] have investigated the possibility of achieving second-order direction focusing in the Mattauch-Herzog type of instrument with first-order double focusing at all masses. They have shown that, in such instruments, it is usually possible to attain second-order direction focusing at one mass near the center of the plate by a suitable choice of the ratio a_e/a_m. But this is not possible for all pairs of deflection angles ϕ_e and ϕ_m which give double focusing of the first order. The Mattauch-Herzog instrument falls just on the border line of the permissible region in the ϕ_e-ϕ_m diagram, while the instrument of Reuterswärd falls outside the permissible region. A new large mass spectrograph of the Mattauch-Herzog type (in the sense that it will provide first-order double focusing for all masses) has been built by Mattauch, Hintenberger, and their coworkers [37]. It possesses, in addition, second-order direction focusing for one mass. The theoretical resolving power for a slit width of 1 μ is 2.7×10^6. Dispersion is 13.9 mm for 1% mass difference. Either photographic plates or an electron multiplier can be used for ion detection.

The Argonne Instrument. Finally, mention should be made of a new large instrument ($a_e = a_m = 254$ cm), constructed at the Argonne National Laboratory by Stevens and his collaborators [54], which also provides second-order direction focusing. Unusual features of this instrument are: (a) the plates of the electrostatic analyzer are spherical in shape, thus providing axial focusing, (b) partial axial focusing is achieved in the magnetic analyzer by employing $\epsilon'_m = 36.5°$, and (c) the second-order direction focusing is achieved by giving the entrance boundary of the magnetic analyzer a slight concave curvature. In preliminary work a resolving power of 100,000 has been achieved with exceptionally high transmission.

(2) **Second-order double focusing.** The conditions under which a two-field combination instrument can be made double-focusing to the second order has been investigated by Hintenberger and König [46]. The final lateral displacement of the ions emanating from a point source, making an angle α_e with the mean ray (Fig. 7.16), after passing through a radial electric field and a sector-shaped homogeneous magnetic field in tandem, can be shown to be of the form

$$y_B = a_m(B_1\alpha_e + B_2\beta + B_{11}\alpha_e{}^2 + B_{12}\alpha_e\beta + B_{22}\beta^2) \qquad (7.70)$$

The lateral displacement is measured at a distance l''_m from the exit boundary of the magnetic field. The complete expressions for the coefficients B_1, etc., have been given by Hintenberger and König [46, 52]. These are rather lengthy and are not given here.

$B_1 = 0$ gives the condition for direction focusing of the first order. B_2 is the coefficient of velocity dispersion, and hence $B_2 = 0$ is the condition of velocity focusing of the first order.   The simultaneous vanishing

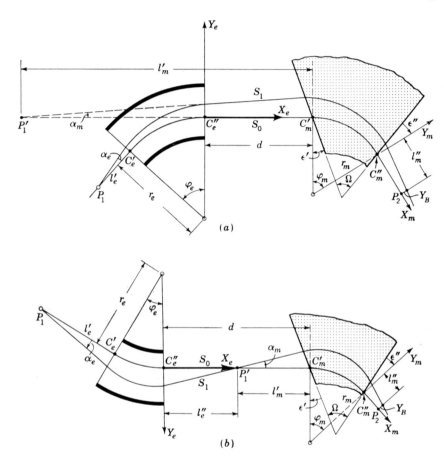

Fig. 7.16.   Paths of ions in a radial electric field and a magnetic field in tandem. Diagram a is for deflections in the same sense in the two fields, while diagram b is for deflections in opposite senses in the two fields.   The position of the intermediate image P_1' depends on α_e' and β.   The mean orbit S_0 is described by ions of velocity v_0 with $\alpha_e' = 0$, while the orbit S_1 is followed by ions with velocity $v = v_0(1 + \beta)$ and $\alpha_e' \neq 0$.

of B_1 and B_2 therefore gives the condition of double focusing of the first order.

B_{11} is the coefficient of the angular aberration (of the second order). B_{12} is the coefficient of *mixed* aberration dependent upon both angular divergence and velocity dispersion $(\alpha_e\beta)$. B_{22} is the coefficient of the velocity-dependent aberration (of the second order).

The condition $B_{11} = 0$, along with the usual double-focusing conditions $(B_1 = 0, B_2 = 0)$, gives the condition of double focusing in the first order with second-order direction focusing, which has been discussed before. In order to achieve second-order double focusing, all the five coefficients must vanish simultaneously; i.e.,

$$B_1 = 0, \ B_2 = 0, \ B_{11} = 0, \ B_{12} = 0, \ B_{22} = 0 \tag{7.71}$$

The above equations represent a series of five nonlinear equations with eight unknowns: ϕ_m, ϵ'_m, ϵ''_m, $\sqrt{2}\ \phi_e$, a_e/a_m, l'_e/a_m, Δ/a_m, and l''_m/a_m, where Δ is the distance between the two fields. The other quantities have their usual significance. Three of these unknowns are to be chosen arbitrarily. Then the other five can be expressed as functions of these three parameters. Hintenberger and König [46] chose the magnetic-field parameters ϕ_m, ϵ'_m, and ϵ''_m arbitrarily. The solutions were carried out numerically with the help of an IBM magnetic calculator. The results are shown in

TABLE 7.1.   Typical Numerical Values of Different Field Parameters for Which Complete Double Focusing of Second Order Is Possible at One Mass

Numbers 1 to 12 represent Solution I, and numbers 13 to 22 represent Solution II. The solutions 5 and 15, 6 and 16, as also 9 and 20, show the multivalued character of the solutions for a given set of arbitrarily chosen values of ϕ_m, ϵ'_m, and ϵ''_m. The table refers to electric and magnetic deflections in the *same* sense. (Hintenberger and König [46].)

Serial no.	ϕ_m, deg	ϵ'_m	ϵ''_m	$\sqrt{2}\ \phi_e$	ϕ_e	Δ/a_m	a_e/a_m	l'_e/a_m	l''_m/a_m
1	55	0	0	115.272	81.510	3.4364	1.2762	0.4768	1.4239
2	60	0	−3	128.692	90.999	2.9930	1.2203	0.2835	1.2145
3	60	0	−15	80.000	56.569	4.5200	1.8143	1.8204	0.8377
4	60	−15	−15	138.380	97.849	2.8459	1.0081	0.0879	0.6637
5	70	15	−15	70.761	50.036	4.9392	2.6663	4.1875	0.9116
6	72	15	−15	83.582	59.101	4.0507	2.2773	2.5560	0.9082
7	80	15	−15	129.381	91.486	2.3494	1.5454	0.5406	0.8768
8	90	27	−15	133.855	94.650	1.5955	1.5938	0.7084	0.9477
9	105	32	−35	65.392	46.239	3.8878	4.5927	45.8907	0.4426
10	110	31	−35	154.540	109.276	1.3432	2.0185	0.7614	0.4424
11	111	31	−35	167.058	118.128	1.2136	1.8983	0.4473	0.4392
12	116	35.9	−45	182.770	129.238	1.1190	2.5988	0.7416	0.3174
13	60	15	−15	183.220	129.556	0.9270	3.8756	5.8805	0.9434
14	60	23	−15	227.282	160.713	0.9449	4.5896	2.8751	0.9870
15	70	15	−15	140.724	99.507	0.6908	3.6209	22.1367	0.9092
16	72	15	−15	133.451	94.364	0.6499	3.6208	45.2327	0.9017
17	80	33	−15	183.292	129.607	0.7162	3.3174	2.9046	0.9528
18	90	33	−15	128.778	91.060	0.7056	2.6338	6.3650	0.9789
19	90	33	−30	250.548	177.164	0.7458	3.8714	0.1440	0.5230
20	105	32	−35	144.110	101.901	1.0577	3.1496	3.8777	0.4463
21	110	33	−45	158.727	112.237	1.2991	3.6810	3.1829	0.3210
22	115	34	−40	184.544	130.492	1.0553	2.0840	0.3143	0.3714

Tables 7.1 and 7.2. As can be seen, there is a large range of possible values of the various parameters for which second-order double focusing can be achieved. For a given set of ϕ_m, ϵ'_m, and ϵ''_m values, two sets of solutions for the other five quantities are possible, which are designated in the tables as Solution I and Solution II, respectively.

TABLE 7.2. Typical Numerical Values of Different Field Parameters for Which Complete Double Focusing of Second Order Is Possible at One Mass

Numbers 1 to 8 represent Solution I, and numbers 9 to 16 represent Solution II. The table refers to electric and magnetic deflections in *opposite* senses. (Hintenberger and König [46].)

Serial no.	ϕ_m	ϵ'_m	ϵ''_m	$\sqrt{2}\,\phi_\epsilon$	ϕ_e	Δ/a_m	a_e/a_m	l'_e/a_m	l''_m/a_m
1	196	−19.5	0	196.853	139.196	6.6206	0.9686	4.6671	6.9003
2	197	−19.5	0	197.690	139.788	6.5723	1.0147	4.6555	6.4834
3	200	−19.5	0	200.108	141.498	6.4211	1.1434	4.6071	5.4829
4	210	−20	0	207.349	146.618	5.7939	1.4569	4.2861	3.5846
5	210	−26	0	207.958	147.049	5.0098	1.3098	3.7867	3.5099
6	210	−32	0	208.252	147.256	4.4167	1.1894	3.4205	3.4278
7	210	−32	8	195.167	138.004	3.5530	4.6347	2.4993	7.4693
8	211	−19.5	0	207.929	147.028	5.8144	1.4954	4.3029	3.4681
9	177	−19.5	0	213.447	150.930	6.8342	8.6135	10.4646	3.2699
10	180	−19	0	205.263	145.143	5.6462	6.8927	9.4120	2.9668
11	180	−34	0	155.254	109.781	1.7517	2.8409	8.7832	2.2471
12	180	−35	0.5	151.429	107.076	1.6115	2.6709	9.7907	2.2973
13	180	−35	−12.5	239.683	169.481	2.4853	4.1779	1.3164	1.1265
14	189	−19.5	0	178.108	125.941	3.1831	3.7284	8.2417	2.2943
15	195	−19.5	0	165.716	117.179	2.4910	2.8218	8.9651	2.0287
16	197	−19.5	0	162.302	114.765	2.3215	2.6020	9.4402	1.9547

It has also been shown by Hintenberger and König by actual numerical calculation that considerable tolerances in the design values or the parameters are admissible ($\sim2°$ in the angles and a few per cent in the lengths) without causing an appreciable deterioration in the aberration correction.

Hintenberger and his coworkers [46] have further shown, exceeding their earlier hopes (Hintenberger, Wende, and König [51]), that it is possible to attain complete second-order double focusing simultaneously with first-order double focusing for all masses.

Ewald and his coworkers [18, 55, 56] have discussed the possibility of second-order focusing in an instrument with stigmatic-image formation (Secs. 7.2C and 7.3C) which is double-focusing for all masses in the first order. The second-order aberration has now four terms dependent on α_r^2, $\alpha_r\beta$, β^2, and α_z^2, where α_r and α_z are the initial angular divergence of the beam in the radial and axial directions. They have shown that it is not possible to correct for all the four aberrations simultaneously. However, it is possible to find an arrangement in which the two aberrations dependent on α_r^2 and $\alpha_r\beta$ can be corrected for, while the aberration dependent on α_z^2 can be made to vanish independently of the other requirements if the faces of the condenser electrodes are made cylindrically curved on the incident and emergent sides.   The fourth aberration, dependent on β^2, which cannot be completely eliminated, can, however, be reduced to a very small magnitude.   The combinations proposed by Liebl and Ewald [56] therefore give *practically* double focusing of second order.

E. Wien filter—Jordan's apparatus

The instruments described so far use a curved electrostatic field to produce the velocity dispersion of the incident ions.   There is, however, another way of producing velocity dispersion.   This is done with the help of the so-called Wien *velocity filter*.   It consists of crossed homogeneous electric and magnetic fields such that the electric force is exactly counterbalanced by the magnetic force for the ions traveling along the axis of the filter.   It can be easily seen that the condition required for this is given by

$$v = \frac{E}{B} \tag{7.72}$$

where v = velocity of ions
E, B = electric and magnetic field strength, respectively
The electric field acts between two parallel plates.

Herzog [13] investigated the focusing properties of the Wien filter. The general focusing effect of a crossed *radial* electric field and a homogeneous magnetic field can be deduced in a similar manner as that given in Sec. 7.2B for a purely radial electric field.   The lateral displacement of the ions of mass $M = M_0(1 + \gamma)$ and initial velocity $v = v_0(1 + \beta)$ at a distance x'' after emergence from the field region is given by [13]

$$y'' = a \left[-\frac{\alpha'}{\tau} \sin \tau\phi + \delta(1 - \cos \tau\phi) + \frac{y_1}{a} \cos \tau\phi \right] \\ + x'' \left[-\alpha' \cos \tau\phi + \delta\tau \sin \tau\phi - \frac{y_1}{a} \tau \sin \tau\phi \right] \tag{7.73}$$

α' is the initial angle of divergence of the ion beam from a point source which enters the field at the ordinate y_1. ϕ is the angle of deflection of the ions. a, the radius of curvature of the central ion beam (of mass M_0 and mean velocity v_0) is given by

$$\frac{1}{a} = \frac{1}{a_e} + \frac{1}{a_m} \tag{7.74}$$

where $$a_e = \frac{M_0 v_0^2 a \ln (R_1/R_2)}{eX} \tag{7.75}$$

and $$a_m = \frac{M_0 v_0}{Be} \tag{7.76}$$

R_1 and R_2 are the radii of curvature of the cylindrical condenser electrodes, and X is the difference of potential between them. δ and τ are given by

$$\tau^2 = 1 + \left(1 - \frac{a}{a_m}\right)^2 = 1 + \left(\frac{a}{a_e}\right)^2$$

$$\tau^2\delta = \gamma + \beta\left(2 - \frac{a}{a_m}\right) \tag{7.77}$$

The condition for direction focusing is that y'' be independent of α' at a distance $x'' = l''$. This, along with the expression $y_1 = b' - \alpha'l'$ (l' and b' being, respectively, the axial distance of the point source of ions in front of the field region and its ordinate normal to the axis), gives us

$$\frac{l'l'' - a^2/\tau^2}{l' + l''} = \frac{a}{\tau}\cot\tau\phi \tag{7.78}$$

Equation (7.78) cannot be directly applied to the case of the Wien filter since, in this case, $a \to \infty$ and $a_m = -a_e$. However, for large a, we can write $L = a\phi$, where L is the length of the condenser plates. Also, from Eq. (7.77), $\tau = a/a_m$ in this case. Then, for a Wien filter, Eq. (7.78) reduces to

$$\frac{l'_w l''_w - a_m{}^2}{l'_w + l''_w} = a_m \cot\frac{L}{a_m} \tag{7.79}$$

Equation (7.79) can be easily transformed to

$$(l'_w - g_w)(l''_w - g_w) = f_w{}^2 \tag{7.80}$$

where $\qquad g_w = a_m \cot\dfrac{L}{a_m} \qquad f_w = \dfrac{a_m}{\sin(L/a_m)} \tag{7.81}$

It should be noted that the Wien filter is the complete analog of the optical cylindrical lens, since the central rays pass through the filter, undeviated. All other fields work as a lens combined with a prism.

From Eq. (7.77), we get

$$\delta a = -a_m\beta$$

The image width at the focal point due to an entrance slit of width b'_w can then be obtained from Eq. (7.73) and is given by

$$b''_w = -a_m\beta\left(1 + \frac{f_w}{l'_w - g_w}\right) - b'_w\frac{f_w}{l'_w - g_w} \tag{7.82}$$

from which it will be seen that the Wien filter is fundamentally a device for producing a velocity dispersion.

The special case ($l'_w = l''_w = 0$), in which the object and image are located at the entrance and exit boundaries, respectively, requires

$$\cos\frac{L}{a_m} = \pm 1$$

whence $L = n\pi a_m$, $n = 1, 2, \ldots$.

On the other hand, for a velocity filter with a length $L = (n - \frac{1}{2})\pi a_m$, a parallel beam will emerge if the object is located at the entrance boundary.

Bainbridge [57] was the first to use a Wien velocity filter with a mass spectrometer employing a homogeneous magnetic field of equal strength.

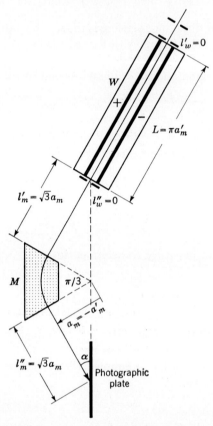

Fig. 7.17. Jordan's double-focusing mass spectrograph using a Wien velocity filter W followed by a sector-shaped homogeneous magnetic field M.

This arrangement antedated Herzog's discovery of the direction focusing properties of the Wien filter.

Jordan [58], at the University of Illinois, combined a 60° magnetic analyzer with a Wien filter for which $l'_w = l''_w = 0$. His arrangement, shown in Fig. 7.17, is a double-focusing one. The condition for velocity focusing can be easily deduced by putting the velocity-dispersion coefficient of the Wien filter equal to that for the subsequent magnetic field (cf. Sec. 7.3A). The velocity-dispersion coefficient of the Wien filter is deduced from Eq. (7.82):

$$K_w'' = -a_m \left[1 - \cos\left(\frac{L}{a_m}\right) + \frac{l_w''}{a_m} \sin\frac{L}{a_m} \right] \qquad (7.83)$$

This reduces to $K_w'' = -2a_m$ for the case $l_w' = l_w'' = 0$ and $L = \pi a_m$. For the subsequent symmetric magnetic field, the velocity-dispersion coefficient is $2a_m'$ [cf. Eq. (7.7a)], where a_m' is the radius of curvature of the central beam in the magnetic analyzer. Hence the condition of velocity focusing is

$$a_m' = -a_m \qquad (7.84)$$

Thus the magnetic fields of the Wien filter and the magnetic analyzer must be equal in magnitude and opposite in direction.

The resolving power, mass dispersion, and magnification of such an instrument are a_m/S, $a_m/25$ cm per 1% mass difference, and unity, respectively.

In Jordan's apparatus, L was 115 cm and a_m was 36.6 cm; the resolving power and dispersion were, respectively, 28,000 and 1.4 cm per 1% mass difference. The instrument was used for a short time in mass-comparison work.

F. Focusing by crossed electric and magnetic fields: trochoidal focusing instruments

The focusing conditions so far discussed are all approximate, either to the first order or to the second order of small quantities. Bleakney and Hipple [59] showed that a combination of crossed electric and magnetic fields possesses perfect double focusing in the plane normal to the magnetic field.

Referring to Fig. 7.18, we note that the equations of motion of the ions in such a crossed-field combination can be written as

$$\begin{aligned} M\ddot{x} &= eB\dot{y} \\ M\ddot{y} &= eE - eB\dot{x} \end{aligned} \qquad (7.85)$$

where B points in the Z, and E in the Y, direction. The solutions of these equations are given by

$$\begin{aligned} x &= a\omega t + \frac{\dot{y}_0}{\omega}(1 - \cos\omega t) - \left(a - \frac{\dot{x}_0}{\omega}\right)\sin\omega t \\ y &= \frac{\dot{y}_0}{\omega}\sin\omega t + \left(a - \frac{\dot{x}_0}{\omega}\right)(1 - \cos\omega t) \end{aligned} \qquad (7.86)$$

where $\dot{x}_0$ and $\dot{y}_0$ are the initial velocity components of the ions of mass M

and charge e in the X and Y directions, respectively. ω and a are given by

$$\omega = \frac{eB}{M} \tag{7.87}$$

$$a = \frac{ME}{eB^2} \tag{7.88}$$

The ions emerge from the point $x = 0$, $y = 0$.

Equations (7.86) represent a trochoid. Hence the ions follow trochoidal paths, the nature of which depends upon the initial conditions.

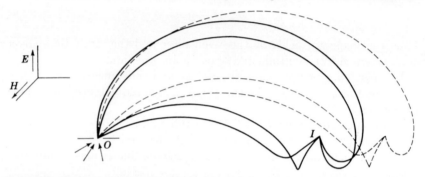

Fig. 7.18. Focusing by crossed uniform electric (E) and magnetic (H) fields. (*From Hipple and Sommer* [63].)

As is well known, such trochoids are generated by a point fixed to a circle of radius a which rolls on a straight line with an angular velocity ω. The distance of the fixed point from the center of the circle is given by

$$\rho = \left[\left(\frac{\dot{y}_0}{\omega} \right)^2 - \left(a - \frac{\dot{x}_0}{\omega} \right)^2 \right]^{\frac{1}{2}} \tag{7.89}$$

For $\rho < a$, the orbits are *curtate* trochoids, generated by ions emerging obliquely from the object slit. For $\rho > a$, the orbits are *prolate* trochoids, generated by ions emerging more or less normally from the object slit.

Figure 7.18 shows typical examples of both curtate and prolate orbits. The ordinary cycloidal path from O to I will be followed by the ions emerging normally with zero velocity.

From Eqs. (7.86) it follows that, after a time $t = 2\pi n/\omega = 2\pi nM/eB$, n being a positive integer, we have

$$x_n = 2\pi na = \frac{2\pi nME}{eB^2} \qquad y_n = 0 \tag{7.90}$$

Thus ions of mass M and charge e emerging from a point $(0, 0)$ are all focused at a distance $b_n = 2\pi nME/eB^2$ from the origin along the x axis, regardless of the initial direction or velocity of the ions. It is thus a case of perfect double focusing.

Obviously, there will be different points of focus, depending on the value of n. The first focusing point I corresponds to one complete revolution of the generating circle. Its distance from the object slit O is

$$b = \frac{2\pi M E}{eB^2} \tag{7.91}$$

Since b is directly proportional to M/e of the ions, the mass scale is linear. For an object slit of width S_0 and photographic detection, the resolving power is given by

$$\frac{M}{\Delta M} = \frac{b}{\Delta b} = \frac{b}{S_0} \tag{7.92}$$

For a given value of b, the resolution will of course be better if the focusing point selected corresponds to higher number (n) of complete revolutions of the generating circle.

Bleakney and Hipple [59] first demonstrated the practicability of this focusing scheme. They investigated both curtate and prolate orbits. For a given analyzer, the curtate orbit gives higher resolution than the prolate. This is due to the fact that the focal point in the former is at the end of the orbit. Hence, for given dimensions of the apparatus, b is larger. However, in practice, it was found that instruments using prolate orbits are the more satisfactory. In them the ions enter almost normally to the entrance slit, in which case it is easier to produce a suitably intense ion beam than it is with oblique incidence. Also, and more important, in this arrangement there is no possibility of the occurrence of cycloidal orbits, which, because of sharp cusps, could give rise to space-charge problems and consequent defocusing. Ordinary cycloidal orbits as well as the curtate ones which would lead to such space-charge defocusing are baffled out, as are also those followed by multiply charged ions of the species under study. This stopping down does not materially affect the high ion currents to be expected in instruments of this type.

Several larger instruments have since been constructed, all of which employ prolate orbits [60–62].

In the Mariner and Bleakney [60] apparatus, constructed at Princeton University about 1940, b was 262 mm, while the object and image slit widths were 0.05 and 3.17 mm. The theoretical resolving power of about 80 was realized in practice. The peaks had sharply rising sides with flat tops, indicating that, with narrower slits, much higher resolution could be attained. This work, interrupted by World War II, was not continued thereafter.

A small instrument of this type, developed by the Consolidated Electrodynamics Corporation [62], is commercially available. This has a resolving power of about 100, with $b = 2.7$ cm, an energy spread of 10%, and a half-angular divergence of about $6°$. The same firm has also experimented with a larger version having $b = 15.2$ cm, and achieved therewith a resolving power of about 4,000.

As mentioned before, trochoidal-path instruments with a larger number of cycles would give better resolution. Hipple and Sommer [63] incorporated this principle in a time-of-flight instrument, which will be discussed in Sec. 7.4D.

7.4. Mass Spectrometers Using the Cyclotron Principle

In this section we shall discuss a group of mass spectrometers which utilize the so-called *cyclotron principle* for their operation. The common characteristic of all such instruments is that the ions describe circular orbits in homogeneous magnetic fields, the period of rotation being

$$T = \frac{2\pi M}{eB} \tag{7.93}$$

This is independent of the velocity of the ions, but depends linearly on the mass.

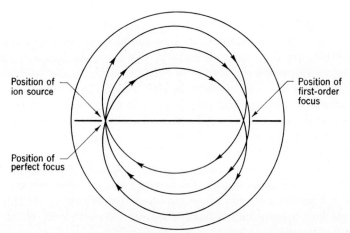

Fig. 7.19.   Trajectories of monoenergetic ions of a given mass, emitted from a point source, in a homogeneous magnetic field. The position of the first-order focus occurs after 180° deflection and depends both on the mass and the energy of the ions. The position of the perfect focus occurs after 360° deflection and is independent of both mass and energy of the ions. (*From Inghram and Hayden* [3].)

Referring to Fig. 7.19, we can see that, if a beam of ions is projected from a point in a magnetic field, without any velocity component parallel to the field, the ions will all describe circular orbits, passing through the starting point, regardless of their initial velocity or the initial direction of motion (perfect double focusing). In the absence of any other deflecting field, these circulating ions will, of course, hit the back of the ion source at the end of their first revolution. However, if either the ions have a velocity component parallel to the field or their energy is changed while they are in circulation, the circular orbits can be changed into helices or spirals, in which case the ions can be made to circulate a large number of times before being finally collected in a suitably located detector. The frequency of rotation—the cyclotron frequency—will remain unchanged in the process. The mass of the ions can be measured either by time of arrival of the ions at the detector (time-of-flight principle) or by frequency of its rotation (or its higher harmonics) by some resonance method. Both these methods have been used to develop a number of mass spectrometers, which are described below.

A. Helical-path mass spectrometer

In this instrument, first proposed by Goudsmit [64] and subsequently constructed by Hays, Richards, and Goudsmit [65], the ions enter the homogeneous magnetic field with a finite velocity component parallel to the field and hence describe a helical path. The perfect spatial double focusing, exemplified in Fig. 7.19, is maintained even when the ion orbits are drawn out into helix. All ions from a point source thus form, after each 360° rotation, a sharp focal line through the source and along the direction of the field. This is true irrespective of the mass of the ions. However, the time or arrival of the ions of different masses will be different, given by Eq. (7.93). An ion collector can be placed above or below the source, assuming the field B to be vertical. Obviously, the measurement of the time of arrival of the ions after one or more complete revolutions gives a measure of the mass of the ions.

In the mass spectrometer constructed by Hays, Richards, and Goudsmit [65] a pulsed ion source was used. Pulses up to 600 volts, lasting 0.25 μsec or longer, could be given to the ion source. The time of seven complete revolutions was measured, which, with their arrangement, changed by 10 μsec per mass unit. Because of the short duration of the pulses, τ, ions heavier than a limiting mass do not have time to get the full acceleration within the time τ. The energy gained by them in this time is inversely proportional to their mass, and hence all of them emerge from the ion source with the same momentum and thus follow a path of definite radius (12 cm). Ions lighter than the limiting mass, which all gain the same energy equal to the full acceleration energy in the ion source during the short duration of the pulse, emerge with lower momenta and are prevented from reaching the collector by a suitable arrangement of baffles.

Ions arriving at the collector after describing different numbers of complete revolutions give rise to pulses of different orders. For time measurement, all these pulses from the detector were displayed on an oscilloscope. Two separately movable electronic markers then selected two pulses, which appeared on two separate fast traces on a second scope. By turning the dials, which control the markers, the two pulses on the second scope could be made to coincide. The dial setting then read directly the time interval between the pulses. Good pulses could be matched to within 0.01 μsec, even if they were as long as $\frac{1}{2}$ μsec. Masses could be measured with a precision of about 1 millimass unit, or 1 mev. The instrument is of special use for heavy ions for which the time intervals to be measured are longer.

Intensities of the higher-order pulses were lower because of the longer paths of ions. However, the intensity was not maximum in the first order. Especially for a large vertical distance l between the source and collector slits, the first-order intensity was very weak. For $l = 12$ cm, the third-order intensity was the maximum for a source temperature of 1000°K, a field of 450 gauss, and a mass of 100 mass units for the ions. In practice, pulses up to the twelfth order could be used for measurement.

Because of the linear relationship between mass and time, the resolving power in this type of instrument is independent of the mass of the ion, and the equipment gives an essentially constant precision at all masses. This property of the instrument is valuable in the study of heavy atoms.

The ion transmission of the machine is low. This is actually a common limitation of all pulsed-beam machines, which are therefore not very suitable for isotopic abundance work.

The magnetic field was produced by a spherical air-core magnet which, because of its dimensions, produced a rather low intensity field. This necessitated the use of low-energy ions (25 ev for $M = 100$), which were strongly influenced by the polarization of the wall of the vacuum chamber. It was not practicable to increase appreciably either the strength or the volume of the magnetic field, the former to reduce the polarization effect, or the latter to allow a larger number of spirals. Although the instrument was, for these reasons, not further exploited, it was used to make a timely and significant contribution to our knowledge of atomic masses.

B. Omegatron

This machine, developed by Hipple, Sommer, and Thomas [66, 67] and shown in Fig. 7.20, is actually nothing but a small cyclotron adapted to sharpen mass discrimination. As will be shown below, the resolution in this type of instrument is improved by making the ions go through a large number of revolutions. This requires that the applied radio-frequency voltage and the resulting radial increment per cycle be small. Since the ordinary cyclotron depends upon r-f focusing as well as magnetic focusing, the r-f voltage cannot be reduced sufficiently to obtain the required resolution. The omegatron effectively overcomes this.

As shown in Fig. 7.20, the ions are produced at the center of the machine by an axial electron beam (in the direction of the magnetic field). A positive trapping voltage on the guard rings produces an electric field which prevents the ions from escaping in the axial direction, so that the r-f field can act on them for a longer interval and the required large number of cycles may be obtained. The ions gain energy, and thus spiral outward, if the frequency of the r-f field is equal to the frequency of revolution of the ions [the reciprocal of the time period given in Eq. (7.93)]. They are then collected by a collector, thus producing an indication of resonance. Unlike for a cyclotron, no D chamber is used, the r-f voltage being applied between two parallel-plate electrodes, and also to the guard rings, to ensure a uniform field.

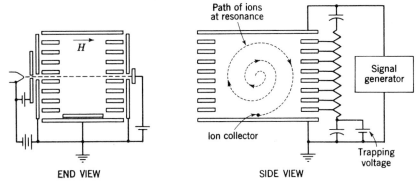

Fig. 7.20. Schematic diagram of the omegatron. (*From Sommer, Thomas, and Hipple* [68].)

An analysis of the motion of an ion of mass M and charge e starting from rest in an r-f electric field $E = E_0 \sin \omega t$, which acts at right angles to a homogeneous and steady magnetic field B, shows that the orbit of the ion is a spiral of radius r, given by

$$r = \frac{E_0}{B\epsilon} \sin \frac{\epsilon t}{2}$$

where $\epsilon = |\omega - \omega_0| \ll \omega_0$, $\omega_0 = eB/M$ being the angular frequency at resonance. The particles will describe the spirals with an angular frequency $(\omega + \omega_0)/2$. For $\epsilon \neq 0$, the radius will *beat*; i.e., it will go through successive maxima and minima with an angular frequency of $(\omega + \omega_0)/2$. At resonance, $\epsilon = 0$, and the radius becomes

$$r_0 = \frac{E_0 t}{2B}$$

If a collector is placed at a distance R_0 from the origin, the ions will not be able to reach it if

$$\frac{E_0}{B\epsilon} < R_0$$

Thus, for a given value of E_0/B, there is a critical value ϵ_c of ϵ for which the ions will just reach the collector.   This is given by

$$\epsilon_c = \frac{E_0}{BR_0}$$

The resolving power will be given by

$$\frac{M}{\Delta M} = \frac{\omega_0}{2\epsilon_c} = \frac{\nu_0}{\Delta \nu} = \frac{B^2 R_0 e}{2M E_0}$$

where $\Delta \nu$ measures the width of the peak at the base, ν_0 being equal to $\omega_0/2\pi$.

The time required by the ions to reach the collector when $\epsilon = \epsilon_c$ is equal to one-quarter of the time period of the beat of r and is thus equal to $t_c = \pi/\epsilon_c$.   Hence we have

$$\frac{M}{\Delta M} = \frac{\omega_0 t_c}{2\pi} = \frac{t_c}{T_0} = n_c$$

where T_0 = time period of circulation of ions at resonance

n_c = number of revolutions ions make before reaching collector when $\epsilon = \epsilon_c$

Time required by the ions to reach the collector at resonance is

$$t_0 = \frac{2}{\epsilon_c} = \frac{2t_c}{\pi}$$

We thus get

$$\frac{M}{\Delta M} = \frac{\pi t_0}{2T_0} = \frac{\pi n}{2} \tag{7.94}$$

where $n = t_0/T_0$ is the number of revolutions made by the ions to reach the collector at resonance.   The resolving power is thus proportional to the number of revolutions made by the ions before reaching the collector.

In practice, the loss in resolution is due to the space charge built up in opposition to the trapping field mentioned above, to inhomogeneities in the magnetic field, and to gas scattering.   To minimize these effects, the machine should be small.   To make the number of revolutions high in a small machine, it is necessary to use an intense magnetic field and very small r-f voltage.

The original omegatron built by Sommer, Thomas, and Hipple [68] consisted of two 3- by 5-cm parallel plates (r-f plates), 2 cm apart, with eight parallel guard rings of the same outer dimensions and equally spaced between the plates.   An ion collector was inserted through one of the r-f plates.   The source was of the electron-bombardment type and was located at the central region.   The whole assembly was enclosed in a glass tube 4.7 cm in diameter.   The resonance peaks could be scanned by varying the frequency of the r-f field. For high-resolution work, the magnetic field was varied. The number of revolutions made by the ions before reaching the collector varied between 3,000 and 7,000. The resolving power was about 10,000 for low masses.

The instrument was used for measuring the mass difference $H_2 - D$, which compared favorably with the measurement of Roberts and Nier [69] by the conventional deflection mass spectrometer. As far as precise mass determination is concerned, the machine is especially useful at low masses ($A < 30$). For heavy masses, the space charge and other difficulties, mentioned above, interfere with the measurements.

The chief application of the machine has been in the determination of the proton moment in nuclear magneton units. This was done by measuring the cyclotron frequency and nuclear resonance frequency for protons in the same field.

Bloch and Jeffries [70, 71] have used a similar method for the measurement of the proton magnetic moment. They used a small cyclotron in which the ions were *decelerated*, which caused them to spiral inward from an outer radius to an ion collector at the center. This arrangement had the advantage that, because the ion source was outside the D chamber, the pressure in the chamber could be kept very low, so that gas collisions and ionization did not affect measurement. Further, the decelerating cyclotron can be operated, not only at the resonant cyclotron frequency ν_0, but also at $\nu = n\nu_0$, where n is an odd integer. This also helped in the attainment of better resolution.

The proton magnetic moment in nuclear magnetons μ_n could be obtained from the relation

$$\frac{\mu_P}{\mu_n} = \frac{\mu_P}{eh/4\pi M_P} = \frac{\nu_N}{\nu_0}$$

where $\nu_N = 2B\mu_P/h$ is the nuclear resonance frequency, for protons.

C. Mass synchrometer

The third type of mass spectrometer employing the cyclotron principle to attain a high resolution was developed by L. G. Smith at the Brookhaven National Laboratory [72]. The essential principle underlying the operation of the instrument is that an electric field of small extent in space and time performs both the functions of forming pulses and of deflecting them, so that they miss hitting the back of the ion source. Ions of a limited mass range from the same slit are focused after half a revolution on a triple slit, forming the *pulser*, of which the two outer ones are grounded, while the inner one is connected to a source of square-wave-voltage pulses. A negative voltage pulse of 1 μsec duration on the latter decelerates the ions so that they revolve in an orbit of smaller radius after emerging from the pulser. After they have executed a number of revolutions in the new orbit, a second pulse, applied to the pulser after an interval of time equal to an integral multiple of the time period of revolution of the ions, further decelerates the ions. The ions, after emerging from the pulser, now follow a still smaller orbit and, after another half turn, enter the slit in front of the detector, which is a 15-stage magnetic electron multiplier using dynodes of beryllium-copper [73].

With this arrangement, 250-volt ions of mass 28 were observed after 90 revolutions. The resolving power was 24,000, but the intensity had reached a very low level after so many revolutions.

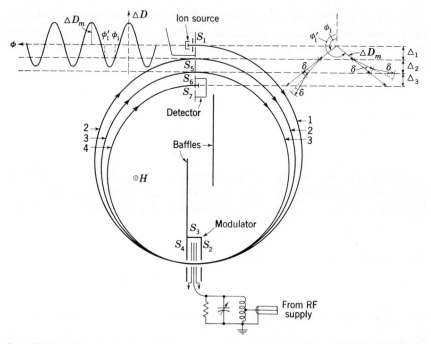

Fig. 7.21. Schematic diagram of the mass synchrometer. (*From Smith and Damm* [49].)

Subsequently, the instrument was converted to an r-f mass spectrometer which provides much higher resolution and intensity, with far greater precision of measurement [48, 49, 74, 78]. The pulser, which now acts as an *orbit-diameter modulator*, has an r-f voltage applied to its central slit S_3, with S_2 and S_4 grounded (Fig. 7.21). The r-f fields in the modulator cause the velocity of the ions to be modulated approximately sinusoidally. Hence the diameter of the subsequent orbit of the ions and the distance from the source slit S_1 of the point through which they will pass after a complete turn are approximately harmonic functions of the phase of the r-f field at the instant when the ions pass S_3.

The ions are actually decelerated by the r-f field of the modulator during each of the three successive transits through it. This decreases the orbit radius during the next cycle after each emergence from the modulator, so that the ions pass through an appropriately placed slit at the end of each of the three cycles (slits S_5, S_6, and S_7 in Fig. 7.21). At the end of the third cycle they reach the collector, placed behind the slit S_7, which consists of a Faraday cup plus a d-c amplifier.

In Fig. 7.21, the change in the diameter ΔD as a harmonic function of the phase of the r-f voltage on S_3 is shown at the upper left. The corresponding vector diagram is shown at the upper right of the same figure. The amplitude of the r-f voltage is usually adjusted so that the amplitude of diameter modulation ΔD_m lies between Δ_1 and $\Delta_1 + \Delta_2$, where Δ_1 and Δ_2 are the distances between S_1 and S_5 and between S_5 and S_6, respectively. As can be seen from Fig. 7.21, under this condition only two short groups of ions are allowed to pass through the baffle. These groups, which pass S_3 at the phases ϕ_1 and ϕ_1' of the r-f voltage, travel along orbit 2 through S_5 back to the modulator. In the vector diagram, they are represented by the two top vectors, ΔD_m, at ϕ_1 and ϕ_1', each of which has a projection Δ_1 in the downward direction.

If each of these two ion groups passes through S_3, the second time at the same phase as the first time, they will receive an approximately equal deceleration and will follow orbit 3 through the slit S_6 back to the modulator again. Since the time for complete revolution of the ions in their orbits in a magnetic field is independent of their speed [Eq. (7.93)], the two ion groups will have the same phase again as previously, when they pass through S_3. They will thus receive nearly the same deceleration again and will follow orbit 4 into the detector. In order that such synchronization of the phase of the r-f voltage with the arrival of the ions at S_3 may be possible, the r-f frequency must be an integral multiple of the cyclotron frequency of the ions in the magnetic field. Sharp peaks of ion currents are actually observed at the resonance frequency of the r-f field, $f = nf_0$, where f_0 is the cyclotron frequency. The resonance frequency is thus inversely proportional to the mass of the ion [Eq. (7.93)].

The resonance condition is represented in the vector diagram by the collinearity of the three vectors ΔD_m corresponding to the transit of the ions through the three slits S_1, S_5, and S_6, for both the phases ϕ_1 and ϕ_1'. Some ions may arrive at the detector even when there is small departure from collinearity of the three vectors, because of the finite widths of the slits S_1, S_5, and S_7. This case is represented in the vector diagram by introducing the small angle δ between the successive vectors ΔD_m.

It has been shown by Smith and Damm [48, 49] that, in general, the frequencies at which the two groups of ions corresponding to the two phases ϕ_1 and ϕ_1' are detected are slightly different from the resonance frequency $f = n f_0$ and are symmetrically located on two sides of the latter. This would cause a splitting of the peaks corresponding to the two groups. However, if the order n has a value n_m such that the corresponding transit angle θ of the r-f field through the modulator has a value of $\theta_m \approx \frac{3}{4}\pi$, then, by a suitable choice of the interslit distances Δ_1, Δ_2, Δ_3, it is possible to reduce the splitting of the peaks, mentioned above, to zero, and the resonance condition $f = n_m f_0$ will be satisfied to the first order of approximation. The transit angle θ is equal to $a\omega/v$, where $2a$ is the thickness of the modulator, ω is the circular frequency of the r-f field, and v is the unperturbed velocity of the ions. The condition $\theta = \theta_m$ gives a maximum change in the velocity of the ions in transit through the modulator and hence also the maximum change in the orbit diameter. The interslit distances Δ_1, Δ_2, and Δ_3 must satisfy the condition

$$D_1 \Delta_1 = D_2 \Delta_2 = D_3 \Delta_3$$

where D_1, D_2, D_3 are the diameters of the successive orbits 1, 2, and 3, respectively.

The resolving power of the apparatus is given by

$$A = \frac{n}{\eta} \tag{7.95}$$

where η = fraction of a cycle represented by the half-width
n = number of r-f cycles per turn
From the vector diagram in Fig. 7.21, it can be seen that, when the angle between the successive vectors is δ, the lateral displacement S of the end of the final vector from the line of the three collinear vectors is given by

$$S = \frac{N(N - 1)\delta \Delta D_m}{2} \tag{7.96}$$

where N is the number of vectors considered. The maximum possible value of S corresponding to the maximum value δ_m of δ (which occurs when the ions from the inner edge of S_1 passing the outer edge of S_5 arrive at the inner edge of S_7) is then given by

$$S_m = \frac{N(N - 1)\delta_m \Delta D_m}{2} \tag{7.97}$$

If W_0, W_1, and W_N are the widths of the slits S_1, S_5, and S_7, respectively, then

$$S_m = \left[\frac{W_1}{2} + (N - 1) \left(\frac{W_1}{2} + \frac{W_0}{2} \right) + \frac{W_N}{2} \right] \operatorname{cosec} \phi_1 \tag{7.98}$$

It may be shown that for a given δ_m the maximum peak intensity corresponds to $(N - 1)W_0 = NW_1 = W_{N1}$ in which case (7.97) and (7.98) yield

$$\frac{\delta_m}{2\pi} = \frac{3W_0}{2\pi N \, \Delta D_m \sin \phi_1} \tag{7.99}$$

Under these conditions η may be shown to be simply related to δ_m,

$$\frac{1}{\eta} = \frac{2\pi}{0.844\delta_m} = 2.48N \frac{\Delta}{W_0} \tan \phi_1 \tag{7.100}$$

from which the resolving power can be calculated. For $N = 3$, (7 . 100) may be rewritten in terms of r-f voltages

$$\frac{1}{\eta} = 7.44 \frac{\Delta}{W_0} \left[\left(\frac{V'}{V_\Delta'} \right)^2 - 1 \right]^{\frac{1}{2}} \tag{7.100a}$$

where V' is the r-f voltage employed and V_Δ' is the r-f voltage for which $\Delta D_m = \Delta_1$. The quantity $1/\eta$ has been measured by Smith and Damm [48, 49] and agrees well with the values calculated from (7.100a).

With their experimental arrangement, $n_m = 103$. Root-mean-square voltages between 180 and 320 volts were used (for a total energy of 2,500 volts). The observed resolving power varied between 10,000 and 25,000 for all masses below 250. The intensity was considerably lower ($\sim 12\%$) than that obtainable in deflection-type instruments with comparable resolving power and the same source, slit height, path length, and resolution.

Smith [78] has described plans for a third version of the mass synchrometer, in which the resolving power will be substantially increased by increasing n (that is, increasing the frequency of the r-f). The material reduction in $\Delta D_m/D$ which this involves is made possible by placing the ion source at a distance from S_1 and conducting ions of desired m/e to it by a Wien filter. A second Wien filter to conduct the analyzed beam from detector slit S_7 to a detector outside the magnetic field further facilitates lowering $\Delta D_m/D$ [78].

The mass synchrometer has been used solely for the determination of atomic masses, and with it results of high precision have been obtained. The mass difference between two masses M_1 and $M_2 = M_1 + \Delta M$ is given by the exact relationships

$$\frac{M_1}{\Delta M} = \frac{f_2}{\Delta f} \quad \text{or} \quad \frac{M_2}{\Delta M} = \frac{f_1}{\Delta f} \tag{7.101}$$

where f_1 and f_2 are the cyclotron frequencies of the two ions in question and $\Delta f = f_1 - f_2$. The frequency difference Δf was measured with high precision by the peak-matching technique, first introduced by Smith and Damm [48, 49] which will be discussed in Sec. 7.5D.

D. Trochoidal-path r-f spectrometer

As mentioned in Sec. 7.3F, Hipple and Sommer [63] have constructed a large trochoidal-path mass spectrometer with five cycles which employs the cyclotron resonance principle for the detection of the ions. The ions, after the first cycle along a prolate orbit (Fig. 7.18), pass through a pair of small closely spaced electrodes to which an r-f voltage is applied. This r-f voltage will, in general, deflect the ions in the direction of the magnetic field, and hence the ions will be unable to reach the gap between a similar pair of electrodes 4 cycles later. However, a small group of ions which pass through the first pair of electrodes when the r-f field is approximately zero will be able to move on undeflected and reach the second pair of electrodes. If the time of transit between the two pairs of electrodes is an exact integral multiple of half the time period of the r-f voltage, then these ions will reach the second pair of electrodes when the r-f field between them is again zero, and the ions will pass on undeflected into the detector, suitably located beyond the electrodes.

The cyclotron frequency in the trochoidal instrument remains unaffected by the presence of the electric field and hence will be determined by the reciprocal of time period given by Eq. (7.93). The time of transit between the two pairs of electrodes is equal to $4/f_0$, where f_0 is the cyclotron frequency. The resonance condition is obviously given by

$$f = \frac{nf_0}{8}$$

where f = frequency of r-f voltage
$\quad n$ = an integer

The resolving power of the instrument increases linearly with n with no loss in intensity. In actual operation, the d-c voltage is first adjusted to get the resonance peak. Similar procedure is adopted for the second member of a doublet   The frequency difference gives the measure of the mass difference. A resolving power of 12,000 has been achieved with $n = 270$.

7.5. Some Factors Affecting the Use of High-resolution Mass Spectroscopes

A. Some basic factors governing the performance

The performance of a double-focusing mass spectrometer is judged primarily by its dispersive power and its resolving power. Besides, such quantities as the line width on the photographic plate, fractional energy filtered, and the total distance traveled by the ion beam from the entrance slit to the photographic plate or the electrical detector are of importance.

The theoretically expected values of these quantities are different for the different types of instruments. These values as functions of the parameters of different types of instruments are listed in Table 7.3. The actual numerical values of the various parameters of the different instruments are included in Table 7.4.

The *mass dispersion* $D(\gamma) = \gamma d$, where γ is the fractional difference between two neighboring masses and d is the dispersion coefficient, has already been given in Sec. 7.2A [Eqs. (7.6d) and (7.6e)]. For an energy filter in front of the magnetic analyzer, we have

$$d = \frac{K_m''}{2 \cos \omega}$$

where ω = angle between the mean direction of beam and the normal to
           the image curve
       K_m'' = velocity dispersion in the magnetic field given by Eq. (7.7)
If, instead of an energy filter, a velocity filter is used, as in Jordan's apparatus [58], the above expression for d will be multiplied by a factor of 2. The mass dispersion is a function of the parameters of the magnetic analyzer alone.

The linear dependence of the mass dispersion on γ is true only for $\gamma \ll 1$. A more exact expression will be given in Sec. 7.5B. As we have remarked earlier, the standard mass dispersion is referred to a value of $\gamma = 1\%$ [Eq. (7.6f)]. This standard figure for the dispersion coefficient d as a function of a_m is included in Table 7.3, for different types of instruments. The actual numerical values of d are listed in Table 7.4. From Eqs. (7.6e) and (7.7) it can be seen that the dispersion increases with increasing radius of the magnetic analyzer. Also, asymmetric arrangement in the magnetic analyzer ($l_m'' > l_m'$) helps improve the mass dispersion. This device has been employed, for example, in the apparatus of Ogata, Matsuda, and Matsumoto [31, 32] (Sec. 7.2B).

The *resolving power* of the double-focusing mass spectrometer, which is given by Eq. (7.51), is determined by the parameters of the electrostatic analyzer alone. It can be increased by reducing the entrance slit width. In some of the instruments attaining very high resolving power (10^4 or more), entrance slit widths of only a few microns have been used (Table 7.4). The proper alignment of these slits presents a problem (Ewald[35]). The trend in recent years has therefore been to use a large radius of curvature of the electrostatic analyzer so that comparatively wide slits can be used, without sacrificing resolution (Table 7.4). From Eq. (7.52) we note that the resolving power can be increased by using an asymmetric arrangement in the electrostatic analyzer ($l_e' > l_e''$). This also has been done by Ogata and Matsuda in their new large instrument.

Another way to improve resolution is to use a toroidal condenser instead of a radial condenser as the energy filter. One gains a factor of

TABLE 7.3

Type of apparatus	Lateral magnification (b''_m/S'_e)	Line width (Δy)	Dispersion coefficient (d)	Resolving power $(M/\Delta M)$	Energy filtered $(\Delta W/W_0)$	Length of ion path (L)	Mass scale
Bainbridge and Jordan	$1.00a_m/a_e = 1$	$2.00S'_e$	$2.00a_m$	$1.00a_e/S'_e$	$1.00S_2/a_e$	$6.733a_e = 6.733a_m$	Quadratic
Dempster	$1.127a_m/a_e = 1.291$	$1.291S'_e$	$1.00a_m$	$0.887a_e/S'_e$	$0.873S_2/a_e$	$5.144a_e = 4.490a_m$	Quadratic
Nier et al.	$0.839a_m/a_e = 0.677$	$0.677S'_e$		$1.00a_e/S'_e$	$1.00S_2/a_e$	$6.059a_e = 7.502a_m$	Linear
Reutterswärd	$0.658a_m/a_e = 0.20\text{–}0.48$	$(0.47\text{–}1.14)S'_e$	$0.776a_m$	$0.500a_e/S'_e$	$1.143S_2/a_e$	$2.21a_e = 7.37a_m$ to $2.95a_e = 4.02a_m$	Linear
Mattauch and Herzog	$1.00a_m/a_e = 0.23\text{–}0.87$	$(0.33\text{–}1.23)S'_e$	$0.707a_m$	$0.500a_e/S'_e$	$1.505S_2/a_e$	$2.67a_e = 11.36a_m$ to $3.74a_e = 4.22a_m$	Quadratic

TABLE 7.4

Apparatus	S'_e (μ)	l'_e (cm)	l''_e (cm)	l'_m (cm)	l''_m (cm)	a_e (cm)	a_m (cm)	ϕ_e	ϕ_m	ϵ'_m	ϵ''_m	$D(\gamma)\tfrac12\%$ (mm)	Resolving power	Mass scale
Dempster	25	1.28	5.17	0	0	8.5	9.8	90°	180°	0	0	0.98	3,000	Quadratic
Duckworth	25.4	3	15.5	0	0	20	23	90°	180°	0	0	2.3	7,000	Quadratic
Duckworth et al.	12	96	96	0	0	274	274	90°	180°	0	0		200,000	Linear
Bainbridge and Jordan	5–8	0	0	44	44	25.4	25.4	127°18′	60°	0	0	5	10,000	Linear
Asada et al.	5	0	0	43.3	43.3	25	25	127°18′	60°	0	0	6.25	60,000	Linear
Demirkhanov et al.	5–10	21.2	21.2	17.3	17.3	30	30	63°36′	60°	0	0		70,000	
Nier et al.	12	6.61	6.61	34.77	20.73	18.87	15.24	90°	60°	0	0		14,000	
Nier et al.	10	17.62	17.62	92.71	55.27	50.31	40.64	90°	60°	0	0		100,000	
Mattauch and Herzog	25	19.8	∞	∞	0	28	4.5–25	31°48′	90°	0	−45°	0.3–1.8	6,500	Quadratic
Mattauch and Bieri	2		∞	∞	0	28	6.58–24.26	31°48′	90°	0	−45°	0.465–1.725	70,000	Quadratic
Ewald			∞	∞	0			31°48′	90°	0	−45°			
Bainbridge		151	∞	∞	0	213	183 (max)	31°48′	90°	0	−45°		200,000+	Quadratic
Mattauch, Hintenberger, et al.†		151	∞	∞	172.5	542.6	100	48°26′	166°59′	−37°58′	−6°31′			
Reutterswärd	34	3	∞	∞	4–10	30	9–22	58°	70°	35°	−55°	0.7–1.7	4,400	Quadratic
Ogata and Matsuda	2.3	134	0	120	448	109.3	120	84°48′	60°	0	0	76	500,000	Quadratic
Ewald and Sauermann	3	28.9	∞	∞	1–3	12	5–20	29°42′	87.5°	−45°	−45°	0.4–1.4	25,000	Quadratic
Stevens et al.		68.1	∞	∞	62.2	254	254	75°	110°	36.5°	−35°			

† Under construction.

$2/(2 - a_e/R_e)$, where a_e and R_e are the principal radii of curvature of the zero potential surface in the toroidal field. As mentioned in Secs. 7.2C and 7.3C, Ewald and coworkers have used this method to improve the resolving power by a factor of 2.67 over what would be obtained with a cylindrical condenser of the same a_e.

In Table 7.3 the resolving power, as a function of the ratio a_e/S_e', has been included for a number of different types of mass spectroscopes. In Table 7.4, the actual numerical values attained are listed. In the case of instruments with electrical detection, these values correspond to the resolving power at half the peak height.

The *line width* depends on the lateral magnification of the final image, which is equal to the product of the individual magnifications G_m and G_e by each field, given by Eqs. (7.5) and (7.27a), respectively.

It can be shown, from the condition of velocity focusing, both when the ions are proceeding in the forward direction [Eq. (7.48)] as also when they are proceeding backward in reversed fields, that the lateral magnification reduces to

$$\frac{2b_m''}{S_e'} = \frac{K_m''}{K_e'}$$

where $2b_m''$ = width of final image

S_e' = width of entrance slit

K_m'' = coefficient of velocity dispersion [Eq. (7.7)]

K_e' = velocity dispersion in electric field if ion beam is assumed to proceed backward [obtained from Eq. (7.29) by interchanging l_e' and l_e'']

The line width Δy is then given by

$$\Delta y = \frac{2b_m''}{\cos \omega} = \frac{K_m'' S_e'}{K_e' \cos \omega}$$

The lateral magnification and the line width as functions of a_m/a_e and S_e', respectively, have been listed in Table 7.3. If the line width is very small, the resolving power of the instrument may be limited by the grain size of the photographic plate or by the difficulty in fabricating a small enough exit slit to exploit the narrow line width.

We next consider the fraction of the energy transmitted by the energy analyzer. If S_2 is the width of the intermediate slit between the two fields, and if $\beta_m v_0$ is the maximum spread in the velocity of the ions which can be transmitted by it, then we can write

$$S_2 = K_e'' \beta_m \qquad \beta_m = \frac{1}{2} \frac{\Delta W}{W_0}$$

where K_e'' = dispersion in electrostatic field [Eq. (7.29)]

W_0 = mean energy of ions

ΔW = spread in energy corresponding to velocity spread $\beta_m v_0$

The fraction of the energy transmitted by the apparatus is then

$$\frac{\Delta W}{W_0} = \frac{2S_2}{K_e''}$$

In the case of double focusing at all masses, the intermediate image is formed at infinity and the slit S_2 is usually placed at the entrance to the magnetic field, so that the dispersion K_e'' has to be evaluated by putting the distance Δ between the fields in place of l_e'' in Eq. (7.29).

The intermediate slit should limit β_m to small enough values so that terms containing β_m^2 and higher powers may be neglected. Usually, this width ranges from a few thousandths to a few hundredths of a_e. The distance between the electrostatic deflecting plates is also usually kept a few per cent of a_e. The entrance slit width should be smaller than S_2 by a factor of about 10^2. $\Delta W/W_0$ as a function of S_2/a_e is listed in Table 7.3.

Finally, from the point of view of the intensity of the ions obtained at the detector or the photographic plate, the length of path L traveled by the ions from the entrance slit to the detector is of some interest. This is given by

$$L = l_e' + a_e\phi_e + \Delta + a_m\phi_m + l_m''$$

where the various quantities have their usual significance. L as a function of a_e and a_m is listed in Table 7.3.

B. Mass scale: theory of dispersion

In the photographic method of detection, the mass difference

$$\Delta M = \gamma M_0$$

between two neighboring masses is computed from the accurately measured distance between the two mass-spectral lines on the plate. In order to do this, the dispersion law in this region of the mass spectrum should be known accurately. In general, if the distance y on the photographic plate of a mass line from a fixed point is known, one can write

$$y = f(M) \tag{7.102}$$

This relation determines the mass scale. In the case of mass spectrometers using electrical methods of detection, the measured quantity may be a voltage (Nier [75, 76], Duckworth et al. [29, 77, 80]), a time, or a frequency (Smith and Damm [48, 49]). In these cases, y could represent any of these quantities. In practice, one measures Δy, the difference between the values of y corresponding to the two closely lying masses. This is a function of the fractional mass difference $\gamma = (M - M_0)/M_0$ and can be expressed as

$$\Delta y = y - y_0 = F(\gamma) = \gamma d(1 + a_1\gamma + a_2\gamma^2 + \cdot\cdot\cdot) \tag{7.103}$$

where d is the dispersion coefficient mentioned in Sec. 7.5A, given by

$$d = M_0 f'(M_0) = \left(\frac{dy}{d\gamma}\right)_0 = F'(0) \tag{7.104}$$

The coefficients in the expansion (7.103) are given by

$$a_i = \frac{M_0{}^i}{(i+1)!} \frac{f^{(i+1)}(M_0)}{f'(M_0)} = \frac{F^{(i+1)}(0)}{(i+1)!\,d} \tag{7.105}$$

In principle, it is possible to calculate the theoretical values of the coefficients for different types of instruments if the mass scale is known. The theoretical values of the coefficients depend on Herzog's theory, which, however, does not take into account the fringing-field effects or the fact that the locations of the recording systems are usually determined empirically for optimum performance. In practice, therefore, a number of different *standard mass lines*—dispersion lines—are used to calibrate the mass scale in the region under study, from which d and a few of the coefficients in the power series (7.103) are determined empirically. Usually, it is sufficient to determine only up to a_1. Within narrow regions of masses ($\gamma \ll 1$), the dispersion relation is, of course, linear.

In a Dempster-type apparatus, the photographic plate is placed along the exit boundary of the magnetic field, and hence the measured quantity y is twice the radius of curvature a_m in the magnetic field. In the mass spectrographs which are double-focusing at all masses, the photographic plate passes through the mean ion entrance point into the magnetic field and the distance y is linearly related to a_m. Hence, for these two types, the mass scales are similar. It can be easily seen that, since the energy of the ions remains constant, $a_m \propto \sqrt{M}$ for both the above types. Such a mass scale is called a *quadratic* mass scale. So we can write $y = K \sqrt{M}$. This immediately gives us $d = y_0/2$ from Eq. (7.104), and the coefficients in the expansion (7.103) can be easily evaluated with the help of Eq. (7.105). The dispersion relation then reduces to

$$\Delta y = y - y_0 = \frac{y_0 \gamma}{2}\left(1 - \frac{\gamma}{4} + \frac{\gamma^2}{8} - \frac{5\gamma^3}{64} + \cdots\right) \tag{7.106}$$

In the Bainbridge and Jordan type of mass spectrograph [25], y is measured from the apex of the 60° sector magnetic field. The mass scale has been shown by Herzog and Hauk [22] to be almost *linear* ($y \propto M$). The linearity is particularly good near the double-focusing point. More exact calculation [42] gives the relation

$$\Delta y = y - y_0 = y_0 \gamma\left(1 + \frac{\gamma}{32} + \frac{\gamma^2}{8} + \cdots\right) \tag{7.107}$$

In the earlier instrument of Bainbridge, using a velocity filter, the mass scale was linear. Hence, in this case, one gets $\Delta y = y_0 \gamma$.

On the other hand, in Jordan's instrument, which also used a velocity filter, the dispersion relation was more complicated [42]:

$$\Delta y = y - y_0 = 2y_0\gamma \left(1 + \frac{9\gamma}{16} + \frac{9\gamma^3}{16} + \cdots \right) \tag{7.108}$$

In the electrical method of detection by Nier [75], the potential difference V between the cylindrical electrodes producing the electric field is varied to bring the different masses into the detector. This is given by [Eqs. (7.12) and (7.13)]

$$V = \frac{Mv^2}{e} \ln \frac{r_a}{r_b} \tag{7.109}$$

where r_a and r_b are the radii of curvature of the two electrodes.

Since the magnetic analyzer accepts ions of a definite momentum ($B = $ constant), the above relation can be transformed to

$$V = \frac{eB_0{}^2 a_m{}^2}{M} \ln \frac{r_a}{r_b} \tag{7.110}$$

where a_m is the radius of the mean orbit in the magnetic field B_0. The above expression shows that the product MV remains constant for all masses. Identifying y with V, one can easily get

$$\Delta y = y_0[(1 + \gamma)^{-1} - 1] = -y_0\gamma(1 - \gamma + \gamma^2 - \gamma^3 + \cdots) \tag{7.111}$$

In practice, Nier et al. measure a resistance (Sec. 7.3D) instead of a voltage. But that does not change the above dispersion law, in which the measured quantity is inversely proportional to mass.

In the helical-path instrument, y is the time of flight τ of the ions from the source to the receiver, while in the omegatron, it is the magnetic field. In both these cases, the mass scale is linear [42].

In the instrument of Smith and Damm [48, 49], y is the resonance frequency f, and the product of M and y is constant, yielding a dispersion law similar to that in Nier's apparatus.

As mentioned above, in actual practice, the dispersion coefficients are determined empirically. One can, for instance, use three symmetrical mass lines—the middle one being a member of the doublet being measured —as, for example, the three consecutive members of the hydride series C_2H_2, C_2H_3, and C_2H_4. Equation (7.103) can be written twice for the mass-differences $C_2H_2 - C_2H_3$ and $C_2H_3 - C_2H_4$, γ having the same absolute value in both cases. From these two equations, one gets the relation

$$\Delta y = 2\gamma d(1 + a_2\gamma^2 + a_4\gamma^4 + \cdots) \tag{7.112}$$

where Δy is the distance between the two extreme mass lines ($C_2H_2 - C_2H_4$). For $M > 10$ one can calculate d from this equation to an accuracy better than 0.01% by retaining up to the quadratic term and using the theoretical value of a_2.

Under this approximation, a_1 is given by Mattauch and Waldman [42] as

$$a_1 = \frac{-(1 + k)}{(1 - k)\gamma}$$

where k is the ratio between the two line separations mentioned above.

Again, for $M > 10$, a_1 can be determined to an accuracy better than 1%.

If, as is often the case, only one dispersion line is available, the doublet, together with its dispersion line, may be placed at different parts of the plate and the dispersion coefficients can be evaluated from the variation of dispersion along the plate [42].

C. Determination of mass differences

The most important application of high-resolution mass spectrometers is, of course, the accurate determination of atomic masses. In this work, the difference between the masses of two ions, whose specific charges differ only slightly, is determined. If the mass of one is known, the other can then be calculated. This method is known as the *doublet* method of mass comparison. The two masses of the doublet should occur at the same mass number. In the photographic method of detection, as explained in the previous section, the mass scale has to be calibrated by photographing at the same time one or two dispersion lines in the neighborhood of the doublet under study.

For example, if we have a doublet consisting of masses M_1 and M_2 and a dispersion line of mass M_0, then, for an instrument with a quadratic mass scale ($y \propto \sqrt{M}$), one can derive the doublet mass difference ($M_2 - M_1$), M_1 being the unknown mass, in terms of the known masses M_0 and M_2, as follows:

$$M_2 - M_1 = M_2 \left[1 - \left(1 - \frac{\delta y}{\Delta y} \frac{\sqrt{M_2} - \sqrt{M_0}}{\sqrt{M_2}} \right)^2 \right]$$

where δy and Δy are the measured line separations between M_2 and M_1 and between M_2 and M_0, respectively. If the doublet separation δy is small compared with Δy, then the product of the two ratios within the parentheses is small compared with unity, and the doublet mass difference ($M_2 - M_1$) can be determined with great accuracy, even without knowing the masses M_0 and M_2 very accurately. If the dispersion line M_0 belongs to a mass number one or two units different from the mass number of the doublet, then in some cases it may be sufficient to use the corresponding mass numbers for M_0 and M_2 in order to evaluate the doublet mass difference.

This was the method used by Aston in determining the masses of H^1, D^2, and C^{12} from the three fundamental doublets H_2^1-D^2, D_3^2-C^{12}, and $C^{12}H_4^1$-O^{16}, using the mass of $O^{16} = 16$ to define the unit of the atomic mass scale.

In the case of mass measurement by mass spectrometers (i.e., using electrical detection), it is not necessary to use a dispersion line. For instance, in Nier's apparatus, we have the relation

$$MV = M_0V_0$$

which gives $\qquad \dfrac{M - M_0}{M_0} = \dfrac{V_0 - V}{V}$

Since the difference between the voltages $(V_0 - V)$, as well as the absolute value of the voltage V, can be measured with great accuracy, the doublet mass difference can be determined accurately without using a dispersion line. However, a dispersion line can also be used, since its use dispenses with the necessity of determining accurately the absolute value of the voltage V (which in some instruments may be quite high).

D. Precision of doublet mass comparison: peak matching

The mass width of a mass-spectral line depends upon the resolution $\Delta M/M$, and hence the accuracy of determining the doublet mass difference depends upon the precision with which the position of a mass-spectral line can be determined. In the photographic method of detection, this is approximately $1/50$ of the line width, so that, with a resolution of 1 in 20,000, the precision achieved in atomic-mass determination is about 1 part in 10^6.

With the electrical method of detection (e.g., using an electron multiplier), coupled with the *peak-matching* technique introduced by Smith and Damm [49], it is possible to locate the peak of the ion current to within $1/1,000$ of the width of the peak, and perhaps even better. Thus precision of the order of 2 parts in 10^8 in atomic-mass measurement is possible, and indeed is achieved in the recent work of Nier et al. [81], Smith [78], and Duckworth et al. [29].

The principle of the technique used by Smith and Damm [49] can be explained with the help of Fig. 7.22. The frequency f of the r-f voltage (which is varied in order to bring ions of different masses to the recorder) is modulated with a saw-tooth voltage in phase with the horizontal sweep of an oscilloscope trace, while the amplified output from the ion detector is applied to the vertical deflecting plates of the oscilloscope. This allows a mass spectrum to be displayed on the oscilloscope screen. At the same time the frequency is allowed to change from f_1 (corresponding to the mass M_1) to f_2 (corresponding to M_2) at the end of each sweep. Thus the two peaks are displayed on the screen in alternate sweeps. Actually, because of persistence of vision, the two peaks corresponding to the two masses are seen by the eye simultaneously, and their positions can be matched (one on the other) with a high degree of accuracy. To improve the precision further, a high-frequency square-wave voltage is applied to the vertical deflecting electrodes in every alternate sweep, so that the peak corresponding to one of the doublet members splits into two, spaced symmetrically above and below the peak corresponding to the other member. With this arrangement, the peaks could be matched to 1 part in 1,000 or better.

Nier [76] has also used the peak-matching technique in his second high-resolution spectrometer (cf. Sec. 7.3D). He modulates the magnetic field in phase with the horizontal sweep of the oscilloscope. At the end

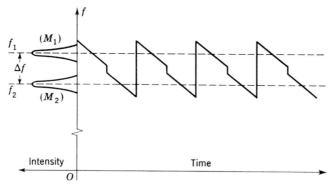

Fig. 7.22. Manner in which the applied frequency f is varied with time in order to match the two peaks of a doublet. (*From Smith and Damm* [49].)

of each sweep, three relay switches operate which change the electrostatic deflecting voltage, accelerating voltage of the ions, and the amplifier gain (to match the heights of the two peaks), respectively, to display, in alternate sweeps, the two peaks corresponding to the doublet. The change in the deflecting voltage is brought about by introducing an additional resistance ΔR in the circuit at the end of each alternate sweep, without otherwise changing the circuit. The mass difference is computed from the relation $\Delta M/M = \Delta R/R$. Bainbridge [38] employs essentially the same method, as also do Duckworth and his collaborators [28], with the difference in the latter case that the ion beam is swept across the collector by applying a sweep voltage to the plates of the electrostatic analyzer.

REFERENCES

1. J. J. Thomson: "Rays of Positive Electricity and Their Application to Chemical Analysis," Longmans, Green & Co., London, 1913.
2. A. J. Dempster: *Proc. Am. Phil. Soc.*, **75,** 755 (1935).

3. M. G. Inghram and R. J. Hayden: A Handbook on Mass Spectroscopy, *Natl. Acad. Sci.-Natl. Res. Council, Nucl. Sci. Ser. Rept.* 14, 1954.
4. K. T. Bainbridge: *Phys. Rev.*, **42**, 1 (1932).
5. F. W. Aston: *Phil. Mag.*, **38**, 707 (1919).
6. F. W. Aston: "Mass Spectra and Isotopes," Edward Arnold (Publishers) Ltd., 1942.
7. J. L. Costa: *Ann. Phys. (Paris)*, **4**, 425 (1925).
8. W. F. G. Swann: *J. Franklin Inst.*, **210**, 751 (1930).
9. W. Wien: *Ann. Phys. Leipsig*, **8**, 244 (1902).
10. N. F. Barber: *Proc. Leeds Phil. Lit. Soc.*, **2**, 427 (1933).
11. W. E. Stephens: *Phys. Rev.*, **45**, 513 (1934).
12. W. Henneberg: *Ann. Phys.* (Leipzig), **19**, 335 (1934).
13. R. Herzog: *Z. Physik*, **89**, 447 (1934).
14. L. Cartan: *J. Phys. Radium*, **8**, 453 (1937).
15. G. V. Schierstedt, H. Ewald, H. Liebl, and G. Sauermann: *Z. Naturforsch.*, **11a**, 216 (1956).
16. R. Herzog: *Z. Naturforsch.*, **8a**, 191 (1953).
17. H. Ewald and H. Liebl: *Z. Naturforsch.*, **10a**, 872 (1955).
18. H. Ewald and G. Sauermann: *Z. Naturforsch.*, **11a**, 173 (1956).
19. H. Ewald, H. Liebl, and G. Sauermann: *Proc. Conf. Nuclear Masses and Their Determination*, 1957, p. 184, Pergamon Press, New York; "Advances in Mass Spectrometry" (J. Waldron, ed.), Pergamon Press, New York, 1959, p. 10; *Proc. Intern. Conf. Nuclidic Masses*, p. 491, University of Toronto Press, Toronto, Canada, 1960.
20. H. Marschall: *Phys. Z.*, **45**, 1 (1944).
21. N. Svartholm: *Arkiv Fysik*, **2**, 195 (1950).
22. R. Herzog and V. Hauk: *Z. Physik*, **108**, 609 (1938); *Ann. Phys.*, **33**, 89 (1938).
23. H. Ewald and H. Hintenberger: "Methoden und Anwendungen der Massenspektroskopie," Verlag Chemie GmbH, Weinheim, Germany, 1953.
24. A. J. Dempster: *Proc. Am. Phil. Soc.*, **75**, 755 (1935).
25. K. T. Bainbridge and E. B. Jordan: *Phys. Rev.*, **50**, 282 (1936).
26. J. Mattauch and R. Herzog: *Z. Physik*, **89**, 786 (1934); *Phys. Rev.*, **50**, 617 (1936).
27. H. E. Duckworth: *Rev. Sci. Instr.*, **4**, 532 (1950).
28. H. E. Duckworth, J. T. Kerr, and G. T. Bainbridge: *Proc. Conf. Nucl. Masses and Their Determination*, p. 218, Pergamon Press, New York, 1957. N. R. Isenor, R. C. Barber, and H. E. Duckworth: *Can. J. Phys.*, **38**, 819 (1960); *Proc. Intern. Conf. Nuclidic Masses*, p. 439, University of Toronto Press, Toronto, Canada, 1960.
29. R. C. Barber, L. A. Cambey, J. H. Ormrod, R. L. Bishop, and H. E. Duckworth: *Phys. Rev. Letters*, **9**, 16 (1962); R. C. Barber, R. L. Bishop, L. A. Cambey, W. McLatchie, and H. E. Duckworth: *Can. J. Phys.*, **40**, 1496 (1962).
30. T. Asada, T. Okuda, K. Ogata, and S. Yoshimoto: *Proc. Phys. Math. Soc. Japan*, **22**, 41 (1940).
31. K. Ogata and H. Matsuda: Mass Spectroscopy in Physics Research, *Natl. Bur. Std. (U.S.) Circ.* 522, p. 217, 1953; *Phys. Rev.*, **89**, 27 (1953); *Z. Naturforsch.*, **10a**, 843 (1955); *Proc. Conf. Nucl. Masses and Their Determination*, p. 202, Pergamon Press, New York, 1957.
32. K. Ogata, H. Matsuda, and S. Matsumoto: *Proc. Intern. Conf. Nuclidic Masses*, p. 474, University of Toronto Press, Toronto, Canada, 1960.
33. R. A. Demirkhanov, T. I. Gutkin, V. V. Dorokhov, and A. D. Rudenko: *Soviet J. At. Energy*, **1**, 163 (1956).
34. J. Mattauch and H. Lichtblau: *Phys. Z.*, **40**, 16 (1939).
35. H. Ewald: *Z. Naturforsch.*, **1**, 131 (1946); Mass Spectroscopy in Physics Research, *Natl. Bur. Std. (U.S.) Circ.* 522, p. 37, 1953.

36. J. Mattauch and R. Bieri: Z. *Naturforsch.*, **9a,** 303 (1954).
37. F. Everling, H. Hintenberger, L. A. König, J. Mattauch, W. Müller-Warmuth, and H. Wende: *Proc. Conf. Nucl. Masses and Their Determination*, p. 221, Pergamon Press, New York, 1957; see also *Proc. Intern. Conf. Nuclidic Masses*, p. 387, University of Toronto Press, Toronto, Canada, 1960.
38. T. L. Collins and K. T. Bainbridge: *Proc. Conf. Nucl. Masses and Their Determination*, p. 213, Pergamon Press, New York, 1957; see also K. T. Bainbridge and P. E. Moreland, Jr.: *Proc. Intern. Conf. Nuclidic Masses*, p. 460, University of Toronto Press, Toronto, Canada, 1960.
39. A. E. Shaw and W. Rall: *Rev. Sci. Instr.*, **18,** 278 (1947).
40. N. B. Hannay: *Rev. Sci. Instr.*, **25,** 644 (1954).
41. C. Reuterswärd: *Arkiv Mat. Astron. Fys.*, **A30,** 7 (1943).
42. J. Mattauch: Mass Spectroscopy in Physics Research, *Natl. Bur. Std. (U.S.) Circ.* 522, p. 1, 1953; J. Mattauch and L. Waldman: Z. *Naturforsch*, **8a,** 293 (1953).
43. H. Hintenberger and L. A. König: Z. *Naturforsch.*, **12a,** 443 (1957).
44. A. O. Nier and T. R. Roberts: *Phys. Rev.*, **81,** 507 (1951).
45. E. G. Johnson and A. O. Nier: *Phys. Rev.*, **91,** 10 (1953).
46. H. Hintenberger and L. A. König: Z. *Naturforsch.*, **12a,** 773 (1957); "Advances in Mass Spectrometry," (J. Waldron, ed.), Pergamon Press, New York, 1959, p. 16.
47. K. S. Quisenberry, T. T. Scolman, and A. O. Nier: *Phys. Rev.*, **102,** 1071 (1956).
48. L. G. Smith and C. C. Damm: *Phys. Rev.*, **90,** 324 (1953).
49. L. G. Smith and C. C. Damm: *Rev. Sci. Instr.*, **27,** 638 (1956).
50. H. Hintenberger, H. Wende, and L. A. König: Z. *Naturforsch.*, **10a,** 605 (1955).
51. H. Hintenberger, H. Wende, and L. A. König: Z. *Naturforsch.*, **12a,** 88 (1957).
52. H. Hintenberger and L. A. König: Z. *Naturforsch.*, **11a,** 1039 (1956); **12a,** 140 (1957).
53. L. A. König: *Proc. Intern. Conf. Nuclidic Masses*, p. 498, University of Toronto Press, Toronto, Canada, 1960.
54. C. Stevens, J. Terany, G. Lobell, J. Wolfe, N. Beyer, and R. Lewis: *Proc. Intern. Conf. Nuclidic Masses*, p. 403, University of Toronto Press, Toronto, Canada, 1960.
55. H. Liebl and H. Ewald: Z. *Naturforsch.*, **12a,** 538 (1957).
56. H. Liebl and H. Ewald: Z. *Naturforsch.*, **12a,** 541 (1957).
57. K. T. Bainbridge: *Phys. Rev.*, **42,** 1 (1932).
58. E. B. Jordan: *Phys. Rev.*, **57,** 1072 (1940).
59. W. Bleakney and J. A. Hipple: *Phys. Rev.*, **53,** 521 (1938).
60. T. Mariner and W. Bleakney: *Rev. Sci. Instr.*, **20,** 297 (1949).
61. G. W. Monk, J. D. Graves, and J. L. Horton: *Rev. Sci. Instr.*, **18,** 796 (1947).
62. C. F. Robinson and L. G. Hall: *Rev. Sci. Instr.*, **27,** 504 (1956); "Advances in Mass Spectrometry" (J. Waldron, ed.), Pergamon Press, New York, 1959, p. 60.
63. J. A. Hipple and H. Sommer: Mass Spectroscopy in Physics Research, *Natl. Bur. Std. (U.S.) Circ.* 522, p. 123, 1953.
64. S. A. Goudsmit: *Phys. Rev.*, **74,** 622 (1948).
65. E. E. Hays, P. I. Richards, and S. Goudsmit: *Phys. Rev.*, **84,** 824 (1951).
66. J. A. Hipple, H. Sommer, and H. A. Thomas: *Phys. Rev.*, **76,** 1877 (1949).
67. J. A. Hipple, H. Sommer, and H. A. Thomas: *Phys. Rev.*, **78,** 332 (1950).
68. H. Sommer, H. A. Thomas, and J. A. Hipple: *Phys. Rev.*, **82,** 697 (1951).
69. T. R. Roberts and A. O. Nier: *Phys. Rev.*, **77,** 746 (1950).
70. F. Bloch and C. D. Jeffries: *Phys. Rev.*, **80,** 305 (1950).
71. C. D. Jeffries: *Phys. Rev.*, **81,** 1040 (1951).
72. L. G. Smith: *Phys. Rev.*, **81,** 295 (1951); *Rev. Sci. Instr.*, **22,** 115 (1951).
73. L. G. Smith: *Rev. Sci. Instr.*, **22,** 166 (1951).

74. L. G. Smith: Mass Spectroscopy in Physics Research, *Natl. Bur. Std.* (*U.S.*) *Circ.*, 522, p. 117, 1953.
75. A. O. Nier: Mass Spectroscopy in Physics Research, *Natl. Bur. Std.* (*U.S.*) *Circ.* 522, p. 29, 1953.
76. A. O. Nier: *Proc. Conf. Nucl. Masses and Their Determination*, p. 185, Pergamon Press, New York, 1957.
77. J. T. Kerr, G. R. Bainbridge, J. W. Dewdney, and H. E. Duckworth: "Applied Mass Spectrometry," Pergamon Press, New York, 1959, p. 1.
78. L. G. Smith: *Proc. Intern. Conf. Nuclidic Masses*, p. 418, University of Toronto Press, Toronto, Canada, 1960.
79. H. E. Duckworth: "Mass Spectroscopy," Cambridge University Press, London, 1958.
80. J. T. Kerr and H. E. Duckworth: *Can. J. Phys.*, **36,** 986 (1958).
81. R. R. Ries, R. A. Damerow, and W. H. Johnson, Jr.; V. B. Bhanot, W. H. Johnson, and A. O. Nier: *Proc. Intern. Conf. Nuclidic Masses*, pp. 446 and 435, University of Toronto Press, Toronto, Canada, 1960.
82. H. Ewald: "Handbuch der Physik," vol. 33, p. 581, 1956.
83. J. Mattauch: private communication.

8

Vacuum Techniques

H. A. Tasman, A. J. H. Boerboom, and J. Kistemaker

Laboratory for Mass Separation (F.O.M.)
Amsterdam, The Netherlands

8.1. Introduction

A. Vacuum requirements in mass spectrometers

As described in Chap. 3, in analyzing a sample with a mass spectrometer, the substance is converted into ions, which are electrically accelerated and passed through some mass discriminating system (a magnetic field, for instance) and measured on a collector. To make the measured intensity at a certain mass proportional to the corresponding concentration of ions near the source, it is essential to eliminate or reduce perturbing effects, caused by possible collisions of the ions with gas molecules in the analyzer tube, on their way from the ion source to the collector.

The gas in a mass-spectrometer tube causes peak broadening, as the ions are diverted from their prescribed trajectories by the collisions. Only ions scattered either at low angles or near the collector slit can reach the collector. For neutral atoms collision cross sections have been measured for low-angle scattering that are about an order of magnitude larger than for corresponding large-angle scattering [1]. Unfortunately, for ions, comparable measurements are not available. Further, the residual gas penetrates the ionization chamber of the source, and the ions produced from it are measured, together with those from the sample. In some cases this is not important, as the influences of the two are separated by the analyzer; in other cases, however, it puts an upper limit to the attainable accuracy in abundance ratio measurements. Thus the vacuum requirements depend on the purpose that is to be served by the mass spectrometer.

Should the instrument be used for accurate measurements of isotope ratios (as it is impossible to exclude coincidence of sample and background peaks), the vacuum must be "as good as possible," unless one can use a type of ion source which will not ionize the residual gas (three-filament source). Under operating conditions a vacuum of 10^{-8} mm Hg may be considered normal. The same requirements, even more stringently, apply to those instruments that are to be used for the analysis of very small sample quantities. Reynolds [2] designed an apparatus for the analysis of small samples of the noble gases, operating at a pressure of 10^{-10} mm Hg.

Sometimes, however, the background from the residual gas will not matter so much. As was mentioned above, the three-filament ion source generally does not ionize the residual gas. Sometimes one requires only a rough determination of the composition of a gas mixture (e.g., in studies of chemical kinetics, with an accuracy of 1 to 5%). In other cases the mass ranges of sample and background can be separated entirely (e.g., in UF_6 analysis, where air background will not interfere). Further, there are instruments with a very high resolving power, capable of resolving ions of the same mass number but of different chemical composition (multiplets). Now, the vacuum should be only good enough to avoid intolerable peak broadening, an operating pressure in the tube of 1×10^{-6} mm Hg being usually sufficient. For high ion energies as in mass spectrography, i.e., about 20 kev, a pressure of 1×10^{-5} mm Hg may be tolerated.

8.2. Construction Materials

For the construction of vacuum systems, materials should generally fulfill three special requirements: (1) they should be sufficiently gastight; (2) they should have a sufficiently low vapor pressure; (3) they should not evolve excessive amounts of adsorbed or dissolved gases. Moreover, for obtaining a good vacuum quickly, there is a fourth requirement: (4) they must withstand a degassing process at elevated temperatures (bake-out).

For very low pressures the fourth requirement is indispensable. For the expulsion of water vapor, etc., a temperature of at least 150°C is needed; for very low pressures, i.e., below 10^{-8} mm Hg, a much higher temperature is essential (300 to 450°C, the higher the better). (See page 304.)

To comply with these demands, two groups of materials are primarily chosen: glass, preferably hard glass, such as Pyrex, and metal.

Ceramic materials have been used successfully [3], especially for seals and terminals which can be subjected to high temperatures. Ceramic-metal seals can stand operating temperatures up to 400°C; their resistance to thermal shock is reported to be excellent (Advac,† Ceramseal‡, Ferranti§).

Some plastics have been tested, like polyethylene, Teflon, etc. [4, 5], but the vacuum properties of polyethylene are not desirable for mass spectrometry, while Teflon may exhibit cold flow to an unpleasant degree and may desorb more gas than is sometimes claimed.

Glass is easy to work with, and vacuumtight joints are readily made. For high-vacuum work glass may be sealed to metal. Glass-Kovar seals and glass-copper seals, e.g., Housekeeper seals, can be made to hold a high vacuum. For experimental work at pressures not below 10^{-7} mm Hg, one may use epoxy-resin (Araldite), which offers a very convenient and simple means of connecting glass to other materials. Pyrex can be baked out at 450°C, and it is therefore extensively used for the construction of ultrahigh-vacuum systems. The electrical insulation of electrodes, etc., in a glass envelope is quite simple with glass-metal seals.

Metal offers the advantage of a higher mechanical strength. The precision in electrode systems, etc., can be improved, at least in a simpler way. For high-vacuum technique, stainless steel and copper are of primary importance. Sometimes nickel alloys such as Monel or Inconel are used. Care should be taken in the use of alloys, as sometimes components may evaporate at elevated temperatures. This occurs with brass, where the zinc is volatile; some hard solders contain volatile cadmium, and soft solders, especially those containing antimony, are to be avoided. The volatile metal may precipitate on electrical insulations and ruin their insulation resistance. Aluminum has been tested [6], but neither welded nor brazed aluminum vessels seem fit for pressures below 10^{-5} mm Hg.

Steel is often used, but it is easily covered with a layer of rust which can evolve enormous quantities of gas [7]. Moreover, the use of a ferromagnetic material is often not permissible. Some types of stainless steel are nonmagnetic. Among these, types 304 (18/8), 347 (18/10 Cb), and 310 (25/20) have excellent vacuum properties. Type 304 is normally supplied in the nonmagnetic austenitic phase, but it can be brought into the magnetic ferritic phase by heat-treatment or by cold-working. In type 310 this transformation is impossible because it is the stable austenitic phase. Therefore, when essentially nonmagnetic parts are subjected to high temperatures (welding), type 310 is to be preferred.

As mentioned above, the alloys Monel and Inconel are also used. Inconel has a good resistivity against corrosion at elevated temperatures. For electrodes, Monel and Nichrome V are suitable.

† Advanced Vacuum Products Inc., Stamford, Conn.
‡ Ceramseal Inc., New Lebanon Center, N.Y.
§ Ferranti Ltd., Hollinwood, Lancashire, England.

In choosing materials for ion-optical systems, one should be careful that no ion beam can hit insulating surfaces, or an insulator may become charged by the impinging particles, causing deflection of the ion beam [8]. A metal surface can become insulating by surface oxidation and contamination, e.g., by an oil film. This can happen with alloys containing a high content of chromium, chromium oxide film being an insulator.

Copper is particularly suitable, as the oxide film is semiconducting; usually polarization phenomena will not cause trouble. Therefore the use of copper mass-spectrometer tubes is recommended. For high vacua of 10^{-8} mm Hg and better, ordinary electrolytic copper may prove porous. This porosity is caused by small amounts of oxide, occluded in the copper. In brazing under reducing conditions, the oxide is reduced, leaving small channels. The best way to avoid this source of trouble is to use OFHC copper exclusively [9, 10].

The formation of an oxide film can be inhibited by silver or gold plating [11]. Gold can be applied electrolytically, or by vacuum-cathodic sputtering. One coating should be applied. A heat-treatment should then be given to allow part of the gold to diffuse into the carrier; finally, another coating should be applied. Electrolytically applied gold sometimes has a rough surface, which may absorb appreciable amounts of gas. This condition can be improved by polishing. In narrow tubes polishing can be done by passing a number of carefully cleaned and degreased small steel balls many times through the tube. However, even a gold-plated surface will not necessarily eliminate all polarization trouble, at least not with high intensities of impinging ions. This was observed in a mass spectrograph, the ion beam hitting the deflector plates of the electro-static analyzer [12], although some of the difficulties may have been due to contamination by diffusion pump oil.

In glass mass-spectrometer tubes, a conducting coating is more or less compulsory. This can be obtained by passing a mixture of $SnCl_2$ vapor and water vapor through the heated tube. A transparent semiconducting film is deposited, consisting of a suboxide of tin [13]. Conducting transparent Pyrex glass may also be obtained commercially (Pyrex E.-C. glass†).

Some metals are somewhat porous to a few gases [14–17]. For instance, copper is permeable to oxygen and hydrogen, and silver is permeable to oxygen. This effect is especially striking at elevated temperatures. At 450°C the diffusion rate per square centimeter surface area, at a wall thickness of 1 mm, is of the order of magnitude of 10^{-7} cm³ NTP per second at a pressure difference of about $\frac{1}{5}$ atm of hydrogen for copper, or oxygen for silver. For a mass-spectrometer tube of 1 mm wall thickness and a surface area of 750 cm², at 450°C this amounts to about 50 liters/sec at 10^{-6} mm Hg. The permeability of copper for oxygen is said to be much

† Made by Corning Glass Works, Corning, N.Y.

lower; however, not many quantitative data are available. At very low pressures most glasses prove to be porous to helium [17, 18].

In using electrical connections of tungsten wire sealed in glass envelopes, trouble can be caused by leakage through the metal and along the crystal faces in the wire. This leakage can be eliminated by covering the wire with a sleeve of glass, a few centimeters long, or silver plating, thus closing straight-through canals in the material from atmosphere to vacuum. When so made, tungsten-wire seals can be used for pressures below 10^{-10} mm Hg [2]. Special care is needed in cutting and bending tungsten wire [9].

8.3. Joints and Connections

A. Fixed joints

Usually, a vacuum system is designed and built in separate parts. These parts have to be joined together in such a way as to leave them vacuumtight, the connection being either permanent or demountable. In a glass system, permanent vacuumtight joints are readily obtained. A radio-frequency spark tester offers a convenient means for testing their vacuum tightness.

In metal systems, permanent joints should be made by brazing or welding. To avoid porosity, all welding should be done in an inert atmosphere, e.g., argonarc or heliarc. For stainless steel and copper, welding can give excellent joints, provided the compositions of the parts are identical or nearly so. However, successful welds have been made between stainless steel and Kovar. Unfortunately, not all parts can be subjected to the rather high temperatures required for welding, as these may produce strains and deformations. For these cases brazing is the obvious method, although larger domains are warmed up in this method of joining metals.

Hard-soldering usually offers considerable advantages: high mechanical strength and fitness for high-temperature bake-out. Provided it is done with care, it will produce vacuumtight joints. Attention must be given, however, to the accumulation of impurities, which can prohibit sealing at the final point of closure, resulting in a pinhole leak. If no high-temperature fitness is needed, soft solders (tin) may be used. In choosing a brazing alloy, one should avoid volatile components. Soft solders containing antimony can cause endless trouble: the antimony evaporates and is very hard to remove. In hard solders care must be taken with alloys containing cadmium or zinc; these can spoil electrical insulators after high-temperature bake-out. An excellent high-vacuum solder is eutectic silver-copper. (Ag/Cu = 72:28, melting point 779°C.) Brazing is best done in a hydrogen atmosphere, without flux. If no hydrogen furnace is available, vacuumtight joints are also possible using a flux. Easyflo can be used; others recommend Handy Flux, Lloyd's No. 7, or Fluxine PDS 8355-1. A somewhat lower working temperature (720°C) can be used for a silver-copper-indium brazing alloy, containing 5 to 15%

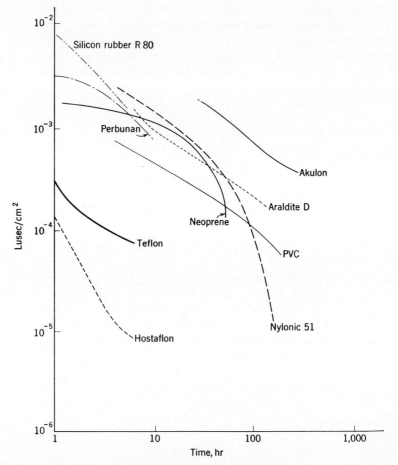

Fig. 8.1. Rates of degassing at room temperature of some sealing materials for high-vacuum purposes, as a function of time. It is emphasized that these data are only semiquantitative, as the degassing rate may vary greatly from one sample to another. (*From Blears, Greer, and Nightingale* [23]; *Diels and Jaeckel* [24].)

In (S1Ag64-In alloy†). This alloy contains no volatile components; good joints can be made with it on stainless steel using Easyflo as a flux, but the alloy oxidizes rapidly if it is exposed to air in the molten state for longer periods. Philips‡ makes a cadmium-free brazing alloy containing

† Manufactured by Dr. E. Durrwachter-Duduco-K.G., Pforzheim, Germany.
‡ Philips Nederland N.V., Eindhoven, The Netherlands.

silicon, melting point 725°C, which is well suited for stainless steel, for torch-brazing with Easyflo as a flux. Hydrogen brazing gives clean parts, free from oxide crusts. When a flux is used, it should be removed thoroughly after brazing. Many fluxes contain corrosive compounds, and moreover they may give off undesirable amounts of gases and vapors. Greasy fluxes should be avoided; borax dissolves in boiling water or hot diluted sulfuric acid. Excellent and extensive information on brazing procedures can be found in the Tube Laboratory Manual [9].

In designing vacuum systems with welded or brazed joints, care should be taken to avoid dirt traps or occluded volumes of gas at the vacuum side. These volumes may release gas or impurities at a sufficient rate to cause trouble, but the evolution can keep on almost indefinitely. Pumping grooves of intermittent welds must serve to evacuate these volumes rapidly [19].

For permanent connections in systems where no pressures below about 10^{-6} mm Hg are required, attention is drawn to the convenience of Araldite.† This is an epoxy resin with very high adherence to nearly all materials, including glass, ceramics, and many metals (for instance, copper, aluminum, stainless steel). For vacuum work, mainly Araldite types 1, 103, B, and D are of interest. Type 1 is a warm-setting resin, without solvent, giving very strong connections. Type 103 has a somewhat lower mechanical strength, but offers the convenience of being cold-setting after addition of a hardening compound. Types D and B are warm-setting after addition of a hardening compound and have excellent mechanical and electrical properties. Data on the gas evolution of Araldite types 1 and D were published by Johnson [20] and by Stivala and Denniger [21]. If the coefficients of thermal expansion of the materials to be joined and the Araldite are similar, Araldite (type 1) sealed joints will stand repeated cooling, down to liquid-air temperature [22].

B. Demountable joints

For demountable connections, *flanges* are mainly used, the sealing between two flanges being effected by a gasket (Fig. 8.2). If in a continuously pumped system pressure below 10^{-6} mm Hg is not required, O-ring seals are permitted. These toroidal rings of highly elastic material possess excellent flexibility. All gaskets should be deformed to make a seal, but for rubber O-rings only moderate pressures are needed to perform the necessary deformation. Gaco rubber, neoprene, and silicone rubber O-rings are to be preferred, as their gas evolution is lower than that of natural rubber. Rubber O-rings may be used repeatedly. Recently Viton O-rings have come on the market.‡ This is a rather expensive

† Manufactured by CIBA A.G., Basle, Switzerland.
‡ Viton V495-7 compound, Parker Seal Company, Culver City, Calif.

fluor-containing synthetic polymer, which degasses considerably less than other elastomers, and can be heated to 250°C continuously. Pressures below 10^{-9} Torr may be obtained with Viton gaskets. These elastomer O-rings may be used repeatedly. Viton shows a seemingly permanent

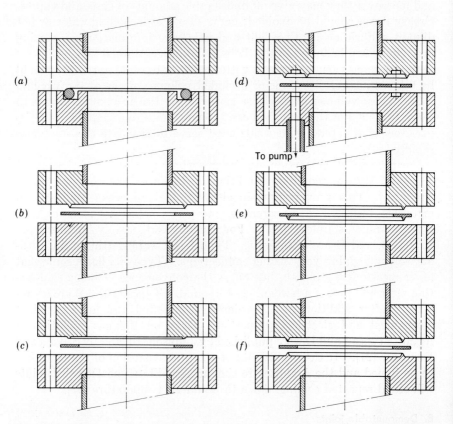

Fig. 8.2. Flanges and gaskets. (*a*) O-ring seal; (*b*) flat metal gasket between ridge and groove; (*c*) flat metal gasket between ridge and flat flange; (*d*) two concentric seals with evacuated interspace; (*e*) cooper gasket with ridge between ridge and flat flange; (*f*) flat metal gasket between two ridges. (*From Von Ardenne* [31]; *Mann* [32]; *Van Heerden* [33, 34]; *Ruthberg and Creedon* [35]; *Wadey* [36].)

deformation after being heated under compression, which recovers on heating in air at 180°C for 30 min.

When pressures below about 10^{-6} mm Hg are required, similar constructions are possible with massive golden O-rings (Fig. 8.3). Gold wire (24-carat gold) is brazed to itself in a small flame to form a ring; after brazing the whole ring is annealed. The ring should be compressed to one-half to one-quarter of its original thickness between scratch-free surfaces to form a good vacuumtight seal that will withstand bake-out at 450°C. The ring can be used several times. Massive aluminum O-rings may also be used [25].

To overcome the difficulty of the cold flow of Teflon, it may be used as a coating on an elastic core. As far as is known to the authors, no practical use of this possibility for high-vacuum sealing has been reported yet, but it seems interesting enough to mention it here. Teflon coatings are available on silicon rubber cores† for temperatures up to 220°C; the Teflon has a C-shaped cross section, the opening of the C pointing to the atmosphere, where contaminants from the rubber will do no harm.

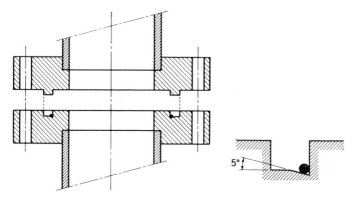

Fig. 8.3. A gold ring seal, which will stand bake-out at 450°C. The gasket is made of 24-carat-gold wire, about ½ to 1 mm thick; the ends are brazed together with a small flame. The dimensions should be chosen so as to compress the gold wire on sealing to one-half to one-quarter of its original thickness. The insert shows details of the groove, in which the ridge on the male flange should preferably fit at the outside for centering. The gold ring is best made slightly smaller than the groove and stretched on mounting.

Another solution, for temperatures up to 315°C, is a thin coating of Teflon (0.05 to 0.10 mm thick) on a hollow pressure-filled metallic O-ring.‡ The Teflon conforms to surface irregularities; the metal provides ruggedness, and the pressure filling enables elastic deformation.

Sensibly used metal gaskets give good seals, evolve little or no gas, and will stand high-temperature bake-out. They can be used in ultrahigh-vacuum systems at pressures below 10^{-9} mm Hg. Besides gold, mainly copper or aluminum gaskets are used. Indium is also an interesting material for making vacuum seals. It is very ductile and has an extremely low vapor pressure. A disadvantage is its low melting point (156.4°C) and its lack of elasticity. A gasket may be made of indium wire (0.8 to 1.6 mm diameter); the mating ends may be welded to each other in a

---

† Eriks Pakking & Rubber N.V., Alkmaar, The Netherlands.
‡ United Metallic O'Ring Corp., United Aircrafts Products, Inc., Dayton, Ohio.

small flame, but this is not even necessary. The material exhibits a tendency to cold-welding to itself, and overlapping of the ends to make a circle will suffice. The material may be put in a trapezoid groove, but tight seals have also been obtained between flat flange faces [26]. Further information may be found in the references [19, 27–29].

For copper or aluminum gaskets, the flanges must be profiled to achieve deformation above the flow limit of the gasket at one radius at least. On one flange a ridge is machined of semicircular, trapeziform or triangular cross section and rounded at the top; the corresponding flange may either be plane or contain a corresponding ridge or groove. A ridge can also be machined on the gasket. Step-type flanges have been proposed ([30]; Fig. 8.2b, c, e, f). Aluminum and copper gaskets can be used only once.

Aluminum gaskets may be cut from either soft 1-mm-thick, or from semihard $\frac{1}{2}$-mm-thick, aluminum sheet [37]. Copper gaskets, about 1 mm thick, are best made from OFHC copper sheet to avoid porosity. They must be annealed in hydrogen after cutting (temperatures from 800 to 950°C are used in annealing).

To reduce drastically any remaining leaks, it is possible to construct a double seal (Fig. 8.2d; [34]). The number of bolts should be doubled. If the space between the two concentric seals is evacuated to 10^{-3} mm Hg, the leakage is reduced by a factor 10^6 as compared with the inner seal alone. This is also a convenient arrangement for applying a test gas in leak testing.

A vacuumtight seal is also possible using foil gaskets. Ruthberg and Creedon [35] use commercial aluminum foil, 0.025 mm thick, between Monel flanges, on one of which is machined a trapeziform ridge of 1.5 mm top width. In using indium foil [27] a ridge seems to be dispensable. However, indium has a rather low melting point (156°C). Foil gaskets are more difficult to remove after use than thicker gaskets; they have to be scraped off carefully; aluminum may be dissolved in alkali.

Construction of flanges. O-ring seals require a groove in one of the flanges, or a spanner is necessary to hold the O-ring in its place. A typical example is shown in Fig. 8.2a. A standard system of sizes has been proposed by Wadey [36]. A good survey of the possibilities is given by Von Ardenne [31]. The sealing surfaces should be sufficiently free from scratches. For metal gaskets the sealing ridge and corresponding surface should be finely polished on the lathe to eliminate all radial scratches as far as possible. Tangential scratches are not so important. The gaskets should be made from reasonably scratch-free material; small irregularities are eliminated by the deformation.

In conventional usage the flanges are either brazed or welded to the tubes which are to be connected, and the seal is effected by a gasket between the flanges. Here the vacuum-tightness of the brazed or welded joint is important. One can get around this requirement by a simple modification, due to Higatsberger and Erbe ([38]; Fig. 8.4). The tube end protrudes about 1 mm through the flange, is finely polished, and acts as the deforming ridge. An analogous idea was proposed by Drowart et al. [39]; these authors use a glass-copper seal (Housekeeper seal), the copper tube of which is threaded externally; a flange is screwed on and the seal is effected by deforming a protruding copper rim, the copper tube itself acting as a gasket. The difficulty of brazing to a Housekeeper seal (because of the high thermal conductivity of copper) is also avoided in this way.

Glass-metal seals. For more rigid vacuum requirements the connection must stand baking out at a high temperature. Optical windows

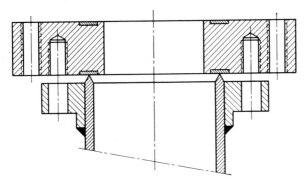

Fig. 8.4. High-vacuum seal. The deforming ridge, which seals against the gasket, is formed by the polished end of the tube itself. The vacuum tightness of the joint between tube and flange is thus rendered unimportant. (*From Higatsberger and Erbe* [38].)

(e.g., for pyrometers) can no longer be mounted in rubber or Teflon, but a demountable construction is possible using a metal gasket. Higatsberger and Erbe [38] describe a seal for these purposes. First a strongly adherent chromium film is evaporated on to the glass, which is then covered with a thicker copper layer by electroplating. A vacuumtight seal is obtained by using an aluminum gasket against a deforming ridge on the corresponding flange. Vacuumtight optical windows made of sapphire brazed to nickel-alloy tubes are made by Ceramseal; they are capable of withstanding high temperatures. For tubular glass-metal seals a powdered-glass technique is possible [3], but usually glass-copper or glass-Kovar seals are used.

Glass-copper seals have been used for a long time, and their construction and properties have been described so extensively in the literature that only a brief discussion seems to be within the scope of this chapter. For further details see Reimann [40], Yarwood [41], and Parr and Hendley [42]. In these seals the glass has a coefficient of thermal expansion considerably different from that of the copper. When the seals are cooled below the softening temperature of the glass, stresses occur in both the glass and the copper. At the seal the copper is machined to a fine edge, and it is sufficiently ductile to accommodate to the differences in thermal expansion by plastic and elastic deformation without cracking the glass. As the copper exhibits a tendency to harden by repeated deformations, there is a danger of cracking in the case of frequent temperature changes, even when the temperature differences are as small as 50°C. Because of the thin wall at the seal, its mechanical strength is rather low, and the danger of porosity of the copper has to be considered [4]. Provided the edge is made very thin, the same type of seal can be made with other metals; even with stainless steel 304 [43]. Also, small-diameter wires of copper, molybdenum, and tungsten may be sealed in a number of glasses for use as terminals [40]. The choice of the glass is more or less unimportant; these seals can be made with both hard and soft glasses.

Kovar-glass seals have been described fully in the literature [40–42, 44]. Contrary to the case of copper-glass seals, Kovar is an alloy with a thermal-expansion curve which is particularly well matched to that of some glasses (Corning 7052†) (Philips 28,‡ Fig. 8.5). Similar alloys, Vacon 10 and Vacon 12, are manufactured by the Vacuumschmelze,§ for which Schott-Jena‖ especially developed their 8234^{III} glass. As the difference in expansion over the whole temperature range between the softening point of the glass and room temperature is very small, much thicker walls, greater temperature changes, and larger tube diameters are possible. High standards of adherence, mechanical strength, and vacuum tightness are attainable, and therefore their use is strongly recommended. For terminals, also, Kovar is better suited than, for instance, tungsten, which is apt to show leakage through the metal.

Kovar surfaces that have to be covered with glass should be polished, and all sharp corners eliminated. The edges should be profiled according to Fig. 8.6a. The Kovar is then degreased thoroughly with ether and annealed in a wet hydrogen atmosphere, at 800°C for at least 30 min. If no hydrogen furnace is available, satisfactory seals can be obtained after annealing in a hydrogen flame (hydrogen only, no air or oxygen fed into the burner!) at a higher temperature (about 1000°C), for about 1 min. The Kovar is then heated in a hydrogen-oxygen flame (no coal gas should be used!), and a piece of Kovar glass (7052 or 28) introduced into it (Fig. 8.6a). The glass is first made to adhere to the inner Kovar surface; it is then bent around the edge with a carbon rod and onto the outer

† Corning Glass Works, Corning, N.Y.
‡ Philips Nederland N.V., Eindhoven, The Netherlands.
§ Vacuumschmelze, Hanau, Germany.
‖ Schott-Jena, Mainz, Germany.

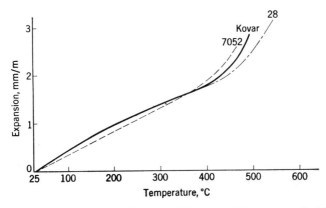

Fig. 8.5. Thermal expansion of Kovar, of Corning 7052 glass, and of Philips 28 glass, as a function of temperature.

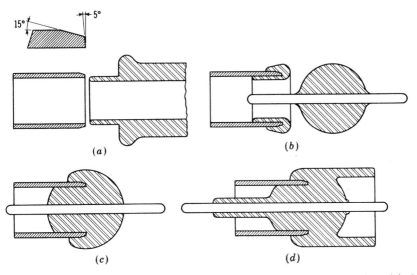

Fig. 8.6. The fabrication of glass-Kovar seals for electrical terminals. (*a*) A piece of 28 glass (Philips) or 7052 glass (Corning) is inserted into the Kovar tube, which is profiled as shown in the insert. (*b*) The glass is made to adhere to the inner surface, bent around the edge, and then melted onto the outer surface. A bead of the same glass is melted onto a Kovar pin, and inserted. (*c*) The finished seal for voltages up to at least 2 kv. (*d*) A modification of *c* that will stand 6 kv. In (*c*) and (*d*) the atmosphere should be at the right-hand side.

surface (Fig. 8.6*b*). For making electrical terminals a Kovar pin (1 to 2 mm), similarly polished, degreased, and annealed, is clad with a bead of the same glass (Fig. 8.6*b*) and sealed to the glass-covered Kovar. A good seal has a light-gray color; the Kovar becomes blackened after annealing.

In making tubular seals between Kovar and Pyrex, the 28 glass cannot be sealed directly to the Pyrex since the difference in thermal expansion causes the seal to crack. Therefore, after having prepared the end of the Kovar tube according to Fig. 8.6b, an intermediate ring of Philips 08 glass is inserted, which can be sealed to Pyrex (Fig. 8.7a). The insulating resistance of a bead of type 28 glass at room temperature is of the order of magnitude of 10^{12} ohms; in some cases (mounting of electrometer collectors), a resistance of at least 10^{15} ohms is desirable without excessive lengths of glass. This may be obtained by inserting a ring of highly insulating glass 18 which can be sealed to 08 glass (Fig. 8.7b).

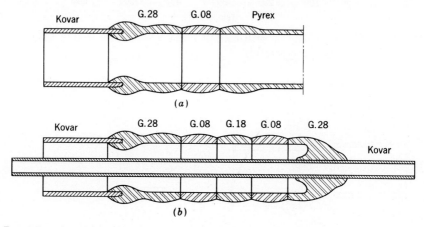

Fig. 8.7. (*a*) To make tubular seals between Kovar and Pyrex, the insertion of an intermediate ring of 08 glass between the 28 glass and the Pyrex is necessary. 08 glass has a thermal expansion intermediate between the expansion of 28 glass and Pyrex. (*b*) Support for electrometer collectors and other arrangements, where an insulating resistance of at least 10^{15} ohms is required. The highly insulating glass is 18 glass, which can be sealed to 08 glass.

The possibility of joining two different types of glass to each other is largely governed by their difference in thermal expansion and to a lesser extent by their difference in softening points and chemical composition. A difference in softening points of 200°C can be tolerated, although such joints are difficult to make. The difference in the thermal expansion of Philips glass 28 and 08, on the one hand, and between 08 and Pyrex, on the other, is rather large; it can be tolerated for smaller diameters when the joints can be made on the bench. For larger diameters and for joints made *in situ*, intermediate glasses will reduce the strains. The sequences, recommended by Philips, are (Kovar)-28-149-08-38-Pyrex, or (Kovar)-28-149-08-155-Pyrex, and (Kovar)-28-149-08-18. Joints with 155 glass should be cooled down rapidly to avoid structural changes in the glass; with 38 glass no such precaution is necessary. 155 glass corresponds almost to Nonex glass. Joints between 08 and 18 glasses are difficult to make (because of the difference in softening points), but the resulting joints are nearly free from stresses (Fig. 8.8).

The peak voltage for terminals, as in Fig. 8.6c, is more than 2 kv for a Kovar tube of 6 mm diameter; for higher voltages one may either choose larger diameters or modify the design as in Fig. 8.6d. Peak voltages of up

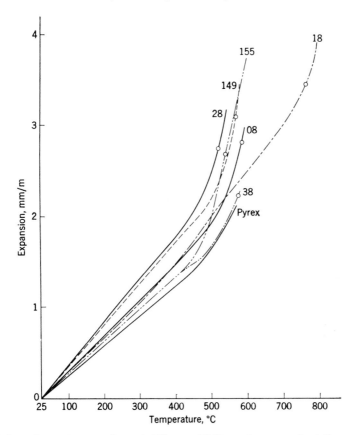

Fig. 8.8. Thermal expansion of different Philips glasses and of Pyrex. The Philips softening points, where the logarithm of the viscosity in poises equals 12.4, are marked with an O.

to 6 kv are then permissible in 6-mm Kovar tubes. All glass surfaces in electrical terminals should be scrupulously cleaned. Contaminations on newly made ones are usually burnt off by the heat; otherwise they should be degreased with acetone and ether. *In vacuo* surface conduction usually does not occur along clean surfaces. Surface conductivity in moist atmospheres can be reduced by orders of magnitude by a silicone treatment [45]. The silicones types MS 997 and MS 804, manufactured by Dow Corning† and by Midland Silicones Ltd.,‡ proved to be equally effective. First the surface to be treated is cleaned with ether, then the silicone fluid is applied, and finally the part is baked at 300°C for 1 hr. Dirt inside a narrow Kovar tube is difficult to remove; therefore terminals are best mounted with the atmosphere at the right side of Fig. 8.6c and d. Strains on the glass should be avoided while bending the Kovar pins, for these may cause cracking. These terminals can be heated up to

TABLE 8.1.   Properties and Composition of Some Glasses

	Pyrex	155 (Nonex)	38	08	149	28	18
Philips softening point,†°C	585	545	580	580	575	525	770
American softening point, °C	......	760	805	770	765	720	955
Specific resistance:							
log ρ (250°C) (ohm-cm)	8.1	8.8	8.6	8.0	8.0	8.8	
log ρ (350°C) (ohm-cm)	6.7	7.4	7.0	6.4	6.4	7.3	10
log ρ (20°C) (ohm-cm)	......	......	......	14.8			
Percentage Composition							
SiO_2	80.5	72.8	77.6	76	77.0	65	57
B_2O_2	12.9	15.3	13.9	16	14.5	22	4
Na_2O	3.8	2.9	5.1	4.5	6.5	4.5	
K_2O	0.4	0.5	1.1	1	0.8	4	0.5
CaO	......	......	......	......	......	......	6.5
BaO	......	......	......	......	......	......	1
MgO	......	......	0.25	......	......	......	9
Al_2O_3	2.2	......	2.0	0.5	0.4	4.5	21.5
Sb_2O_3	......	......	......	2	0.8	......	0.5
PbO	......	7.5					
Li_2O	......	0.05					

† Philips softening point is the temperature at which the logarithm of the viscosity in poises equals 12.4.

450°C, although their insulation resistance falls off rapidly at higher temperatures. Hard-soldering in the flanges, etc., is possible, provided the whole assembly is uniformly heated. Several terminals may be soldered in one flange. The holes in the flange and the flange ends of the Kovar tubes are tapered (1:10) to make a good fit. The flux, Borax or Easyflo, is applied, and a ring of silver solder, melting point about 725°C, is laid around every tube. The whole assembly is heated gradually and uniformly until the flux melts, and the heating is continued to a dull-red heat. With a small hand torch, using a fine flame, heat is applied locally to the brazing points until the solder melts and flows. After cooling

† Midland, Mich.
‡ London.

down gradually to room temperature, the flux is dissolved in boiling water or hot diluted sulfuric acid. If the Kovar tubes fit tightly enough in the tapered holes, neither small amounts of cadmium in the brazing alloy nor the flux will give trouble. The uniform heating procedure will not cause deformation of stainless-steel (type 304) flanges. The whole brazing can be done better in a hydrogen furnace without a flux by choosing brazing alloys free from volatile components. As was mentioned

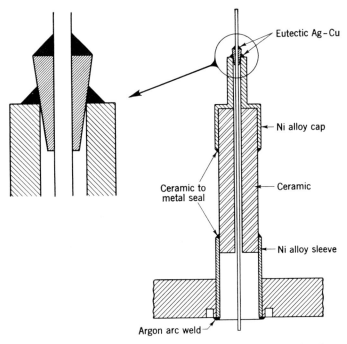

Fig. 8.9. Ceramic-to-metal seal for electrical terminals. The insert shows details of a good way of brazing the lead to the cap, by means of a tapered copper plug and a eutectic silver-copper.

earlier, ceramic materials have been used successfully for terminals (Fig. 8.9). These are now available for tensions up to 100 kv. They will stand operating temperatures up to 400°C, and they are very resistant to thermal shock. They may be brazed or welded in place. Their electrical resistance at 150°C exceeds 10^{14} ohms; at 250° to 300°C it is some 10^{12} ohms.

8.4. Flexible Connections. Introduction of a Movement in a Vacuum

It is often desirable to make a flexible connection between parts of a vacuum system so as to avoid the transmission of mechanical vibrations from, say, a rotary pump, or for making mechanical adjustments, such as the positioning of a mass-spectrometer tube in a magnetic field, the alignment of an ion source, etc. To avoid stresses in complicated glass systems, a flexible connection is often to be recommended.

In backing lines heavy-walled rubber tubing can be used; however, rubber is always somewhat porous, though it can be improved by varnishing. Plastic vacuum tubing, e.g., polyvinylchloride, is better, but still unsuitable for high-vacuum systems. A good high-vacuum flexible connection can be made with bellows, preferably constructed from stainless steel,† attached to flanges. Tombak bellows are not recommended, as they often show leaks after having been in use for some time. Tombak will not stand mercury vapor. The mounting of bellows to flanges can be done using tin soldering [46]. The mechanical strength of

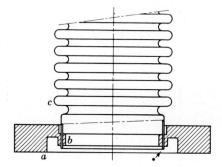

Fig. 8.10. Provisions for argonarc or heliarc welding of a stainless-steel bellows to a stainless-steel flange. The flange (*a*) is profiled to a thin collar and fits snugly over the end of the bellow (*c*). A snugly fitting protecting ring (*b*) is inserted in the bellow. The welding arc should be restricted to the ultimate edge (*) and be kept away from the corrugated surfaces.

soft solder is often too low, however, and therefore the best way to mount bellows is by welding (argonarc or heliarc). By shaping the flange as in Fig. 8.10, an excellent joint can be obtained. The bellow should fit snugly in the flange, as well as over the inner protecting ring. The welding arc should be restricted to the ultimate rim of the collar, and certainly not hit the corrugated surfaces. O-ring mounting of bellows is also possible [47], but without a substantial reinforcement of the wall at the end of the bellow, the radial and tangential forces exerted by the compressed O-rings are likely to cause cracking and leakage. The atmospheric pressure tends to collapse an evacuated bellow. If both end flanges can be rigidly attached to a frame, this tendency will not be objectionable. The collapsing force may be eliminated, however, by mounting two bellows in a counterbalanced arrangement ([48]; Fig. 8.11).

Bellows may also be used for the introduction of small linear displacements in a vacuum for adjustable slits, etc., and even for full-turn rotary motions ([49]; Fig. 8.12). Thus the contamination with rubber or vacuum grease is avoided, and a positively leak-free system can be

† I.W.K. Industrie Werke, Karlsruhe, Germany; Metallschlauchfabrik O. Meyer-Keller A.G., Luzerne, Switzerland.

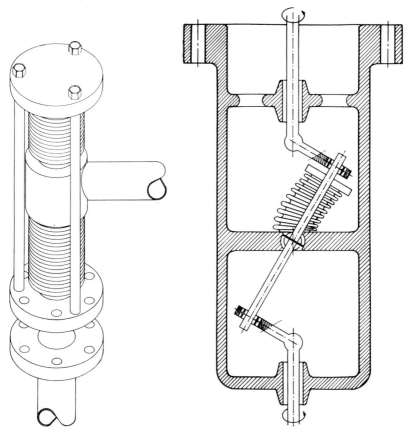

Fig. 8.11. Counterbalanced bellows. The side tube "floats freely" on evacuation; the collapsing forces due to the atmospheric pressure cancel out.

Fig. 8.12. A bellow-sealed device for introduction of a continuous rotary movement into a vacuum. (*From Guthrie and Wakerling* [49].)

obtained. In case of less stringent demands, much simpler constructions are possible with Simmer rings or pieces of vacuum tubing [31, 40, 50, 51].

8.5. Pumps

The high vacuum necessary for mass spectrometry cannot be obtained by one type of pump alone. For pressures below 10^{-4} mm Hg one relies mainly on diffusion pumps. These pumps need a forevacuum that should be of the order of 10^{-2} to 10^{-1} mm Hg, which is usually supplied by a rotary mechanical pump. The necessary backing pressure depends somewhat on the type of diffusion pump; some pumps work against a backing pressure of 1 mm Hg. The pumping speed is usually influenced favorably by a lower backing pressure.

The action of rotary pumps depends on an oil-film seal between the moving and the stationary parts. This oil can cause trouble in pumping condensable vapors, especially when they are readily soluble in the oil. Volatile solutes in the oil increase the vapor pressure and can spoil the final vacuum. Special attention is needed in pumping considerable amounts of organic vapors, e.g., alcohol, acetone, or ether. In these cases the insertion of a cold trap to condense the vapors is strongly recommended. For other vapors that will not dissolve as readily in the oil, e.g., water vapor, a simpler remedy is possible, namely, the use of an air-ballast pump. When the vacuum space in the pump has been separated from the system to be pumped and the compression toward atmospheric pressure begins, air is admitted to the gas that must be driven out, to 1 atm or less. The mixture of gas and air is compressed and expelled, and saturation pressure of vapors is not attained. In case of water vapor, the contamination of the oil is considerably reduced; to a lesser extent the same is true for organic vapors [52]. The ultimate pressure of air-ballast pumps is higher than the pressure achieved by the same pump without air ballast in the absence of condensable vapors. If water or other vapors are present, the air-ballast pump is by far superior. The final vacuum is usually sufficient for backing diffusion pumps, although the backing of an oil diffusion pump requires, then, a double-stage rotary pump. The air-ballast inlet can be shut off by a valve in many designs.

A new type of rotary pump,[†] based on the Rootes principle, has been developed. These pumps operate without any oil-film seal, on the gear-pump principle. The sealing is effected by high mechanical precision. The principal features of these pumps are very high pumping speeds (150 m^3/hr and higher), good final vacuum (10^{-4} to 10^{-5} mm Hg), and total absence of oil. However, they require a forevacuum of a few millimeters of mercury.

Diffusion pumps have already been treated extensively in different books on vacuum techniques (Ref. 16, chaps. 4 and 5; Ref. 14, chap. 1; Ref. 40, chaps. 5 and 6). Therefore only the mentioning of features of particular interest for mass spectrometry seems to fall within the scope of this chapter. The pumping action of a diffusion pump is caused by a vapor stream or jet, which transfers momentum to the gas molecules that have to be pumped out of the vacuum system. Usually, the gas is compressed by a number of successive streams of jets before it is finally handled by the backing pump. Depending on the type of vapor, one may choose between mercury or oil diffusion pumps. Both types have their merits. For mass spectrometry, however, mercury-vapor diffusion pumps are most suited. All diffusion pumps release small amounts of mercury or oil vapor at the high-vacuum side. To avoid penetration of

† Heraeus, Hanau, Germany; E. Leybold's Nachfolger, Cologne-Bayental, Germany.

this vapor into the vacuum system and to achieve pressures below the vapor pressure of the mercury at the temperature of the cooling water, the insertion of a cold trap or baffle between the pump and the system is necessary. The mercury or oil condenses, or is otherwise eliminated, and contamination of the vacuum system is avoided. However, this elimination is never complete.

Considerable progress has been made in reducing the back stream of oil. Now it seems that oil diffusion pumps can be used for mass spectrometers, provided that great care is taken in choosing the type of baffle. The baffle that is mounted directly above the pump should not be cooled for long periods below the stock point of the oil (about 17°C) because then the fluid charge of the pump would be collected at the baffle instead of returning to the boiler. A second baffle, at a much lower temperature, offers better possibilities for reducing the back stream.

The vapor pressure of ordinary diffusion pump oils at room temperature and below is lower than that of mercury. However, in a mass spectrometer, a little mercury penetrating into the source will not interfere, because its mass spectrum is confined to masses 196 to 204. On the other hand, oil will give an extended and strongly fluctuating spectrum, depending on the previous history of the oil and the amount of decomposition. Oil diffusion pumps are to be chosen for those instruments where medium sensitivity and accuracy suffice and a high pumping speed is of primary importance. Because of the low vapor pressure of the oil, a cold trap can sometimes be omitted or substituted by a moderately cooled baffle; water cooling is often sufficient. Without the inconvenience of liquid air or liquid nitrogen a sufficient reduction of oil penetration is possible, the action of the baffle being mainly a mechanical one, consisting largely of collecting small oil droplets. At this point mention should be made of the extensive studies which were performed at Balzers† to elucidate the mechanism of the oil back stream [53, 54]. It was found that the strongest molecular beam originates from the space near the rim of the upper stage cap. A second more diffuse and less intense source is caused by the scattering of the primary beam by collisions with other oil molecules or with residual gas molecules. This secondary beam is scattered again to give rise to a still weaker tertiary beam. A good baffle should screen the primary beam completely and the secondary beam as far as possible (Fig. 8.13). For helium mass spectrometers for leak detection, oil diffusion pumps are the obvious choice, as the hydrocarbon background will not interfere perceptibly at mass 4. A new and very interesting type of pump is the molecular pump, made by Pfeiffer‡ ([55]; Fig. 8.14). It introduces no oil or mercury at the high-vacuum side. Pumping speeds of about 150 liters/sec have been

† Gerätebau-Anstalt Balzers, Liechtenstein.
‡ Arthur Pfeiffer G.m.b.H., Wetzlar, Germany.

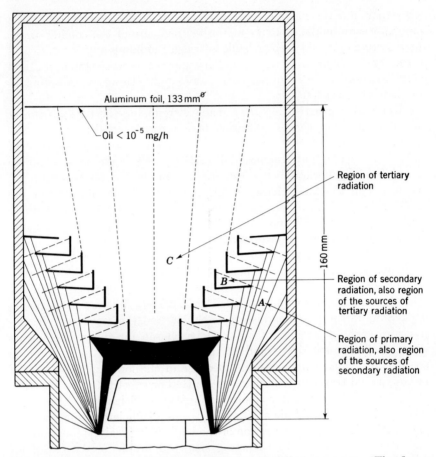

Fig. 8.13. Balzers' type D "Cern baffle" for oil diffusion pumps. The figure shows the arrangement in which the oil back stream of a small diffusion pump was measured. The strongest radiation originates near the rim of the upper jet cap; this is indicated as primary radiation. This primary radiation is scattered by collisions with residual gas molecules, yielding a much weaker secondary radiation, which is scattered again to result in a still much weaker tertiary radiation. This type of baffle is designed so as to eliminate both the primary and the secondary radiation completely. With water cooling only, the back stream is reduced by a factor of 10^5 as compared with the unbaffled pump. A high conductance is still obtained.

measured. The ultimate pressure is somewhat of the order of 10^{-10} mm Hg. Ion vacuum groups which do not require a mechanical pump fore-pressure for continuous operation have been recently made by several companies.[†,‡] Some of these have very high pumping speeds (1,900 liters/sec for nitrogen, 3,000 liters/sec for hydrogen). An essentially

† Consolidated Electrodynamics Corp. Rochester Division, Rochester, N.Y.
‡ Varian Associates, Palo Alto, Calif.

clean vacuum is also furnished by *getter-ion-pumps*.† Here gas is removed by chemical combination with a getter, usually a titanium surface, whereas chemically inert gases are shot into this layer by a discharge and buried by subsequent getter deposition. Unfortunately these pumps

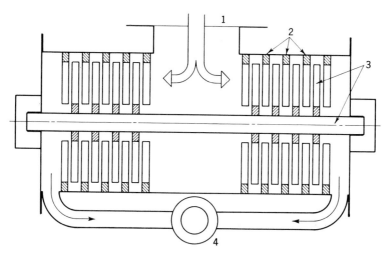

Fig. 8.14. Turbomolecular pump, made by Pfeiffer (schematic). 1, high-vacuum inlet; 2, rotor; 3, stator; 4, forevacuum connection. Both the rotor and the stator consist of a number of slotted disks. The slots are machined obliquely with a small angle to the plane of the disk. The obliqueness of the rotor slits is the reverse of that of the stator slits. If the rotor rotates at a high speed, the relative motion of the inclined surfaces acts as a molecular pump and produces a compression in the axial direction. Pressures as low as 10^{-10} mm Hg are readily obtainable. (*From Becker* [55].)

tend to release some of the previously pumped gas under the ion impact, giving rise to a memory effect in gas analysis.

8.6. Pumping Speeds

From the kinetic theory of gases it follows that, in changing the pressure of a gas, the mean free path of the molecules between two successive collisions changes inversely proportional to the pressure. To give an idea of the order of magnitude, there is a rule of thumb stating that, for nitrogen at room temperature, the mean free path at $\frac{1}{4}$ mm Hg pressure is about $\frac{1}{4}$ mm. The pressures commonly used in mass spectrometry are nearly always so low that the mean free path is long as

† Vac-Ion, Varian Ass., Palo Alto, Calif. Dri-Vac, Consolidated Vacuum Corporation, Rochester 3, N.Y. ULTEK Corp., 920 Commercial Street, Palo Alto, Calif. Hughes Vacuum Equipment, Vacuum Tubes Products Division, 2020 Short Street, Oceanside, Calif.

compared with the dimensions of the vessel. Under these circumstances, i.e., when molecular flow occurs, the resistance to gas flow of tubulations, etc., may be readily calculated.

The gas flow F (in liters per second) is given by the expression

$$F = C \frac{\delta P}{P} \tag{8.1}$$

P = pressure
δP = pressure difference between ends of the tube
C = a constant, the conductance of the tube

For series or parallel arrangements of a number of such elements, there is a relation analogous to electrical circuits. For a series arrangement the total conductance C is given by

$$\frac{1}{C} = \frac{1}{C_1} + \frac{1}{C_2} + \cdots \tag{8.2}$$

For a number of common cases the values of C will be given. These data are valid for air at 300°K under molecular-flow conditions. For other gases and temperatures they are to be multiplied by $0.31 \sqrt{T/M}$ (M = molecular weight).

Long tube of circular cross section $(R \ll L)$:

$$C = \frac{0.98R^3}{L} \cong \frac{R^3}{L}$$

where R = radius, mm
L = length, mm
C = conductance, liters/sec

Short tube of circular cross section (same symbols as above):

$$C = \frac{R^3}{2.7R + L}$$

Circular aperture of radius R (millimeters) terminating a tube of radius R_0 (millimeters) (C in liters per second):

$$C = 0.365 \frac{R_0^2}{R_0^2 - R^2} R^2 \tag{8.3}$$

For $R_0 \gg R$ this simplifies to

$$C = 0.365R^2$$

or
$$C = 0.116 \text{ liter sec}^{-1} \text{ mm}^{-2}$$

(Rule of thumb: conductance of an aperture $C \cong 10$ liters/sec^{-1} cm^{-2}.)

Bends.   The conductance of a bend is equal to that of a tube of the same diameter and length, measured along the center line.

Pumps.   These may be considered as conducting elements with zero pressure, or the ultimate pressure with zero gas flow, at the other side. A pump with a pumping speed S (liters per second) connected to the system by a tube of conductance C acts at the system with an effective pumping speed S' given by

$$\frac{1}{S'} = \frac{1}{S} + \frac{1}{C} \tag{8.4}$$

Cold traps.   If a system contains condensable vapors, a cold trap will act as a pump with a pumping speed given by

$$S = 0.116 \sqrt{\frac{29}{M}} \left(1 - \frac{P_2}{P_1}\right) A \tag{8.5}$$

where S = pumping speed, liters/sec
  M = molecular weight of condensable vapor
  P_1 = partial pressure of condensable vapor in system near trap
  P_2 = partial pressure of same vapor at cold-trap temperature
  A = effective area, mm^2, equal to actual area in case of unobstructed thimble and less than that in case of obstruction by a wall

The same expression gives the pumping speed of a charcoal trap, where P_2 is the equilibrium pressure of the adsorbed gas over the charcoal, this depending on activation of the charcoal, degree of saturation, and temperature. The above data are mentioned only to give an idea for some important cases. More detailed information, also for more complicated cases, will be found in the references [56–58]. One further detail may be mentioned here, namely, the "stopping power" of, for instance, mercury vapor in the tube between diffusion pump and cold trap; this point is treated on page 303.

The pumping speed of a vacuum system can be measured with an arrangement as in Fig. 8.15 (see also Refs. 59 and 60). To the pump a manometer and a variable leak are attached, preferably to a dome of a diameter at least equal to the pump aperture. The leak is a fixed constriction; the gas flow is adjusted by varying the pressure at the high-pressure side of the leak. This pressure is read from the height h of the mercury column, with a correction for the barometric pressure B. If the cross section a of the mercury column and the volume V at the high-pressure side of the leak (with $h = 0$) are known, the flow Q can be determined by measuring the time t required for the meniscus to rise from h_1 to h_2:

$$Q(p) = \frac{(V - ah_1)(B - h_1) - (V - ah_1)(B - h_2)}{t} \tag{8.6}$$

At pressures $p > 10$ mm Hg, the flow Q varies roughly proportional to $p^2 = (B - h)^2$. For low values of Q this method is inaccurate, and the flow is best determined by closing the spherical joint valve (Fig. 8.20)

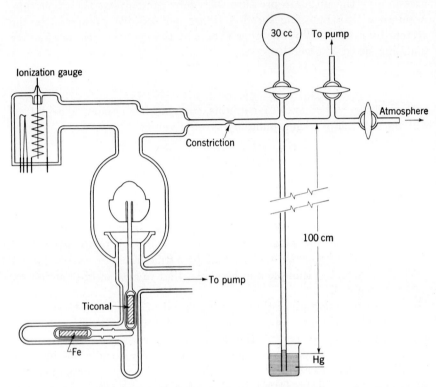

Fig. 8.15.   Arrangement for the introduction of an adjustable gas flow to a pump, for the measurement of pumping speeds.

and measuring the time required for the pressure in the gauge indicated to rise to a higher value (e.g., 10^{-3} mm Hg).

8.7. Cold Traps and Baffles

For a diffusion pump alone, there is a fundamental limit to the best vacuum that can be achieved; this can never be better than that corresponding to the vapor pressure of the working fluid at the temperature at the high-vacuum side.   Earlier mention was made of a better vacuum that can be achieved by insertion of a cold trap or (for oil diffusion pumps) at least a baffle.   The vapor of the working fluid is condensed to a lower saturation pressure, and penetration in the vacuum system is avoided. With oil diffusion pumps a substantial improvement can be achieved by a baffle, cooled only with tapwater or not cooled at all, where small oil droplets are gathered and "condensed" to larger ones instead of pene-trating into the system. Besides the inconvenience of a necessary refrigerant (solid CO_2, or better, liquid air or liquid nitrogen), all cold traps introduce the drawback of an extra (and sometimes quite con-siderable) resistance to the gas flow. The conductance diminishes proportional to $T^{+\frac{1}{2}}$. The spacing between the walls may be increased, but this will lower the efficiency.

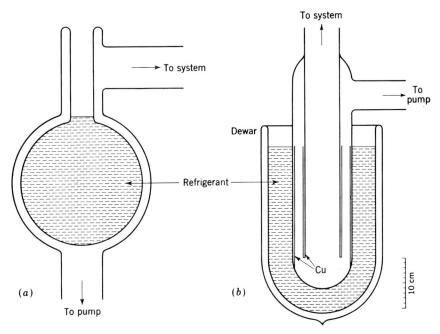

Fig. 8.16. Two different designs of cold traps for glass systems. (*a*) Cold trap for installation directly above the pump, suitable for mechanical refrigeration; (*b*) cold trap suitable for cooling with liquid air. The insertion of copper-foil cylinders results in a nearly constant condensing surface, irrespective of the refrigerant level, and improves the trapping efficiency for mercury. The copper should be activated by repeated oxidation and reduction (see text).

For glass systems two designs of cold traps are suitable (Fig. 8.16*a* and *b*). The first design offers the advantage that the trapped mercury, etc., is returned to the pump when the trap is warmed up. To recover the mercury distilled over from the design shown in Fig. 8.16*b*, the glass must be cut off. However, the second design is much more economical in the consumption of refrigerant; the Dewar flask can hold sufficient liquid air for several days of pumping. An elegant combination consists of a cold trap as in Fig. 8.16*a* directly above the pump, cooled to, for example, $-38°C$, by a Freon machine, in series with a trap as in Fig. 8.16*b*, cooled with liquid air, close to the system.

The trapping efficiency of mercury vapor on a cooled glass surface is much lower than unity, and a glass trap at −180°C will not condense mercury completely. On a copper surface, mercury amalgamates and the accommodation coefficient approaches unity. Insertion of copper-foil cylinders in a glass trap, as in Fig. 8.16b, results not only in a greatly improved trapping efficiency for mercury, but the good thermal conductivity of the copper results in a nearly constant condensing surface. The copper surface should be activated by repeated oxidation and reduction at 400 to 450°C, by admitting alternatively air and hydrogen gas at

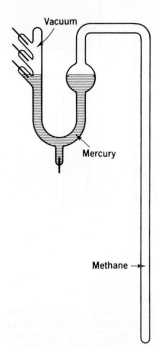

Fig. 8.17. Manometer relay for automatic refilling of cold traps with refrigerant. When the methane-filled limb is submerged in liquid air, the liquid-air level falls below the lower end of the methane-filled limb, and the methane reevaporates and pushes the mercury up in the left-hand limb, making an electrical contact. The methane may be prepared from ordinary coal gas by condensing a sufficient amount in the limb, pumping off the hydrogen, letting off the excess methane after warming up to room temperature, and sealing.

approximately atmospheric pressure. Copper at room temperature is ineffective for trapping mercury [61].

The insertion of copper-foil cylinders does not impair the conductance of the trap to an appreciable extent. This is not true for a variation on this theme, the "corrugated copper," introduced by Alpert [62]. Here a strip of OFHC copper sheet is corrugated by passing it between two gear wheels; the depth of the corrugations may be approximately 6 mm. The copper is activated by repeated oxidation and reduction as indicated above and coiled up to fill the trap (Figs. 8.18 and 8.19). The volume of the trap is thus divided into a large number of small tubes. The cooling surface and the trapping efficiency are increased enormously, but the conductance should be expected to decrease proportional to r/R, where r is the radius of a small tube, and R is the original radius of the trap volume. The original conductance can be restored by increasing the radius of the trap by the factor $(R/r)^{1/2}$ (see also Refs. 63 and 64). A copper-foil trap exhibits a chemical trapping action, notably for oil vapor, and a trap as in Fig. 8.18c may suffice to eliminate penetration of diffusion pump oil in a glass system, without the necessity of cooling the trap below room temperature.

When working with considerable amounts of condensable gases, a cold trap will exert a pumping action. Therefore it is advantageous to make

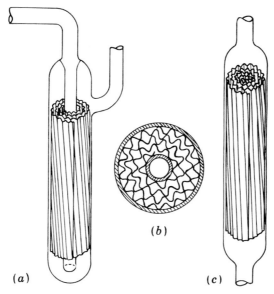

(b)

(a)   (c)

Fig. 8.18. "Corrugated-copper" traps. Copper sheet is corrugated by passing it between two gear wheels and coiled up to fill the annular space in a trap of the reentrant type (a and b). A straight, uncooled trap with corrugated copper (c) exerts a chemical trapping action on pump oil. (*From Alpert* [62].)

the connection between mass-spectrometer tube and cold trap wide and short. In calculating conductances, attention is directed to the "stopping action" of mercury vapor in the tube between diffusion pump and cold trap. In this section, the mean free path between two collisions with a mercury atom is only about 2 cm [65]. In a tube of 2 cm diameter collisions with mercury atoms in the gas phase are as frequent as collisions with the walls. On increasing the diameter, the mean free path remains essentially constant, and the conductance will thus increase proportional to R^2 instead of to R^3. Moreover, there is a steady flow of mercury from the pump to the cold trap, transferring momentum to the gas molecules that are to be pumped away. This corresponds to some kind of counter-pressure that will reduce the effective conductance at room temperature by the amount δC, where $\delta C/C = R/10$ (R = radius in centimeters). (See McLeod-gauge correction.)

Fig. 8.19. Corrugated-stainless-steel sheet for filling a cold trap. Instead of copper, stainless steel was used to make the trap resistant to mercury vapor. By coiling up a flat sheet together with a corrugated one, a more regular spacing of the small tubes is achieved. (*From Thomas, Destappes, and Dupont* [64].)

8.8. Pumping Procedure and Bake-out Provisions for Obtaining Very Low Pressures

For the attainment of pressures below 10^{-7} mm Hg in a glass system, degassing by high-temperature bake-out at, at least, 300°C is imperative. For vacua better than 10^{-9} mm Hg, a bake-out at 400 to 450°C is strongly advisable [66–68]. The whole system should be heated, including valves, flange connections, manometers, and preferably also one cold trap. Rubber gaskets can no longer be used, and all-metal valves should be chosen [67, 69–71]. In these valves a conical metal nonrotating nose (Kovar or Monel) is driven down into a soft metal seat (OFHC copper or silver). The malleable metal yields and fills small scratches and irregularities. The movable nose is sealed to the housing by means of a Kovar or Monel diaphragm. All joints are brazed in hydrogen atmosphere, without flux, with eutectic copper-silver or some other nonvolatile alloy. Considerable forces are required for closing such a valve, which are best provided by some differential screw mechanism. These valves have a continuously variable conductance of from 1 liter/sec to less than

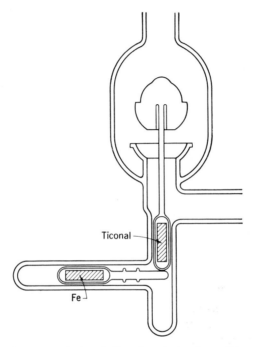

Fig. 8.20. Greaseless ground-glass ball-and-socket valve. The valve is shown in the open position. It is operated with two small magnets, one for lifting the ball and the other for moving the iron slug. In the closed position (ball in the socket), the conductance is very small if the pressure at both sides is in the molecular-flow region. The valve will not seal against one atmosphere. (*From Vogl and Evans* [73].)

10^{-10} liter/sec. Bills and Allen [71] report a minimum conductance of as little as 10^{-14} liter/sec; the same high standard is reached by Granville-Phillips valves, which are commercially available.† These valves are also excellent for dosing very small quantities of gases, and they are reported to remain satisfactory after extensive use.

In cases where the bake-out requirements are similar but a higher minimum conductance is permitted, greaseless ground-glass valves can be used [73]. A finely ground ball-and-socket joint, mounted as in Fig. 8.20, will provide a very small leakage rate if the pressure at both sides is in

† Granville-Phillips Co., Pullman, Washington, D.C.

the molecular-flow range. The device is thus suitable for isolating a system from the diffusion pump, for leak tests by the pressure-rise method. However, it has been demonstrated by Venema and Bandringa [74, 75] that extremely low pressures can be achieved with a mercury diffusion pump, provided that the system is carefully degassed and that the trapping of mercury vapor is very efficient. For this purpose, Venema installed several cold traps in series and reached vacua of the order of 10^{-10} mm Hg. Attention is also drawn to the results achieved with the molecular pump of Pfeiffer [55], where pressures of the order of 10^{-10} mm Hg were readily obtained.

Liquid metal cutoffs are also suitable, if the pressure difference between both sides of the cutoff does not exceed a few centimeters of mercury. Mercury is unsuitable because of its high vapor pressure; molten indium [76] or gallium [77] or In-Ga-Sn-eutectica [78] is satisfactory. Pure indium provides a leak-free seal because it "wets" glass, melts at 156.4°C, and has a vapor pressure of 10^{-7} mm Hg at 540°C, 10^{-6} mm Hg at 600°C. Gallium melts at 29.5°C and has a vapor pressure of 10^{-5} mm Hg at 711°C, 10^{-4} mm Hg at 859°C. The alloy of 62% Ga, 25% In, and 13% Sn melts at about 5°C and has a negligible vapor pressure.

On evacuating a system from atmospheric pressure, this is first done with the rotary pump alone; mercury diffusion pumps with sufficient conductance need not be short-circuited; for oil diffusion pumps or pumps which offer considerable obstruction to the gas flow, a bypass is advisable. When the pressure has fallen below 10^{-1} mm Hg, the diffusion pump is switched on, and the whole assembly, including one cold trap near the system, is heated to as high a temperature as is tolerable (300 to 450°C). The heating may be accomplished by placing the whole system under a hood [67] or by surrounding each element by a small furnace in which a temperature of 400 to 500°C is maintained. Insulating walls may be made of asbestos or Uralite. The power required is about 35 to 40 watts/dm² [66].

The heating is maintained for at least three hours, with the diffusion pump functioning. The cold trap is then filled with refrigerant, and filaments, etc., are heated for degassing. Thorough degassing of metals may require a much higher temperature than needed for glass. Metal can evolve tremendous quantities of gas. According to Dushman [79], tungsten may give off ten times its own volume of gas NTP. Besides filaments, also other electrodes that are exposed to either heat or to electron or ion impact are to be degassed [67]. After cooling down the cold trap, it is still advisable to keep large sections of the system (for instance, the mass-spectrometer tube) at an elevated temperature. This will reduce gas adsorption and memory effects when changing samples. About 200°C is sufficient. Some authors [2] obtain excellent vacua by a sequence of alternating high and low temperatures.

8.9. Manometers

The simple U-tube manometer has a lower limit of sensitivity which makes it unsuitable for pressures below about 0.5 mm Hg with mercury filling; with oil one can measure somewhat lower pressures. This pressure range is met with only in gas-handling systems, etc. The U tube is quite suitable for measuring sample reservoir pressures in systems with a viscous leak, provided the mercury vapor is not objectionable. In many systems the pressure of the sample is adjusted by raising or lowering a mercury level, and then mercury vapors are present already. Only a small model is needed, as the working pressures rarely exceed 150 mm Hg. A convenient method for making these things is

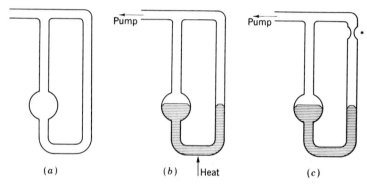

(a)   (b) ↑Heat   (c)

Fig. 8.21. A convenient design for small U-tube manometers. A glass tube as shown in (a) is filled with mercury and heated under vacuum to expel dissolved gases (b) and finally sealed under vacuum at * (c).

shown in Fig. 8.21. A tube as shown in Fig. 8.21a is filled with mercury to the appropriate level; then it is evacuated and heated cautiously under vacuum to expel all dissolved gases (Fig. 8.21b); and finally the mercury is boiled under vacuum and the tube sealed off as in Fig. 8.21c. This method of sealing contributes greatly to the mechanical strength of the device.

A delicate, but otherwise ideal, instrument for the same pressure range and below is the diaphragm manometer [80–85]. The pressure difference between the two sides of a thin flexible membrane causes a small displacement, and this displacement is detected electronically. Opposite the membrane an electrode is mounted, and the capacity between the two is measured and taken as an indication of pressure or brought back to some original value by displacing the electrode [80] or by applying a polarization voltage to the electrode that pulls the diaphragm back [81, 82]. Diaphragm manometers can be baked out, together with the rest of the system, and are independent of the composition of the gas and its saturation pressure. They can be adapted for measurement of lower pressures, down to 10^{-5} mm Hg [81, 82], and may then be used for the measurement of sample inlet pressures in molecular-leak inlet systems, and even for measuring the ionization-chamber pressure.

However, for lower pressures, the principal gauge is the McLeod gauge. Its principle is well known: a comparatively large volume V of gas at the pressure to be measured is compressed into the smaller volume of a capillary (cross section a). The pressure of the compressed gas is measured with essentially an ordinary U-tube manometer or a Thermistor manometer [86]. To balance capillary depression, both limbs should have the same internal diameter. If the mercury in the "open" limb is raised to the level of the sealed end of the other limb, the pressure P of the system to be measured is given by

$$P = \frac{a(\Delta H)^2}{V} \tag{8.7}$$

where P = pressure, mm
$\quad\ \ a$ = cross section of capillary, mm^2
$\quad \Delta H$ = difference in height of menisci, mm
$\quad\ \ V$ = volume, mm^3

The raising of the mercury can best be accomplished by the atmospheric pressure and the lowering by a rough vacuum (a few centimeters of mercury). The variations in barometric pressure can be very nearly eliminated by an appropriate design (Fig. 8.22).

Although the McLeod gauge offers the considerable advantage of indicating pressures independent of the nature of the gas, condensable vapors are not registered, e.g., water vapor. The vapor pressure of its own mercury is not indicated. Insertion of a cold trap in the connection to the system to be measured is necessary. The lower limit is about 10^{-6} mm Hg. Therefore, and because of the discontinuity of the measurement, the McLeod gauge is mainly used for calibration of other continuously indicating gauges, which may extend the measurable pressure range to much lower pressures.

Another standard gauge for the range of 10^{-4} to 10^{-6} mm Hg is the Knudsen gauge. It measures absolute gas pressures, independent of the ease of condensation of the gas and of its nature. It depends essentially on the transfer of momentum by the gas molecules from a heated surface to a movable vane. It does not release vapors, and it exhibits no pumping action. Knudsen gauges are commercially available.†

Even for permanent gases a McLeod gauge indicates a pressure that is systematically too low; the error is small, however. It is caused by the steady flow of mercury atoms from the gauge to the cold trap, which transfers momentum to the molecules of the permanent gas. The effect

† Edwards High Vacuum Ltd., Crawley, Sussex, England.

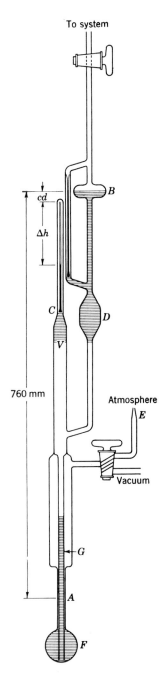

Fig. 8.22. Self-adjusting McLeod gauge. As shown, the mercury is raised; the volume V at a pressure P has been compressed to a volume $ax = h$ at a pressure h. cd = capillary depression. The influence of variations in barometric pressure is reduced by the small ratio of the mercury surface areas at A and B. When raising the mercury, breakage may occur at C if the mercury arrives at the entrance of the capillary at a tool high speed. Therefore the narrowing at C is made gradual, and the larger diameter at D serves to reduce the velocity of the mercury at C. The air-admitting point at E should be drawn out to a convenient size. The bulb at F serves to reduce the velocity of the mercury before it streams upward; thus air bubbles are not swept along into V. The volume G is not critical; it should hold the mercury from V, D, and B.

is independent of length of the tube between the gauge and cold trap, and for small diameters is proportional to the radius. At room temperature the error is given by [87] $\delta P/P = R/10$ (R in centimeters).

For practical purposes other types of gauges are used. The choice from a wide variety of working principles is chiefly governed by the pressure range to be measured (Fig. 8.23). A survey of instruments for pressures down to about 10^{-4} mm Hg is given by Roberts and O'Hara [88] and by Lawrence [89]. An almost complete survey of low-pressure gauges is that compiled by Schwarz [90]. For mass spectrometry only a few types are in general use.

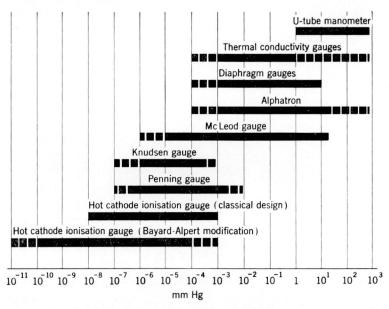

Fig. 8.23. Ranges of different vacuum gauges.

For pressures from about 20 mm Hg down to about 10^{-4} mm Hg, a number of gauge types exist, based on the pressure-dependent heat conductivity of the gas, e.g., Pirani gauges and modifications. A small filament, or a thermistor or an NTC resistor, is heated, and either the change in resistance or the change in current needed to keep the temperature constant is taken as a measure of pressure. Instead of resistance measurements for temperature indication, thermocouple devices may also be used. Besides the general reviews [88–90], special articles on the Pirani gauge have been published by Glockler and Horst [91], Leck [92], Leck and Martin [93], Von Ubisch [94], and Seiden [95]. Sensitivity increases with wire temperature. Leck [92] describes a Pirani gauge with a quartz-coated wire that can be run at 1050°K nearly free from drift. Cooling the walls will also increase the sensitivity; immersion in liquid air will render it about three times more sensitive. All thermal-conductivity devices are somewhat sensitive to fluctuations in the wall temperatures. These effects may be compensated by mounting an identical element in the opposite arm of a Wheatstone bridge. The second element is mounted in an evacuated and sealed tube, maintained at the same temperature as the pressure-indicating one, by placing both in a metal block.

Lower pressures are usually measured with ionization gauges. The gas is ionized, and the ion current, which is proportional to the pressure, is measured. The ionization may be caused by alpha particles (alphatron) or by electrons (thermionic and cold-cathode ionization gauge). The alphatron is very solid and particularly suited for industrial use; in

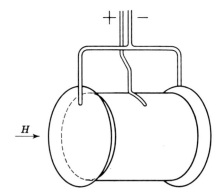

Fig. 8.24. Penning gauge (cold-cathode ionization gauge, Philips gauge) in the most sensitive modification. H = direction of the magnetic field.

mass spectrometry chiefly cold cathode (Penning) ionization gauges (for pressures from 10^{-3} to 10^{-7} mm Hg) and thermionic ionization gauges (for pressures from 10^{-3} to 10^{-8} mm Hg or 10^{-10} mm Hg, depending on the design) are chosen.

The cold-cathode ionization gauge (Penning gauge, Philips gauge), in the most sensitive modification, consists of a cylindrical anode, at both ends closed by disk-shaped cathodes (Fig. 8.24). A magnetic field is applied along the cylinder axis. A voltage of a few kilovolts is applied to the electrodes. Electrons, originating from cosmic rays or from field emission at a sharp point at one of the electrodes, are accelerated toward the anode, but their paths are curled up by the magnetic field, and the field shape at the cylinder ends forces them to oscillate parallel to the axis. Along this extended path the gas is ionized, and the ions are accelerated parallel to the magnetic field and withdrawn by the cathodes. The ion current is proportional to the pressure, but depends on the nature of the gas. In mass spectrometry pressures often occur which fall below the range of the Penning gauge, but still this gauge offers outstanding advantages in its range. Its construction, as well as electronic circuitry, is simple, and the device is virtually indestructible. There is no filament, and the gauge is not damaged by a sudden inrush of air. The Penning gauge is therefore particularly suitable for safety devices which will switch off thermionic ionization gauge and ion-source filaments and diffusion-pump heaters in case of pressure rise above a safe level. Extensive investigations on the Penning gauge were carried out by Conn and Daglish: a general survey [96], the influence of electrode geometry [97], the influence of the ballast resistance [98].

Conn and Daglish also describe a modification [99] which is essentially a thermionic gauge derived from the Penning gauge. The electrode geometry is similar to that of the Penning gauge, but a filament is mounted along the cylinder axis. The magnetic field is retained. This device covers about the same range as the original Penning gauge, but because of the much higher number of ionizing electrons, the sensitivity is increased by a factor of about 100. The ion current is about 1 μa at a pressure of 2×10^{-7} mm Hg. The arrangement is much more complicated than the common Penning gauge, and the vulnerability is comparable with that of an "ordinary" thermionic ionization gauge, which usually is to be preferred because of its more extended range.

The electrodes of a Penning gauge can evolve considerable amounts of gas on electron or ion impact and should therefore be degassed for working pressures below about 10^{-4} mm Hg. When the permanent magnet is removed, the glass envelope can be degassed by ordinary furnace bakeout, together with the rest of the apparatus. For the electrodes, much higher temperatures are necessary. After cooling down the walls, this may be accomplished by r-f induction heating; a much cheaper alternative, resistance heating, is proposed by Bobenrieth [100].

When a Penning gauge is to be used only as a rather rough indication of pressure (for instance, for safety devices), it will function satisfactorily with a voltage of 2 to 4 kv and a magnetic field of about 500 gauss. If accurate pressure measurements are desired, Phillips [101] demonstrates that a magnetic field of about 1,600 gauss and a voltage of 10 kv are to be preferred (Fig. 8.25). The particular values are valid for the arrangement of Phillips with the dimensions ring anode 15 mm diameter and cathode spacing 15 mm, but the trend should be general. In fact, this had been observed by Guthrie and Wakerling [102].

The thermionic ionization gauge is the most suitable gauge for pressures between 10^{-3} and 10^{-8} mm Hg or 10^{-3} and 10^{-10} to 10^{-11} mm Hg, depending on the type. It is believed to be the most accurate direct-reading instrument for this range. In the "classical" arrangement, the gauge consists of a central filament, serving as a hot cathode, surrounded by a positive grid (electron collector) and a negative plate (ion collector) (Fig. 8.26a). The electrons, emitted from the cathode, are accelerated toward the grid and ionize the gas molecules. The ion current is proportional to the pressure and to the electron current; the proportionality constant depends on the nature of the gas. The calibration was done for different gases by Dushman and Young [103], Wagener and Johnson [104], and Nöller [105]; the values relative to nitrogen = 1 are given in Table 8.2 (see also Vick [106]).

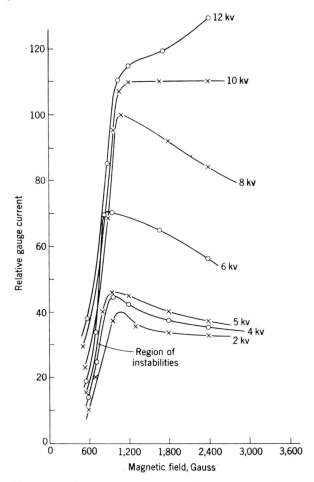

Fig. 8.25. Characteristics of a Penning gauge (Philips 101). The gauge current is plotted as a function of the magnetic field strength for different voltages. The pressure was 1×10^{-6} mm Hg. To avoid instabilities in the discharge, a magnetic field of at least 1,000 gauss is desirable. Operating at 1,600 gauss and 10 kv offers the advantages of a considerably increased gauge current and insensitivity for small changes in the magnetic field strength.

The values given in the table are influenced by the pumping action of the gauge and thus depend somewhat on the conductance of the tubulation, on the degassing condition of the electrodes, and on the electron current.

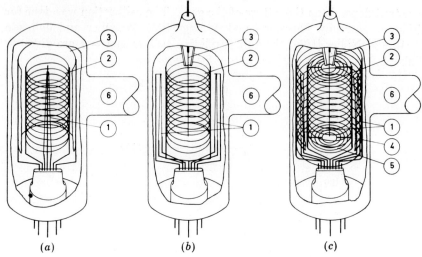

Fig. 8.26. Different types of thermionic-ionization gauges. (*a*) "Classical" design with axial filament and surrounding plate ion collector, for the pressure range of 10^{-3} to 10^{-8} mm Hg. (*b*) Bayard-Alpert modification, with thin central ion-collector wire, for the pressure range of a few times 10^{-4} to 10^{-10} or 10^{-11} mm Hg, depending on the diameter of the ion-collector wire. As shown, there is a separate filament, which can also be used for the flash-filament technique (p. 316). (*c*) The modification due to Nottingham. In addition to (*b*), there is a screening grid surrounding the whole arrangement. There are gridlike covers on the anode. Instead of a screening grid, a conductive coating may be applied at the inside of the glass envelope. Pressure range: from a few times 10^{-3} to 10^{-10} or 10^{-11} mm Hg. 1, filament; 2, anode = electron collector; 3, ion collector; 4, screening grid; 5, gridlike covers; 6, connection to vacuum system.

TABLE 8.2.   Thermionic-ionization-gauge Sensitivity, Relative to That for Nitrogen

Gas	Relative sensitivity	References
Hydrogen..............	0.47	103
	0.53	104
Helium................	0.16	103
Neon..................	0.24	103
Nitrogen..............	1.00	103, 104
Argon.................	1.19	103
Carbon monoxide........	1.07	104
Carbon dioxide..........	1.37	104
Water vapor............	0.89	104
Oxygen................	0.85	104
Krypton...............	1.9	103
Xenon.................	2.7	103
Mercury vapor	3.4	103
Narcoil-40 vapor........	13.0	105

The pressure is calculated from the ratio of ion current to electron current. To make a measurement, the electron current is adjusted to a constant value, either manually or by an electronic stabilization, which should meet high standards; or one may use a circuit that measures the desired ratio directly. Examples of such circuits are the simple logarithmic differential amplifier of Hariharan and Bhalla [107], the arrangement of Osborne [108], and the more complicated design of Schutten [109]. The gauge is calibrated, for example, for nitrogen or argon, by comparing the indication at a pressure of about 10^{-4} mm Hg with that of a McLeod gauge. For lower pressures the reading is taken to be proportional to the pressure. Once such a calibration has been done, it will be valid in a good approximation for all gauge tubes of the same type.

When the pressure falls below 10^{-8} mm Hg, the classical ionization gauge (Fig. 8.26a) gives an ion current that will not decrease any further; this constant value corresponds to a pressure of about 10^{-8} mm Hg. This current originates no longer from ions, but rather from electrons that are emitted from the ion collector by soft X rays. These X rays originate from the electrons hitting the grid. This effect is also proportional to the electron current and adds to the ion current. To get round this limitation of the range, Bayard and Alpert [110] inverted the arrangement of the electrodes. Now a very thin wire is mounted along the axis of the cylindrical grid and serves as an ion collector; the filament is mounted outside the grid (Fig. 8.26b). The solid angle for X-ray capture of this thin-wire ion collector is about 1,000 times smaller than in the classical arrangement of Fig. 8.26a; a corresponding decrease of the minimum measurable pressure results. The sensitivity (microamperes ion current per milliamperes electron current per millimeter hydrogen) is comparable with that of the older design.

A further modification of the design was given by Aisenberg et al. [111], Nottingham [112] (see also Ref. 113), and Redhead and McNarry [114]. The cylindrical grid is closed here with gridlike covers, and the whole arrangement is surrounded by a screening grid maintained at a negative potential (Fig. 8.24c). The supporting wire of the ion collector is screened by a glass tube, not in contact with it, to avoid photoelectron emission from this thicker wire. The aim of the screening grid is to force the ionizing electrons back in the collection space within the electron collector. This results in a further linear extension of the response at higher pressures and in an increased sensitivity over the whole range. The gridlike covers serve to prevent the ions from escaping the collection volume at lower pressures. When an ion, at its point of origin, has a tangential velocity component, from the thermal energy, it will describe something like an elliptical path under the influence of the radial electrical field. If the ion originates at 10 mm from a wire collector of 0.075 mm diameter, with an initial tangential velocity corresponding to 0.03 ev, it follows from the second theorem of Kepler that its path will reach the collector surface directly only if its total acceleration exceeds about 500 volts. As such high voltages are rarely applied, it is evident that part of the ions will turn around the collector until a collision alters their angular momentum. These collisions become more rare as the pressure is lowered, and any axial-velocity component of the ion will thus permit it to escape from the collection volume. Because of this incomplete collection, the ratio of ion current to electron current will fall off more rapidly than the pressure when the latter is lowered. The gridlike covers, electrically connected to the electron collector, should keep the collection complete, even at very low pressure.

To diminish the X-ray effect, the ion-collector wire is made as thin as possible, and the heavier mechanical supporting wire is screened by a small glass tube. A further improvement can be achieved by deliberately choosing the optimum voltages to be applied to the electrodes. The ionization efficiency depends on the electron energy, and the maximum sensitivity of the gauge is obtained with the grid about 100 volts positive in respect to the filament. The spurious X-ray current is also strongly dependent on the electron energy; when lowering the filament-grid voltage from 100 to about 45 volts, this spurious photoelectron current decreases further than the ion current. To extend the pressure range as far as possible, the gauge should be operated at about 45 ev energy, and the lower sensitivity compensated otherwise.

Flash-filament technique. Even without an extremely refined ionization gauge, pressure measurements are possible at pressures considerably below the lower limit of the instrument's range. This can be done by using a property of tungsten to adsorb gases to form a monomolecular layer and to give off that layer quantitatively on heating to about 2500°K in a time of the order of 1 sec [112, 115]. After cooling down to room temperature, the layer is formed again, but this will take some time. The time constant for nitrogen on tungsten, at a pressure of 10^{-8} mm Hg, is about 100 sec. With a flash-filament surface area of 0.4 cm² in a gauge volume of 500 cm³, the flashing of a complete monolayer will correspond to a pressure peak of about 10^{-5} mm Hg. If one flashes such a filament once every 100 sec at a pressure of, for instance, 10^{-10} mm Hg, the pressure peaks are still 10^{-7} mm Hg, and this may be recorded even with a "classical" ionization gauge. For lower pressures one may also increase the time interval; the accuracy attainable decreases somewhat, but there still remains some indication of pressures far below the lower limit of the instrument range. In many thermionic-ionization-gauge tubes, a second, or spare, filament is already installed, electrically insulated from the first filament. To suppress electron emission from the flash filament, it should be connected to the most positive electrode, i.e., the grid.

The electrodes of a thermionic ionization gauge should be degassed at a considerably higher temperature than the glass envelope. The method of Alpert [67] consists of connecting grid and ion collector to one side of a 750-volt a-c source, the filament to the other side; the filament current is adjusted so as to heat the electrodes by electron impact to about 1200°C. A dissipation of 25 to 75 watts is necessary.

A hot tungsten filament will not stand oxygen pressures higher than about 10^{-3} mm Hg. Therefore a safety device is desirable to switch off the filament in case of pressure rise, for instance, by means of a Penning gauge. VEECO† and Edwards‡ bring models on the market with, respectively, a thoria-coated iridium filament or a barium-strontium-coated platinum alloy filament that will stand air at atmospheric pressure without damage to later readings. The VEECO gauge has a double-ended grid that can be degassed by resistance heating and heats the other electrodes by radiation. In addition, tungsten will not stand small concentrations of ammonia, because of the formation of nitrides that drastically reduce the filament life. For this case, Hall [116] reports favorable results with rhenium cathodes.

It should be realized that, although all gauges measure some property of the gas, which depends on the pressure, only a few respond to the pressure directly. Of the types mentioned in this section, only the U-tube manometer, the diaphragm manometer, and the McLeod gauge indicate pressure; the McLeod indicates only the pressure of the permanent gases in the large volume V. However, thermal-conductivity gauges respond to mean free path and molecular concentration rather than to pressure; ionization gauges respond to molecular concentration. When a fixed quantity of gas is heated without change in volume, its pressure will increase proportional to T, but the molecular concentration remains constant, and the reading of, for example, an ionization gauge, will essentially not increase proportional to the pressure.

However, one should bear in mind that what usually counts in vacuum technique and mass spectrometry is not pressure itself. For collision-free passage of an ion beam, the mean free path is the important thing; for the background from residual gas, the molecular concentration and the rate of penetration into the ionization chamber are determining factors.

All ionization gauges exhibit a pumping action. The ions impinging on the collector are removed nearly completely from the system, and, in addition, a hot filament may dissociate molecules to chemically active fragments which are trapped chemically by the walls. Alpert [67] used this effect to obtain pressures below 10^{-8} mm Hg. Supposing a complete removal of all collected ions, Alpert calculates, for a gauge sensitivity of 10^{-n-1} amp ion current at 10^{-n} mm Hg and an electron current of 10 ma, a pumping speed of 0.02 liter/sec. However, Young [117] demonstrates

† VEECO Vacuum-Electronics Engineering Corp., New Hyde Park, N.Y.
‡ Edwards High Vacuum Ltd., Crawley, Sussex, England.

that the ion current to the glass wall can be five to ten times that to the collector and gives a pumping speed of 0.1 to 0.2 liter/sec. The chemical trapping depends on the filament surface area and its temperature and may amount to 0.2 liter/sec. A connection between gauge and system of 1 cm diameter and 25 cm long has a conductance of about 0.5 liter/sec; the ion pumping alone will produce errors of 20 to 40%, the readings being generally too low. To reduce these errors, the connection must be made wide and short. Reducing the electron emission will reduce the ion pumping speed proportionally, whereas the chemical pumping speed remains essentially unaltered. An analogous effect is also exhibited by Penning gauges; as the ion current is about ten times that in a hot-cathode ionization gauge at 10 ma emission, a pumping speed of about 0.2 liter/sec may be expected. The capacity of electrodes to hold the removed material can be considerable, especially at low pressures. A well-degassed surface will remove material by ion pumping up to a few monolayers. At a pressure of 10^{-9} mm Hg, a gauge can be expected to pump for a time of the order of a year without saturation.

8.10. Sample Introduction Systems

For manipulation of gases at relatively high pressures, say, 1 mm Hg or higher, greased glass stopcocks may be used. Even low-vapor-pressure stopcock greases, however, will release dissolved gas and solve organic vapors, causing annoying memory effects. Greased stopcocks cannot be baked out and are unsuitable to meet high demands of gas purity; all-metal valves of the Alpert type will meet the severest requirements. For somewhat less rigorous demands there are several types of valves, suitable for gas-handling systems, made by Hoke.† These valves are much cheaper and easier to manipulate than Alpert valves. They should be ordered mass-spectrometer leak-tested.

For the analysis of liquids, a heatable inlet system is desirable for obtaining a convenient vapor pressure of the sample. The elevated temperature will assist the removal of the sample after analysis and reduce pump-out times, but for some strongly adherent compounds merely baking out is not sufficient to expel the compounds completely and to avoid interference and memory effect with succeeding similar samples. Hageman and Van Katwijk [118] investigated oxygenated organic compounds and found it necessary to flush the inlet system repeatedly with a sample to drive off the previous sample from the walls.

In most ion sources for gaseous samples, the optimum pressure in the ionization chamber is about 10^{-4} mm Hg. Between the sample container and the ionization chamber some type of constriction is necessary to make possible the use of higher sample pressures. Two types of

† Hoke Inc., Cresskill, N.J.

constriction are commonly used [119]. The first type is the *molecular leak*. At the entrance side of the leak the sample pressure is here so low that the mean free path of the molecules is large as compared with the constriction diameter. If the sample consists of a mixture of several components, then the flow of each component is determined by its mass. Two components *a* and *b* pass the leak in quantities that are related by the expression

$$\frac{N_a}{N_b} = \sqrt{\frac{M_b}{M_a}} \frac{p_a}{p_b} \tag{8.8}$$

(N_j = number of molecules passing; M_j = molecular weight; p_j = partial pressure.)

However, the conditions for molecular flow are satisfied also in the ionization chamber, and hence the flow to the pump will be governed by the same expression. In the stationary state the ratio of the partial pressures in the ionization chamber, p_{ai}/p_{bi}, is equal to that in the sample, p_a/p_b. The components, however, are not withdrawn from the supply in the same proportion as they are present in the sample. When the sample quantity is finite, its composition will change with time; the heavier component will be enriched in respect to the lighter one. Because molecular flow requires a relatively low sample pressure, usually about 0.15 mm Hg, rather large storage bulbs are necessary for accurate abundance determinations and should have a volume of at least a few liters; a small volume at a higher pressure is usually expanded into it.

The second type of leak is the *viscous leak*. It is operated at higher inlet pressures, and the mean free path is small as compared with the constriction dimensions. Now the flow is governed by the laws of viscous flow. The gas mixture moves as a whole, and the flow depends on the viscosity of the mixture as a whole. However, because of the pressure drop in the constriction, there is a region where viscous flow blends into molecular flow. In the flow equation two separate terms represent both types of flow. Again, the molecular flow discriminates proportional to $M^{-\frac{1}{2}}$. The relative importance of the discriminating molecular-flow contribution as compared with the nondiscriminating viscous-flow term depends on the geometry and the pressure. The total fractionation depends on a number of conditions in a not readily calculable manner, and these conditions can be used in practice only if a capillary connection is inserted between sample supply and constriction. Because of the discrimination in the constriction, the composition at its high-pressure side will change. By choosing a long and narrow capillary connection, the influence by back diffusion of this composition change on the sample supply can be made very small. All components are then withdrawn from the supply in their original proportions, and consequently high-precision abundance-ratio determinations are possible with

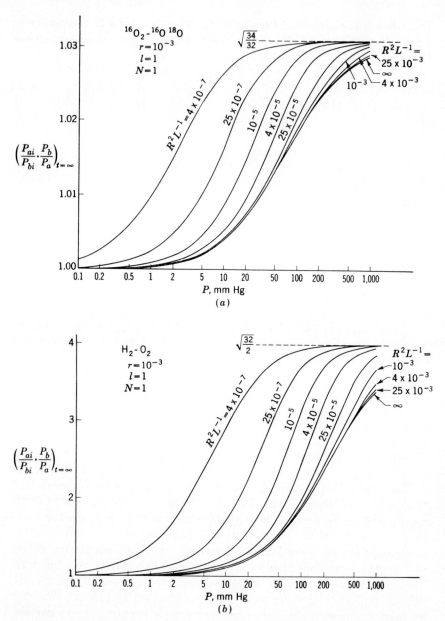

Fig. 8.27. Fractionation with a viscous leak, as a function of sample pressure. The curves represent the total fractionation in the stationary state, from sample reservoir to ionization chamber, with R^2/L of the supply capillary as a parameter (in centimeters). The constriction is supposed to consist of one tube of 10^{-3} cm radius, 1 cm long. (a) 1:1 mixture of $^{16}O_2$-$^{16}O^{18}O$; (b) 1:1 mixture of H_2-O_2.

much smaller sample containers. The flow from ionization chamber to pump is molecular, and in the stationary state, supposing a very narrow and long capillary connection, there is the relation

$$\frac{p_{ai}}{p_{bi}} = \sqrt{\frac{M_a}{M_b}}\frac{p_a}{p_b} \tag{8.9}$$

Quantitative calculations on the viscous leak have been published by Halsted and Nier [120], by Kistemaker [121], and by Boerboom and Tasman [122].

The conditions for satisfactory dimensions of the supply capillary may be estimated from the following considerations. The back diffusion through the capillary is negligible if

$$\frac{R^2}{L} \ll \frac{Q}{D}\frac{1+\delta}{1-\delta} \tag{8.10}$$

where R = radius of capillary
L = capillary length
Q = gas flow at sample pressure
D = diffusion constant at sample pressure
δ = fractionation just in front of the leak, compared with the sample composition, defined by

$$\delta = \left(\frac{p_a}{p_b}\right)_1 \left(\frac{p_b}{p_a}\right)_2 \tag{8.11}$$

where the subscript 1 represents pressures just in front of the constriction, and the subscript 2, pressures in the sample.

If the condition (8.10) is fulfilled, the total fractionation from the sample reservoir to the ionization chamber is given by (8.9). In Fig. 8.27a and b, $(p_{ai}/p_{bi})(p_b/p_a)$ is plotted against the sample pressure, for the cases $^{16}O_2$-$^{16}O^{18}O$ (1:1) and H_2-$^{16}O_2$ (1:1). Curves are drawn for different values of R^2/L. The radius of the constriction was assumed to be 10^{-3} cm, its length 1 cm. The gas flow at $p = 100$ mm Hg corresponds to about 0.2 liter/sec at 10^{-4} mm Hg.

The time necessary to obtain the stationary state is strongly dependent on the supply capillary dimensions [121, 122]. The time in which the discrepancy with the stationary state at $t = \infty$ is fallen down to 10% of its original value is given in Fig. 8.28a for $^{16}O_2$-$^{16}O^{18}O$ (1:1) and in Fig. 8.28b for H_2-$^{16}O_2$ (1:1). Curves are drawn for $L = 10$ cm and for different values of R. The constriction is the same as assumed in Fig. 8.27a and b. The time at which the discussed discrepancy has decreased to 1% is double the "10% time"; the "0.1% time" is three times the "10% time," and so on.

For isotopic-abundance-ratio determinations the viscous leak is particularly suitable. Relatively small sample quantities and containers

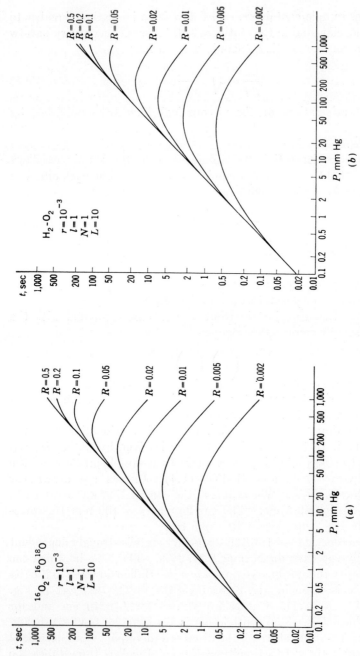

Fig. 8.28. Time taken for the discrepancy with the stationary state to diminish to 10% of its original value, as a function of sample pressure. The curves refer to the same viscous leak as was assumed in Fig. 8.27. The length L of the supply capillary was taken to be 10 cm; its radius R is used as a parameter. (*a*) 1:1 mixture of $^{16}O_2$-$^{16}O^{18}O$; (*b*) 1:1 mixture of H_2-O_2. (*From Boerboom and Tasman* [122].)

suffice; the sample can be at a rather high pressure; while any fractionation can be eliminated by comparison with a standard sample. The sample pressure for the viscous leak can be relatively high, e.g., about 100 mm Hg, and thus this type of leak is suitable for continuous sampling, for instance, from a process stream. However, the total gas flow through a viscous leak depends, among other things, on the viscosity of the mixture, and this viscosity depends on the composition. The peak heights are not linear functions of the partial pressures in the sample. Inghram [123] reported a 50% change in the helium peak height if 1% benzene ($M = 78$) was added to helium ($M = 4$), using a viscous leak. The change in viscosity, however, will not suffice to explain this effect. This property makes the viscous leak not very suitable for following the course of a chemical reaction. This is especially the case when the sample is at a low pressure. Under these conditions, the molecular leak is more appropriate; the discrimination is known, the peak heights are linearly dependent on the partial pressures in the sample, and the change in sample composition is accepted.

For the *construction of a molecular leak* the dimensions should be small as compared with the mean free path. At a sample pressure of 0.15 mm Hg, the mean free path of CO_2 is about 0.2 mm; for most other gases it is somewhat higher. The leak dimensions should certainly not exceed 0.02 mm, and for pure molecular flow even a hole four times smaller, or 0.005 mm in diameter, is advisable. The flow through a 0.02-mm orifice in a thin membrane amounts to about 4.5×10^{-2} liter/sec at 10^{-4} mm Hg. This is somewhat too small for most ion sources, and therefore, in a thin foil, e.g., gold foil, a number of perforations are arranged in parallel. One may also insert a sintered-glass disk or a plug of unglazed earthenware, which contains a large number of small capillaries. A quartz thimble [124] or platinum plug [125] may be used. The drilling of these very small holes requires a technique in itself and may be accomplished by a conical pin or a spinning needle [126, 127].

For the *construction of a viscous leak* the simplest method is to constrict the long narrow connection capillary between sample supply and ionization chamber, not too far from the latter. The amount of constriction should be chosen so as to obtain the desired gas flow at a convenient sample pressure of, for instance, 100 mm Hg. Nickel capillary of 0.1 mm internal diameter† may be compressed somewhat reversibly; when using copper capillary, care should be taken not to compress further than desired, because the elastic recovery after removal of the constricting force is very small. Nickel should not be used for the analysis of H_2-HD-D_2 samples, as it exerts a catalytic action on the establishment of the equilibrium

$$H_2 + D_2 \rightleftharpoons 2HD$$

† Manufactured by Accles and Pollock Ltd., Birmingham, England.

and the three species can thus no longer be determined independently of each other. For this case a stainless-steel capillary must be used. All leaks of the compressed capillary type should be set once, but are not easily adjustable.

Adjustable leaks can be of the needle-valve type. A number of designs of these rather complicated constructions have been published [119, 128]. Simpler to construct are thermal-expansion leaks, where the flow is

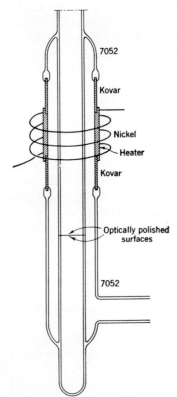

Fig. 8.29. Thermal-expansion leak. Two optically polished glass surfaces are driven apart by heating a nickel cylinder. (*From Forman* [130].)

adjusted by a difference in thermal expansion of two parts. Harrison [129] gives a survey of a number of such devices; there the throughput between two ground-glass surfaces is varied by the different expansion of two different kinds of glass. Forman [130] uses a nickel cylinder, which is heated and drives apart two optically polished glass-tube ends (Fig. 8.29). Perhaps the simplest design is that of Nester ([131]; Fig. 8.30). A gold plug is inserted into a Nonex capillary. The glass and gold are then drawn out together to a fine point, assuring a perfect fit. On cooling, the gold shrinks more than the glass and the leak opens; on heating, it closes again. Excellent reproducibility is claimed; remote control is possible.

Smither [132] makes use of the temperature dependence of the viscosity of a gas. By heating a narrow Pyrex capillary, the viscosity is increased and the throughput decreased. The flow can be varied by a factor 6; remote control is possible.

For accurate abundance ratio determinations the dead space between the high-pressure side of the leak and the supply capillary should be kept as small as possible to avoid unduly long times to establish the stationary state.

Solid samples may be placed in a small furnace, or painted on a filament, and this furnace or filament may be changed for changing samples. This requires breaking of the vacuum. If the atmospheric pressure is

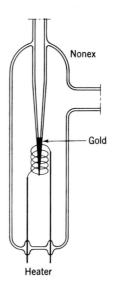

Fig. 8.30. Thermal-expansion leak of Nester [131]. The glass and the gold plug are drawn out together. On cooling, the gold shrinks more than the glass, and the leak opens. On heating, it closes again.

also admitted to the diffusion pumps, these should be allowed to cool first. In mercury diffusion pumps it is sufficient when the mercury ceased boiling; the pressure rise will exclude further boiling immediately. In oil diffusion pumps, however, the hot oil will not stand air at atmospheric pressures because of the possibility of oxidation, and therefore it should be allowed to cool down to below 100°C before admitting air.

To avoid this time-consuming cooling cycle, a gate valve between the ion source and the diffusion pump is desirable. Thus the diffusion pumps may be kept hot while changing samples. The system is then pumped down to about 10^{-2} mm Hg with the mechanical pump via a bypass line and evacuated further with the diffusion pumps. The time needed for evacuation may be reduced considerably if, instead of air, dry argon or helium is admitted during sample change. Gas adsorption at the walls is thus largely avoided.

For the construction of these gate valves numerous suggestions have been published. If a working pressure of 10^{-6} mm Hg can be tolerated (three-filament source), O-ring seals may be used. A survey of the possibilities is given by Von Ardenne [133] and by Yarwood [134]. Such a survey is beyond the scope of this chapter. A perfect seal across the seat is not required; the leakage needs only to be restricted so far as to avoid damage to the diffusion pump. Special attention is called to the danger of leakage from the atmosphere along the stem. More rigid requirements are posed by demands for better vacua, especially for small samples that should fill the tube with the pumps shut off. Reynolds [2] uses this technique. Glass valves as designed by Deckel [72] and by Vogl and Evans [73] may also be used; the leakage when exposed to pressure differences of 1 atm is here no drawback.

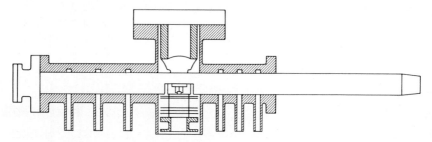

Fig. 8.31. Stainless-steel vacuum lock for the introduction of solid samples to a 180° mass spectrometer. The precision-ground bar, which carries the sample, moves in the channel with a radial clearance of 0.001 mm. By evacuating the grooves in the channel walls to successively better vacua, the bar can be moved freely without breaking the vacuum.

For routine analysis of solid samples a device is desirable that allows changing samples without breaking the vacuum. A vacuum lock for this purpose was developed by Stevens [135]. A rod passes through a channel, in the middle of which the mass-spectrometer tube is mounted with the appropriate ion accelerating and focusing slits. At both sides of the tube three grooves are machined in the channel wall, and the air that leaks in along the rod is pumped away by connecting the grooves in succession to a rotary pump, a diffusion pump, and a second diffusion pump backed by the first. In the original design of Stevens, the sealing between rod and channel was effected by O-rings; in a later version all rubber was eliminated, and the seal effected by a narrow clearance (0.012 ± 0.003 mm for the first stage between atmosphere and rotary pump, 0.025 ± 0.005 mm for the other stages). At the Laboratorium voor Massascheiding in Amsterdam, a similar vacuum lock was constructed for a 180° mass spectrometer (Fig. 8.31). The material had to be nonmagnetic; type 304 stainless steel was chosen. Space was limited, and to maintain the vacuum, the clearance was reduced to

0.002 mm over the whole length; the precision was 0.001 mm. The grooves were evacuated by a rough vacuum, a rotary pump, and a diffusion pump. To avoid seizing, the rod was chromium-plated; lubrication was accomplished by Molykote (microsize powder, MoS_2), of which the excellent properties for vacuum lubrication are known [136]. A pressure of 10^{-6} mm Hg in the tube is easily obtained. The sample is applied to a tube base with three filaments. If magnetic materials are permitted in the construction, the use of nitriding steel offers the possibility of very high surface hardness (see also Turnbull [137]).

Changing samples without breaking the vacuum may also be accomplished by means of valve constructions (Roberts and Walsh [138]; Fultz and Wilson [139]). The time needed for changing samples can be reduced to about 25 min with a valve construction; perhaps even shorter with a vacuum lock of the Stevens type. A device for the continuous introduction of solid samples was patented by Parsegian and Stevens in 1952 [140]; the sample is applied to a conveyer belt, which passes through three pumping chambers at successively lower pressures, analogous to the rod carrier in the Stevens vacuum lock [135].

8.11. Leak Hunting

Among the most troublesome difficulties in vacuum techniques is the presence of leaks. The size of a leak that can be tolerated depends on the vacuum requirements and the effective pumping speed, including the resistance of connecting tubes, valves, cold traps, etc.

When one starts to evacuate a vacuum apparatus that has been exposed to the atmosphere for a considerable time, it is not impossible that even in the absence of leaks the pressure will not drop below about 10^{-4} mm Hg after the diffusion pump has been switched on. This effect is caused by gas or vapor desorption from the walls. In the presence of water, for instance, the pressure may not be reduced for hours. A direct, but not always unambiguous, method for discerning between this gas desorption and the presence of a real leak is isolation from the pump and plotting the pressure rise against time. In the presence of a real leak the pressure will rise linearly with time; if the pressure rise is caused only by gas or vapor desorption or by evaporation of contaminants which can be regarded as a virtual leak, then the rate of pressure rise will decrease gradually, and the pressure will finally tend to a saturation value. The degassing process may be speeded up considerably by high-temperature bake-out, but this procedure remains questionable as long as the presence of important leaks is possible.

As in most things in life, here also prevention is better than cure. Maximum care in making joints requires far less extra effort than the time expended tracing and repairing later leaks. If a system is assembled from different parts, it is advisable to test these parts separately before assembling. If they stand a pressure of a few atmospheres, this test may be accomplished with compressed air, which must, of course, be clean and oil-free, or compressed nitrogen at a few atmospheres over-pressure; the leaks are seen by submerging the whole part in water. If submersion is impossible, leaks may be found by painting the joints with soap solution. Submersion in water provides more pronounced detection, especially when strong illumination is applied from the side. There is, however, a minimum for the leaks that can be detected; not only will the formation of an air bubble take an unduly long time, but it may even be prevented by the surface tension of the water or soap solution. Pressurizing with Freon (see below) is much more sensitive and may as well be applied to stationary mountings. The alternative way of finding a leak is to evacuate the system and trace the influence of the leak on the vacuum by different methods.

A simple and straightforward, but time-consuming, method consists in isolating the system from the pumps after evacuation, dividing it up into as many parts as possible by closing valves, etc., and locating the leaky part by the pressure rise. Suspected areas are then painted with plastic lacquer, e.g., soluble polyvinyl-chloride-acetate, Vinyline.† A leak is closed when the rate of pressure rise decreases. For permanent repair the lacquer coating is not suitable, being delicate and unfit for bake-out; as a purely temporary measure, it may do.

In an apparatus comprising glass parts, an r-f spark tester may be used. This is a device, commercially available from several high-vacuum firms, which generates an r-f high voltage. If the electrode is brought near to a glass system, the residual gas is excited to fluorescence, provided it is at a pressure between 10 and 10^{-3} mm Hg. If a little alcohol or acetone is applied to the leak with a piece of cotton wool, it may either clog the leak temporarily, causing the pressure and fluorescence to decrease, or it penetrates, causing a change in color of the discharge. When the sparker is brought near to a leak in glass, it is found as a brilliant spot. Care should be taken, as thin glass walls are easily perforated by the spark. This is an actual method used for making fine holes in glass [141]. In some Penning gauges the discharge is visible, e.g., in the older models, and its color may be taken as an indication. Leaks may be expected at joints in glass tubing, at welded or brazed joints in metal parts, and at badly sealing gaskets. Instead of alcohol or acetone, hydrogen (or helium) may be used as a test gas. The detection may be accomplished by different means, which are mostly present in usual systems.

The pressure reading of ionization gauges depends on the nature of the gas, but the low viscosity of hydrogen has a larger influence. Thus spraying hydrogen over the leak causes the pressure indication to rise. For thermal-conductivity gauges (Pirani, thermocouple, etc.), the effect

† Supplied by Draka N.V., Amsterdam, The Netherlands.

is larger because of the good heat conduction in hydrogen. The sensitivity may be increased considerably by insertion of a selective trap between gauge and system, which absorbs air but lets through hydrogen. Such a trap with charcoal has been described by Kent [142]. A similar device was used by Van Leeuwen and Oskam [143]; these authors used a Penning gauge and a trap with silica gel. The gauge and the trap are degassed by heating; for leak detection the silica gel is cooled in liquid air, and then it adsorbs air, but hydrogen passes through. Silica gel can adsorb remarkable quantities of air without saturation. The cleanup action of the Penning gauge removes the transmitted hydrogen.

The mass spectrometer itself offers another, and usually even more sensitive, means of detection. The instrument is then adjusted to mass 2 (for hydrogen) or mass 4 (helium), and the peak height is used as an indication. Argon (mass = 40) may also be used as a test gas. Especially at mass 4, the background is very low and the detection can therefore be made extremely sensitive. Leak detectors, consisting of a mass spectrometer, permanently tuned to mass 4, are on the market. These are the most sensitive leak detectors existing at the moment, but their price is rather high.

Another detector, approaching the helium leak detector in sensitivity, is the halogen leak detector. The positive-ion emission from a hot platinum surface is strongly dependent on small concentrations of halogenated hydrocarbons such as Freon compounds, carbon tetrachloride, etc. These detectors are commercially available.† For finding leaks, Freon, etc., may be introduced into the system to an excess pressure (1 atm of air with, for instance, 10 to 20 cm additional Freon pressure), and the outer surface is then traced with a "sniffler," containing the platinum element; or alternatively, the system may be evacuated, the sensing element placed in the vacuum, and Freon sprayed over the leak. If greased stopcocks are present, the vacuum method is to be preferred; then only little Freon enters the system, and cleaning and regreasing of the stopcocks can usually be omitted.

When searching for leaks with a vacuum method, it is of particular importance where the sensing element is mounted. As was pointed out by Ziock [144], the essential factor which determines the minimum size of a detectable leak is the partial pressure of the test gas in the sensing device rather than the size of the leak alone. A leak of size L mm liter^{-1} sec^{-1} (1 mm liter^{-1} sec^{-1} is 1 liter gas at 1 mm Hg per second), in a vessel of V liters, connected to a pump with an effective speed of S liters/ sec, causes a partial pressure of the test gas p mm Hg after spraying for t sec:

$$p = \frac{L}{S} \left(1 - e^{-(S/V)t}\right) \tag{8.12}$$

† Among others from General Electric Company (United States), from B.T.H., Rugby, England, and from Leybold's, Cologne, Germany.

Reducing the pumping speed S will increase the partial pressure (Alers et al. [145]), but the response time is increased by the same factor; at times short as compared with S/V, the two effects cancel.

If the system is evacuated with a diffusion pump with speed S_D liters/sec and the detector is mounted in the backing line to the rotary pump with speed S_V, the partial test-gas pressure in the backing line, supposing its volume to be zero, is given by

$$p = \frac{L}{S_V} \left(1 - e^{-(S_D/V)t}\right) \tag{8.13}$$

S_D is usually of the order of the volume V sec^{-1}, causing a response time of the order of 1 sec. By mounting the sensing element in the backing line instead of in the high vacuum, the sensitivity of detection is increased by a factor S_D/S_V. S_V can be reduced so far that it just maintains the required backing pressure.

The foregoing reasoning is valid only for detectors that respond to the partial pressure of the test gas, irrespective of the total pressure; this is the case with Freon detectors, but not with mass-spectrometer methods.

The methods and constructions that have been described in this chapter, which was written in March, 1959, do not pretend completeness, emphasis having been placed on principles. The authors' present opinions may differ in some details. Where manufacturers' or suppliers' names were given, they serve only to indicate what materials or instruments were actually used or are usable; it was not intended to indicate their superiority.

REFERENCES

1. W. Jawtusch, R. Jaeckel, and G. Schuster: *Z. Naturforsch.*, **9a,** 475 (1954).
2. J. H. Reynolds: *Rev. Sci. Instr.*, **27,** 928 (1956).
3. N. Anton: *Ceram. Age*, **63,** 15 (1954).
4. J. F. Duncan and D. T. Warren: *Brit. J. Appl. Phys.*, **5,** 66 (1954).
5. J. R. Sites and R. Baldock: *U.S. At. Energy Comm. Rept.* ORNL-1405.
6. M. A. Miller and A. S. Russell: *Welding J.*, February, 1953, p. 116.
7. F. J. Norton: *J. Appl. Phys.*, **28,** 34 (1957).
8. G. P. Barnard: "Modern Mass Spectroscopy," p. 67, The Institute of Physics, London, 1953.
9. Tube Laboratory Manual, Research Laboratory of Electronics, MIT, Cambridge, Mass., 1951.
10. W. Espe: *Nachrichtentechnik*, **6,** 355, 401 (1956).
11. D. Fischer: *Z. Physik*, **133,** 471 (1952).
12. G. M. Delfosse; private communication, Lovain, Belgium.
13. R. Gomer: *Rev. Sci. Instr.*, **24,** 993 (1953).
14. J. Yarwood: "High Vacuum Technique," pp. 144, 165, John Wiley & Sons, Inc., New York, 1955.
15. E. L. Jossem: *Rev. Sci. Instr.*, **11,** 164 (1940).

16. S. Dushman: "Vacuum Technique," p. 547, John Wiley & Sons, Inc., New York, 1949; S. Dushman and J. M. Lafferty, "Scientific Foundations of Vacuum Technique," p. 516, John Wiley & Sons, Inc., New York, 1962.
17. F. J. Norton: *Vacuum Symp. Trans.*, 1954, p. 47, Committee on Vacuum Techniques, Boston.
18. D. Alpert and R. S. Buritz: *J. Appl. Phys.*, **25**, 202 (1954).
19. H. Kronberger: "Vacuum Technique in the Atomic Energy Industry," Institute of Mechanical Engineering, London, 1957.
20. J. E. Johnson: *At. Energy Res. Est. (G. Brit.)*, G/R 486, January, 1950.
21. S. S. Stivala and V. L. Denniger: *Ind. Eng. Chem.*, **49**, 1106 (1957).
22. J. E. Quarrington: *J. Sci. Instr.*, **31**, 387 (1954).
23. J. Blears, E. J. Greer, and J. Nightingale: *Proc. First Intern. Congr. Vacuum Techniques*, Namur, Belgium, 1958.
24. K. Diels and R. Jaeckel: Leybold Vakuum Taschenbuch, p. 168, Berlin, 1958.
25. H. G. Nöller (Leybold's, Cologne): private communication.
26. K. Ridley (Edwards High Vacuum Ltd., Crawley, England): private communication.
27. S. L. Martin: *Chem. Proc. Eng.*, **34**, 276 (1953).
28. H. A. Adam, S. Kaufman, and B. S. Liley: *J. Sci. Instr.*, **34**, 123 (1957).
29. F. L. Reynolds: *U.S. At. Energy Comm. Rept.* UCRL-2989 (23-5-1953).
30. W. J. Lange and D. Alpert: *Rev. Sci. Instr.*, **28**, 726 (1957).
31. M. von Ardenne: Tabellen der Elektronenphysik, Ionenphysik und Uebermikroskopie, p. 743, VEB Deutscher Verlag der Wissenschaften, Berlin, 1956.
32. J. B. Mann: *Rev. Sci. Instr.*, **27**, 1083 (1956).
33. P. J. van Heerden: *Rev. Sci. Instr.*, **26**, 1130 (1955).
34. P. J. van Heerden: *Rev. Sci. Instr.*, **27**, 410 (1956).
35. S. Ruthberg and J. E. Creedon: *Rev. Sci. Instr.*, **26**, 1208 (1955).
36. W. G. Wadey: *Rev. Sci. Instr.*, **27**, 322 (1956).
    W. G. Wadey: *Vacuum*, **4**, 53 (1954).
37. M. J. Higatsberger: private communication, Vienna.
38. M. J. Higatsberger and W. W. Erbe: *Rev. Sci. Instr.*, **27**, 110 (1956).
39. J. Drowart, P. Goldfinger, and R. van Steenwinkel: *J. Sci. Instr.*, **34**, 248 (1957).
40. A. L. Reimann: "Vacuum Technique," p. 265, Chapman & Hall, Ltd., London, 1952.
41. J. Yarwood: Ref. 14, pp. 62, 182.
42. L. M. Parr and C. A. Hendley: "Laboratory Glass Blowing," p. 101, G. Newnes, Ltd., London, 1953.
43. H. H. Pattee, Jr.: *Rev. Sci. Instr.*, **25**, 1132 (1954).
44. W. Duesing: *Telef. Ztg.*, **26**, 111 (1953).
    W. Duesing: *Glastech. Ber.*, **26**, 232 (1953).
45. M. W. Riley: *Mater. & Methods*, **132**, 139 (1956).
46. M. von Ardenne: Ref. 31, p. 763.
47. M. L. Randolph, M. Slater, and D. L. Parrish: *Rev. Sci. Instr.*, **27**, 878 (1956).
48. C. B. Leffet, W. H. Kasner, and T. M. Donahue: *Rev. Sci. Instr.*, **27**, 1084 (1956).
49. A. Guthrie and R. K. Wakerling: "Vacuum Equipment and Techniques," p. 169, McGraw-Hill Book Co., Inc., New York, 1949.
50. C. C. Lauritsen and T. Lauritsen: *Rev. Sci. Instr.*, **19**, 919 (1948).
51. E. W. Webster: *Electron. Eng.*, **17**, 53 (1944).
52. B. D. Power and R. A. Kenna: *Vacuum*, **5**, 34 (1955).
53. J. Ruf and O. Winkler: in M. Auwärter (ed.), "Ergebnisse der Hochvakuumtechnik und der Technik dünner Schichten," p. 207, Wissenschafteliche Verlag G.m.b.H., Stuttgart, Germany, 1957.

54. G. Zinsmeister: *Proc. First Intern. Congr. Vacuum Techniques,* 1958, Namur, Belgium.

55. W. Becker: *Proc. First Intern. Congr. Vacuum Techniques,* 1958, Namur, Belgium; *Vakuumtechnik,* vol. 7, no. 7 (1958).

56. A. H. Turnbull: "Vacuum Technique for Beginners," *At. Energy Res. Estab. (G. Brit.),* 1951.

57. W. P. Dryer: "Calculating High Vacuum Systems," U.S. *At. Energy Comm.,* M.D.D.C., 459, 1946.

58. A. Guthrie and R. K. Wakerling: Ref. 49, p. 12.

59. J. Yarwood: Ref. 14, p. 108.

60. S. Dushman: Ref. 16, p. 159.

61. R. F. McFarland and D. G. McDonald: *Rev. Sci. Instr.,* **29,** 530 (1958).

62. D. Alpert: *Rev. Sci. Instr.,* **24,** 1004 (1953).

63. J. Burns: *Rev. Sci. Instr.,* **28,** 469 (1957).

64. E. Thomas, J. Destappes, and J. Dupont: *Vacuum,* **3,** 413 (1953).

65. A. L. Reimann: Ref. 40, p. 108.

66. A. L. Reimann: Ref. 40, p. 176.

67. D. Alpert: *J. Appl. Phys.* **24,** 860 (1953).

68. J. Yarwood: *J. Sci. Instr.,* **34,** 297 (1957).

69. D. Alpert: *Rev. Sci. Instr.,* **22,** 536 (1951).

70. S. C. Brown and J. E. Coyle: *Rev. Sci. Instr.,* **23,** 570 (1952).

71. D. G. Bills and F. G. Allen: *Rev. Sci. Instr.,* **26,** 654 (1955).

72. R. W. Deckel: *J. Appl. Phys.,* **25,** 1441 (1954).

73. T. P. Vogl and H. D. Evans: *Rev. Sci. Instr.,* **27,** 657 (1956).

74. A. Venema: *Proc. First Intern. Congr. Vacuum Techniques,* Namur, Belgium, 1958.

75. A. Venema and M. Bandringa: *Philips Tech. Rev.,* **20,** 97 (1958).

76. L. Pátý and P. Schürer: *Rev. Sci. Instr.,* **28,** 654 (1957).

77. M. J. O'Neal: "Applied Mass Spectrometry," p. 27, The Institute of Petroleum, London, 1954.

78. N. Milleron: *Vacuum Symp. Trans.,* 1957, p. 38.

79. S. Dushman: *J. Franklin Inst.,* **211,** 689 (1931).

80. D. Alpert, C. G. Matland, and A. O. McCourbey: *Rev. Sci. Instr.,* **22,** 370 (1951).

81. J. J. Opstelten and N. Warmoltz: *Appl. Sci. Res.,* **B4,** 329 (1955).

82. J. J. Opstelten, N. Warmoltz, and J. J. Zaalberg van Zelst: *Appl. Sci. Res.,* **B6,** 129 (1956).

83. T. A. Perls, W. H. Kaechele, and D. S. Goalwin: *Natl. Bur. Std. (U.S.) Ann. Rept.* 2165, January, 1953.

84. D. B. Cook and C. J. Danby: *J. Sci. Instr.,* **30,** 238 (1953).

85. D. C. Pressey: *J. Sci. Instr.,* **30,** 20 (1953).

86. R. S. Bradley: *J. Sci. Instr.,* **31,** 129 (1954).

87. C. J. Zilverschoon: Thesis, Amsterdam, 1954.

88. C. C. Roberts and W. W. O'Hara: *Prod. Eng.,* **24,** 190 (1953).

89. R. B. Lawrence: *Chem. Eng. Progr.,* **50,** 155 (1954).

90. H. Schwarz: *Arch. Tech. Messen,* t96 (1951); 7, 53, 99 (1952).

91. G. Glockler and H. V. Horst: *Science,* **116,** 364 (1952).

92. J. H. Leck: *Sci. Instr.,* **31,** 226 (1954).

93. J. H. Leck and C. S. Martin: *Rev. Sci. Instr.,* **28,** 119 (1957).

94. H. von Ubisch: *Vakuum-Tech.,* **6,** 175 (1957).

95. P. E. Seiden: *Rev. Sci. Instr.,* **28,** 657 (1957).

96. G. K. T. Conn and H. N. Daglish: *Vacuum,* **3,** 24 (1953).

97. G. K. T. Conn and H. N. Daglish: *Vacuum,* **4,** 136 (1954).

98. G. K. T. Conn and H. N. Daglish: *J. Sci. Instr.,* **31,** 433 (1954).

99. G. K. T. Conn and H. N. Daglish: *J. Sci. Instr.,* **31,** 412 (1954).

100. A. Bobenrieth: *Vide*, **8**, 1302 (1953).
101. K. Phillips: *J. Sci. Instr.*, **31**, 110 (1954).
102. A. Guthrie and R. K. Wakerling: Ref. 49, p. 134.
103. S. Dushman and A. H. Young: *Phys. Rev.*, **68**, 275 (1945).
104. S. Wagener and C. B. Johnson: *J. Sci. Instr.*, **28**, 278 (1951).
105. H. G. Nöller: *Vacuum*, **5**, 59 (1955).
106. E. A. Vick: *Sci. Progr.*, **42**, 462 (1954).
107. P. Hariharan and M. S. Bhalla: *Rev. Sci. Instr.*, **27**, 448 (1956).
108. F. J. Fitz Osborne: *Can. J. Phys.*, **31**, 11 (1953).
109. J. Schutten: *Appl. Sci. Res.*, **B6**, 176 (1957).
110. R. T. Bayard and D. Alpert: *Rev. Sci. Instr.*, **21**, 571 (1950).
111. S. Aisenberg, W. J. Lange, L. E. Sprague, and W. B. Nottingham: *MIT. Res. Lab. Electronics Tech. Rept.*, July, 1951, and January, 1952.
112. W. B. Nottingham: *Vacuum Symp. Trans.*, 1954, p. 76.
113. F. A. Baker and J. Yarwood: *Vakuum-Tech.*, **6**, 147 (1957).
114. P. A. Redhead and L. R. McNarry: *Can. J. Phys.*, **32**, 267 (1954).
115. H. D. Hagstrum: *Rev. Sci. Instr.*, **24**, 1123 (1953).
116. L. D. Hall: *Rev. Sci. Instr.*, **28**, 653 (1957).
117. J. R. Young: *J. Appl. Phys.*, **27**, 926 (1956).
118. F. Hageman and J. Van Katwijk: "Applied Mass Spectrometry," p. 59, The Institute of Petroleum, London, 1954.
119. G. P. Barnard: Ref. 8, p. 107.
120. R. E. Halsted and A. O. Nier: *Rev. Sci. Instr.*, **21**, 1019 (1950).
121. J. Kistemaker: *Physica*, **18**, 163 (1952).
122. A. J. H. Boerboom and H. A. Tasman: *Physica*, **24**, 683 (1958).
123. M. G. Inghram: *Advan. Electron. Electron Phys.*, **1**, 219, 232 (1948).
124. E. G. Leger: *Can. J. Phys.*, **33**, 74 (1955).
125. G. Nief: *J. Chim. Phys.*, **49**, 49 (1952).
126. S. Shinozuka: *Rev. Sci. Instr.*, **27**, 542 (1956).
127. R. Marks: *Rev. Sci. Instr.*, **28**, 381 (1957).
128. E. F. Babelay and L. A. Smith: *Rev. Sci. Instr.*, **24**, 508 (1953).
129. E. R. Harrison: *J. Sci. Instr.*, **30**, 170 (1953).
130. R. Forman: *Rev. Sci. Instr.*, **24**, 326 (1953).
131. R. G. Nester: *Rev. Sci. Instr.*, **27**, 874 (1956).
132. R. K. Smither: *Rev. Sci. Instr.*, **27**, 964 (1956).
133. M. von Ardenne: Ref. 31, p. 776.
134. J. Yarwood: *Vacuum*, **3**, 398 (1953).
135. C. M. Stevens: *Rev. Sci. Instr.*, **24**, 148 (1953).
136. V. R. Johnson and G. W. Vaughn: *J. Appl. Phys.*, **27**, 1173 (1956).
137. A. H. Turnbull: *Vacuum*, **5**, 131 (1955, publ. 1957).
138. R. H. Roberts and J. V. Walsh: *Rev. Sci. Instr.*, **26**, 890 (1955).
139. C. R. Fultz and L. V. Wilson: *Carbide Carbon Chem. Corp. Instr. Rept.* Y-990, July, 1953.
140. V. L. Parsegian and C. M. Stevens (and U.S. Atomic Energy Commission): U.S. Patent 2,618,750 (1952).
141. R. J. Muson: *Rev. Sci. Instr.*, **26**, 236 (1955).
142. T. B. Kent: *J. Sci. Instr.*, **32**, 132 (1955).
143. J. A. van Leeuwen and H. J. Oskam: *Rev. Sci. Instr.*, **27**, 328 (1956).
144. K. Ziock: *Glas u. Hochvakuumtech.*, **2**, 292 (1953).
145. G. A. Alers, J. A. Jacobs, and P. R. Malmberg: *Rev. Sci. Instr.*, **24**, 399 (1953).

9

Chemical Analysis by Mass Spectrometry

V. H. Dibeler

Mass Spectrometry Section, National Bureau of Standards
Washington, D.C.

9.1. Introduction

The use of the mass spectrometer in many kinds of research in physics and chemistry is fundamentally an analytical one. Therefore knowledge and experience in the techniques and methods of analysis are frequently as important to those investigating basic properties of matter as to the chemist concerned with the synthesis of organic material, to the kineticist interested in reaction mechanisms, or to the staff of a control laboratory in a large chemical company or refinery.

The principles of mass spectrometry as described in detail elsewhere in this book have been applied to the determination of precise mass measurements, atomic weights, isotope abundances, and other nuclear properties, bond strengths, and heats of formation of molecules, radicals, and ions. In addition, the principles have been applied to the solution of problems in qualitative and quantitative analysis of mixtures containing nearly every known type of compound. In such application it has frequently surpassed previously existing chemical or physical methods.

As early as 1931, Smyth [1] reviewed the work of various investigators who had adapted Dempster's 180°-magnetic-deflection instrument to the study of ionization processes by electron impact. Many of the studies reported at that time indicated the characteristic features of mass spectra that were to be of subsequent use in identifying compounds. However, it was not until 1940–1941 that Hoover and Washburn [2] proposed the mass spectrometer as a tool for chemical analysis of multi-component mixtures. Furthermore, broad application of the method awaited the commercial development and production of quasi-standardized techniques and apparatus so necessary for reproducible and reliable analyses. As a result of these developments and through a remarkably cooperative effort of industrial, governmental, and university laboratories, mass-spectrometric analysis is not only applicable to the simple gases and volatile liquids, but also to lubricating oils and waxes, polymers, and other nonvolatile organic materials. Present research and development in the fields of inorganic solids and refractory materials indicate a similarly broad application for mass spectrometry.

The literature relating to analytical mass spectrometry is already very extensive. For background information, the reader is referred to the literature cited in other chapters of this volume and to a number of monographs and collected works [3–8a] that specifically include discussions of analytical applications. In addition, attention is drawn to the various semiannual [9] and other periodic reviews of analytical methods that include mass spectrometry.

9.2. Instrumentation and Techniques

A. Commercial mass spectrometers

Numerous types of mass spectrometers have been designed and built, including some for very specific applications. At the present time, however, the great majority of mass spectrometers used for analytical work are obtained from commercial sources. Of these instruments, one of the most widely used in the United States is the large general-purpose mass spectrometer designated as the 21-103 series by the Consolidated Electrodynamics Corporation. This type of instrument has a 5-in. radius of curvature, 180° magnetic deflection, and direction focusing, with a nominal sensitivity of better than 1 part in 10^4, a maximum resolving power of about 500, and a mass range of 1 to 600 or more. It is a versatile instrument that gives good performance on routine organic and other mixture analyses of materials volatile at room temperature, but because of design characteristics of value as an analytical machine, it is not most easily adapted to some kinds of original research investigations.

The general-purpose mass spectrometer can be modified to obtain spectra of less volatile materials by adding a heated sample-introduction system permitting temperatures up to 350°C and by increasing the resolving power. Such modifications are generally made by individual laboratories and are patterned after the instrument developed by O'Neal and Wier [10]. As laboratory requirements for high-molecular-weight materials are not standardized and problems of chemical reaction, pressure measurement, and solid sample introduction are not completely resolved, commercially available inlet systems operating above 250°C are still in the development stage in the United States, though some are already available.

Other analytical mass spectrometers available from Consolidated include:

A small magnetic mass spectrometer of limited mass range, Model 21-6111, for the analysis of light gases [11]. It is easily portable and is also useful as a process monitor [12, 13].

Fig. 9.1. A group of gas-analysis instruments showing, in left foreground, crossed electric and magnetic field instrument; in center foreground, a portable light-compound analyzer; and right background, the general-purpose instrument with a heated gas-handling system. (*Consolidated Electrodynamics Corp.*)

A small magnetic analyzer, Model 21-610, which is a leak detector fitted for light-gas analysis.

A small crossed electric and magnetic field (cycloidal) mass spectrometer, Model 21-620, which is based on the principle devised by Bleakney and Hipple [14]. The instrument is described by Robinson and Hall [15].

A Mattauch-type [16] instrument intended for a wide variety of uses by virtue of a module construction permitting interchange of specially designed ion sources and collectors, is offered for solids as well as gases or liquids.

Figure 9.1 illustrates several of the above instruments. In the right background is an instrument of the 21-103 series, but provided with a heated inlet system. In the left foreground is a Model 21-620, with a programming unit above the recorder such as would be applicable when using the instrument as a process monitor. In the center foreground is the portable Model 21-6111.

A cycloidal mass spectrometer designed for the study of reaction kinetics is also available [17]. Scan speeds are adjustable in decade steps from 1,000 to 0.1 sec for mass ranges up to 150. Strip chart,

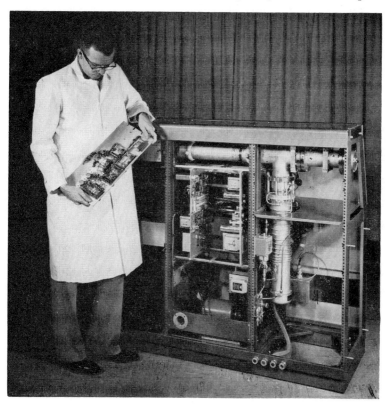

Fig. 9.2. Rear view of time-of-flight instrument showing drift tube, vacuum system, and chassis. (*Bendix Aviation Corp.*)

recording oscillograph, or oscilloscope can be used for data recording. Several instruments specifically designed for isotope abundance measurements and leak detection are also available, as well as a "dual-purpose" mass spectrometer useful both for isotope ratio measurements and routine gas analysis in the range of about 2 to 100.

The Cincinnati Division of the Bendix Aviation Corporation manufactures a pulsed-beam time-of-flight mass spectrometer [18, 19]. Figure 9.2 shows a rear view of Model 12-100 displaying the horizontal drift tube, vertical connection to the pumping system, and electronic racks.

Provisions for supplemental ion sources and alternative choices for sample introduction, ion acceleration, and ion detection make this a highly versatile instrument [20]. The main advantage of the time-of-flight principle lies in its ability to make very high speed qualitative identification, particularly of short-lived species. Nominal characteristics include a useful resolution of about 1 in 400 and a highly repetitive rate of scan of 1 spectrum per 100 μsec. Thus it is particularly applicable to the analysis of systems in which composition changes with time.

Fig. 9.3. Multipurpose 12-in. radius of curvature sector field instrument. (*Nuclide Analysis Associates.*)

This includes the analysis of gaseous fractions from vapor-phase chromatographic columns [21], as well as the observation of reaction intermediates and flash photochemical-reaction studies [22].

Nuclide Analysis Associates, State College, Pa., manufactures a multipurpose, 12-in.-radius, 60°-sector magnetic-field mass spectrometer. Their model 12-60 is similar in appearance to that shown in Fig. 9.3. The basic unit, designed by Professor Mark Inghram, makes use of an analyzer terminated by flanges to which various units can be bolted. Thus source, collector, and pumping units can be interchanged to meet demands of analytical and research programs. Also available is a general-purpose 6-in.-radius, 60°-sector magnetic-field mass spectrometer, as well as instruments specifically designed for various purposes, including precise isotope abundance measurements.

In Great Britain, the Metropolitan-Vickers Electrical Co., Ltd., Manchester, makes available several 90°-sector magnetic-deflection and double-focusing instruments. The MS2 series includes instruments of 6-in. radius adaptable to a wide variety of systems. In general, the

Fig. 9.4. Sector field instrument for gas, liquid, and solid analysis. Dual-inlet system permits comparison of unknown and standard mixtures. Heated inlet system (right rear chassis) handles materials of low volatility. (*Associated Electrical Industries Ltd., Manchester, England.*)

mass range is from 2 to 450, except for MS2-H. The latter, shown in Fig. 9.4, is provided with a heated inlet system of glass and metal, but with a vitreous-enamel lining to reduce catalytic effects. Operation at temperatures up to 300°C is feasible. The nominal resolution is 600, but useful information can be obtained up to mass 1,000. Type MS3 is a 4-in.-radius instrument with a mass range of 2 to 130. It is useful for general chemical analysis of gases and light liquids. The normal sample size is about 1 standard cubic centimeter, and concentrations of components as low as 0.01% are measurable. Type MS4 is a completely portable instrument designed primarily for use in medical research and clinical practice. It has a 2-in. radius and 180° magnetic deflection, with provision for continuous probe sampling and simultaneous presentation of up to four components. The mass range is limited to 18 to 48 mass units.

A double-focusing Mattauch-type instrument, type MS7, is also available, with a mass range of 7 to 240 mass units. The instrument

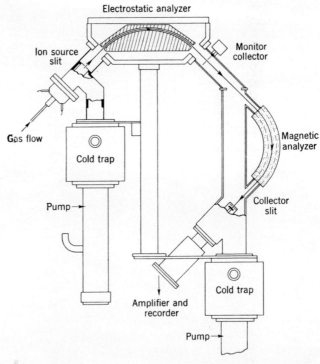

Fig. 9.5. Schematic diagram of a double-focusing mass spectrometer used to attain sufficient resolution for "mass-defect" analysis. (*Associated Electrical Industries Ltd., Manchester, England.*)

includes a vacuum-spark source and photographic detection. The maximum radius in the magnetic field is 8.5 in., and the electrostatic analyzer is 15 in. Its use for the determination of impurities in solids has been described [23].

Recently, Metropolitan-Vickers has produced type MS8, for analytical application [24]. A schematic diagram of this instrument is shown in Fig. 9.5. A precision of 1 part in 100,000 is attainable, with a resolving power of 10,000. Details of the construction have been reported by Craig and Errock [24], and examples of its application to the identification of organic compounds have been discussed by Beynon [25]. Other instruments specifically designed for leak detection or isotope abundance measurements are also available.

In Germany, the Atlas Werke A.G. of Bremen has recently announced completion of its model CH$_4$. This is a large-sector field instrument designed for analytical work. Other specialized instruments are also manufactured.

Instruments of limited application such as mass-spectrometer leak detectors and residual gas analyzers for high vacuum work are available from several instrument manufacturers. Other domestic and overseas groups in France, Italy, and Russia manufacture general-purpose and isotope-ratio instruments on a special-order or restricted basis. However, the intention is to provide a representative list of instrument sources primarily intended for routine chemical analysis.

B. Sample-introduction systems

A typical gas-handling system and the relation of its component parts to a mass-spectrometer analyzer are shown in Fig. 9.6. Samples may be introduced into the spectrometer in any of several ways. A gas sample tube is attached to a manifold with a suitable joint. A small volume of the gas is metered into the measuring volume between two valves and then expanded into the reservoir bottle. A liquid sample is introduced rapidly and reproducibly through a mercury orifice [26], or volatile and not easily adsorbed materials through a mercury-covered sintered disk [27].

One of the important aspects of the routine analytical mass spectrometer is the small size of the sample required for an analysis, e.g., of the order of 0.1 cm^3-atm of gas at room temperature, or 0.0005 ml liquid, or a milligram or less of solid sample. This is at once a great advantage and a source of possible difficulties. The problem of obtaining a truly representative sample is more critical for small samples. Furthermore, changes in composition occur more readily in small samples by sorption and desorption on the interior walls of the sample-handling system. For these reasons, it is often very important to know the pressure of the sample (in the micron region) after expansion into the reservoir. Diaphragm-type micromanometers such as supplied by Consolidated or as described by Dibeler and Cordero [28] are particularly useful for this purpose.

A major problem in the analysis of high-molecular-weight substances is the introduction of a precisely known quantity of pure material for calibration purposes. One means of solving this problem is discussed in a later section on type analysis.

The introduction of nonvolatile solid samples requires special devices such as furnaces or filaments for evaporating the sample within the ion source. Vacuum port openings with provision for changing samples rapidly have been developed.   One type has been designed by Stevens [29].   Such devices permit analysis of solid samples in the amounts of 10^{-9} g or less and consequently are of great advantage in handling dangerously radioactive or extremely rare materials.

The problem of representative sampling may be subdivided according to whether the components of a mixture are (1) completely gaseous, (2)

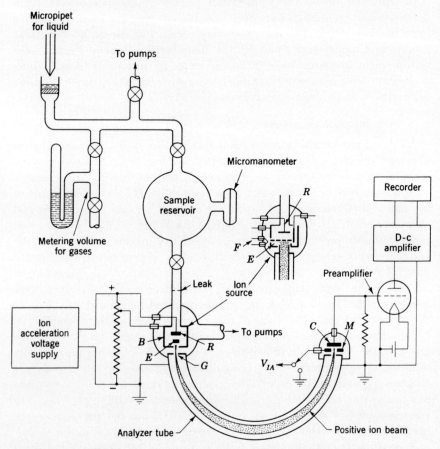

Fig. 9.6. Schematic diagram of gas- and liquid-introduction system and other principal components of an analytical mass spectrometer.

liquids boiling above room temperature at atmospheric pressure, (3) liquids boiling below room temperature at atmospheric pressure, or (4) solids.

The first case is a relatively straightforward one that includes samples commonly expanded from a storage bulb, a reaction vessel, or a compressed-gas cylinder into a manifold attached directly to the evacuated sample-introduction system or to an evacuated sample bulb for subsequent attachment to the mass spectrometer. The sample transfer is usually accomplished under conditions of viscous flow, and no mass fractionation occurs in the process. It will be seen that the nature of the mixture to be analyzed often requires careful consideration of the design of a sample container.

For the second case, the use of a micropipet and mercury orifice, mentioned above, is greatly preferred. The method of introducing a liquid-mixture sample from a break-seal ampule is not considered good practice. It is a well-known fact that the composition of the vapor above the liquid is not representative of the sample and varies with the liquid and vapor volumes. Furthermore, if the size of the liquid sample were limited to that which would completely vaporize upon expansion, adequate mixing of the vapor throughout the inlet system would still remain an uncertainty. Careful consideration is required even in the case of "pure" samples used for calibration purposes. Traces of more volatile impurities may be relatively abundant in the vapor phase, and the apparent percentage impurity greatly magnified in the mass spectrum.

In the third case, the problem of withdrawing a representative sample from a cylinder or pipeline containing liquids boiling below room temperature at atmospheric pressure is a difficult one. Shepherd [30] has discussed the factors that influence the composition of a sample obtained by flashing the sample from a valve located below the liquid level. He also proposed a procedure for obtaining a small representative sample without fractionation as follows: (1) make the liquid phase homogeneous; (2) isolate an appropriate portion of the liquid without change in composition (i.e., without boiling); (3) completely vaporize the isolated sample; and (4) make the vapor phase homogeneous.

For the analysis of compounds with vapor pressures less than 1 mm at room temperature but greater than 1 mm at 350°C, the sample-introduction system is heated as previously indicated. The use of a dual-inlet system (one at room temperature and the other heated) has been described [31, 32] that permits the analysis of materials of both high and low volatility on the same instrument. Such a system is now available on some types of commercial mass spectrometers.

The introduction of corrosive gases such as the halogens, hydrogen halides, and some volatile metallic halides to the mass spectrometer presents unique and troublesome problems. These have been solved to varying degrees, particularly by laboratories associated with gaseous-diffusion plants. One such laboratory has described [33, 34] some recent developments in this field, and Fig. 9.7 shows a typical form of an inlet system for corrosive gases. The unit is kept as simple and compact as possible. The basis of construction is a single block of nickel into which four valves are built. Monel valve diaphragms, gold gaskets, and nickel construction permit bake-out at 300°C. Diffusion welding is apparently prevented by heavy fluorination of the valve seats. Irsa and Friedman [35] have described a metal system for the introduction of halogen fluorides. However, their baking temperature was limited to 100°C because of the presence of polytetrafluoroethylene gaskets and valve seats. This may have accounted for some minor difficulties encountered in their work.

Although noncorrosive in the usual sense, compounds such as water, ammonia, amines, alcohols, and carboxylic acids are so strongly sorbed on walls that accurate analysis of mixtures containing these compounds

Fig. 9.7. Sample-handling system for corrosive gases showing all-metal block and valves.

is virtually impossible without unusual consideration and special techniques. The latter commonly include heated-sample inlet lines and analyzer tubes, placement of the leak in close proximity to the ion source, elution, and repetitive sample flushing [36, 37]. By these means satisfactory analyses have been made on samples containing up to 99% water, including the analysis of the water content.

The importance of sorption effects is clearly demonstrated by the introduction of compounds containing exchangeable deuterium atoms into a conventional metal or glass inlet system. The mass spectrum of the first sample admitted usually shows only a small fraction of the original isotopic abundance, and for a single analysis the amount of sample flushing required to condition the system completely is usually prohibitive. However, by minimizing sorption effects and other instrumental difficulties, Washburn, Berry, and Hall [38] report the successful isotopic analysis of water vapor on a routine basis.

C. Ion source and analyzer region

The typical arrangement shown in Fig. 9.6 indicates the ionization region, the momentum analyzer, the ion collecting and recording regions, and the vacuum envelope. In the analytical mass spectrometer, gases enter the ionization chamber by means of a molecular or effusive flow leak. The use of such leaks and the preferred arrangement for gas flow within the mass spectrometer have been described by Honig [39] and by Zemany [40]. Briefly, however, the aim is to design pumping systems, regulate gas flow rates, and choose sample reservoir dimensions such that the analyzer pressure will be sufficiently low (10^{-6} to 10^{-7} mm Hg for an instrument with a 5-in. radius) to minimize gas scattering of the ion beam and to permit a pressure within the ionization region greater by about two orders of magnitude (10^{-4} to 10^{-5} mm Hg) to reduce background effects and to prevent gases from reentering the ionization region after passing over the hot filament (Chap. 8).

The principles incorporated in ion sources, the desiderata of ion optics, the ionization-dissociation processes characteristic of molecules, and operating characteristics and properties of detecting and recording systems are presented elsewhere in this book (Chaps. 4, 5, and 12). With regard to the latter, however, a brief description is included in the following paragraphs.

Nearly all the analytical instruments for routine chemical analysis employ either d-c or vibrating-reed amplifiers. The latter is particularly well noted for its favorable signal-to-noise ratio and stable base line. However, the property of high zero stability is apparently being matched by recent developments in preamplifier circuitry in which a pair of electrometer tubes is arranged so that drift due to the effects of variation in operating voltages is largely canceled (Chap. 6).

One of the principal requirements of an analytical spectrometer is that of rapid and accurate recording of mass spectra. Although these are relative terms, within the usual meaning for routine analysis, the requirements have been met in several ways. In one instance, the amplifier output is fed to a bank of five galvanometers appropriately shunted for relative sensitivities of 1, $\frac{1}{3}$, $\frac{1}{10}$, $\frac{1}{30}$, and $\frac{1}{100}$. By means of a light beam and mirrored suspensions, a plot of ion current vs. mass is traced on photographic paper. When traces of the more sensitive galvanometers are off the paper, one of the less sensitive galvanometers is in proper range to record the peak. Horizontal grid lines are traced simultaneously to obviate possible effects of variation in paper width as a result of processing. The advantage of this system lies in the simultaneous recording of peaks representing different degrees of attenuation. A dynamic range of 10^5 with a precision of about 0.1% is obtained in this manner.

Slide-wire recorders using a single pen trace are also in common use. However, these require manual or electromechanical shunt-selecting devices in order to record the necessary range of ion intensities. One method of accomplishing this automatically is first to scan the peak on a nonlinear scale to select the proper shunt, and then to rescan the peak on a linear scale [41]. A second method of attenuation requires a pre-collector with a separate amplifier circuit. In this device an ion beam approaching the collector slit strikes a sensing electrode that operates the attenuation controls before the ion beam strikes the collector.

Another detector for the measurement of very low ion currents, under favorable conditions as low as 10^{-20} amp, is the electron multiplier. Although the response of this type of detector is fast and the gain extremely high, it is subject to fluctuations related to kinetic energy and the chemical and physical nature of the ions which impose limitations for its use as a quantitative device [42].

Systems designed for very rapid and repetitive scanning of mass spectra have already been mentioned. However, in these cases precision, sacrificed for speed, is frequently not comparable with the more conventional recording systems.

9.3. Principles of Analysis

A. Basic requirements

The fundamental requirements that must be met for successful chemical analysis by mass spectrometry are threefold: (1) components must exhibit spectra significantly different from each other in mass or relative abundance of ions; (2) spectra and sensitivity (ion current per unit of pressure) must be reproducible and constant within allowable limits; and (3) mass peaks in a mixture spectrum must result from linear additivity of the contributions from each component. When the latter requirement is met, ion currents of a given mixture can be expressed mathematically as follows:

$$k_{11}d_1 + k_{12}d_2 + \cdots + k_{1n}d_n = D_1$$
$$k_{21}d_1 + k_{22}d_2 + \cdots + k_{2n}d_n = D_2$$
$$k_{31}d_1 + k_{32}d_2 + \cdots + k_{3n}d_n = D_3$$
$$\cdots \cdots \cdots \cdots \cdots \cdots \cdots \cdots \cdots \cdots \cdots$$
$$k_{m1}d_1 + k_{m2}d_2 + \cdots + k_{mn}d_n = D_m$$

where k_{mn} is the ion current at mass m due to component n, d_n is the partial pressure of component n, and D is the ion current at mass m in the mixture spectrum.   As peak heights are usually measured in arbitrary units such as recorder chart divisions, one procedure is to employ pattern coefficients for the k_{mn} values.   These coefficients are relative peak heights, usually referred to the largest peak in a spectrum, and are previously obtained for a given component from a calibration spectrum of that material.   If the D_m values are expressed as recorder chart divisions, the solution to the equations yields component peak heights in chart divisions.   These values must be converted to partial pressures by dividing component peak height by sensitivity.   The latter, expressed in chart divisions per unit of pressure, are also obtained from the calibration spectrum.   Percentage composition is then obtained by dividing each partial pressure by the computed sum or by the originally measured sample pressure.

Computational methods.   The computation of a simple hydrocarbon mixture containing C_4 paraffins, olefins, a diolefin, and an acetylene is used to illustrate the technique and method.   Significant portions of the calibration spectra of the components appearing in the mixture are tabulated in Table 9.1.   These spectra are taken from the API Tables

TABLE 9.1.   Partial Calibration Spectra of Some C_4 Hydrocarbons

m/e	$n\text{-}C_4H_{10}$	$iso\text{-}C_4H_{10}$	$n\text{-}C_4H_8$	$1,3\text{-}C_4H_6$	2-Butyne
39	12.5	16.5	35.2	100.	25.7
40	1.63	2.37	6.49	3.42	0.84
41	27.8	38.1	100.	0.10	0.03
42	12.2	33.5	3.41		
43	100.	100.	0.07		
44	3.33	3.33			
52	0.26	0.15	1.46	10.6	10.2
53	0.74	0.50	2.28	59.3	44.8
54	0.19	0.07	3.12	85.9	100.
55	0.93	0.42	19.9	3.76	4.34
56	0.72	0.34	42.5	0.09	0.07
57	2.42	3.00	1.86		
58	12.3	2.73	0.03		
59	0.54	0.11			
Sensitivity (d/μ) of underlined peak	39.6	46.4	33.1	19.4	36.4

of Mass Spectral Data.   Here the most abundant ion in each spectrum is given the value 100.   The abundances of other ions are relative.   Sensitivities in chart divisions per micron of sample pressure are given at the bottom of the table.

Column 2 of Table 9.2 lists peak heights in chart divisions for a hypothetical mixture.   In the illustration, the components are known, and, as

TABLE 9.2. Analysis of a Simple Hypothetical Hydrocarbon Mixture

m/e	Mixture spectra (chart division)	n-C_4H_{10}	iso-C_4H_{10}	N-C_4H_8	1,3-C_4H_6	2-Butyne	Residual
39	2,131	37.1	120.9	262.1	1,427	288.0	−4.0
40	128.9	4.8	17.4	48.8	48.8	9.4	0.1
41	1,108	82.6	279.1	745.0	1.4	0.3	−0.4
42	307.1	36.3	245.5	25.4	....	....	−0.1
43	1,030	297.0	733.0	0.5	....	....	−0.5
44	34.3	9.9	24.4				
52	278.8	0.8	1.1	10.9	151.3	114.3	0.4
53	1,370	2.2	3.7	17.0	846	502.0	−0.9
54	2,369	0.6	0.5	23.2	1,225	1,120	0.3
55	256.6	2.8	3.1	148.0	53.7	48.6	0.4
56	323.9	2.1	2.5	316.5	1.3	.9	0.6
57	43.2	7.2	22.0	13.9	....	....	0.1
58	56.4	36.2	20.0	0.2			
59	2.4	1.6	0.8				
$p(\mu)$	....	7.5	15.8	22.5	73.6	30.8	= 150.2
Mole %	....	5.0	10.5	15.0	49.0	20.5	= 100

indicated in the previous section, one can immediately write a set of five equations stating the contributions of each component to selected mixture peaks. For example, the $m/e = 39$ peak is given by

$$0.125(n\text{-B}) + 0.165(i\text{-B}) + 0.352(n\text{-Bu}) + 1.00(1,3\text{-Bd}) + 0.257(2\text{-Bt}) = 2,131$$

The simultaneous solution of the set is then accomplished by a method such as Crout's [42a], which is rapid and accurate and requires only a desk calculator. However, several simplifications suggest themselves. Except for the very small contributions of the n-butene to $m/e = 43$ and 58, the n- and isobutane composition could be determined directly from a set of two simultaneous equations. From Table 9.1, column 4, it is apparent that the maximum possible contributions of the butene to the 43 and 58 peaks is equal to 0.07 and 0.03% of the 41 peak, respectively. As these are very nearly trivial corrections in this case, it can be made without further consideration. Thus the butane equations become:

$$m/e = 43, \ 1.00(n\text{-B}) + 1.00(i\text{-B}) = 1029.5$$

and

$$m/e = 58, \ 0.123(n\text{-B}) + 0.0273(i\text{-B}) = 56.2$$

The solution gives the calculated number of divisions of n-butane and isobutane on the 43 peak, and these are entered on the appropriate line of columns 3 and 4 of Table 9.2. The n- and isobutane contributions to all peaks in the mixture are calculated from the calibration spectra in Table 9.1 and tabulated. Next, the butadiene and butyne contributions to the $m/e = 41$ peak are estimated to be 0.08% of the $m/e = 39$ peak. This contribution (1.8 divisions) plus the butane contributions is subtracted from the $m/e = 41$ peak, and the remainder is entered in Table 9.2, column 5. Butene contributions to all peaks are computed with the aid

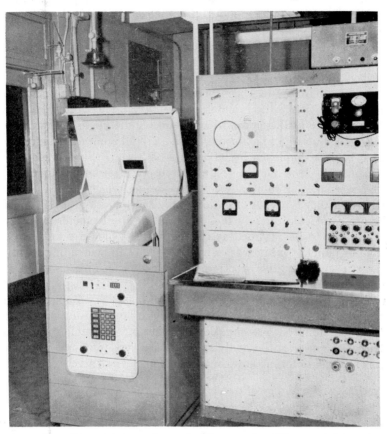

Fig. 9.8. Consolidated Electrodynamics Corporation Mascot digitizer in conjunction with a general-purpose gas analyzer. The chassis include mass-number and peak-amplitude digitizers, binary-code display, and a Clary printer with serial read-out to an IBM 523 summary punch. (*Esso Research Engineering Co.*)

of Table 9.1 and tabulated. Finally, the butadiene and butyne are separated by two simultaneous equations in a manner similar to the butane computation, except using the net $m/e = 39$ and $m/e = 54$. The respective contributions to all mass numbers are again computed. All calculated contributions to each mass number are summed and subtracted from the original mixture peak heights. The residuals, tabulated in the last column, indicate a satisfactory solution of the problem. Therefore partial pressures are computed by dividing the underlined peak heights in Table 9.2 by the sensitivities at the bottom of Table 9.1. Normalizing each to give a total pressure of 100 results in the mole-percentage composition.

Solutions such as the above and the Crout matrix solution are usually most useful when only a limited number of analyses of a particular kind is anticipated. When a large number of analyses is required, using one set of calibration spectra, the final solutions are greatly simplified if an inverse or reciprocal matrix is first calculated. Details of this somewhat more complicated, but routinely advantageous, method have been described [43].

Considerable effort has been directed to the application of automatic computing devices to do calculations rapidly and accurately. The first computers were analog types which could handle up to 12 simultaneous equations [44]. Recently, medium-sized digital computers, such as the IBM 650, were employed by laboratories with large sample loads. For these purposes the manual reduction of data from photographic or similar records is prohibitively time-consuming. One means of alleviating this difficulty is the use of analog-to-digital converters [45, 46]. These units are connected directly to the output of spectrometers and permit peak heights to be recorded as digital values which are automatically printed or punched into cards or paper tape [47]. Such a device, used in conjunction with a mass spectrometer, is shown in Fig. 9.8. Details of recent developments and experiences in digitizing mass-spectral data have been reported by Dudenbostel and Klaas [48]. Such mechanical computing aids are ideally suited to obtain quick solutions to large systems of simultaneous equations or for using previously prepared matrices. Consequently, analyses which formerly required one or more hours of calculation time now can be accomplished in a very few minutes.

B. Precision and accuracy of results

The commercial analytical mass spectrometer was originally designed for application to light-hydrocarbon analysis. Most of the information on precision and accuracy of specific instruments and the general method is available for this class of materials. Probably the most satisfactory means of determining the accuracy of an analytical method is by comparison of analyses with the known composition of a synthetic blend. This has been done in a number of instances, but mostly for relatively simple mixtures. Washburn [49] has tabulated the results of running a synthetic mixture on 24 instruments. The mixture was independently synthesized for each instrument and run three or more times for a total of 93 analyses. Table 9.3 gives the average error in the determination of each component. For all components except the butenes, 90% of the errors were ±0.5 mole % or less. In 1-butene, 2-butene, and isobutene,

90% of the errors were less than ± 1.2 and ± 0.8 mole %, for the n- and isobutenes respectively. Total butenes were determined with greater accuracy.

TABLE 9.3. Average Error in Determination of Each Component in 93 Analyses of a Synthetic C_1–C_4 Paraffin-Olefin Mixture

Component	Composition, mole %	Average error, mole %
Methane................	15	±0.14
Ethane.................	20	±0.22
Propene................	10	±0.17
Propane................	20	±0.21
n-Butane..............	8	±0.18
Isobutane..............	10	±0.19
1-Butene...............	5	±0.83
2-Butene...............	5	±0.77
Isobutene..............	7	±0.33
Total butenes........	17	±0.27

Barnard [50] quotes data communicated by Blears on a very similar mixture analyzed with a Metropolitan-Vickers mass spectrometer. A brief summary is given in Table 9.4.

TABLE 9.4. Errors in Analyses of a Synthetic Hydrocarbon Blend

Components	Composition, mole %	Actual errors, mole %			
		Run 1	Run 2	Run 3	Run 4
Methane.............	15.1	+0.61	+0.39	+0.17	+0.09
Ethane..............	19.1	+0.27	+0.23	+0.07	−0.09
Propene.............	9.9	+0.39	+0.44	+0.42	+0.34
Propane.............	20.1	−0.25	+0.02	+0.06	0.00
n-Butane...........	8.1	−0.51	−0.47	−0.45	−0.33
Isobutane...........	9.9	−0.23	−0.16	−0.21	−0.03
1-Butene............	5.0	−0.60	−1.12	−0.84	−0.57
2-Butene............	5.0	−0.54	+0.67	+0.24	−0.05
Isobutene...........	7.0	+0.85	+0.01	+0.44	+0.64
Total butenes......	17.0	−0.29	−0.44	−0.16	+0.02

The importance of conditioning the mass spectrometer to attain accurate analyses is well illustrated by the data of Table 9.4. In general, the deviation between calculated and known compositions decreases markedly with successive samples of the blend run through the mass spectrometer. This is illustrative of the generally accepted fact that instruments continually analyzing hydrocarbon mixtures usually attain a greater degree of reproducibility and accuracy than those operating intermittently or analyzing various types of compounds. Apparently, this observation is related to the phenomenon known as "interference," an undesirable condition in which mass spectra and sensitivities of mixture components are not independent and mass spectra cease to be linearly additive. Its cause has been related to the surface condition of the tungsten filament [51], commonly used in analytical instruments. Rather specific treatments for eliminating the effect have been suggested, but one of the best seems to be the use of rhenium [52] as a filament in place of tungsten. In this case, treatment of the filament is not necessary.

Interesting indications of the recent state of the mass-spectrometric method of analysis of light gases and a comparison with volumetric chemical methods were obtained by Shepherd in a series of cooperative analyses of standard samples of a natural gas [53] and of a carburetted water-gas [54]. In one of the latter studies, 27 mass-spectrometer laboratories cooperated. These represented both chemical industries and government. In addition to tabulating the analytical data reported by the various laboratories, Shepherd constructed a set of frequency-distribution plots in which each point represented a single determination of the component plotted. The number of points placed equidistant along the ordinate indicated the number of times a particular mole percentage plotted as the abscissa was reported. The plots also included average values for each laboratory.

Plots of the results for methane and for hydrogen are given in Fig. 9.9. For methane, a gaussian distribution is evident, and the mean (three low values omitted) was 8.0 ± 0.4 mole %. The composition previously determined by volumetric chemical means was 8.4 ± 0.3 mole % [54].

Frequency-distribution plots were also obtained for ethane and ethylene. The saturated hydrocarbon was determined with a relatively narrow spread of values (± 0.15 mole %), whereas the ethylene analysis was less definitive (± 0.4 mole %). Carbon dioxide was analyzed with good reproducibility (± 0.14) and good agreement with the chemical value. However, large variations were observed for hydrogen and for carbon monoxide and nitrogen. The latter two might be expected from interference with each other and with ethylene.

Blears [54a] reports the results of a cooperative analysis program organized by the Institute of Petroleum for sector-type mass spectrometers. In all, 140 samples of four C_1–C_4 hydrocarbon mixtures were analyzed by eight laboratories. A comparison with a similar program carried out entirely on 180° mass spectrometers and reported by Starr and Lane [55] indicated a remarkable degree of similarity in analytical

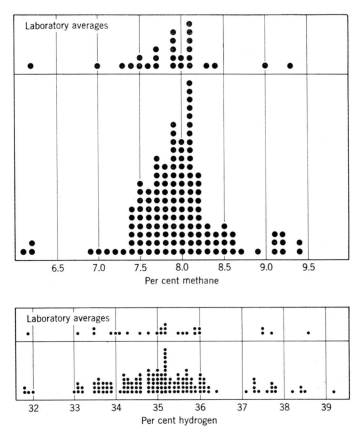

Fig. 9.9. Data-distribution plot for methane.   Each filled circle represents one laboratory determination.

errors between the two groups in spite of some reservations concerning the direct comparison of the two programs.

C. Laboratory requirements

Commercial analytical mass spectrometers are almost completely automatic in operation.   Instrumental variations that commonly affect the accuracy of the analysis, such as the temperature of the ion source, electron energy and emission, and other ion-source conditions, are generally removed from manual control.   Design factors usually limit major variations in pumping speeds, pressure in the ion source, scanning rate, amplifier sensitivity, and recording speed.   Furthermore, the increasing use of digitization of mass-spectral data for routine analysis ensures still more rigorously standardized conditions.   Thus the established programs are easily followed by personnel with a minimum of technical background.   However, an interest in electronics and laboratory procedure, as well as a moderate degree of dexterity, is advantageous.

A brief description of typical space, maintenance, sample preparation, and personnel requirements is given by Barnard [3] for mass-spectrometer laboratories operating at several activity levels.   One published study [56] of laboratory cost, in terms of man-hours per sample analyzed, is now seriously out of date because of technical advances.   In 1946, however, the cost varied from 1.3 man-hours per 3- or 4-component samples to 1.9 man-hours for 10- to 20-component samples.   These estimates included time of operating, computing, supervisory, maintenance, and supply personnel.

9.4. Qualitative Analysis

In a completely unknown mixture, qualitative analysis is a necessary preliminary to quantitative analysis, and a successful analysis depends on previous knowledge of the characteristics of mass spectra.   Such information includes the specific mass numbers given or omitted from the spectrum of a given compound, the approximate relative abundances of the principal ions, the normal isotopic abundances of elements comprising the components, the recognition of meta-stable and multiply charged ions, the presence of rearrangement peaks and ions characteristic of particular functional groups, and so forth.   Fortunately, all these data are included in a single spectrum if a suitable mass range has been scanned.   When components of a mixture have been qualitatively identified, estimates or precise determinations of abundances usually follow with little difficulty. A possible exception to this arises in the analysis of inorganic or refractory compounds using photographic techniques.   This will be discussed in a later section.

A. Identification by mass number and mass defect

In a nearly pure compound or simple mixture, the positions and relative abundances of mass peaks are often sufficient to identify several or all components.   The absence of mixture peaks in particular mass ranges is also important information and, as in the case of optical spectrometry, is used to establish an upper limit to the presence of possible components. An indispensable aid in these identifications is a cumulative library of mass-spectral data either in the form of tables [57] or punch cards.   Sets of the latter, usually constructed by individual laboratories, can be arranged for sorting on the basis of physical and chemical properties, as well as mass number and ion abundance.   In addition, a very useful summary of the spectra of 279 compounds, including many nonhydrocarbons, has been published [58] in which peaks are roughly classified according to magnitude and mass-charge ratio.

Beynon [8a, 59] has pointed out the usefulness of instruments with resolving powers of the order of 1 in 10,000 in identifying components producing ions of the same nominal mass number but different mass defects. With such resolving power, for example, mass peaks of the ions CO^+, $C_2H_4^+$, CH_2N^+, and N_2^+, all with nominal $m/e = 28$, are easily resolved, and the mass intervals used to confirm identification. Obviously, this technique is also of importance in the following section.

B. Functional group and structure analysis

Although remarkable progress has been made [60], there is yet no complete theoretical basis for predicting the mass spectra of any but the simplest molecules [61]. Consequently, the qualitative determination of functional groups and of group-type analysis is based entirely on empirical principles established by means of correlation of mass-spectral data. The scope of this relatively new and rapidly expanding field of study is too great for detailed discussion in the present work. However, some brief references to source material are included.

Probably the most intensive study has been applied to oxygenated compounds, as the large number of possibilities makes functional-group identification particularly valuable. Friedel, Shultz, and Sharkey [62] have studied alcohols in the C_1–C_{11} range, and Brown, Young, and Nicolaides [63] have examined heavier alcohols. Friedel et al. conclude that spectral characteristics identify various subclassifications within the general (i.e., primary, secondary, tertiary) alcohol types. Sharkey, Shultz, and Friedel [64] have also correlated ketone spectra with structure.

Other correlation studies of oxygenated compounds include aldehydes [65], acetals [66], carboxylic acids [67], esters [68], and ethers [69].

With regard to hydrocarbons, no detailed study of paraffins has been published, although characteristic spectra of normal- and branched-chain compounds are well known. Monoolefins [70] and alkyl benzenes [71] have received considerable study.

Work has also been reported on sulfur compounds [72]. Kinney and Cook [73] have reported the feasibility of identifying any thiophene homolog through C_9 and certain higher homologs. They also reported the possibility of identifying any mono- or disubstituted benzene homolog through C_{11} and certain higher homologs. Although trialkyl and higher substituted benzene homologs and homologs of benzenes and thiophenes with molecular weight greater than 154 were not identifiable, correlations were used to eliminate some possible structures from consideration. Biemann and coworkers have successfully applied mass-spectrometric techniques to the study of naturally occurring compounds. These studies include determination of the structure of sarpagine [73a], proline derivatives from apples [73b], and various alkaloid derivatives

[73c].   Studies of sterols and bile acids are reported by Stenhagen and coworkers [73d].

C. Isotopic structure analysis

Knowledge of the natural isotopic composition of the elements can be very useful in qualitative analysis.   The presence of C^{13} and D atoms in natural abundance, although complicating the spectrum of a hydrocarbon,

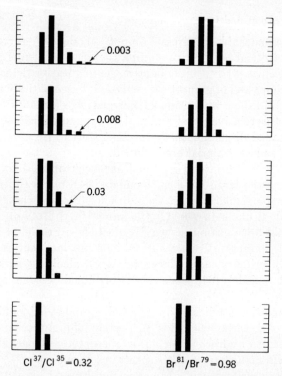

$$Cl^{37}/Cl^{35} = 0.32 \qquad Br^{81}/Br^{79} = 0.98$$

Fig. 9.10. Graphical representation of the statistical distribution of the isotopes of chlorine and bromine for atoms and for molecules containing two to five atoms.

also serves to establish the nature of the ion.   This is particularly true for the identification of components of semiorganic and nonhydrocarbon mixtures.   For example, an ion containing one Cl^{35} atom appearing at mass $(m/e)_x$ will be accompanied by an ion approximately one-third the abundance containing a Cl^{37} atom appearing at $(m/e)_{x+2}$.   Similarly, an ion containing two chlorine atoms appearing at $(m/e)_y$ will be accompanied by two other ions in the ratios of $1:0.6:0.1$ appearing at $(m/e)_{y+2}$ and $(m/e)_{y+4}$.   Ions containing bromine atoms are characterized in a similar fashion.

Figure 9.10 illustrates the statistical distribution of the halogen isotopes ($Cl^{37}/Cl^{35} = 0.32$, and $Br^{81}/Br^{79} = 0.98$) in ions containing one to five chlorine or bromine atoms.   Each peak is separated by two mass units, and the abundances are relative to the most abundant species in each group.   A similar chart can be constructed for ions containing both bromine and chlorine atoms.   The addition of fluorine or iodine does not change the distribution, as these halogens are anisotopic.   By the use of such charts, fragment ions containing these atoms are often readily identified.

D. Molecular weights by microeffusiometry

As previously noted for analytical spectrometers, the gas or vapor sample enters the ion source by molecular flow through an effusive leak. Thus the rate at which molecules of molecular weight M enter (and leave) the ion source is proportional to $1/\sqrt{M}$.   By proper choice of sample reservoir volume, a convenient rate of pressure drop can be achieved.   As peak heights are directly proportional to the partial pressure of a component, the mass spectrometer can be used to measure the rate at which the partial pressure of an unknown decreases in the reservoir compared with the rate of pressure decrease for a standard.   In practice, a calibration curve is determined which relates molecular weight with leak rates.   Furthermore, as the peak heights for all ions produced from a component decrease at the same rate, any convenient peak can be selected.   It is not necessary for the molecule ion to be identified or even observed.   Eden, Burr, and Pratt [74] have determined molecular weights of mixture components by this means, and Zemany [75] reports the successful application of this method in a number of instances, including the identification of unusual compounds such as CH_2ClCOF. The technique is also useful in distinguishing between unicomponent peaks and those representing superpositions of two or more ion currents, thus aiding in establishing the presence of components with overlapping spectra.

E. Other special techniques

Analytical problems have been encountered which could not be directly and completely solved from mass-spectral data.   The use of supplementary methods such as infrared or ultraviolet spectrometry, distillation or crystallization techniques, vapor-phase chromatography, and so forth, has already been mentioned.   Occasionally, analysts have resorted to chemical treatment.   For example, isobutene has been analyzed as *tert*-butyl chloride [76], alcohols as trimethylsilyl derivatives [77, 78], and the skeletal-structure analysis of ketones by conversion to corresponding hydrocarbons [79].

A very useful, purely physical technique is that of low-electron-energy analysis.   Complex mass spectra can usually be simplified by reducing the energy of the bombarding electrons.   In the most favorable cases, it is feasible to operate at energies that ionize the parent molecule without causing dissociation. This is particularly advantageous in the analysis of mixtures of some isotopically substituted molecules [80].

Field and Hastings [81] have utilized the above to improve the determination of olefinic and aromatic hydrocarbons. They found measurabe parent ions for olefins and aromatics to the exclusion of ions from aliphatic paraffins. Examples of other applications to complex mixtures have been reported [82–84].

Another device for obtaining analytical data is the isotope-dilution principle, originally introduced by Hevesy and Paneth for analyzing the radioactive isotopes of lead. This principle has been used by Grosse, Hinden, and Kirshenbaum [85] to determine oxygen, carbon, and nitrogen in organic compounds. This technique will be discussed more fully in a later section.

9.5. Quantitative Analysis

A. Gases and light liquids

By far the most extensive application of the mass spectrometer has been to the field of gas and light-liquid analysis. Examples of several of these have already been mentioned. In general, analysis of these mixtures yields satisfactory results without fractionation because of the relatively small number of isomers present. In fact, one of the outstanding advantages of the mass spectrometer as applied to this type of sample is the large number of components that can be analyzed in a single mixture without supplementary distillation or fractionation procedures. With regard to refinery fractionating tower samples, Washburn [86] observes that tower overheads, including depropanizers, debutanizers, and depentanizers, can be analyzed directly and completely by mass spectrometer. Although Crowe [86a] states that it is impossible to quote a general figure for the accuracy of analyzing such samples, an estimate of usual ability is shown in Table 9.5.

The ease with which mixtures of hydrocarbon gases can be analyzed decreases with increasing similarity of the component spectra. However, Brown and coworkers [87] have shown that individual paraffin isomers in the C_5–C_8 range can be analyzed. For example, Table 9.6 summarizes results for a mixture containing all nine heptane isomers and one octane.

Naturally, the analytical problem is simplified if only selected components need be determined. An example of such a procedure has been reported by Meyerson [88] for the analysis of benzene and benzene-forming naphthenes (cylclohexane and methyl cyclopentane) in naphthas. The necessary conditions for accurate analysis of these components have been established, and Meyerson gives comparative results for three samples analyzed by mass spectrometer and by ultraviolet absorption and

TABLE 9.5. Analysis of Hydrocarbon Gas Separation Plant Streams

Stream	Major component	Trace component	Detection limit, mole %	Average component	Maximum error (±), mole %
Demethanizer tops	H_2			65.0	1.0
	C_1			25.0	1.0
		$C_2^=$	0.05	7.0	0.1
		C_2	0.05	3.0	0.1
De-ethanizer tops	C_2			45.0	1.0
	$C_2^=$			45.0	1.0
		C_1	0.05	4.0	0.2
		C_3	0.05	4.0	0.1
		C_3	0.1	2.0	0.1
Depropanizer tops	C_3			45.0	1.0
	C_3			45.0	1.0
		C_2	0.1	5.0	0.2
		C_4	0.1	5.0	0.2
Ethylene splitter tops	$C_2^=$			90.0	0.3
		C_1	0.05	5.0	0.2
		C_2	0.05	5.0	0.1

Table 9.6. Analysis of Mixture Containing C_7 Isomers

Compound	Synthetic composition, mole %	Analysis, mole % Run 1	Analysis, mole % Run 2	Mean difference (±)
2,2-Dimethyl pentane	3.5 } 5.7	5.8	5.7	0.1
2,2,3-Trimethyl butane	2.2 } 5.7			
2,4-Dimethyl pentane	50.7	48.9	49.3	1.6
3,3-Dimethyl pentane	1.9	1.9	1.8	0.1
2,3-Dimethyl pentane	31.7	33.0	33.6	1.6
2-Methyl hexane	1.7	1.3	0.9	0.6
3-Methyl hexane	3.8	4.6	3.0	0.8
3-Ethyl pentane	1.8	1.6	2.5	0.5
n-Heptane	1.3	1.6	2.0	0.5
2,2,4-Trimethyl pentane	1.4	1.3	1.2	0.2

for three samples analyzed by mass spectrometer and by refractive-index measurements. The mass-spectrometric method offers the important advantages of being fast, reliable, and more informative than the other procedures. Sobcov [89] has reported a study aimed at determining the smallest number of distillation cuts necessary to obtain sufficiently accurate analyses of olefin-free mixtures found in the boiling range from isopentane to toluene. By careful selection of matrices and IBM programming, a minimum number of four cuts was chosen and four overlapping matrices of order 9, 10, 12, and 10 were used to search for 28 compounds. Benzene and toluene were independently calculated because of the specific natures of their spectra.

B. Light-type analysis

Probably one of the most significant advances in the mass-spectrometric method of hydrocarbon analysis is the development of compound-type analysis such as that described by Brown [90]. This method, which resulted from a correlation study of C_4–C_{16} compounds, is based on the following simple observations of data published in the API Tables of Mass Spectral Data:

1. Ions at masses 43, 57, 71, 85, and 99 were generally most abundant in paraffins.

2. Relatively large peaks at 41, 55, 69, 83, and 97 appeared to characterize cycloparaffins and monoolefins.

3. Ions at masses 67, 68, 81, 82, 95, and 96 are generally characteristic of cycloolefins, diolefins, and acetylenes, known as the "coda" group.

4. Alkylbenzene fragments are usually observed at masses 77, 78, 79, 91, 92, 105, 106, 119, 120, 133, and 134.

When peaks in each of the above groups are arithmetically treated as 1, it is found that pattern coefficients of the combined peaks are similar for compounds of the same type but are markedly different for compounds of other types. Thus it is possible to consider gasoline as consisting of only four components which could be resolved with four simultaneous equations. Cycloparaffins and olefins, considered as one component, are subsequently resolved by chemical determination of the olefins.

Numerous modifications of the method have been applied to determinations of the major-compound types in the gasoline boiling range. For instance, Lumpkin, Thomas, and Elliott [91] describe a procedure designed to handle streams with low olefin content, no "coda" group, but including condensed-ring naphthenes. In this method, olefins are first determined by bromide-bromate titration and then removed in a small-scale bromination-steam distillation apparatus before mass-spectrometric analysis. Paraffins, naphthenes, aromatics, and condensed-ring naphthenes are then determined by a single mass-spectrometer analysis on a 70 to 200°C boiling-range sample involving the use of distinctive m/e ratios in the mass spectra of the hydrocarbon types. The method is rapid and is accurate to $\pm 10\%$ of the type present.

A rapid and accurate determination of total aromatics and aromatic-molecular-weight groups is frequently necessary in refineries producing motor fuel and aromatic solvents. The distillation, silica gel, acid absorption, and specific dispersion methods often become laborious. As mass spectra of aromatic compounds are sufficiently different from the other compound types in a hydrocarbon mixture to allow determination of aromatics only, Lumpkin and Thomas [92] have developed a rapid analytical procedure with an accuracy of about $\pm 1\%$ for the different aromatic-molecular-weight groups and for total aromatic contents.

Table 9.7 illustrates the wide range of total aromatic content that can be analyzed by this procedure and also shows the extent of agreement between the mass-spectrometric results and those obtained by acid absorption and by specific dispersion methods.

TABLE 9.7. Analysis of Total Aromatic Content by Mass Spectrometer Compared with Acid Absorption and Specific Dispersion

Compound	Volume %						
Benzene	81.6	73.0	12.1	1.1	0.8	0.2	0.0
Toluene	2.2	1.1	8.8	55.3	12.9	2.7	6.0
C_8 aromatics	0.2	0.1	0.8	2.5	32.2	34.5	45.5
C_9 aromatics	0.0	0.0	0.0	1.0	2.4	2.1	2.0
Total, mass-spectrometric	84.0	74.2	21.7	59.9	48.3	39.5	53.5
Acid absorption	84.0	73.9	22.2	60.3	48.1	39.2	55.7
Specific dispersion	85.4	69.7	24.1	62.9	48.7	39.2	54.7

A novel technique applied in conjunction with a type analysis has been described by Melpolder et al. [93]. A thermal-diffusion process is first used to separate a light lubricating oil into specific hydrocarbon types. Then mass and ultraviolet spectrometry are used to identify 16 different hydrocarbon types in the thermal-diffusion fractions.

Recently, compound-type methods have been based on total-ionization measurements [94]. This method essentially makes use of some early observations by Mohler and coworkers [95], who found that total ionization (1) increased with molecular weight within a particular class, (2) increased with the number of hydrogen atoms of hydrocarbons, for a given number of carbon atoms, and (3) is essentially constant for structural isomers. Further work [96, 97] indicates that similar observations hold for compounds containing as many as 36 carbon atoms. Molar volume relationships for these compounds are such that for high-molecular-weight compounds the total ion intensity is approximately proportional to the liquid volume of the sample charged to the mass spectrometer. Spectra are thus roughly standardized with respect to both liquid volume and instrument sensitivity.

C. Heavy-type analysis

The chemical constitution of petroleum fractions boiling above the gasoline range has recently become of considerable interest to the petroleum industry. In order to obtain this information, O'Neal and Wier

[10] greatly extended the range of application of the mass spectrometer to include spectra of compounds having molecular weights up to about 600. A heated sample-introduction system referred to above was devised so that hydrocarbons up to C_{40} could be completely vaporized into the reservoir.

Because of the complex nature of molecular structures, nearly all analysis of high-mass compounds is type analysis. The method is similar to, but necessarily less specific than, that for the lighter fractions. Nevertheless, the modification and extension of the principles of type analysis to heavy hydrocarbon mixtures has provided probably the most useful analytical method presently available to high-molecular-weight hydrocarbon chemistry. As an example, O'Neal [98] describes analytical methods that permit the direct determination of the isoalkane content of lubricating oils, the composition and molecular-weight distribution in a low-melting paraffin wax, and the analysis of the aromatic portion of the 400°C-boiling-range material from a catalytically cracked, clarified slurry oil. Other recent work has indicated practical application of this technique to the analysis of petrochemicals, coal products, and other related materials [99, 100, 100a].

One very specialized technique for high-mass studies should be mentioned at this time. Bradt and Mohler [101] have studied the spectrum of polymeric substances obtained by heating p-$C_6F_4Br_2$ and p-$C_6F_4I_2$ in the presence of metallic copper. Samples of the products containing various lengths of the perfluorophenyl chains Br-$(C_6F_4)_x$-Br and I-$(C_6F_4)_Y$-I were evaporated from a small oven into the ion source with little or no decomposition at temperatures up to 260 and 431°C, respectively. As expected from the aromatic structure of the molecules, the molecule ions were relatively abundant. In the iodine compound, ions containing 6 or 7 phenyl rings were most abundant, and ions containing up to 11 rings with molecular weights of 1,755 were observed. This represents some of the heaviest ions ever reported by mass spectrometry. As noted elsewhere in this chapter, unit resolution at this mass is not required for satisfactory identification of ions in this case.

D. Pollution and trace-molecule-impurity analysis

The recent application of mass spectrometry to problems of air and stream pollution illustrates, perhaps better than any other, the amazing compound and sensitivity range of this method. For example, the mass-spectrometric determination of solvent vapors of the order of parts per million in air was reported by Happ, Stewart, and Brockmyre [102]. In the examination of smog samples, Shepherd and coworkers [103] exceeded this limit by several orders of magnitude. The method employed by Shepherd was capable of determining as little as 10^{-4} ppm of some pollutants from a 100-liter sample of air and 10^{-6} ppm of some substances from larger samples. The gaseous phase of some smog samples was found to be of the order of 0.5 ppm of air. Approximately 60 compounds or types were positively or tentatively identified, and the amounts of some of these determined. The authors showed that the gaseous phase of smog was primarily a mixture of hydrocarbons and of hydrocarbon derivatives containing oxygen, nitrogen, and chlorine. The means of formation of some of these and their relation to eye and respiratory irritants were also investigated.

Other determinations of air contaminants are described by Quiram and coworkers [104] in which sample mixtures were analyzed containing methane, propane, and isopentane of the order of 3 to 8 ppm. Newton [105] describes techniques for the inverse case of noncondensable impurities in condensable gases.

Melpolder, Warfield, and Headington [106] have developed a method for qualitative and quantitative determination of traces of volatile contaminants which affect the taste and odor of water. These include such petroleum products as natural gas, gasoline, kerosene, and furnace oil. Hydrocarbon contaminants boiling below 200°C were determined at concentrations as low as 0.01 ppm.

E. Volatile inorganic and semiorganic analysis

The versatility of the mass spectrometer in analyzing other than hydrocarbons and substituted hydrocarbons has been noted previously. This section includes a few interesting examples.

The erratic behavior of nitrogen dioxide in the mass spectrometer has long made difficult the analysis of mixtures containing this material. The results of a detailed study of this problem have been reported by Friedel et al. [107]. For mixtures containing nitrogen dioxide, nitric oxide, nitrous oxide, carbon monoxide, carbon dioxide, nitrogen, oxygen hydrogen, and water vapor, the analysis is routine for all components except the NO_2. The latter requires prior conditioning of the mass spectrometer and particular attention to instrumental conditions if the partial pressure of the NO_2 is less than 15 μ. Other components required no preconditioning, and the presence of these compounds had no effect on the spectral behavior of NO_2. The absence of chemical reaction between nitric oxide and oxygen was clearly demonstrated at total pressures less than 0.09 mm Hg. Thus, in the analysis of synthetic blends, reaction between the various components was avoided by mixing at pressures of microns rather than millimeters. The results of the analysis of two synthetic blends prepared directly in the mass-spectrometer reservoir are summarized in Table 9.8.

Bernstein, Semeluk, and Arends [108] have compared mass-spectrometric and infrared methods for determining some chlorine-substituted methanes and ethanes. Reference mass-spectral patterns and relative ionization efficiencies were measured for the pure compounds, hexa-chloroethane, pentachloroethane, *sym*-tetrachloroethane, tetrachloro-ethylene, methylene chloride, carbon tetrachloride, and chloroform. A matrix suitable for analysis of mixtures of these compounds was devised. Apparently the sensitivity of the method is good for pentachloroethane and carbon tetrachloride, but because of extensive superposition, it is difficult to detect small amounts of methylene chloride in these mixtures.

The continued interest in fluorocarbon chemistry and technology has made mass-spectral studies of these compounds of primary importance. Although details of analyses of fluorocarbon mixtures have not been published, a summary of spectra for a variety of compounds is available

TABLE 9.8. Analysis of Synthetic Blends Containing Oxides of Nitrogen

Blend	Components, mole %								
	NO_2	NO	N_2O	CO_2	CO	N_2	O_2	H_2O	H_2
Analysis.........	5.3	3.9	....	13.1	13.1	16.2	4.3	44.1	
Synthetic 1.......	5.1	4.0	....	13.2	13.3	16.3	4.3	43.1	
Analysis.........	9.4	10.1	12.1	12.4	12.0	25.2	6.7	....	12.1
Synthetic 2.......	10.2	9.6	12.0	12.0	12.1	25.5	6.8	....	11.9

[109]. With regard to the saturated fluorocarbon compounds, the most abundant positive ion is the CF_3^+ ion, and molecule ions are frequently not observed. Observations of negative molecule ions in such compounds [110] suggest the possibility of identification on that basis. However, a considerable amount of systematic study of negative ions yet remains before this is a practical consideration.

Unsaturated fluorocarbon molecules and cyclics have very distinctive mass spectra, and possibilities of analysis of mixtures of these are favora-ble. As there are gaps of 12 or 19 mass units between various groups of peaks in fluorocarbon spectra, impurities other than fluorocarbons are usually conspicuous, e.g., compounds containing hydrogen atoms or other halogen atoms.

Perhaps one of the most intriguing of recent analytical applications is that of the mass-spectrometric study of the upper atmosphere. In these studies, simple, lightly constructed mass spectrometers—usually of the Bennett type [111]—are sent into the upper atmosphere in sounding rockets. Townsend [112] and Holmes and Johnson [113] have described such compact rocket-borne models, which have been used successfully to detect positive and negative ions and to report gas composition at alti-tudes of 150 miles.

Analyses of other simple gaseous systems have been reported in some detail. These include the determination of respiratory dead space [114], the evaluation of accumulated carbon dioxide in anesthetized patients [115], the study of respiratory metabolism [116], and metabolic studies of cells, tissues, and complete organisms [117].

9.6. Analysis of Nonvolatile Materials

A. Organic compounds

It has been known for many years that many classes of organic and semiorganic polymers, when pyrolyzed or otherwise degraded in vacuum, yielded products of varying degrees of volatility that were sufficiently characteristic of the parent material to serve as a means of identifying that material. Although of less concern in this chapter, a study of the products also leads to information on the more fundamental question of kinetics and mechanisms of the degradation process [118–120].

Two principal methods have evolved for carrying out pyrolytic experiments. In the *indirect method*, the material to be studied is pyrolyzed in a specially designed vacuum system and the products collected, sometimes with fractionation, and transported in sample bulbs to the mass spectrometer. In the *direct method*, very small samples are placed in a furnace usually mounted within the vacuum envelope of the instrument, but always as close as possible to the ion source. The decomposition products evolved by heating pass more or less directly into the electron beam and analyzer tube.

Using the indirect method, Zemany [121] describes a simple apparatus in which samples weighing a few tenths of a milligram are coated on or placed inside the coils of standard radio receiving-tube filaments and are then heated rapidly in vacuum to above 800°C. The pyrolysis vessel was attached to the inlet system of a mass spectrometer and cooled to liquid nitrogen temperatures, and the volatile gases analyzed and pumped away. The vessel was then warmed to room temperature, and another spectrum obtained. The nature and relative amounts of pyrolysis products were rather sensitive functions of the experimental conditions, and reproducibility was not as high as that obtained for electron-impact dissociation patterns. For the simpler materials such as linear homopolymers, however, the reproducibility approached that of the mass spectra of molecules.

Patterns were obtained for copolymers, alkyds, proteins, and numerous commercial materials. The distinctly different spectra obtained for most of these constituted an empirical method of recognizing the original material similar to that used for identifying compounds.

The more elaborate apparatus and method used by Madorsky and coworkers in the pyrolysis and fractionation of pyrolysis products of polymers have been described in detail in a study of polystyrene [118].

It consists essentially of heating a 25- to 50-mg sample of a polymer, spread as a thin film on a platinum tray, to temperatures of 300 to 500°C in a high vacuum. The separable fractions obtained were (1) a solid residue, (2) a waxlike portion volatile only at the temperatures of the pyrolysis, (3) a liquid fraction volatile at room temperature, and (4) a fraction volatile at liquid-nitrogen temperatures. Fractions 3 and 4 were subjected to mass-spectrometric analysis. Fraction 4 was principally carbon monoxide. Average results of several runs for fraction 3 showed 94% styrene monomer, 5.5% toluene, and traces of ethylbenzene and methyl styrene.

Madorsky and coworkers [119] have studied other polymers, including polyisobutylene, polyisoprene, polybutadiene, polyethylene, the copolymer, GR-S, polytetrafluoroethylene, polyhydrofluoroethylene, and others.

An example of the newer technique of direct pyrolysis is by Bradt, Dibeler, and Mohler [122]. In this experiment, polystyrene samples were pyrolyzed, and products evaporated directly into the ion source after comparatively few collisions. Identified ions corresponded to the monomer, dimer, trimer, tetramer, and pentamer. Mohler [123] has given a review of recent pyrolysis studies covering a wide variety of materials.

B. Trace-element and inorganic-solids analysis

Until recently, relatively few laboratories had developed means for the analysis of inorganic materials having negligible vapor pressure at room temperature. This in part resulted from the lack of suitable commercially available equipment and in part from the very specific nature of any method for nonvolatile materials and the lack of general application of the equipment and techniques.

A major share of the development of trace-element methods was contributed by Professor A. J. Dempster's laboratory at the University of Chicago. Much of the present application and work of establishing potentialities and limitations of the methods is continuing at the University of Chicago under the direction of Professor M. G. Inghram, at Bell Laboratories under Dr. A. J. Ahearn, at Westinghouse under Dr. W. M. Hickam, and at the RCA Laboratories under Dr. R. E. Honig. Honig [123a] has recently prepared an excellent review of this field.

There are four basic methods currently in use for trace-element analysis: (1) Dempster vacuum spark, (2) ion bombardment, (3) isotope dilution, and (4) total evaporation. The first two, like gaseous mass-spectrometric analysis, are analogous to optical-spectroscopic methods in that all components present are analyzed in a single measurement. They differ, however, in the range of sensitivities with which elements are detected. The last two methods detect only a very limited number (usually one) of elements per measurement. However, they have the advantage of much higher sensitivity.

Recently, Hannay and Ahearn [124], using a Mattauch-type instrument, have applied the vacuum-spark method to the routine analysis of boron

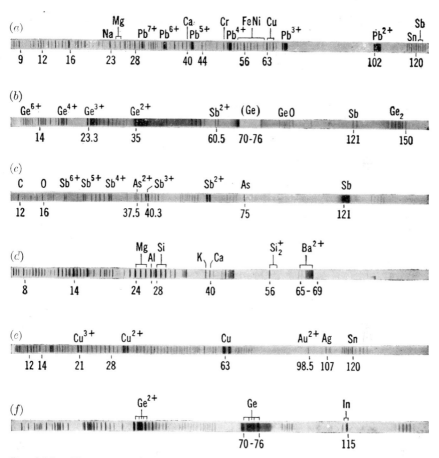

Fig. 9.11. Vacuum-spark spectra. From top to bottom, they are identified as (a) high-purity lead; (b) germanium with 6 ppm added antimony; (c) antimony with 100 ppm of arsenic; (d) steatite, using high-purity silicon reference electrodes; (e) copper, with one monolayer of indium deposited on the surface; and (f) germanium, with one monolayer of indium on the surface. (*Obtained by Hannay, Bell Telephone Laboratories.*)

in silicon and antimony in germanium. In each case they attained sensitivities of about 1 part in 10^7. An illustration of the application of the method is given in Fig. 9.11, which shows six vacuum-spark spectra obtained by Hannay. These are identified in the caption. Hannay and Ahearn [124] report sample requirements of about 10^{11} atoms for detection.

The secondary positive ions produced by bombardment of a surface by electrons are useless for analyzing solid materials [125]. However, ion-beam bombardment of surfaces apparently does give usable secondary ion beams, and several such ion sources have been described [126]. Preliminary studies by Inghram and coworkers indicate a potentially reliable method for several elements in concentrations of 1 ppm and less.

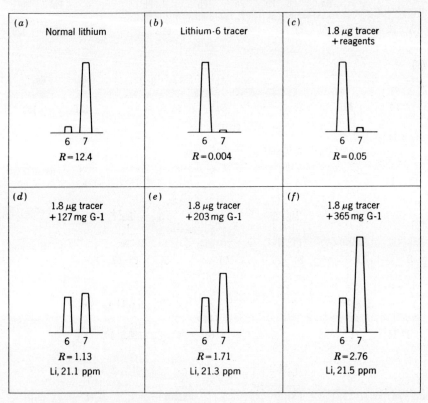

Fig. 9.12.   Determination of lithium in granite using lithium-6 tracer.

Honig [127] has bombarded silver, germanium, and germanium-silicon-alloy surfaces with the inert-gas ions Ne^+, Ar^+, Kr^+, and Xe^+ at energies between 30 and 400 ev.   Of the sputtered secondary particles, neutrals as well as positive and negative ions, some were due to gases adsorbed on the surfaces and others were characteristic of the material being examined.   Honig concluded that, provided surface contamination could be reduced, the method was useful for the analysis of the solid surfaces.

The isotope-dilution method, originally proposed by G. Hevesy and used for many years to determine the elements hydrogen, nitrogen, and carbon and their compounds, was limited to those elements for which separated isotopes were available. The recent availability, through the U.S. Atomic Energy Commission, of many additional stable and long-lived isotopes has greatly expanded the usefulness of this technique. Inghram [128] has outlined the methods of trace analysis and given typical examples of the isotopic-dilution technique; therefore the details will not be repeated here. Several examples of recent applications, however, are presented.

The completely specific application to the determination of lithium in a granite is described by Webster [129] and shown in Fig. 9.12. Sections *a* and *b* show the spectra of normal lithium and lithium-6 tracer,

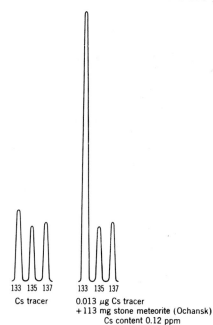

Fig. 9.13. Determination of cesium in a stony meteorite using fission-product cesium.

133 135 137
Cs tracer

133 135 137
0.013 µg Cs tracer
+113 mg stone meteorite (Ochansk)
Cs content 0.12 ppm

respectively. Section *c* is a blank determination of lithium in reagents used to dissolve the rock samples. The increase in the lithium-7 peak causes a small correction of about 0.1 µg. Sections *d* to *f* show the spectra obtained when 1.8 µg of tracer is added to 127, 203, and 365 mg of sample, respectively, and the calculated lithium content.

In principle, the method is limited to elements that are polyisotopic, but the use of long-lived radioactive tracers permits analysis of some anisotopic elements. One such example [130] is shown in Fig. 9.13, in which fission cesium (mainly Cs-133, Cs-135, and Cs-137) is used to determine the cesium content of a stony meteorite.

The isotope-dilution method is applicable to about 80% of the elements. Although sensitivity varies with the element, applications to as little as 10^{-12} g are feasible. There are few interference effects, and accuracies of a few per cent are reasonable. A limitation of the method is the practical one of restriction to one or two elements at a time. However, the major limitation is the possibility of contamination. Depending on the particular application, extreme precautions are resorted to in selecting reagents and even in laboratory construction.

The total-vaporization method has been reported useful for trace analysis by Hickam [131], for a determination of silver in copper, and by Honig [132], for the determination of impurities in germanium. In this method, the sample is placed in a crucible very close to the ion source, and the peak corresponding to the element in question is monitored as the temperature is increased. The concentration of the element under investigation is proportional to the integrated peak height. Unfortunately, the accuracy of this method is presently not comparable with that obtainable from any of the three previous methods. This results in part from possible chemical complications and in part from the physical nature of the evaporation process. However, the use of calibration standards and other developments should make this method reliable for a variety of trace-element applications.

The foregoing discussion, providing only a brief introduction to methods and techniques that for the most part have received intensive study and broad application by those responsible for their development, is intended only as a guide to more detailed information. Moreover, as new types of instruments become commercially available, and as new supplementary techniques are developed for use on present instruments, the application of mass spectrometry will undoubtedly improve on current uses and extend to fields not yet considered feasible.

REFERENCES

1. H. D. Smyth: *Rev. Mod. Phys.*, **3**, 347 (1931).
2. H. H. Hoover, Jr., and H. W. Washburn: *Proc. Calif. Nat. Gas Assoc.*, 16th Annual Fall Meeting, 1941.
3. G. P. Barnard: "Modern Mass Spectrometry," The Institute of Physics, London, 1953.
4. "Mass Spectrometry," The Institute of Petroleum, London, 1952.
5. "Applied Mass Spectrometry," The Institute of Petroleum, London, 1954.
6. *Advan. Mass Spectrometry Proc.*, 1959.
7. H. E. Duckwork: "Mass Spectroscopy," Cambridge University Press, London, 1958.
8. A. J. B. Robertson: "Mass Spectrometry," John Wiley & Sons, Inc., New York, 1954.
8a. J. H. Beynon: "Mass Spectrometry and Its Applications to Organic Chemistry," Elsevier Publishing Company, Amsterdam, 1960.
9. For example, see Analytical Reviews: 1958 Review of Fundamental Developments in Analysis, *Anal. Chem.*, **30**(4), 604 (1958); also **34**(4), 243R (1962).
10. M. J. O'Neal, Jr., and T. P. Wier, Jr.: *Anal. Chem.*, **23**, 830 (1951); see also *Advan. Mass Spectrometry Proc.*, 1959, p. 175.

11. R. F. Wall: *Ind. Eng. Chem.*, **49**, 59A (August, 1957).
12. J. K. Walker, A. P. Gifford, and R. Nelson: *Ind. Eng. Chem.*, **46**, 1400 (1954).
13. T. C. Wherry and F. W. Karasek: *J. Appl. Phys.*, **26**, 682 (1955).
14. W. Bleakney and J. A. Hipple: *Phys. Rev.*, **53**, 521 (1938).
15. C. F. Robinson and L. G. Hall: *Rev. Sci. Instr.*, **27**, 504 (1956).
16. J. Mattauch and R. Herzog: *Z. Phys.*, **89**, 786 (1934).
17. L. G. Hall, C. K. Hines, and J. E. Slay: A High Speed Cycloidal Mass Spectrometer, *Advan. Mass Spectrometry Proc.*, 1959, p. 266.
18. W. C. Wiley: *Science*, **124**, 817 (1956).
19. W. C. Wiley and I. H. McLaren: *Rev. Sci. Instr.*, **26**, 1150 (1955).
20. D. B. Harrington: *Advan. Mass Spectrometry Proc.*, 1959, p. 249.
21. R. S. Gohlke: *Anal. Chem.*, **31**, 535 (1959).
22. G. B. Kistiakowsky and P. H. Kydd: *J. Am. Chem. Soc.*, **79**, 4825 (1957).
23. R. D. Craig, G. A. Errock, and J. D. Waldron: *Advan. Mass Spectrometry Proc.*, 1959, p. 136.
24. R. D. Craig and G. A. Errock: *Advan. Mass Spectrometry Proc.*, 1959, p. 66.
25. J. H. Beynon: *Advan. Mass Spectrometry Proc.*, 1959, p. 328.
26. E. M. Charlet: *Consolidated Eng. Corp. Group Rept.* 74, Pasadena, Calif.
27. R. C. Taylor and W. S. Young: *Anal. Chem.*, **17**, 811 (1955).
28. V. H. Dibeler and F. Cordero: *J. Res. Natl. Bur. Std.*, **46**, 1 (1951).
29. C. M. Stevens: *Rev. Sci. Instr.*, **24**, 148 (1953).
30. M. Shepherd: *J. Res. Natl. Bur. Std.*, **38**, 351 (1947); **36**, 313 (1946); **44**, 509 (1950).
31. R. A. Brown, F. W. Melpolder, and W. S. Young: *Petrol. Processing*, **7**, 204 (1952).
32. G. L. Cook, R. A. Meyer, and D. G. Earnshaw: ASTM Committee E-14 Meeting, New Orleans, La., June, 1958.
33. P. G. Bentley, A. N. Hamer, and P. B. F. Evans: *Advan. Mass Spectrometry Proc.*, 1959, p. 209.
34. P. G. Bentley, J. R. Fothergill, and A. N. Hamer: *At. Energy Res. Est. (G. Brit.) Rept.* IG 132 (RD/CA).
35. A. P. Irsa and L. Friedman: *J. Inorg. Nucl. Chem.*, **6**, 77 (1958).
36. A. P. Gifford, S. M. Rock, and D. J. Comaford: *Anal. Chem.*, **21**, 1026 (1949).
37. H. M. Kelley: *Anal. Chem.*, **23**, 1081 (1951).
38. H. W. Washburn, C. E. Berry, and L. G. Hall: *Anal. Chem.*, **25**, 130 (1953).
39. R. E. Honig: *J. Appl. Phys.*, **16**, 646 (1945).
40. P. D. Zemany: *J. Appl. Phys.*, **23**, 924 (1952).
41. D. Grove and J. A. Hipple: *Rev. Sci. Instr.*, **18**, 837 (1947).
42. M. G. Inghram and R. J. Hayden: A Handbook on Mass Spectroscopy, *Natl. Acad. Sci.-Natl. Res. Council, Nucl. Sci. Ser., Rept.* 14, 1954.
42a. P. D. Crout: *Trans. Am. Inst. Elec. Engrs.*, **60**, 1235 (1941).
43. E. C. Daigle and H. A. Young: *Anal. Chem.*, **24**, 1190 (1952).
44. C. E. Berry, D. E. Wilcox, S. M. Rock, and H. W. Washburn: *J. Appl. Phys.*, **17**, 262 (1946).
45. B. F. Dudenbostel, Jr., and W. Priestley, Jr.: *Anal. Chem.*, **26**, 1275 (1954).
46. B. K. Fritts and G. Peattie: *Anal. Chem.*, **28**, 10 (1956).
47. W. H. King and W. Priestley: *J. Anal. Chem.*, **23**, 1418 (1951).
48. B. F. Dudenbostel, Jr., and P. J. Klaas: *Advan. Mass Spectrometry Proc.*, 1959, p. 232.
49. H. W. Washburn: Mass Spectrometry, in W. G. Berl (ed.), "Physical Methods in Chemical Analysis," vol. 1, Academic Press Inc., New York, 1950.
50. See Ref. 3, p. 212.

51. A. G. Sharkey, Jr., E. H. Bean, R. A. Friedel, and L. J. E. Hofer: Meeting of ASTM E-14 Committee on Mass Spectrometry, New York, May, 1957.
52. C. F. Robinson and A. G. Sharkey, Jr.: *Rev. Sci. Instr.*, **29**, 250 (1958).
53. M. Shepherd: *J. Res. Natl. Bur. Std.*, **38**, 19, 491 (1947); *Anal. Chem.*, **19**, 635 (1947).
54. M. Shepherd: *J. Res. Natl. Bur. Std.*, **36**, 313 (1946); **44**, 509 (1950).
54a. J. Blears: Cooperative Mass Spectrometric Analysis of C_1-C_4 Hydrocarbon Mixtures, *Metropolitan-Vickers Res. Ser.* 6, June, 1955, Metropolitan-Vickers Electrical Co., Trafford Park, Manchester, England.
55. C. E. Starr and T. Lane: *Anal. Chem.*, **21**, 572 (1949).
56. W. S Young: *Natl. Petrol. News, Tech. Sec.*, March, 1946.
57. American Petroleum Institute Tables of Mass Spectral Data, Carnegie Institution, Pittsburgh.
58. S. M. Rock: *Anal. Chem.*, **23**, 261 (1951).
59. J. H. Beynon: *Nature*, **174**, 735 (1954); *Mikrochim. Acta*, **1**, 437 (1956); *Advan. Mass Spectrometry Proc.*, 1959, p. 328.
60. M. B. Wallenstein, A. L. Wahrhaftig, H. Rosenstock, and H. Eyring: *Proc. Symp. Radiobiol.*, p. 20, Oberlin College, 1950; see also H. M. Rosenstock, M. B. Wallenstein, A. L. Wahrhaftig, and H. Eyring: *Proc. Natl. Acad. Sci. (U.S.)*, **38**, 667 (1952).
61. D. P. Stevenson: *J. Chem. Phys.*, **15**, 409 (1947).
62. R. A. Friedel, J. L. Shultz, and A. G. Sharkey, Jr.: *Anal. Chem.*, **28**, 926 (1956).
63. R. A. Brown, W. S. Young, and N. Nicolaides: *Anal. Chem.*, **26**, 1653 (1954).
64. A. G. Sharkey, Jr., J. L. Shultz, and R. A. Friedel: *Anal. Chem.*, **28**, 934 (1956).
65. J. A. Gilpin and F. W. McLafferty: *Anal. Chem.*, **29**, 990 (1957).
66. R. A. Friedel and A. G. Sharkey, Jr.: *Anal. Chem.*, **28**, 940 (1956).
67. G. P. Happ and D. W. Stewart: *J. Am. Chem. Soc.*, **74**, 4404 (1952).
68. A. G. Sharkey, Jr., J. L. Shultz, and R. A. Friedel: *Anal. Chem.*, **31**, 87 (1959).
69. F. W. McLafferty: *Anal. Chem.*, **29**, 1782 (1957).
70. R. A. Brown and E. Gilliams: Mass Spectra of Monoolefins, Meeting of ASTM E-14 Committee on Mass Spectrometry, New Orleans, La., May, 1954.
71. S. Meyerson: *Appl. Spectry.*, **9**, 120 (1955).
72. E. J. Levy and W. H. Stahl: Meeting of ASTM E-14 Committee on Mass Spectrometry, New York, May, 1957.
73. I. W. Kinney, Jr., and G. L. Cook: *Anal. Chem.*, **24**, 1391 (1952).
73a. K. Biemann: *J. Am. Chem. Soc.*, **83**, 4801 (1961).
73b. K. Biemann, G. G. J. Deffner, and F. C. Steward: *Nature*, **191**, 380 (1961).
73c. K. Biemann, M. Friedmann-Spiteller, and G. Spiteller: Tetrahedron Letters, 1961, p. 485.
73d. S. Bergstroem, R. Ryhage, and E. Stenhagen: *Svensk Kem. Tidskr.*, **73**, 566 (1961).
74. M. Eden, B. E. Burr, and A. W. Pratt: *Anal. Chem.*, **23**, 1735 (1951).
75. P. D. Zemany: *J. Appl. Phys.*, **23**, 924 (1952).
76. F. W. Melpolder and R. A. Brown: *Anal. Chem.*, **20**, 139 (1948).
77. S. H. Langer, R. A. Friedel, I. Wender, and A. G. Sharkey, Jr.: *Anal. Chem.*, **30**, 1353 (1958).
78. A. G. Sharkey, Jr., R. A. Friedel, and S. H. Langer: *Anal. Chem.*, **29**, 770 (1957).
79. H. Siegel and D. O. Schissler: *Anal. Chem.*, **28**, 1646 (1956).
80. A. W. Tickner, W. A. Bryce, and F. P. Lossing: *J. Am. Chem. Soc.*, **73**, 5001 (1951).
81. F. H. Field and S. H. Hastings: *Anal. Chem.*, **28**, 1248 (1956).
82. C. J. Varsel, F. A. Morrell, F. E. Resnik, and W. A. Powell: The Use of Low Voltage Mass Spectroscopy in the Analysis of Multicomponent Mixtures,

Meeting of ASTM E-14 Committee on Mass Spectrometry, New Orleans, La., June, 1958.

83. H. E. Lumpkin: *Anal. Chem.*, **30**, 321 (1958).

84. R. J. Gordon, R. J. Moore, and C. E. Muller: *Anal. Chem.*, **30**, 1221 (1958).

85. A. V. Grosse, S. G. Hinden, and A. D. Kirshenbaum: *Anal. Chem.*, **21**, 386 (1949).

86. H. W. Washburn: Mass Spectrometry, in W. G. Berl (ed.), "Physical Methods in Chemical Analysis," vol. 1, Academic Press Inc., New York, 1950.

86a. See Ref. 54a.

87. R. A. Brown, R. C. Taylor, F. W. Melpolder, and W. S. Young: *Anal. Chem.*, **20**, 5 (1948).

88. S. Meyerson: *Anal. Chem.*, **25**, 338 (1953).

89. H. Sobcov: *Anal. Chem.*, **24**, 1386 (1952).

90. R. A. Brown: *Anal. Chem.*, **23**, 430 (1951).

91. H. E. Lumpkin, B. W. Thomas, and A. Elliott: *Anal. Chem.*, **24**, 1389 (1952).

92. H. E. Lumpkin and B. W. Thomas: *Anal. Chem.*, **23**, 1738 (1951).

93. F. W. Melpolder, R. A. Brown, T. A. Washall, W. Doherty, and W. A. Young: *Anal. Chem.*, **26**, 1904 (1954).

94. G. F. Crable and N. D. Coggeshall: *Anal. Chem.*, **30**, 310 (1958).

95. F. L. Mohler, L. Williamson, and H. M. Dean: *J. Res. Natl. Bur. Std.*, **45**, 235 (1950).

96. A. Hood: *Anal. Chem.*, **30**, 1218 (1958).

97. J. W. Otvos and D. P. Stevenson: *J. Am. Chem. Soc.*, **78**, 546 (1956).

98. M. J. O'Neal, Jr.: "Applied Mass Spectrometry," p. 27, The Institute of Petroleum, London, 1954; see also Ref. 10.

99. F. W. Melpolder, R. A. Brown, T. A. Washall, W. Doherty, and C. E. Headington: *Anal. Chem.*, **28**, 1936 (1956).

100. A. Hood and M. J. O'Neal: *Advan. Mass Spectrometry Proc.*, 1959, p. 175.

100a. R. J. Clerc and M. J. O'Neal: *Anal. Chem.*, **33**, 380 (1961).

101. P. Bradt and F. L. Mohler: *Anal. Chem.*, **27**, 875 (1955).

102. G. P. Happ, D. W. Stewart, and H. F. Brockmyre: *Anal. Chem.*, **22**, 1224 (1950).

103. M. Shepherd, S. M. Rock, R. Howard, and J. Stormes: *Anal. Chem.*, **23**, 1431 (1951).

104. E. R. Quiram, S. J. Metro, and J. B. Lewis: *Anal. Chem.*, **26**, 352 (1954).

105. A. S. Newton: *Anal. Chem.*, **25**, 1746 (1953).

106. F. W. Melpolder, C. W. Warfield, and C. E. Headington: *Anal. Chem.*, **25**, 1453, (1953).

107. R. A. Friedel, A. G. Sharkey, Jr., J. L. Shultz, and C. R. Humbert: *ibid.*, **25**, 1314 (1953).

108. R. B. Bernstein, G. P. Semeluk, and C. B. Arends: *Anal. Chem.*, **25**, 139 (1953).

109. F. L. Mohler, V. H. Dibeler, and R. M. Reese: *J. Res. Natl Bur. Std.*, **49**, 343 (1952).

110. R. M. Reese, V. H. Dibeler, and F. L. Mohler: *J. Res. Natl. Bur. Std.*, **57**, 367 (1956).

111. W. H. Bennett: *J. Appl. Phys.*, **21**, 143 (1950).

112. J. W. Townsend, Jr.: *Rev. Sci. Instr.*, **23**, 538 (1952).

113. J. C. Holmes and C. Y. Johnson: *Anal. Chem.*, **30**(9), 19A (1958).

114. J. Bartels, J. W. Severinghaus, R. E. Foster, W. A. Briscol, and D. V. Bates: *J. Clin. Invest.*, **33**, 41 (1954).

115. F. A. Miller, A. Hemingway, E. B. Brown, A. O. C. Nier, R. Knight, and R. L. Varco: *Surg. Forum Proc. 36th Congr. Am. Coll. Surgeons*, 1950, p. 602.

116. A. W. Pratt, B. E. Burr, M. Eden, and E. Lorenz: *Rev. Sci. Instr.*, **22**, 694 (1951).

117. A. H. Brown, A. O. C. Nier, and R. W. Van Norman: *Plant Physiol.*, **27**, 320 (1952).

118. S. L. Madorsky and S. Straus: *J. Res. Natl. Bur. Std.*, **40**, 417 (1948).
119. S. L. Madorsky, S. Straus, D. I. Thompson, and L. Williamson: *ibid.*, **42**, 499 (1949).
120. L. A. Wall: *ibid.*, **41**, 315 (1948).
121. P. D. Zemany: *Anal. Chem.*, **24**, 1709 (1952).
122. P. Bradt, V. H. Dibeler, and F. L. Mohler: *J. Res. Natl. Bur. Std.*, **50**, 201 (1953).
123. F. L. Mohler: Meeting of ASTM E-14 Committee, Los Angeles, June, 1959.
123a. R. E. Honig: "Trace Analysis in Semi-Conductor Materials" (J. P. Coli, ed.), Pergamon Press, New York (in press).
124. N. B. Hannay and A. J. Ahearn: *Anal. Chem.*, **26**, 1056 (1954).
125. M. G. Inghram: Mass Spectroscopy, in Yoe and Koch (eds.), "Trace Analysis," John Wiley & Sons, Inc., New York, 1957.
126. M. G. Inghram and R. J. Hayden: A Handbook on Mass Spectroscopy, *Natl. Res. Council Publ.* 211, 1954.
127. R. E. Honig: *J. Appl. Phys.*, **29**, 549 (1958).
128. M. G. Inghram: *Ann. Rev. Nucl. Sci.*, **4**, 81 (1954); *J. Phys. Chem.*, **57**, 809 (1953).
129. R. K. Webster: *Advan. Mass Spectrometry Proc.*, 1959, p. 103.
130. R. K. Webster, J. W. Morgan, and A. A. Smales: *Geochim. Cosmochim. Acta.*; see also R. D. Russell and R. M. Farquhar: "Isotopes in Geology," Interscience Publishers, Inc., New York, 1960.
131. W. M. Hickam: *ASTM Spec. Tech. Publ.*, **149**, 17 (1951).
132. R. E. Honig: *Anal. Chem.*, **25**, 1530 (1953).

10

Isotope Abundance Measurements and Their Application to Chemistry

C. C. McMullen and H. G. Thode

Departments of Chemistry and Physics
McMaster University
Hamilton, Ontario

10.1. Introduction

The conventional mass spectrometer still retains its preeminence as regards speed and accuracy of analysis of stable isotopes. In fact, it is used almost exclusively for the determination of absolute and relative isotope abundances, with the exception of the wide use of infrared methods for the determination of hydrogen-deuterium ratios.

The existence of isotopes was clearly established during the early years of this century, when intensive studies of the radiations of naturally occurring radioactive species indicated that elements of different atomic weights possessed identical chemical properties. Using the parabola method for positive-ion analysis, J. J. Thomson [1] showed conclusively, without recognizing the significance of his discovery, that neon consisted of at least two types of atoms, one of mass 20 and another of mass 22. A third stable isotope of mass 21 was later shown to exist.

In 1919 F. W. Aston [2] designed the first "mass spectrograph" which could be used to determine isotopic abundances. Although the photometric determination of the density of the traces produced by ions on the photographic plates introduced a number of errors, Aston succeeded in making the first abundance determinations of the isotopes of a number of elements using mass-spectrographic techniques.

At the time that Aston was engaged in the development of his mass spectrograph, A. J. Dempster [3] constructed a new type of apparatus employing a method used earlier by Classen [4] to determine the electron charge-mass ratio. This instrument utilized the geometrical focusing properties of a homogeneous magnetic field on a stream of charged particles. With this instrument Dempster measured the abundances of the isotopes of a number of the light elements.

Following the pioneer work of Aston and Dempster, there was little development work done in the field of mass spectrometry during the next twenty years. In fact, until after 1935, when A. O. Nier [5] introduced a number of refinements to mass-spectrometer design, rather limited advances were made in the field of isotopic abundance measurements. In more recent years, many more refinements have been made, particularly with regard to more efficient ion sources and to more sensitive ion detectors, so much so that the mass spectrometer has become an extremely sensitive instrument for the detection of practically all the elements and remains second only in sensitivity to radioactive-counting techniques.

10.2. Identification of Natural Isotopes

As a research tool the mass spectrometer has been put to very wide use for the identification of natural isotopes. Since 1920, when Aston and Dempster first published results of investigations of the isotopic composition of neon and of magnesium, all the natural elements have undergone close scrutiny in an effort to establish which isotopes exist in nature. As pointed out in the previous section, mass-spectrometric techniques saw little development prior to 1935. The sensitivity of all the early instruments was very low. As an example, in the work of Aston [6] on the isotopes of ruthenium, he was unable to establish the existence of Ru^{98}, which was later found to have a natural abundance of 1.9%. Similarly, Ni^{64}, which represents 1.16% of natural nickel, was not measured by the same observer [7]. When sector-type instruments were developed, which incorporated more efficient ion sources, the limits of detection were extended by another order of magnitude, so that isotopes which represented 0.1% of the abundance of an element could be determined with relative ease.

In recent years considerable scientific effort has been directed toward the elucidation of the properties of nuclei. The isotopic constitution of matter provides a basis upon which certain hypotheses of nuclear structure can be tested. In addition, considerable interest has been shown in the radiative-capture cross sections of many elements for neutrons of various energies. On this account it was desirable to have a method whereby it would be possible to measure the build-up of a relatively small number of nuclei.

With the development of the electron multiplier by Bay [8] and Allen [9] and its subsequent incorporation into a mass spectrometer as an ion detector [10], it has been possible to increase the sensitivity of a standard mass spectrometer, using an electrometer ion-detecting system, by as much as 10^3 to 10^4. With this increased sensitivity it was possible for many workers to set upper limits on the abundance of many isotopes of 1 part in 50,000 of the most abundant isotope.

By modifying a surface-ionization source to incorporate two filaments, Inghram and Chupka [11] have succeeded in analyzing samples of the heavy elements as small as 10^{-12} g, when the source is used in conjunction with an electron multiplier detector. The overall efficiency of a surface-ionization source depends upon the number of neutral molecules impinging on the ionizing surface at the operating temperature of the filament. In the case of many elements which have high ionization potentials, it is found that the vapor pressure of the compound containing the element reaches a high value before a suitable ionization temperature can be achieved. In such cases the sample evaporates from the surface before the atoms are ionized. The multiple-filament source circumvents this difficulty when the sample is dried on one of two sample filaments which is in close proximity to a high-temperature ionizing filament. By controlling the current in the sample filament, it is possible to set the sample evaporation rate at any given level. The neutral atoms which then impinge on the hot ionization filament are then ionized and accelerated into the magnetic analyzer.

White, Collins, and Rourke [12] have employed a filament which is folded in the shape of a V, into which the sample is placed. Neutral atoms which are evaporated from the hot tungsten are given an additional opportunity to become ionized on the walls of the tungsten V before they reach the first slit of the ion source. Both of the above methods prove successful in increasing the efficiency of an ion source for elements having relatively high ionization potentials.

Moreover, White et al. [13] have made isotopic ratio determinations of adjacent masses whose abundances are different by 10^4 to 1 or greater. Employing a two-stage 180° magnetic analyzer in which a double-slit system was incorporated between analyzers, they showed that Ta^{180} existed in nature with an abundance of about 1 part in 10^4 of the abundance of Ta^{181}. Other isotopes of tantalum had abundances less than 3 parts in 10^6 of the Ta^{181}.

Using an inflection-type mass spectrometer employing an electron multiplier as a detector, Kerwin [14, 15] and his coworkers [16, 17] have studied a number of elements in a search for new stable isotopes. In general, he has placed upper limits up to 1 part in 10^4 on the abundance of the stable isotopes of phosphorous, argon, manganese, cadmium, neon, chlorine, gallium, carbon, oxygen, and zinc relative to the most abundant isotope of the element.

10.3. Isotope Abundances

With the discovery in 1914 of the isotopic nature of stable elements by J. J. Thomson [1], an intense interest developed in learning more of the physical composition of matter. Today, some three hundred stable nuclides have been shown to exist in nature and their natural abundances have been determined.

A. Absolute abundances

As early as 1910 Soddy [18] suggested the existence of isotopes of the heavy elements when he found that radioactive materials having easily distinguishable radioactive radiations could not be separated by chemical means. The irrefutable evidence acquired from studies of the stable elements by means of the mass spectograph showed beyond doubt the isotopic structure of nearly all the elements. Moreover, the isotopic abundances of the elements appeared to be constant, and their chemical properties similar, as suggested by Soddy [18]. There were exceptions, of course, in instances where the heavy elements were known to be of radiogenic origin. Subsequently, it was shown that, in the light-element region, argon found in sedimentary rock was enriched in Ar^{40}, as compared with atmospheric argon, due to the decay of K^{40}.

Needless to say, it is now known that the isotopes of an element do vary in their chemical properties and that some fractionation of isotopes does occur in geological and biological processes. For example, the S^{32}/S^{34} ratio in terrestrial sources shows variations up to 10% depending upon the history of the source of the sulfur. Also, the C^{13} content of limestones is found to be enriched relative to that found in other sources. A more complete discussion of these effects will follow later, in sections dealing with isotopic variations.

With the discovery of nearly three hundred stable nuclides in recent years, some very interesting features of nuclear structure have come to light. For example, stable nuclides having an even number of both protons and neutrons (even-even) are about forty times greater in number than nuclides having odd numbers of protons and neutrons (odd-odd). In addition, the number of even-even nuclei is about three times greater than either even-odd or odd-even nuclides. Nuclear-shell effects are indicated by the larger number of stable isotopes for $Z = 50$ than in other regions and by the large number of stable nuclides for $N = 50$ and $N = 82$. Also, it should be pointed out that the even-even isotopes of an element, with but few exceptions, account for between 70 and 100% of the total element.

Before discussing the mass-spectrometric methods for determining isotopic abundances, two important reasons for having accurate isotopic abundance measurements should be mentioned. First, any detailed test of postulates of element formation would require a comparison of the abundance distribution of individual nuclides. Second, since there is considerable interest in the radiative-capture process of many elements, accurate upper limits on the abundance of the isotopes of these elements are essential in order that a small build-up of a given nuclide can be accurately determined.

Absolute isotopic abundances have been determined chiefly by the use of two types of mass spectrometers. These are gas-analysis instruments using electron-impact-source assemblies and solid-source instruments using surface-ionization and crucible-type sources.

B. Calibration methods

In 1950 Nier [19] reported the redetermination of the relative abundances of the isotopes of a number of elements. These measurements were made using two 60°-type mass spectrometers, one employing an electron-impact source for gas analysis, the other incorporating a furnace that was used for the analysis of substances which were not volatile at room temperature. For specific details of the source arrangements, the reader is referred to the publications of Nier and his coworkers. By taking into account the mass-discrimination effects which are inherent in the mass spectrometer, Nier was able to report absolute isotopic abundances with an accuracy of 0.1%. Prior to the publication of these improved data, the absolute accuracy of abundance measurements was limited to about 1%, although the experimental precision in most cases was very often an order of magnitude better than this figure.

In order to determine the value of the discrimination corrections, essentially pure samples of Ar^{36} and Ar^{40} were produced in thermal-diffusion columns [20] and then used to produce synthetic-argon mixtures which had a very accurate Ar^{36}/Ar^{40} ratio. These standard mixtures were then utilized to calibrate the mass spectrometers for errors which were dependent on the relative masses of the ions under study.

For the mass-spectrometric analysis of gaseous elements, the sample is introduced into the ionization chamber through what is known as a "leak." Molecular flow will occur through the leak when the mean free path of the molecules is at least ten times greater than the diameter of the leak. Thus the velocity of flow will be inversely proportional to the square root of the isotopic mass. At the same time, molecular flow will, in general, occur when the gases are pumped from the ionization chamber. Provided the volume of the gas reservoir behind the leak is large, so that the depletion of the light isotopes is small during an analysis, a cancellation of the discrimination in the source occurs. Thus the gas in the ionization chamber will have the same composition as that in the reservoir.

On the other hand, viscous flow will dominate in the leak when conditions are such that the mean free path is small compared with the leak dimensions. In this instance the flow rate is independent of the isotopic mass, and a correction which is proportional to $1/\sqrt{M}$ is necessary to compensate for molecular flow from the ionization chamber.

When analyzing elements which are ionized by evaporation from an oven or from a heated surface, isotopic fractionation is once again a serious problem. This fractionation, which again depends upon $1/\sqrt{M}$, is large when the isotopic masses are small. Lithium is an example where this effect has been quite noticeable in the many investigations of the isotopic abundance of this element. It is therefore essential that each ionization process be investigated to determine the amount of discrimination which occurs during the analysis of a particular element.

When electron-impact sources are employed, some mass discrimination can be attributed to space charges which build up in the ionization chamber. Since the ion current is a function of the magnitude of the ionizing electron beam, corrections for this effect can be made by observing the size of the ion currents at various electron currents and then extrapolating to zero ionizing current.

In addition, isotope-fractionation effects can occur in the dissociation of molecules by electron impact. Since there is no satisfactory explanation for this phenomenon, care must be exercised when mass-spectrometric studies of molecular species are undertaken.

Other effects which are of a minor nature but still require attention when making absolute abundance measurements are those which occur in the analyzer tube and at the collector. Residual gases in the instrument can provide a source of error when weak ion currents are being measured. These can usually be minimized by baking the spectrometer at temperatures between 300 and 400°C in order to outgas the entire source-and-analyzer system before the sample is introduced. Small-angle ion scattering can in general be overcome to a large extent by proper design of the instrument. The tandem analyzer design of Inghram and Hayden [21] and White and Collins [22] practically eliminates this effect, and it is then possible to measure the ion currents of adjacent masses which have an abundance ratio of approximately 10^8.

At the collector, secondary electron emission produced when the energetic ions are stopped is prevented by incorporation in the detector assembly of a suppressor grid. A less elegant method is to collect the ions in a long V-shaped electrode from which very few secondary electrons will escape. If an electron multiplier is employed to detect the ions, a mass discrimination will occur when the integrated electron current from the multiplier is used as a measure of the ion current. If pulse-counting techniques are used to detect the ions, this discrimination disappears, provided each ion produces an electron pulse which is of sufficient magnitude to be counted by some rate-measuring device.

The high ohmic resistances which are used in d-c amplifiers and vibrating-reed electrometers very often have a positive or a negative voltage coefficient. Calibration procedures for this effect can in general be made with little difficulty using various methods which introduce a known current into the detection system.

It is obvious that accurate absolute abundance measurements require very careful studies of a number of discrimination effects which occur in the mass spectrometer if absolute accuracy of the order of 0.1% is to be achieved.

10.4. Applications to Nuclear Chemistry

A. Natural radioactive isotopes

One of the first applications of the mass spectrometer to nuclear problems was in the identification of naturally occurring radioactive isotopes. Stable isobars, which are nuclides having the same mass number, do not exist when their atomic number differs by unity. Therefore, when adjacent stable isobaric nuclides are found in nature, it is to be expected that one of the two species will exhibit radioactivity. Of the 10 groups which can be found in the table of isotopes, 7 have shown radioactivity in one of their members. In some cases the natural activity of an element was observed prior to the discovery of the isotope responsible for this activity. Two such cases are K^{40} and Lu^{176}. In the case of K^{40}, which constitutes 0.0119% of natural potassium, radioactivity was conclusively demonstrated with the aid of a high-intensity mass spectrometer [23] which was capable of separating the isotopes of this element. Subsequent study revealed that the activity arose in K^{40}.

Studies of a suspected daughter element of a radioactive parent have in two other instances resulted in identifying the source of the radioactivity. These are Rb^{87} (Mattauch [24]) and Re^{187} (Hintenberger, Herr, and Voshage [25]). By analyzing mass-spectrometrically the strontium which had been extracted from a rubidium-rich Manitoba mica, Mattauch observed that the Sr^{87} made up 99.7% of the total strontium. Since normal strontium contains 7.02% Sr^{87}, it was apparent that the enhancement of the Sr^{87} was due to the natural beta decay of Rb^{87}.

Isobaric triplets make up three of the groups in which natural radioactivity has been observed. Because of the odd proton–odd neutron character of the central member in each of these groups, the activity had been tentatively assigned to this member. The discovery of La^{138} by Inghram, Hayden, and Hess [26] led to the detection of its activity by a number of observers (Pringle, Standil, and Roulston [27]; Pringle, Standil, Taylor, and Fryer [28]; Mulholland and Kohman [29]). Ta^{180}, which was discovered by White, Collins, and Rourke [13] in 1955, using a two-stage magnetic analyzer, has been assigned a half-life greater than 10^7 years. Although the determination of the three isobaric masses in the V^{50} group (Johnson [30]) show V^{50} to be unstable against decay to both Cr^{50} and Ti^{50}, several investigations have not detected any activity in this nuclide [31]. This work followed the discovery of V^{50} in 1949 (Hess and Inghram [32]; Leland [33]), using 60°-magnetic-deflection mass spectrometers.

B. Half-life determinations

The use of a mass spectrometer to determine the half-life of a radioactive nuclide has proved to be a very elegant method in a number of instances. This can be achieved in two ways, either by observing the rate of disappearance of the parent nuclide or by the growth of the daughter nuclide. Nier [34] first employed the latter method to determine the half-life of U^{235}. He determined mass-spectrometrically the amount of Pb^{207} (the stable end product of the actinium series) in uranium minerals which had different geologic ages. His value of 7.13×10^8 years is the current accepted value for this half-life.

A second example of the daughter-growth method has been used by Inghram and his collaborators [35] to determine the half-life of Pu^{240}. By mass-spectrometrically determining the isotopic ratios of uranium and plutonium and employing the half-life of Pu^{239} determined by Westrum et al. [36], the half-life of $6,580 \pm 40$ years for Pu^{240} can be found.

Both of these examples which involve long half-lives use the daughter-growth method. When the half-lives are relatively short, the parent-decay approach is a more accurate method. Thode and his coworkers [37, 38] have assigned a half-life of 10.27 ± 0.18 years to the fission product Kr^{85}. This was determined by examining, over a period of seven years, the decay of the Kr^{85} relative to stable Kr^{86}, which was originally extracted by Thode and Graham [37] from irradiated U^{235}. Using similar techniques, Macnamara, Collins, and Thode [39] have shown the half-life of Xe^{133} to be 5.271 ± 0.002 days.

Half-lives in this range, which have been determined by counting techniques, are subject to numerous corrections and, in general, accuracies of 10% can be expected. However, the unambiguous results obtained using mass-spectrometric techniques give half-lives which have accuracies that are better than 1%.

By utilizing the rate of disintegration of a known mass of an element, the half-lives of several nuclides have been determined. These include Ni^{63}, Sr^{90}, Cs^{137}, and U^{233}. In this method the absolute disintegration rate of a given element is determined, using counting techniques, and then the amount of the element is determined, using isotope-dilution methods in conjunction with a mass spectrometer. The equation

$$\frac{dN}{dt} = -\lambda N$$

where dN/dt = rate of disintegration
λ = decay constant
N = number of atoms present
can then be used to evaluate the half-life $T_{1/2} = 0.693/\lambda$. Values for half-lives ranging from about 20 years up to about 100 years can be determined in this manner with an accuracy of about 1%.

C. Neutron-capture cross sections

When an element is irradiated by neutrons, the rate of absorption of a particular stable isotope is given by

$$\frac{dN}{dt} = -\sigma\phi N$$

where σ = isotopic-absorption cross section
ϕ = neutron flux
N = number of atoms of isotope
The isotopic abundance of the nuclide which absorbs the neutrons will after a time t be modified according to the relation

$$N = N_0 e^{-\sigma\phi t}$$

where N_0 is the number of atoms of the nuclide which were present at $t = 0$. Therefore, provided the product of $\sigma\phi t$ is of sufficient magnitude to produce a measurable change in N_0, it is possible by mass-spectrometric means to determine the change in the isotopic abundance, and hence evaluate the value of the isotopic cross section, having first obtained some knowledge of the integrated flux ϕt.

Provided the neutron irradiation produces a stable nuclide, the above simple relationships are valid. However, when the resultant nuclide is a radioactive species, the equations become more complex because due consideration must be taken of the decay of the nuclide produced.

Dempster [40] analyzed a sample of cadmium in his mass spectrograph which had undergone a long exposure to neutrons. He found that the abundance of Cd^{113} had been depleted by approximately the amount that the Cd^{114} abundance had been enhanced. This observation showed that Cd^{113} is the isotope which is mainly responsible for the large-neutron-absorption cross section of elemental cadmium. In addition, Dempster and his collaborators [41], by similar means, ascribed the large absorption in samarium and gadolinium to Sm^{149}, Gd^{155}, and Gd^{157}. Figure 10.1 shows the isotopic pattern of gadolinium before and after irradiation with thermal neutrons. This technique was further employed by Inghram,

Thode, and their coworkers [42–46] to determine the neutron-absorption cross sections for the isotopes of europium, mercury, neodymium, xenon, and krypton. This method gives results which are consistent with those obtained by the use of counting techniques.

In work of this kind the uncertainty in the value of the integrated flux introduces the most serious experimental problem. Flux monitors such as BF_3 and Co^{60} have been used with success by Tomlinson and his workers [47, 48] to obtain an accurate determination of this parameter.

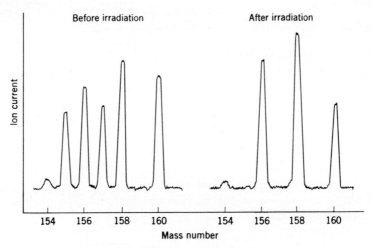

Fig. 10.1. Mass spectrogram of the element gadolinium before and after irradiation with thermal neutrons. [*From W. H. Walker, Ph.D. Thesis, McMaster University, Hamilton, Ontario, Canada (1954).*]

This is made possible since the cross sections for boron and for cobalt as a function of neutron energy have received considerable study and are well known.

D. Mass fission yields

The application of the mass spectrometer to the study of fission yields was first reported in 1947 when Thode and Graham [37] analyzed the rare gases extracted from an irradiated sample of uranium. Prior to this, radiochemical studies of the fission products had shown the general nature of the double-humped fission-yield curve. Although no satisfactory theory for the fission act has as yet been proposed, accurate mass-spectrometric data, in conjunction with radiochemical determinations, have made it possible to make a number of postulates and hypotheses regarding the fission process.

It is common knowledge that the fission of heavy elements with slow neutrons results in the production of fission products which are distributed over two broad mass ranges with centers in the region $A \sim 95$ and

$A \sim 140$. The mass equation for the fission process can be written in the following manner:

$$M(A, Z) + M(1, 0) \rightarrow M^*(A + 1, Z) \rightarrow M(A_1, Z_1) + M(A_2, Z_2)$$
$$+ \, vM(1, 0) + \gamma + Q$$

$M(A, Z)$ is the mass of the fissioning nucleus, $M(A_1, Z_1)$ and $M(A_2, Z_2)$ are the masses of the primary fission fragments, and v the total number of prompt neutrons. Q is the total kinetic energy of the particles released in the fission process, and γ is the energy released at the moment of fission in the form of prompt gamma radiation. Since the fission fragments are neutron-rich, they are unstable toward beta decay. Consequently, by means of a series of beta transitions, the unstable fission products proceed toward stable end products which lie near the Bohr-Wheeler stability line. A typical fission-product decay chain is shown in the following example:

If sufficient time is allowed to elapse so that the primary products can decay to stable nuclides, relative-abundance measurements can be made which will give the relative fission yields of the fission-product decay chains. It was early studies of this kind which led to the discovery of the fission product Kr^{85} [37], which does not appear in nature, and permitted the assignment of the long-lived radioactivity found in rare-gas fission products to this isotope [49]. In this manner the mass-fission-yield data have been obtained for Th^{232}, U^{233}, U^{235}, U^{238}, and Pu^{239}, which have undergone neutron-induced fission, and U^{235} and U^{238}, which have undergone spontaneous fission.

At the outset, the slow neutron fission of U^{235} was studied very extensively. Study of the mass yields of the rare gases krypton and xenon received considerable attention from Thode and his collaborators, and in this instance the mass spectrometer lent itself most effectively. Figure 10.2 shows a recording of the isotope patterns of normal krypton and fission-product krypton obtained using a mass spectrometer. The general asymmetric shape of the fission-yield curve had been previously established by radiochemical methods. This asymmetry of fission is believed to be determined by the extra stability of the 50 and 82 neutron configurations [50]. The latter configuration is believed to determine the fine structure in one of two ways. Either there is, in the initial fission act, a preference for the 82 neutron configuration (Wiles [51]; Wiles, Smith, Horsley, and Thode [52]; Glendenin, Steinberg, Inghram, and Hess [53]) or the extra neutron in fragments containing 83 neutrons

(Glendenin [54]) or 85, 87, or 89 neutrons (Pappas [55]) is loosely bound and may "boil off," producing higher yields in adjacent decay chains. Relative-abundance determinations of the xenon isotopes produced in the fission of U^{235} by slow neutrons showed that Xe^{133} and Xe^{134} lie well above the smooth radiochemical curve. The abnormal yields of Xe^{133} and Xe^{134} are shown in Fig. 10.3. This "fine structure," as it was called, was discovered also in the light-mass region at the mass which is complementary to the mass of the heavy fragment.

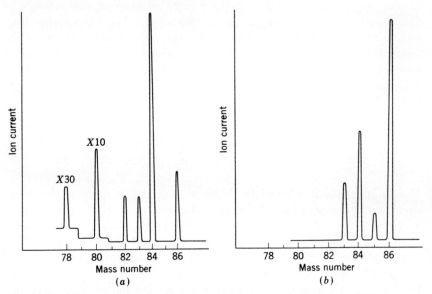

Fig. 10.2. Mass spectrograms of (a) normal krypton and (b) fission-product krypton. [*From Thode, Nucleonics*, **3**, 14 (1948).]

An extensive study of the nongaseous elements produced in fission has been carried out by Tomlinson and his associates. By determining the relative yields of rubidium, strontium, cesium, cerium, neodymium, and samarium isotopes, he was able to construct a mass-yield curve for the fission of U^{235} (Petruska, Thode, and Tomlinson [56]). These results along with the time-of-flight data of Milton and Fraser [209] indicate that most of the fine structure occurs as a result of variations in neutron emission (Farrar and Tomlinson [210]). Katcoff [57] has reviewed the fission-yield data for the neutron fission of uranium, thorium, and plutonium, and the reader is referred to this author for a tabulation of fission yields.

Another aspect of fission that presents greater experimental difficulty but which yields valuable information regarding the fission process is that of primary yields. In cumulative-yield studies the decay chains are allowed to reach their stable end products before studies are made of the chain yields.

Primary-yield studies permit the observer to gain some knowledge of the distribution of charge in the fission act. It has been pointed out that the primary products are unstable toward beta decay, and moreover these products usually have short half-lives. Chemical separations and analysis of the samples must therefore of necessity be carried out rapidly before the primary product decays to its daughter or is enhanced by the decay of its precursor.

Toward the end of the decay chains the half-lives are generally long, but then the primary yields are so small that they are usually swamped

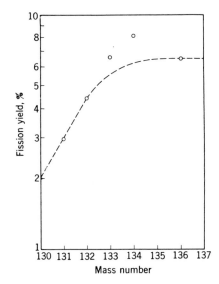

Fig. 10.3. Abnormal cumulative fission yields for Xe^{133} and Xe^{134} in the thermal neutron fission of U^{235}.

by the much higher yields of their predecessors. In a few instances where decay chains have shielded nuclides, such as at Br^{80}, Br^{82}, I^{128}, and I^{130}, the decay chains are terminated at the shielding nuclides of selenium and tellurium. Therefore mass-spectrometer yield determinations of the nuclides Kr^{80}, Kr^{82}, Xe^{128}, and Xe^{130} will give a measure of the primary yields of the parent bromine and iodine nuclides. Moreover, included in such mass-spectrometric measurements will be a contribution of about 1% from the primary yields of the krypton and xenon isotopes, and appropriate corrections must be made to the isotopic abundance values to compensate for this effect. Kennett and Thode [58] have measured the primary yields of Br^{80}, Br^{82}, I^{128}, and I^{130} through their rare-gas daughters for the fission of U^{235}, U^{238}, and Pu^{239}, using high-sensitivity mass-spectrometric techniques. They have shown that the primary yields of these isotopes are about 10^3 to 10^4 less than the yields of the most probable primary products, which have about three additional units of charge at the time of formation.

Figure 10.4 shows the xenon pattern for the thermal fission of Pu^{239}. The mass spectrogram illustrates how high sensitivity and high resolution have been combined to separate the low-yield Xe^{128} and Xe^{130} peaks from the hydrocarbon contaminants at the same mass position.

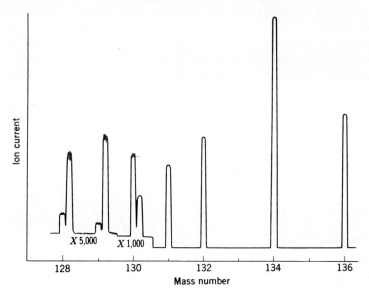

Fig. 10.4.   Mass spectrogram of the xenon pattern for the thermal neutron fission of Pu^{239}, showing how the isotopic mass is resolved from the hydrocarbon with the same mass number.   (*From Kennett and Thode* [58].)

Any satisfactory theory of fission will have to explain both the distribution in mass and the distribution in charge of the fission products. Primary yields such as those determined by Kennett and Thode are therefore of considerable importance for the experimental verification of any such theory.

E. Isotope-dilution techniques

Isotope dilution as an analytical tool (Inghram [59]) has made it possible to determine absolute fission yield over effectively the entire mass-yield curve.   By adding a known number of atoms of an isotope of an element under study and then measuring by means of a mass spectrometer the change in the isotopic abundances of the element, it is possible to calculate the absolute number of atoms which were present in the original sample.   For example, the amount of Xe^{128} produced from the neutron absorption of I^{127} can be measured and then mixed with a fission-product sample of xenon to make absolute yield determinations of the isotopes of xenon. In a similar manner, strontium and cesium, with normal isotopic content, can be used as tracers to determine the yields of the strontium and cesium isotopes produced in fission. This technique has been used very effectively by Petruska, Thode, and Tomlinson [56] to make quantitative determinations of the fission yields of 28 mass chains in the thermal neutron fission of U^{235}. Fritze, McMullen, and Thode [60] have used the Xe^{128} "spiking" technique to measure the absolute fission yields of the xenon and krypton isotopes in the neutron fission of Pu^{239}.

The isotope-dilution procedure can, in addition, be employed to determine the half-lives of long-lived isotopes. In essence, the daughter nuclide, from a given amount of a long-lived radioactive species, is extracted by chemical means. The amount of the daughter nuclide, which has appeared in a given period of time, is then determined by adding a known amount of another isotope of the daughter element. Half-lives which can be determined by the use of this technique are accurate to within a few per cent.

The half-life determination of Sr^{90} and Cs^{137} mentioned earlier was determined by Wiles and Tomlinson [61, 62], using isotope-dilution methods. A 4π proportional counter was used to measure the disintegration rate of the long-lived isotope. In the case of Sr^{90}, the Sr^{90}/Sr^{88} ratio was then measured, using a mass spectrometer. A known amount of Sr^{88} was added, and the ratio remeasured. The mass-spectrometer ratios and the weight of the Sr^{88} added were then used to obtain the initial amount of Sr^{90}.

F. Trace analysis

The use of the mass spectrometer for trace-element analysis has, during the last ten years, become a reality. In fact, the commercial use of this technique would seem to be limited only by the availability of suitable mass-spectrometric equipment.

There are four methods of positive-ion production, each fundamentally different, which have been used for trace analysis in solid materials, and one method of ion production which has been applied specifically to gas analysis. The methods used for solids are Dempster vacuum spark, isotope dilution, ion bombardment, and total vaporization. The first two methods have been employed more extensively than the last two because of their higher sensitivity and accuracy and will therefore be discussed here. For gas analysis, the electron-impact source used in conjunction with high-sensitivity mass-spectrometric techniques has been of good service in the detection of trace impurities in gaseous samples.

The use of the vacuum-spark method in mass spectroscopy was first reported by Dempster [63] in 1935. This method is analogous to the optical-spectroscopic method in that all elements present are analyzed in a single experiment. It is found that all elements are detected with equal sensitivity to within about one order of magnitude. However, two important disadvantages are inherent in this method. First, the ion emission is unstable because of the pulsing procedure required for the high voltage which is applied to the electrodes. Second, when a spark voltage of 40,000 volts is used, an ion spread of 1,000 ev may be expected. The former difficulty is circumvented by using either the photographic-detection technique or ratio recording. In the latter method the ratio of the resolved to unresolved ion beams is measured, rather than the magnitude of the resolved beam alone. The energy spread in the ion beam can be removed by utilizing a double-focusing spectrometer in which an electrostatic velocity selector is combined with a magnetic analyzer.

The maximum sensitivity of the vacuum-spark method for a detectable effect has been reported to be about 10^{11} atoms (Hannay and Ahearn [64]) for elements of medium mass. This sensitivity is limited by instrumental background and by gas scattering. At some mass positions such as at mass 12, where hydrocarbons can interfere, the background level is of ultimate importance, while at other mass positions, little interference is encountered. The use of modern high-vacuum techniques will tend to minimize this difficulty. The problem of gas scattering of an ion current, by the residual gas, to the position where the impurity is to be observed can also be reduced by the use of high-vacuum procedures.

Trace-element analysis based on the vacuum-spark method has been effectively used for some elements. It is destined to be used more widely when double-focusing instruments employing this method are made available on a commercial basis.

The isotope-dilution method which was first proposed by G. Hevesy [65] has been used extensively for the determination of the elements and compounds of hydrogen, carbon, and nitrogen (Rittenberg [66]). Now that separated isotopes of many of the elements are more readily available, the method has been greatly extended, and the isotope-dilution technique has become a very effective tool in quantitative trace analysis.

This method differs from the one previously discussed since it permits the determination of only a few elements at one time. However, the tremendous increase in sensitivity of several orders of magnitude makes it a very attractive method. The methods of trace analysis using this technique have been outlined in the previous section. Since a known amount of tracer is added prior to the processing of the sample by chemical means, a quantitative recovery of the tracer and the unknown is not necessary. Thus the trace-element concentration can be calculated even though but a small fraction of the sample is recovered. A hypothetical isotope-dilution measurement of lanthanum in a solid sample [59] is shown in Fig. 10.5.

One serious difficulty which may be encountered in this method is that of contamination from reagents. By analyzing samples of different size, it is possible to evaluate the extent of this interference. In general, extreme care is essential in the preparation of reagents and distilled water to ensure that contamination by trace quantities of an element is reduced to a minimum.

Since isotope dilution depends upon the mass-spectrometric determination of isotopic ratios, the method is restricted to elements which have at least two stable elements. In cases of monoisotopic elements such as cesium, long-lived fission products Cs^{135} and Cs^{137} can be used

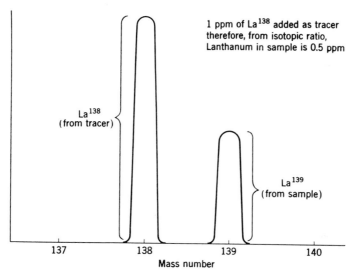

Fig. 10.5. Hypothetical isotope-dilution measurement of lanthanum in a solid sample. (*From Inghram* [59].)

as tracers for analysis of this element. In all, some 68 elements can be analyzed by this means [59, 67].

Tilton et al. [68] have analyzed the reagent ammonium chloride to determine the uranium content in this reagent. They have found it to be 75 parts in 10^{12}, with an accuracy of 4 parts in 10^{12}. This gives some indication of the power of this method for trace-element determinations.

Modern mass spectrometers which use relatively efficient sources and an electron multiplier to detect the ion current can detect as little as 10^{-12} g of most elements. Therefore, as long as the impurity in a sample equals 10^{-12} g, it can be detected.

For gas analysis it has been possible, using high-vacuum techniques and pressures of 10^{-9} mm, to measure 1 part in 10^8 of krypton in xenon (Wetherill [69]). By eliminating the difficulty of ion scattering, which was mentioned in connection with the Dempster vacuum-spark method, it is conceivable that the extension of gas techniques should permit sensitivities of 1 part in 10^{10}.

The application of trace-element techniques has brought fruitful results in the area of solid-state physics. Honig [70] found impurities of arsenic, lead, tin, iron, phosphorus, antimony, and cadmium in germanium. In reality, studies of the abundance of low-yield fission products and the primary fission yields of Br^{78}, Br^{80}, I^{128}, and I^{130} [58] can be classified as employing trace-analysis technique since ultrahigh sensitivities are required in conjunction with good vacuum techniques.

Trace analysis as an analytical tool will undoubtedly become more widely used when high-sensitivity commercial mass spectrometers become available.

10.5. Isotope Chemistry

A. Measurement of small variations in isotope abundances

In most chemical applications only relative isotope abundance measurements are involved. Here we are concerned with measuring small changes in isotope ratios that occur in chemical reactions where such isotope ratios need only be measured relative to an arbitrary standard.

In work with the light elements changes in the isotopic ratio of $\pm 0.1\%$ are just detectable with the best available instruments in which ion currents are collected and measured singly. However, by collecting the two ion beams for the isotopes in question at the same time and measuring the isotope ratio directly, a much higher precision may be obtained. With the simultaneous-collection scheme, fluctuations in the ion source will, for the most part, leave the isotope ratio unchanged. Isotope ratios of two samples can be compared quickly under more nearly identical conditions by this means.

Aston [71] was the first to suggest the use of simultaneous collection of two ion beams. However, the method was not employed until 1941, when Straus [72] applied the technique to the study of the nickel isotopes. Since then, the method has been developed for routine isotope measurements by Nier [73]. Figure 10.6 illustrates Nier's method of measurement. Each ion current passes through a resistor, and the resulting voltages are amplified and compared by a put-and-take potentiometer system, using the galvanometer as a null indicator. Variations of this scheme have been reported by Nier, Ney, and Inghram [73, 74], McKinney et al. [75], and Wanless and Thode [76]. Figure 10.7 shows an amplifier interconnection circuit used by the latter workers.

The ion current i_1, due to the most abundant isotopic species falling on collector 1, is amplified by d-c amplifier A. The output of this amplifier is then applied across a put-and-take potentiometer. The

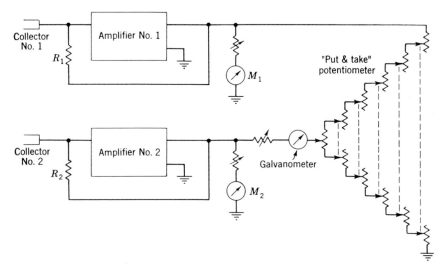

Fig. 10.6. Nier's method of comparing two ion currents. [*From Thode and Shields, Rept. Progr. Phys.*, **12**, 1 (1949).]

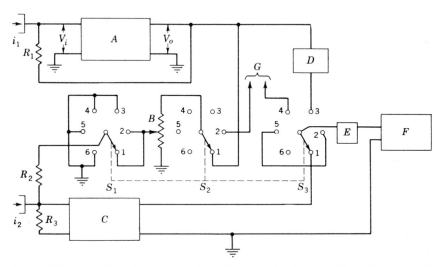

Fig. 10.7. Amplifier interconnection circuit. *A*, d-c amplifier; *B*, put-and-take potentiometer; *C*, vibrating-reed electrometer; *D*, filter circuit; *E*, shunt selector; *F*, pen recorder; *G*, to No. 2 d-c amplifier. (*From Wanless and Thode* [76]).

current i_2, flowing through resistor R_2, is balanced against some fraction X of the voltage developed across the potentiometer. The potential at the junction of resistors R_2 and R_3 is zero when the system is balanced. Under these conditions, a zero signal is applied to the d-c amplifier, and the recorder pen will remain at the null position of a zero-center record-

ing instrument. When a second sample of gas with a slightly different
isotopic ratio is introduced into the mass spectrometer, a potential
difference will develop at the junction of resistors R_2 and R_3. This
difference will be amplified by amplifier 2, and the resulting signal will
drive the recorder pen away from central balance position.

The displacement of the recorder pen is proportional to the difference
in ratio i_1/i_2 in the two samples being analyzed.   Figure 10.8 illustrates
the shift in balance point resulting when two samples with slightly
different isotopic content are compared.

McKinney et al. [75] developed an elegant method for switching
rapidly from one sample to another, thereby increasing the precision

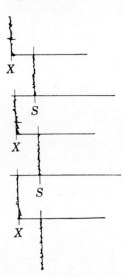

Fig. 10.8.   Recorder   trace. S = standard, X = unknown.   (*From Wanless and Thode* [76].)

by which the isotope ratio of the two samples may be compared.   The
mass spectrometer is provided with a double sample line, double leak,
and magnetic-valve system such that, by means of an electric switch,
either the gas sample from leak 1 or from leak 2 may flow into the mass
spectrometer.   An example of the magnetic-valve system developed
by McKinney et al. [75], which permits rapid alternate introduction of
the reference and unknown samples, is shown in Fig. 10.9.   By the
careful adjustment of the gas pressures of the two samples behind the
two identical leak systems, it is possible to switch back and forth from
one sample to the other rapidly, without changing pressures and temperatures in the source.   The recorder tracing shown in Fig. 10.8 was
obtained in this way.

By means of a dual ion-collection system and a dual sample-handling
system, the limit of detection of small differences in isotope ratios for
isotopes of the light elements has been reduced to 0.01%.   Actually,

the mass-spectrometer precision is so good that sample preparation is now the limiting factor [208]. These high-precision isotope-ratio instruments make possible the direct measurement of simple-process isotope-fractionation factors in isotope-separation work and the accurate measurement of equilibrium and kinetic isotope effects in chemical reactions, as well as provide a means of following the small changes in isotope

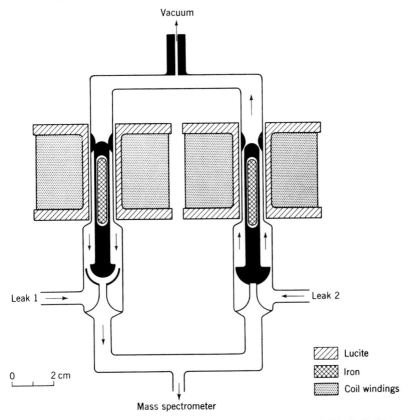

Fig. 10.9. Magnetic sample valves. (*From Wanless and Thode* [76].)

ratios which occur and have occurred in chemical and biological processes in nature.

B. Equilibrium properties of isotopic substances

In the twenty years following the discovery of isotopes by Soddy [18] and Fajans [77], considerable evidence was built up to show that isotopes were identical in their chemical properties and that their abundances were constant in nature, isotopes of radiogenic origin being an exception in this latter connection. This evidence was based on mass-spectrographic measurements, atomic-weight determinations, and unsuccessful attempts to separate isotopes by chemical means. By 1932, however, the rare isotopes of the light elements deuterium, carbon, nitrogen, and oxygen had been discovered by spectroscopic methods (King and Birge [78]; Giauque and Johnston [79]; Naude [80]; Urey et al. [81]). Additional wavelengths in the molecular spectra of compounds containing these elements could be explained only by the existence of these isotopes. Early in the investigations of the properties of hydrogen and deuterium compounds, Lewis and Cornish [82] and Washburn, Smith, and Frandsen [83] observed that there were differences in the vapor pressures of H_2O and D_2O, and it was also observed that there were slight differences in the vapor pressures of O^{16} and O^{18} waters. Since then, partial separation of the hydrogen and oxygen isotopes has been effected by the distillation of water, using very efficient fractionation columns (Huffman and Urey [84]; Randall and Webb [85]; Thode, Smith, and Walkling [86]). In distillation separation methods the simple-process fractionation factor may be defined by the relation

$$\alpha = \frac{p_{H_2O^{16}}}{p_{H_2O^{18}}} = \frac{n/(1 - n)}{N/(1 - N)} \qquad (10.1)$$

where N and n represent the mole fraction of one constituent in the liquid and vapor, respectively.

Then the total fractionation secured by a column is given by the formula

$$\frac{N_t/(1 - N_t)}{N_b/(1 - N_b)} = \alpha^{kz} \qquad (10.2)$$

where N_t, N_b = mole fractions of one constituent at top and bottom of the column, respectively

α = simple-process factor defined above

z = length of column

k = a constant

The quantity kz is referred to as the number of theoretical plates in the column.

Theory. Differences in the chemical properties of the hydrogen isotopes were to be expected in view of the large percentage mass difference. Rittenberg and Urey [87] showed, from theoretical calculations, that the equilibrium constants for the isotopic reactions

$$H_2 + 2DI \rightleftarrows 2HI + D_2$$
$$H_2 + D_2 \rightleftarrows 2HD$$

differed from each other and differed from the constants obtained, assuming a simple statistical distribution of the atoms among the various molecules, a situation which would prevail if the isotopes had identical chemical properties. Farkas and Farkas [88] made similar studies of the hydrogen-water exchange. In each case the constants calculated were beautifully confirmed by experiment. The literature on hydrogen-isotope-exchange equilibrium is now very extensive (Urey and Teal [89]; Kirshenbaum [90]).

In 1935 Urey and Greiff [91] became interested in the equilibrium properties of the isotopic compounds of boron, carbon, oxygen, nitrogen, and chlorine.

Using the well-known methods of statistical mechanics, they calculated equilibrium constants for many isotopic reactions. For example, they studied the oxygen isotope exchange between carbon dioxide and water. The calculated equilibrium constant K for the isotopic reaction

$$\tfrac{1}{2}CO_2{}^{16} + H_2O^{18} \rightleftarrows \tfrac{1}{2}CO_2{}^{18} + H_2O^{16} \tag{10.3}$$

was found to be 1.044 at 25°C. When the exchange reaction is written as in (10.3), then K may be expressed as

$$K = \frac{[CO_2{}^{18}]^{1/2}/[CO_2{}^{16}]^{1/2}}{[H_2O^{18}]/[H_2O^{16}]}$$

The enrichment factor, or separation factor α for an exchange process, is defined as the ratio of the heavy and light atoms in the one compound divided by the same ratio in the other compound. For the carbon dioxide–water exchange reaction,

$$\alpha = \frac{(O^{18}/O^{16})_{CO_2}}{(O^{18}/O^{16})_{H_2O}}$$

Thus α will be given by the equation

$$\alpha = \frac{2[CO_2{}^{18}] + [CO^{16}O^{18}]/2[CO_2{}^{16}] + [CO^{16}O^{18}]}{[H_2O^{18}]/[H_2O^{16}]} \tag{10.4}$$

Since the two isotopes O^{16} and O^{18} will be distributed between the molecules $CO_2{}^{16}$, $CO_2{}^{18}$, and $CO^{16}O^{18}$ in a random manner, the equilibrium constant K' for the reaction

$$CO_2{}^{16} + CO_2{}^{18} \rightleftarrows 2CO^{16}O^{18}$$

is $\qquad K' = \dfrac{[CO^{16}O^{18}]^2}{[CO_2{}^{16}][CO_2{}^{18}]} = 4$

Consequently, Eq. (10.4) reduces to

$$\alpha = \frac{[CO_2{}^{18}]^{1/2}/[CO_2{}^{16}]^{1/2}}{[H_2O^{18}]/[H_2O^{16}]} = K$$

It should be pointed out that, had Eq. (10.3) for the water–carbon dioxide exchange been written

$$CO_2{}^{16} + 2H_2O^{18} \overset{K''}{\rightleftharpoons} CO_2{}^{18} + 2H_2O^{16}$$

then the equilibrium constant K'' would be expressed as

$$K'' = \frac{[CO_2{}^{18}]/[CO_2{}^{16}]}{[H_2O^{18}]^2/[H_2O^{16}]^2}$$

and the enrichment factor α would then be

$$\alpha = \sqrt{K''} = \frac{[CO_2{}^{18}]^{1/2}/[CO_2{}^{16}]^{1/2}}{H_2O^{18}/H_2O^{16}}$$

Therefore a calculated equilibrium constant K for reaction (10.3) above of 1.044 at 25°C means that, at equilibrium, there will be 4.4% more O^{18} in the CO_2 than in the water. This equilibrium isotope effect of 4.4% was beautifully confirmed by experiment (Weber, Wahl, and Urey [92]).

Since then calculations of equilibrium constants or equilibrium isotope effects have been extended to many other isotopic reactions by Urey [93], Bigeleisen and Mayer [94], and Tudge and Thode [95].

Equilibrium constants for isotope-exchange reactions may be calculated quite unambiguously once the characteristic vibrational frequencies for the various isotopic species are known. A typical isotope-exchange reaction may be written

$$aA_1 + bB_2 \leftrightarrows aA_2 + bB_1$$

where the subscripts 1 and 2 indicate that the molecules A and B with one element as a common constituent contain only the light or the heavy isotope, respectively. For gaseous substances the standard free energy is related to the partition function by the relation

$$F = -RT \ln \frac{Q}{N}$$

the partition function being defined by the formula

$$Q = \sum_i g_i e^{-\epsilon_i/kT}$$

Here the summation is over all the allowed energy levels of the molecules, and g_i is the statistical weight of the ith level. Then, since

$$\Delta F^\circ = -RT \ln K$$

it is seen that the equilibrium constant reduces to

$$K = \left(\frac{Q_{A_2}}{Q_{A_1}}\right)^a \left(\frac{Q_{B_1}}{Q_{B_2}}\right)^b$$

where the energy values are measured from a suitable reference state.

The calculation of equilibrium constants therefore involves the direct calculation of partition-function ratios for two isotopic species and is the product of two such ratios. Since the translational partition function is the same as the classical one for all temperatures and the rotational partition function is classical at room temperature in all cases except hydrogen, these cancel out in the partition-function ratio and only the vibrational partition function need be considered. The ratio of partition functions therefore depends upon the vibrations only. Bigeleisen and Mayer [94] have given this ratio in a convenient form as follows:

$$\frac{Q_2}{Q_1} = \frac{S_1}{S_2}\left[1 + \sum_i \left(\frac{1}{2} - \frac{1}{u_i} + \frac{1}{e^{u_i} - 1}\right)\Delta u_i\right] = \frac{S_1}{S_2}\left[1 + \sum_i G(u_i)\,\Delta u_i\right]$$

where S_1 and S_2 are the symmetry numbers of the two isotopic species, $u_1 = h\nu_i/kT$, and the summation is to be taken over the different vibrational frequencies $\nu_i/2$ of the molecule, counting a g-fold degenerate frequency g times. These authors have tabulated the function

$$G(u) = \frac{1}{2} - \frac{1}{u} + \frac{1}{(e^u - 1)}$$

for values of u from 0 to 25.

Therefore, although the calculation of a single partition function is difficult, the calculation of the partition-function ratio for two isotopic species can be made from a knowledge of the vibrational frequencies of the molecule.

Table 10.1 gives a number of interesting exchange reactions for which equilibrium isotope effects have been calculated and measured.

TABLE 10.1.   Equilibrium Isotope Effects

Eq.	Reaction	T, °K	α (exp)	α (calc)	Ref.
1	$HC^{12}O_4{}^-(aq) + C^{13}O_2(g) = HC^{12}O_3{}^-(aq) + C^{12}O_2(g)$	298	1.014	.....	a
2	$N^{15}H_3(g) + NH_4{}^+(aq) = N^{14}H_3(g) + N^{15}H_4{}^+(aq)$	298.1	1.034	1.035	b
3	$HC^{12}N(g) + C^{13}N^-(aq) = HC^{13}N(g) + C^{12}N^-(aq)$	295	1.026	1.030	c
4	$S^{34}O_2 + S^{32}O_3{}^- = S^{32}O_2 + S^{34}O_3{}^-$	298.1	....	1.014	d
5	$S^{34}O_2(g) + HS^{32}O_3{}^-(aq) = S^{32}O_2(g) + HS^{34}O_3{}^-(aq)$	298.1	1.019	.....	e
6	$\tfrac{1}{2}CO_2(g) + H_2O^{18}(l) = \tfrac{1}{2}CO_2{}^{18}(g) + H_2O^{16}(l)$	273.1	1.04070	1.044	f
7	$C^{12}O_3{}^-(aq) + C^{13}O_2(g) = C^{13}O_3{}^-(aq) + C^{12}O_2(g)$	273.1	1.017	1.016	f
8	$\tfrac{1}{3}CO_3{}^{16-}(aq) + H_2O^{18}(l) = \tfrac{1}{3}CO_3{}^{18-}(aq) + H_2O^{16}(l)$	273.1	1.036	1.033	g
9	$N^{14}O_3{}^-(aq) + N^{15}O(g) = N^{15}O_3{}^-(aq) + N^{14}O(g)$	298	....	1.09	h

a H. C. Urey, A. H. W. Aten, Jr., and A. S. Keston, *J. Chem. Phys.*, **4**, 622 (1936).

b H. G. Thode, R. L. Graham, and J. Ziegler, *Can. J. Res.*, **B23**, 40 (1945).

c K. Cohen, *J. Chem. Phys.*, **8**, 588 (1940).

d A. Harrison, Ph.D. Thesis, McMaster University, Hamilton, Ontario, Canada, 1957.

e W. Compston and S. Epstein, Abstract, American Geophysical Union Meeting, May, 1958.

f B. F. Murphy, and A. O. Nier, *Phys. Rev.*, **59**, 771 (1941).

g M. Dole and R. L. Slobod, *J. Am. Chem. Soc.*, **62**, 471 (1940).

h W. Spindel, *J. Chem. Phys.*, **22**, 1271 (1954).

The physical basis for the differences in equilibrium properties of isotopic substances is therefore well established. Reasonably satisfactory agreement between theory and experiment is obtained when sufficient spectroscopic data are available to calculate the partition-function ratios for the isotopic molecules involved. For more complicated molecules, it is usually necessary to assume some model, and the accuracy of the calculations is, of course, limited. For example, in the case of reactions involving condensed phases, calculations give only the order of magnitude at best. In these cases, a direct measurement must be made, using high-precision isotope-ratio techniques in order to obtain accurate values.

C. Isotope separation

These small isotope-exchange factors need to be multiplied many times to effect an appreciable separation of the isotopes. This is most conveniently done where countercurrent exchange between liquid and vapor is possible. For example, exchange reaction 2, Table 10.1, has been used successfully to separate the nitrogen isotopes in apparatus of the distillation-column type (Urey et al. [96]). $NH_4NO_2(sol)$ is pumped into the top of a column, flows to the bottom, where $NH_3(g)$ is liberated, and flows back up in a countercurrent fashion. For such a countercurrent system the theory of distillation columns applies. Since N^{15} is favored in the liquid phase, there will be a net transport of this isotope toward the bottom of the column, where it will concentrate. Under conditions of total reflux, the maximum separation which may be obtained is given by the relation

$$\frac{N_t/(1 - N_t)}{N_b/(1 - N_b)} = \alpha^{kx}$$

where α = sample-process separation factor for exchange
N_b, N_t = mole fractions of heavy constituent N^{15} at bottom and top of the column, respectively

Using a three-stage cascade, Thode and Urey [97] were able to produce nitrogen containing $\sim 75\%$ N^{15}. Since then large quantities of N^{15} have been produced on a commercial basis.

The isotopes of hydrogen, carbon, and sulfur have been successfully separated by similar chemical-exchange methods (Hutchison, Stewart, and Urey [98]; Stewart and Cohen [99]; Smyth [100]). Several good reviews have been written on isotope separation by chemical-exchange methods (Urey [101, 102]).

Spindel and Taylor [103] have developed a combination thermal-diffusion isotope-exchange method of separating the isotopes of nitrogen. In their experiment, NO and NO_2 were contacted in a thermal-diffusion column. The descending cold stream was principally NO_2, while the ascending hot stream was $NO + O_2$. It was found that roughly half the separation could be attributed to the thermal-diffusion effect and half to the gaseous-exchange action, $N^{14}O_2 + N^{15}O \leftrightarrows N^{15}O_2 + N^{14}O$. However, the same authors found the chemical exchange between NO and concentrated nitric acid to be more effective (Table 10.1, Eq. 9). The experimentally determined exchange constant in this case was 1.045 as compared with 1.023 for the NH_3-NH_4^+ system (Table 10.1, Eq. 2). It should be pointed out that these are practical separation factors and are usually less than the simple-process factors given in Table 10.1. For example, in the NH_3-NH_4^+ exchange system, some of the nitrogen in the liquid phase is present as dissolved ammonia rather than NH_4^+ ion. This results in a somewhat lower overall separation factor, depending on the percentage of dissolved ammonia, in view of the low separation factor of 1.006 for the reaction

$$NH_3(g) + NH_2(sol) \leftrightarrows NH_3(g) + NH_3(sol)$$

A similar situation prevails in the NO-$HNO_3(conc)$ exchange process.

There is still considerable interest in the chemical-exchange method of separating isotopes, and high-precision isotope-ratio mass spectrometers are used to measure exchange constants directly and to follow the separations obtained in practical processes.

D. Applications of equilibrium isotope effects

Since equilibrium isotope effects depend on the vibrational frequencies of the various molecular isotopic species in the initial and final states, the measurement of these small isotope effects gives information concerning the nature and strength of certain chemical bonds. Feder and Taube [104] studied the exchange of the oxygen isotopes between hydrated cation and water. They showed that appreciable oxygen isotope fractionation occurs through exchange equilibria of the type

$$A(H_2O^{16})_n + H_2O^{18} = A(H_2O^{16})_{n-1}(H_2O^{18}) + H_2O^{16} \qquad (10.5)$$

The isotope effects found indicate that the heavier species is favored in the ionic hydrates according to theoretical expectations. These isotope effects are determined by measuring the decrease in the relative activity of the heavier species H_2O^{18} which occurs with the addition of a cation to pure water. This is conveniently accomplished by comparing the O^{18}/O^{16} ratio of CO_2 in isotopic equilibrium with the solution with that of CO_2 in isotopic equilibrium with pure water (Table 10.1, Eq. 6). The decrease in the O^{18} content of the CO_2, as hydrate-forming cation is added to pure water, reflects the extent to which H_2O^{18} is favored in the hydrate according to reaction (10.5) above.

The results indicate that, for a number of cations, the isotope-enrichment factor increases with the molality of the solute in a linear fashion, suggesting the constancy of the hydration numbers of the ions up to fairly high concentrations. A trivalent ion with completely filled inner sphere $[Co(en)_3]^{3+}$ did not show an oxygen-isotope effect when added to water. This result indicates the probable identity of the coordination number and the number of water molecules in the inner sphere. The equilibrium constant for the reaction

$$[Co(NH_3)_5H_2O^{16}]^{3+} + H_2O^{18} \rightleftarrows [Co(NH_3)_5H_2O^{18}]^{3+} + H_2O^{16}$$

was found to be 1.019 at 25°C by Rutenberg and Taube [105].

In general, the equilibrium fractionation of the oxygen isotopes which occurs when salts are added to water provides a means of studying the interaction between ions and solvent.

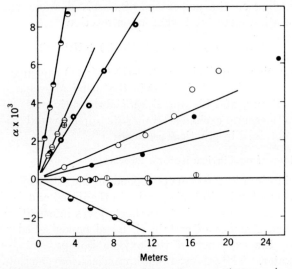

Fig. 10.10. Effect of salts on fractionation of oxygen isotopes in water. $\alpha \equiv R_0/R - 1$; temperature 25° except for CsCl data for which, at 4°: ◐, AlCl$_3$; ⊖, MgCl$_2$; ⊕, Mg(ClO$_4$)$_2$; ○, HCl; ◯, LiCl; ●, AgClO$_4$; ◑, NaI; ◑, NaClO$_4$; ◖, NaCl: ◓, CsCl. *(From Taube [106].)*

The effect studied in the equilibrium method is the change in the O^{18} activity of a solution as a function of salt concentration. The ratio $H_2O^{18}(l)/H_2O^{16}(l)$ will be altered by removal from the solution of one isotopic form in preference to the other when salts interact with solvents. Figure 10.10 shows the change in this ratio with molality of various salts in terms of α, where

$$\alpha \equiv \left(\frac{R_0}{R} - 1\right)$$

and R_0 and R are the $H_2O^{18}(l)/H_2O^{16}(l)$ ratios for the pure water and solutions, respectively (Taube [106]). It is seen that, whereas α/m is very sensitive to the nature of the cation, it is almost insensitive to the nature of the anion. The ratio α/m is essentially a property of the cation only. Taube [106] points out that isotope fractionation exerted by cations in constructing the hydration sphere is in competition with that exerted as a result of the mutual interaction of the water molecules themselves in the bulk solvent. Therefore, when the ion-solvent interactions are stronger (bond-force constants greater) than the solvent-solvent interaction, then the O^{18} activity of the solution will decrease with increasing concentration of the cation, or α/m will be positive (Fig. 10.10).

The value of $\alpha/m = 0$ for sodium can be regarded as resulting from the coincidence that H_2O^{18} is equally favored in solvent-ion and solvent-solvent bonding.

Sheppard and Bourns [107] studied the sulfur isotope exchange between bisulfite ion and the bisulfite addition product of eight aldehydes and ketones. The equilibrium constants for the reactions

$$\text{HS}^{34}\text{O}_3^- + \underset{\substack{|\\ \text{C}}}{\text{R}_2\text{C}}\text{—C—S}^{32}\text{O}_3^- \overset{\substack{\text{OH}}}{\underset{K}{\rightleftharpoons}} \text{HS}^{32}\text{O}_3^- + \underset{\substack{|\\ \text{C}}}{\text{R}_2\text{C}}\text{—S}^{34}\text{O}_3^-$$

were found to range from 1.021 for acetone to 1.010 for anisaldehyde. Since constants of this magnitude are to be expected only if bonding of the sulfur atom is different in the bisulfite ion and the addition product, these results confirm the C—S bond structure of the latter.

The above applications of isotope-exchange equilibrium to chemistry involve the direct determination of isotope-exchange constants. There are many applications, however, in which the rate of exchange of isotopes between two substances under certain conditions is important. These studies give information concerning the nature of substance and reaction mechanisms. The work of Taube and others on the O^{18} exchange between oxycations and solvent are noteworthy [108].

E. Kinetic isotope effects

Soon after the discovery of deuterium, experiments were carried out to show that isotopic atoms and molecules of hydrogen react at different rates. Washburn and Urey [109] showed that, as water is electrolyzed, it grows richer in deuterium, a process which was immediately used to separate the hydrogen isotopes. In 1932, Eyring [110] pointed out the effect of the zero-point energy on reaction rates and showed that the greater zero-point energy of the lighter isotope would effectually aid it in passing over the potential-energy barrier in the surface reaction:

$$2\text{H}^+ + 2 \text{ electrons} = \text{H}_2$$

The importance of kinetic isotope effects in the study of rate theory was, of course, recognized, and many reactions involving hydrogen isotopes were studied. However, there has been a greatly increased interest in kinetic isotope effects since 1948.   This interest was stimulated by the development of high-precision mass spectrometers and new differential techniques for isotope-ratio measurements.   These made possible the direct measurement of kinetic isotope effects for the isotopes of the light elements other than hydrogen.   Perhaps most important, there has been an increased interest in the theory of kinetic isotope effects and their use in the elucidation of reaction mechanisms.

Kinetic isotope effects of the order of 1 to 10% have been reported for the isotopes of carbon, oxygen, nitrogen, and sulfur.   These isotope effects, although small, can be measured with a precision of 0.02% since, as will be seen, only comparative isotope ratios are required and the highly precise differential techniques discussed above may be used.

In general, if we have competitive isotopic reactions of the type

$$A_1 + B \xrightarrow{k_1} X_1 + Y$$
$$A_2 + B \xrightarrow{k_2} X_2 + Y$$

where the subscripts 1 and 2 refer to the light and heavy isotopic species, respectively, assuming the reaction to be first-order in A_1 and A_2, we may write

$$\frac{dX_1}{dt} = k_1(A_{01} - X_1)(B)$$

$$\frac{dX_2}{dt} = k_2(A_{02} - X_2)(B)$$

Then, since X_1 and $X_2 = 0$ at $t = 0$, we have

$$\frac{k_1}{k_2} = \frac{\log\left(\dfrac{A_{01}}{A_{01} - X_1}\right)}{\log\left(\dfrac{A_{02}}{A_{02} - X_2}\right)} = \frac{\log\left(1 - f\,\dfrac{1 + A_{02}/A_{01}}{1 + X_2/X_1}\right)}{\log\left(1 - rf\,\dfrac{1 + A_{02}/A_{01}}{1 + X_2/X_1}\right)} \tag{10.6}$$

where f is the fraction of molecules which have reacted and is given by

$$f = \frac{X_1 + X_2}{A_{01} + A_{02}} = \frac{X_1}{A_{01}}\,\frac{1 + X_2/X_1}{1 + A_{02}/A_{01}}$$

and

$$r = \frac{X_2/X_1}{A_{02}/A_{01}}$$

The ratio of specific rate constants for competitive isotopic reactions is therefore obtained from the determination of isotopic ratios in the initial reactant A_{02}/A_{01} and in the product X_2/X_1 after fraction f of the molecules have reacted.   Since the heavy isotopes of carbon, oxygen, nitrogen, and sulfur are relatively rare,

$$1 + \frac{X_2}{X_1} \quad \text{and} \quad 1 + \frac{A_{02}}{A_{01}} \cong 1$$

where experiments are carried out at the normal levels.

In these cases, Eq. (10.6) reduces to

$$\frac{k_1}{k_2} = \frac{\log (1 - f)}{\log (1 - rf)}$$

There are two important advantages in working with the isotopes of the light elements at normal isotopic levels. First, problems of synthesizing labeled compounds are avoided, and second, errors due to normal contamination are minimized. These were recognized by Lindsay, McElcheran, and Thode [111], Bigeleisen and Friedman [112], and Lindsay, Bourns, and Thode [113, 114], who studied carbon isotope effects in the decomposition of oxalic acid and in the decarboxylation of malonic acids. These reactions are of particular interest in view of the symmetry of the molecules and the two possible modes of reaction for each molecule. At normal carbon-isotope levels we need consider only those isotopic species with zero or one C^{13} atom present, as follows:

(a)

$$\overset{12}{C} \underset{\overset{12}{C}}{\overset{O}{\diagup}} OH \xrightarrow[\substack{H_2SO_4 \\ conc}]{k_1} C^{12}O_2 + C^{12}O + H_2O \qquad (a1)$$

(b)

$$\overset{12}{C} \underset{\overset{13}{C}}{\overset{O}{\diagup}} OH \xrightarrow[\substack{H_2SO_4 \\ conc}]{k_2} C^{13}O_2 + C^{12}O + H_2O \qquad (b2)$$

$$OH \xrightarrow[\substack{H_2SO_4 \\ conc}]{k_3} C^{12}O_2 + C^{13}O + H_2O \qquad (b3)$$

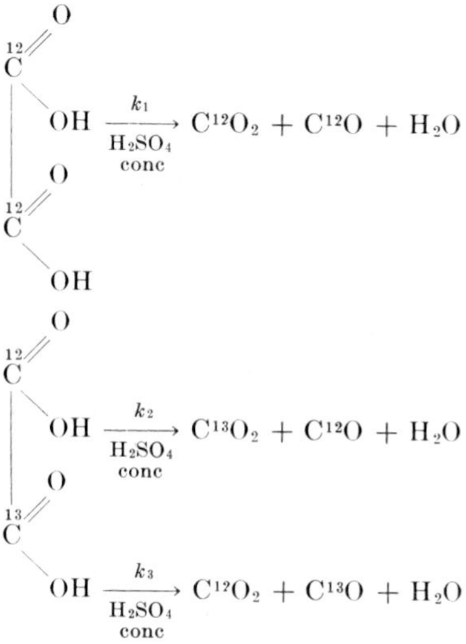

Lindsay, McElcheran, and Thode [111] studied the decomposition of normal oxalic acid in the presence of concentrated H_2SO_4 at 100°C. The products of the reaction, CO_2 and CO, were collected and separated, first, after a small fraction f of the molecules had decomposed, and second, after complete reaction. These samples were analyzed with a mass spectrometer as CO_2 and compared with a sample CO_2 prepared by the complete oxidation of the initial oxalic acid. Two isotope effects were observed. First, species (a) reacts 3.4% faster than species (b), or $k_1/(k_2 + k_3) = 1.034$, and second, in the decomposition of species (b), mode 2 favored over mode 3 by 3.2% or $k_2/k_3 = 1.032$. These isotope effects are referred to as inter- and intramolecular isotope effects, respectively [113]. Yankwich and Promislow [115] confirmed the large intra-

kinetic isotope effect and suggested that perhaps the first step in the reaction involved protonation of the oxalic acid and that an equilibrium isotope effect might therefore be superimposed on the kinetic isotope effect in the decomposition.

The decarboxylation of malonic acid does not involve this difficulty, and a great deal of work has been done to determine the intra- and intermolecular C^{12}—C^{13} isotope effects in this reaction (Bigeleisen and Friedman [112]; Lindsay, Bourns, and Thode [113, 114]; Yankwich and Promislow [115]; Yankwich and Stivers [116]). The thermal decomposition of normal malonic acid may be represented by the following equations:

$$
C^{12}H_2 \Big\langle {}^{C^{12}OOH}_{C^{12}OOH} \xrightarrow{k_1} C^{12}O_2 + C^{12}H_3C^{12}OOH
$$

$$
C^{13}H_2 \Big\langle {}^{C^{12}OOH}_{C^{12}OOH} \xrightarrow{k_2} C^{12}O_2 + C^{13}H_3C^{12}OOH
$$

$$
C^{12}H_2 \Big\langle {}^{C^{13}OOH}_{C^{12}OOH} \begin{array}{l} \xrightarrow{k_3} C^{13}O_2 + C^{12}H_3C^{12}OOH \\ \xrightarrow{k_4} C^{12}O_2 + C^{12}H_3C^{13}OOH \end{array}
$$

As in the case of the oxalic acid decomposition, the carbon isotope ratios are determined for the products, first, after a small fraction has reacted, and second, after complete reaction. In each case the samples are analyzed as CO_2 and compared with CO_2 prepared from initial acid.

Since the decomposition of malonic acid is first-order, the rate equations may be set up and the ratio of rate constants determined. For complete reaction ($t = \infty$), it can be shown that

$$
\frac{k_4}{k_3} = 2 \frac{M_0C^{13}}{M_0C^{12}} \left[\frac{C^{12}O_2}{C^{13}O_2} \right]_C - 1
$$

where M_0C^{13}/M_0C^{12} equals the atom fraction of C^{13} in initial malonic acid (complete homogeneity of malonic acid in carbon isotopes assumed), and for t small (small fraction of reaction),

$$
\frac{k_1}{2k_3} = \frac{M_0C^{13}}{M_0C^{12}} \left[\frac{C^{12}O_2}{C^{13}O_2} \right]_f
$$

Again, both the intra- and intermolecular isotope effects may be calculated from a carbon-isotope-ratio determination for the product CO_2 after complete reaction and after a small fraction of reaction, respectively.

Values of the intramolecular isotope effect, $100(k_4/k_3 - 1)$, reported by different workers at 140°C, vary from 2.0 to 2.7%. Some of this discrepancy is due to lack of isotope homogeneity. This intramolecular isotope effect was found to be independent of temperature within the precision of experiment (Lindsay, Bourns, and Thode [113, 114]). The intermolecular isotope effect, $100(k_1/2k_3 - 1)$, was found to be somewhat larger than the intramolecular isotope effect, values reported by the different works ranging from 3.4 to 3.7%. Here again, some of the discrepancy is due to the lack of isotope homogeneity in the malonic acid, for which a correction is necessary.

From a theoretical point of view, intramolecular isotope effects are of great interest, since the two competing isotopic reactions proceed from the same initial state and the whole effect is due to slightly different paths over the barrier on the potential energy surface for two modes of reaction.

F. Theory of kinetic isotope effects

The transition-state-theory reaction kinetics of Marcelin [117], Polanyi and Evans [118], and Eyring [119] is widely accepted and provides the framework for any theory of kinetic isotope effects. This theory is based on the idea that a chemical reaction proceeds from some initial to a final configuration by a continuous change in the coordinates and that there is some critical intermediate configuration called the *activated complex*, or *transition state*. This critical configuration is in general situated on the highest point of the most favorable reaction path and is considered to be stable along all degrees of freedom except along the path of decomposition. The theory then assumes that there are a small number of activated molecules, molecules having the critical configuration in equilibrium with the reacting species, and that the rate of reaction is controlled by the rate of decomposition of the activated species. Eyring, using a statistical treatment of reaction rates and making only the assumptions inherent in the transition-state hypothesis, developed a formula by which the ratio of rate constants for isotopic reactions can, in principle, be calculated explicitly. More recently, Bigeleisen [120] has developed an expression for the ratio of isotopic rate constants which is more convenient and gives greater insight into the significant factors.

If we consider a process involving reactants A and B which form the activated complex $M^{\ddagger}$ in the reaction

$$A + B \rightleftarrows M^{\ddagger} \rightarrow products$$

then the rate of this reaction is given by the expression

$$Rate = k'_1 C_A C_B \qquad (10.7)$$

where k'_1 = velocity constant

C_A = concentration of reactant A

C_B = concentration of reactant B

The transition-state theory then leads to the relation

$$\text{Rate} = C^{\ddagger} \frac{kT}{h} \tag{10.8}$$

where $C^{\ddagger}$ = concentration of the activated complex
$\quad k$ = Boltzmann's constant
$\quad h$ = Planck's constant
$\quad T$ = absolute temperature in degrees Kelvin
Then, by combining (10.7) and (10.8), we get the well-known equation

$$k_1' = \frac{kT}{h} K^{\ddagger}$$

where $K^{\ddagger}$ is the constant for the equilibrium

$$\text{A} + \text{B} \rightleftarrows \text{M}^{\ddagger}$$

Finally, a factor K, the transmission coefficient, is included to take care of the quantum-mechanical tunnel effect, and we obtain

$$k_1' = \frac{kT}{h} KK^{\ddagger} \tag{10.9}$$

The equilibrium constant can, as before, be expressed in terms of the partition functions of the molecules concerned, so that Eq. (10.9) may be written

$$k_1' = \frac{kT}{h} \frac{Q^{\ddagger}}{Q_A Q_B} K$$

and finally, for two competing isotopic reactions, the ratio of rate constants becomes

$$\frac{k_1'}{k_2'} = \frac{K_1}{K_2} \frac{Q_1^{\ddagger}}{Q_2^{\ddagger}} \frac{Q_{A_2}}{Q_{A_1}} \frac{Q_{B_2}}{Q_{B_1}} \tag{10.10}$$

where the subscripts 1 and 2 denote the two isotopic species.

It is seen that the evaluation of kinetic isotope effects as in the case of equilibrium isotope effects involves the calculation of partition-function ratios for isotopic species. However, the calculation of the kinetic isotope is more difficult, since both K and the partition-function ratio for the two activated complex species involve a knowledge of the potential energy surface for the reaction. However, some terms in the ratio of rate constants will cancel since the potential energy surfaces will be invariant with respect to isotope substitution. In the calculation of kinetic isotope effects, the assumption is usually made that $K_1 = K_2$. This assumption, which may lead to difficulties, has been discussed by Bigeleisen and Wolfsberg [121].

Bigeleisen [120–122] has reduced the ratio of rate constants given by Eq. (10.10) to a form which gives considerably more insight into significant factors and which facilitates computation. By the application of the method developed by Bigeleisen and Mayer [94] for the calculation of isotopic equilibria, he obtains

$$\frac{k_1'}{k_2'} = S\, \frac{\nu_{1L}{}^{\ddagger}}{\nu_{2L}{}^{\ddagger}} \left[1 + \sum_i^{3n-6} G(u_i)\, \Delta u_i - \sum_i^{3n-7} G(u_i)^{\ddagger}\, \Delta u_i{}^{\ddagger} \right] \qquad (10.11)$$

where, as before, $G(u)$ is the free-energy function; $3n - 6$ is the number of vibrational modes in the molecule; $\Delta u_i = hc(\omega_1 - \omega_2)/kT$; 1 and 2 refer to light and heavy molecules, respectively; S is a statistical factor and depends on the symmetry numbers of molecules; and finally, ν_L is the imaginary frequency along the reaction coordinates in the transition state. The factor $\nu_{1L}{}^{\ddagger}/\nu_{2L}{}^{\ddagger}$ gives the ratio of the number of light and heavy activated complexes which decompose in unit time. The quantity in the brackets gives a quantitative description of the effect of the different zero-point energy of light and heavy molecules in the normal and transition states. It should be noted that of the $3n - 6$ vibrational modes in the molecule, one of these vibrations is missing in the free-energy term for the activated complex. This is the imaginary frequency along the decomposition coordinate which gives rise to the frequency factor $\nu_{1L}{}^{\ddagger}/\nu_{2L}{}^{\ddagger}$. The high-temperature limit of the isotope effect does not, therefore, go to unity, but is given by $\nu_{1L}{}^{\ddagger}/\nu_{2L}{}^{\ddagger}$, the ratio of the frequencies along the path of decomposition.

According to a theorem of Slater [123], developed for unimolecular reactions, the frequency factor $\nu_{1L}{}^{\ddagger}/\nu_{2L}{}^{\ddagger}$ of Eq. (10.11) may be replaced by $(\mu_2/\mu_1)^{\frac{1}{2}}$, where μ is the reduced mass across the bond being ruptured. This is based on the premise that the reaction coordinate involves only motions of those atoms whose internuclear separations have to reach a certain critical value. The limiting high-temperature isotope effect is then given by the simple relation

$$\frac{k_1'}{k_2'} = \left(\frac{\mu_2}{\mu_1} \right)^{\frac{1}{2}}$$

where for bond A—B, $\mu = M_A M_B/(M_A + M_B)$. According to the Eyring method [124], the vibrational modes of the activated complex may in principle be found, provided that the potential energy surface for the reaction is known. This surface involves $3n - 6$ vibrations. Of these $3n - 7$ are real vibrations and contribute to the zero-point energy, and 1 is usually imaginary. In the solution of the potential problem, the motion along the reaction coordinate is taken as the imaginary-nongenuine mode. The ratio of the nongenuine frequencies then gives the limiting-high-temperature isotope effect $\nu_{1L}{}^{\ddagger}/\nu_{2L}{}^{\ddagger}$. This approach will, of course, take into account the motion of all atoms in the passage over the barrier from reactants to products along the reaction coordinate. If only the motions of two atoms across the ruptured bond are involved, then the Slater and Eyring methods will give the same result in the classical limit.

Bigeleisen and Wolfsberg [121] have suggested a reaction coordinate q which tears the two decomposition fragments apart. Such a coordinate would lead roughly to

$$\frac{\nu_{1L}^{\ddagger}}{\nu_{2L}^{\ddagger}} = \left(\frac{\mu_2^*}{\mu_1^*}\right)^{1/2}$$

where μ, the reduced mass across the bond ruptured, is evaluated by using the masses of the two fragments and is given by

$$\mu^* = \frac{M_\alpha M_\beta}{M_\alpha + M_\beta}$$

where M_α and M_β are the masses of two fragments.

In principle, it should be possible to determine whether the Slater coordinate or the coordinate suggested by Bigeleisen and Wolfsberg should be used since the high-temperature limit of the isotope effect or the temperature-independent factor of Eq. (10.11) will be very much smaller in the latter case. However, experimental results are inconclusive to date. In the case of the malonic acid decomposition, the agreement between theory and experiment is improved by the use of the mass-fragment approach for the intramolecular isotope effect, but the opposite is true for the intermolecular isotope effect. In general, the choice of the coordinate that tears the two mass fragments apart would seem to be a better approximation, although it of course depends on the nature of the potential energy surface for a reaction. The fact that many isotope effects have been measured which are smaller than the limiting value calculated using the Slater coordinate seems to favor the mass-fragment approach.

G. Kinetic isotope effects and reaction mechanisms

It is clear from the theory of kinetic isotope effects that, in competing isotopic reactions, a significant isotope effect can result only if the bond associated with the isotopic atom is broken in the rate-determining step.

This fact became apparent almost immediately following the isolation of water enriched in the heavy isotope of hydrogen and the realization that compounds of hydrogen and deuterium could react at different rates. Since then, deuterium isotope studies have played an important role in the study of the mechanisms of chemical reactions and in the development of theories of rate processes. Wiberg [125] and Melander [206] have prepared valuable reviews on the theory and application of deuterium and tritium isotope effects. One or two typical examples of hydrogen isotope studies will suffice to illustrate the type of information which this approach to mechanism studies will provide.

It is generally agreed that electrophilic aromatic substitution involves the attack of a positively charged ion, in nitration the nitronium ion, on the aromatic ring. Melander [126–128], in a now classical investigation, observed no significant isotope effect in nitration of tritium-labeled benzene, nitrobenzene, the possible monotritium-substituted toluenes, and naphthalene-α-t_1. From this he concluded that the reaction proceeds in two steps, the first involving the formation of an intermediate addition complex in which the C—H bond is still intact, being rate-determining:

$$\text{⟨benzene⟩} + NO_2^+ \longrightarrow \text{⟨intermediate⟩}^+ \overset{H}{-}NO_2 \longrightarrow \text{⟨benzene⟩}-NO_2 + H^+$$

On the other hand, sulfonation of bromobenzene-4-t_1 did give an isotope effect, $k_H/k_T = 1.82$ (Berglund-Larson and Melander [129]). It was suggested that this result indicated that, in contrast to the nitration, the intermediate formed in the sulfonation reaction is being partitioned between reactants and products.

Hammond [130] has criticized Melander's conclusions that the absence of an isotope effect in nitration constitutes proof for the existence of a two-step process. What this result does show, he points out, is that the C—H bond rupture cannot have made much progress in the transition state. Whether or not there is a nonisolable intermediate of high energy is a question for which the answer is not provided by an isotope-effect study alone.

Zollinger [131, 132], in a brilliant investigation, has clearly demonstrated that the diazo-coupling reaction, a typical electrophilic aromatic substitution reaction, does indeed proceed through an intermediate addition complex, the decomposition of which to products may or may not be rate-determining, depending upon the structure of the compound undergoing coupling or the presence or absence of a base catalyst.

Normally, no isotope effect is observed in diazo-coupling reactions. Thus, for example, 2-methoxydiazobenzene reacts with 1-naphthol-4-sulfonic acid at the same rate as with 2-D-1-naphthol-4-sulfonic acid. The situation is quite different in the coupling of 4-chlorodiazobenzene with 2-naphthol-6, 8-disulfonic acid. In this case a normal isotope effect is observed, $k_H/k_D = 6.55$. This latter reaction is strongly catalyzed by bases such as pyridene, but the reaction rate is not a linear function of base concentration. Furthermore, the magnitude of the isotope effect decreases as the concentration of the base increases.

All experimental observations are accommodated by the following mechanism:

$$X^+ + Ar\text{---}H \underset{k_{-1}}{\overset{k_1}{\rightleftharpoons}} \overset{\overset{\displaystyle H}{\diagup}}{Ar^+}_{\diagdown X}$$

$$\overset{\overset{\displaystyle H}{\diagup}}{Ar^+}_{\diagdown X} + B \overset{k_2}{\rightarrow} Ar\text{---}X + HB^+$$

Assuming a steady-state concentration of the intermediate, the rate equation for the reaction system is as follows:

$$\frac{d(Ar\text{---}X)}{dt} = \frac{k_1 \dfrac{k_2}{k_{-1}} [B]}{1 + \dfrac{k_2}{k_{-1}} [B]} [Ar\text{---}H][X^+]$$

If $k_2 \gg k_{-1}$, the expression becomes

$$\frac{d(Ar\text{---}X)}{dt} = k_1[Ar\text{---}H][X^+]$$

and the rate is independent of the presence of base and shows no isotope effect. On the other hand, when $k_{-1} \gg k_2$, the rate becomes linearly dependent upon concentration of base and the constant k_2. A normal isotope effect corresponding to that of the second step of the reaction should be observed. By varying the three values k_2, k_{-1}, and [B], all possible transitions between a zero and a normal isotope effect and between a zero and a linear dependence on [B] can be expected, and this has been found.

Using the steady-state expression, Zollinger has calculated from the observed isotope-effect values, at intermediate base concentration, the isotope effect k_{2H}/k_{2D} for the bond-rupture process. In all cases this calculated ratio was practically constant, although the observed effects varied over a wide range, and furthermore, this ratio corresponded to theory.

H. Application of carbon and sulfur isotope effects to reaction-mechanism problems

That kinetic isotope effects for elements such as carbon, nitrogen, oxygen, and sulfur were of sufficient magnitude to permit their application to mechanism problems first became apparent in the late 1940s. A number of interesting applications have since been reported, and although this approach has been slow in developing, partly because of the great care required in the measurement of effects of the order of a few per cent, nevertheless it can be expected to play an increasingly important role in reaction-mechanism studies. Two typical applications will be discussed here.

On the basis of kinetic studies, Grovenstein and Henderson [133] suggested that the bromodecarboxylation reaction of 3,5-dibromo-4-hydroxybenzoic acid, a special case of electrophilic aromatic substitution, proceeds by the following mechanism:

$$\text{ArH}_2 \overset{\text{fast}}{\rightleftharpoons} \text{ArH}^- + \text{H}^+ \qquad \text{(i)}$$
$$\text{ArH}^- + \text{Br}_2 \rightleftharpoons \text{ArHBr} + \text{Br}^- \qquad \text{(ii)}$$
$$\text{ArH}_2 + \text{Br}_2 \rightleftharpoons \text{ArHBr} + \text{H}^+ + \text{Br}^- \qquad \text{(iii)}$$
$$\text{ArHBr} \overset{\text{fast}}{\rightleftharpoons} \text{ArBr}^- + \text{H}^+ \qquad \text{(iv)}$$
$$\text{ArBr}^- \longrightarrow \text{Ar'Br}^- + \text{CO}_2 \qquad \text{(v)}$$

where ArH_2 is the dibromohydroxybenzoic acid, Ar'Br^- is the anion of tribromophenol, and ArHBr is the reactive intermediate, presumably of the following structure:

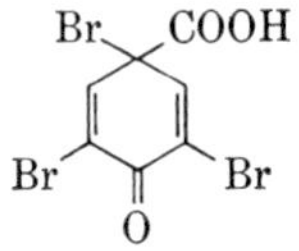

Convincing evidence for this mechanism has been furnished by Grovenstein and Ropp [134]. They found that, in the absence of added Br^- ion, there was essentially a zero isotope effect with respect to the carbon of the carboxyl group. This is to be expected for reaction by the above mechanism when steps (ii) and (iii) are rate-determining, since in these steps the bond between the carbonyl carbon and the ring remains intact. At high bromide-ion concentrations, however, a normal carbon 13 effect of 4.5% was observed. This is readily accounted for on the basis of the suggested mechanism since, under these conditions, a large part of the intermediate, ArHBr, would revert to reactant. In the limiting case, this intermediate would be in rapid, reversible equilibrium with reactants, and the rate-controlling step would be reaction (v), which involves carbon-carbon bond rupture.

I. The Tschugaeff reaction

The thermal decomposition of a xanthate ester, a reaction named after its discoverer Tschugaeff, constitutes a reliable method for the conversion of an alcohol to an alkine. On the basis of both stereochemical (Hückel et al. [135]; Alexander and Mudrak [136, 137]; Cram [138]) and kinetic (O'Connor and Nace [139, 140]) evidence which has been built up over the past fifteen years, it is now generally agreed that this reaction proceeds via a cyclic transition state with a cis elimination of the xanthate group and a hydrogen joined to a β-carbon atom.   Two mechanisms have been proposed.

Mechanism I, originally suggested by Hückel [135], formulates the reaction as a single-stage process in which the hydrogen on the β-carbon is abstracted by the thio ether sulfur atom, while at the same time the bond of this sulfur to the carbonyl carbon is broken.   Mechanism II,

proposed first by Stevens and Richmond [141] and given more explicitly by Cram [138] and Alexander [136, 137], shows the reaction proceeding in two stages.   In the initial rate-determining stage, hydrogen is abstracted by the doubly bonded or thion sulfur atom giving the olefin and an unstable xanthic acid.   The latter then decomposes in a rapid second step to carbon oxysulfide and the mercapton:

$$CH_3-S-\overset{\overset{\displaystyle O}{\|}}{C}-S-H \xrightarrow{\text{fast}} CH_3SH + COS$$

Bourns and Bader [142] have applied sulfur and carbon isotope effects to the problem of deciding between these two possible mechanisms.   A single-stage process, in which hydrogen is abstracted by thio ether sulfur, would proceed through a transition state in which the bond of this sulfur to the carbonyl carbon is partially broken, while the thion sulfur double bond remains intact.   One would predict a S^{32}/S^{34} effect for the thio ether of about 1.3%, based on the simple diatomic model and the application of Eq. (10.11) at usual xanthate decomposition temperatures and a very small or possibly zero sulfur isotope effect for the thion sulfur atom, the bonding of which is essentially unchanged.

In the two-stage mechanism, on the other hand, the bonding of the thio ether sulfur is essentially unchanged in the rate-determining step. A zero effect, therefore, would be predicted, or, if there is some C—S bond weakening in the xanthic acid, then an effect of a few tenths of a per cent at the most. The thion sulfur double bond, however, is converted into a single bond, a transformation which should give rise to an appreciable effect for this atom. Bourns and Bader predicted this effect to be about 1%. This prediction was based again on a simple diatomic model, assuming a C—S bond for the initial state and a C—S bond for the transition state in evaluating the two free-energy terms of Eq. (10.11). A difference in the carbonyl carbon isotope effect for the two mechanisms is also predicted. In the single-stage mechanism the single bond of this carbon to sulfur is broken. This should result in a C^{12}/C^{13} effect slightly greater than 3%, which is the value of the temperature-independent term of Eq. (10.11) obtained by either the diatomic or three-center model. In the rate-determining step of the two-stage mechanism, however, no bond involving the carbonyl carbon is completely broken, but rather a carbon-sulfur double bond is converted to a single bond and a carbon-oxygen single bond to a double. The free-energy summations for the initial and transition states [Eq. (10.11)] might be expected effectively to cancel each other, and since there is no separation of atoms, the effective mass term should also be small. The net result then would be a zero, or at the most, a very small effect for the carbonyl carbon atom.

Distinctly different isotope effects for the three atoms are therefore predicted for the two mechanisms.

The system studied by Bourns and Bader [142] was the thermal decomposition of the xanthate of trans-2-methyl-1-indanyl, the configuration of which will permit only cis elimination. The isotopic ratios in the

reactant were determined by carrying out a number of decomposition reactions to completion, separating the carbon oxysulfide and mercapton and converting each of the sulfurs to sulfur dioxide and the carbonyl carbon to carbon dioxide for mass-spectrometer analysis. Partial reactions were carried to 5 to 10% completion at 80°C, and the isotopic ratios in the products compared with those in the products of the complete decompositions.

Table 10.2 summarizes the predictions made for each of the two mechanisms and gives the experimentally observed isotope effects.

It may be concluded from these results that the one-step mechanism I is clearly eliminated. On the other hand, the isotope-effect values are nicely accommodated by the two-step mechanism II, in which the thion sulfur atom is the hydrogen-abstracting species and a xanthic acid is formed as an intermediate.

TABLE 10.2.   Sulfur and Carbon Isotope Effects in the Decomposition of
Trans-2-methyl-1-indanyl Xanthate

	Per cent isotope effect, $100(k/k' - 1)$		
	Thioether sulfur S^{32}/S^{34}	Thion sulfur S^{32}/S^{34}	Carbonyl carbon C^{12}/C^{13}
Predicted:			
Mechanism I.............	1.3	0	3.5
Mechanism II............	0	1.0	0
Observed................	0.21 ± 0.06	0.86 ± 0.12	0.04 ± 0.04

10.6. Variations in the Abundances of Isotopes in Nature

A. Geochemistry of the oxygen isotopes

Soon after the discovery of deuterium, it was demonstrated that the concentration of this isotope was not constant in the surface of the earth. Later, Dole [143–146] showed that the ratio of the oxygen isotopes varied in nature and that atmospheric oxygen was enriched by 3.1% over that of fresh water.   Also, the ratio of the carbon isotopes was found to vary from inorganic and organic sources (Nier and Gulbransen [147]; Murphy and Nier [148]).   Later, variations in the abundances of the boron isotopes were indicated (Thode, Macnamara, Lossing, and Collins [149]), and finally Thode, Macnamara, and Collins [150] demonstrated large variations in the abundances of the sulfur isotopes in materials from natural sources.   The subject of isotope abundance variations in nature was reviewed by Thode [151], Ingerson [152], and Rankama [153].   Since then interest in the field of isotope geochemistry has continued at an ever-increasing rate.

Differences in equilibrium properties such as vapor pressures and equilibrium isotope effects in exchange reactions are at least partially responsible for the variations in the isotope abundances that occur. Certainly, variations in the hydrogen and oxygen isotopes in fresh waters of the earth as compared with sea water are due to differences in the vapor pressures of the varieties of water containing hydrogen and deuterium or the three oxygen isotopes O^{16}, O^{17}, and O^{18}.   For example, $p_{H_2O^{16}}/p_{H_2O^{18}} = 1.011$ and 1.008 at 0 and 25°C, respectively (Riesenfeld and Chang [154]).   Greene and Voshuyl [155] have shown that

$$\frac{[H_2O^{16}]/[H_2O^{18}] \quad \text{Great Lakes}}{[H_2O^{16}]/[H_2O^{18}] \quad \text{ocean}} = 1.009$$

In the same way, surface or fresh water is depleted in deuterium compared with ocean water. Surface waters are precipitated from the atmosphere in equilibrium with water vapor depleted in deuterium and O^{18} to varying degrees by multiple-stage fractional-distillation processes and are therefore labeled by a roughly constant ratio of deuterium enrichment to O^{18} enrichment, both taken with respect to average ocean water. Friedman [156] demonstrated that a single linear relationship exists between the deuterium and O^{18} content of both fresh and marine waters, with a slope corresponding to the isotope–vapor-pressure ratio at ordinary temperatures. Therefore the fresh waters have been derived from vapor which has attained isotopic equilibrium in the evaporation process. Boato and Craig [157] suggest that the large surface area of the ocean spray, the great travel distances, and the high-humidity conditions provide a mechanism for the equilibration. On the other hand, nonequilibrium evaporation takes place from certain lakes. For example, the waters of Great Salt Lake show a considerable excess enrichment in O^{18} over the concentration of this isotope in meteoric or surface waters of equivalent deuterium content and fall distinctly off a plot of deuterium versus O^{18} concentration for unevaporated surface meteoric waters. This makes possible the identification of unevaporated waters.

Epstein [158] has considered the evaporation of water from the equatorial regions and its condensation as rain at the various latitudes as a multistage distillation column. At each latitude rain falls which is in isotopic equilibrium with atmospheric vapors. Therefore, as the vapors travel north and south from the equator, they become further and further depleted in the heavy isotopes deuterium and O^{18}. Snow falling in the arctic regions should therefore be very low in these isotopes. Epstein also predicted that the extent of depletion of arctic precipitation in O^{18} would depend, as in a distillation column, on the temperature gradient between the equator and the arctic. Actually, since this temperature gradient would be greater in the winter, seasonal variations in O^{18} content of arctic precipitation might be expected. His studies of the O^{18} content of ice from the Greenland icecap confirmed the prediction of a very low O^{18} content of arctic precipitation and showed a variation with depth in O^{18} content, which is attributed to a seasonal variation in arctic temperature. In other words, the O^{18} concentration-depth curve for the ice shows up the seasons much like the rings in a tree show up the annual growth. These studies make possible quantitative measurements on glaciers and give information about climatic conditions under which snow accumulated. Further, the extent of ring distortion in the ice may give information about glacier flow.

B. Isotopic thermometry

In 1947, Urey [93] suggested that the temperature coefficient of some isotopic reactions might be large enough to make possible the determination of temperature by the measurement of isotopic fractionation. In particular, he suggested that the carbonate-water exchange might be used and that the accurate measurement of the O^{18} content of suitable marine carbonate sediments would indicate the temperature of the seas at the time of their deposition. The calcium carbonate and water system has been studied extensively, and a paleo temperature scale established (Urey [159]; Urey et al. [160]; McCrea [161]; Epstein et al. [162]). The equilibrium constant K for the reaction

$$\underset{\text{Calcite}}{\tfrac{1}{3}CO_3^{16-}} + H_2O^{18}(l) \overset{K}{\rightleftarrows} \underset{\text{Calcite}}{\tfrac{1}{3}CO_3^{18-}} + H_2O^{16}(l)$$

has been redetermined by Clayton [163] and is given as 1.0301 at 25°C. This means that limestone and calcium carbonate shell material will be enriched in O^{18} over that of sea water by about 3%. The exact extent of this enrichment will depend on the temperature of the water at the time of deposition. Clayton has shown that the temperature behavior of this constant over the range 0°C to infinite temperature can be described by the equation

$$\ln K = 2{,}725T^{-2} \tag{10.12}$$

To arrive at values of K, two other constants need to be known, in view of the analytical procedure. In the treatment of $CaCO_3$ with H_3PO_4, only two-thirds of the oxygen is liberated, in accordance with the following reaction:

$$CaCO_3 + 2H^+ \rightarrow CO_2 + Ca^{++} + H_2O \tag{i}$$

The CO_2 liberated at 25° will not, therefore, have the same O^{18} content as the carbonate under study, and there will be a certain constant isotope fractionation α, where

$$\alpha = \frac{(O^{18}/O^{16})CO_2 \text{ liberated}}{(O^{18}/O^{16}) \text{ carbonate}}$$

This means that, in the comparison of the O^{18} content of carbonates and water using the above procedure, we actually determine the quantity

$$\frac{(O^{18}/O^{16})CO_2 \text{ liberated}}{(O^{18}/O^{16}) \text{ water}} = K\alpha$$

The constant α has been determined to be 1.00750 at 25°C.

Also, in the determination of the O^{18}/O^{16} ratio for water, what is actually measured is the O^{18}/O^{16} ratio for CO_2 in equilibrium with water at 25°C. To determine the O^{18}/O^{16} ratio for water from the measurements, it is therefore necessary to know the equilibrium constant for the reaction

$$H_2O^{18}(l) + \tfrac{1}{2}CO_2^{16}(g) \rightleftarrows H_2O^{16}(l) + \tfrac{1}{2}CO_2^{18}(g)$$

This constant was redetermined by Compston and Epstein [164] and is given as 1.04070 at 25.2°C.

The O^{18}/O^{16} ratio obtained for CO_2 must then be divided by 1.04070 to determine the O^{18}/O^{16} ratio for the water in isotopic equilibrium with the CO_2. By the use of this constant and α for reaction (i) above, the equilibrium constants for the water-carbonate exchange may be determined. The temperature at which the exchange took place is then obtained from Eq. (10.12).

In early temperature determinations, equations were established experimentally which take into account the temperature coefficient of the isotope exchange constant, but do not permit the determination of the exchange constant itself. The following equation established by Epstein et al. [162] can be used.

$$t°C = 16.5 - 4.3\delta + 0.14\delta^2$$

where $\quad \delta(‰) = \left(\dfrac{O^{18}/O^{16} \text{ sample}}{O^{18}/O^{16} \text{ standard}} - 1\right) 1,000$

Here O^{18}/O^{16} sample is the O^{18}/O^{16} ratio for CO_2 liberated from the carbonate sample by treatment with H_3PO_4, and O^{18}/O^{16} standard is the O^{18}/O^{16} ratio for the standard CO_2. The standard CO_2 for which the constant 16.5 in the above equation applies is CO_2 gas liberated from calcium carbonate shell material classified as Belemnetella Americana Peewee Formation, South Carolina. The standard on this scale $\delta = 0$ is CO_2 gas in equilibrium with average marine water of 38% salinity at 25°C.

The temperature coefficient of the equilibrium constant for the O^{18} exchange between carbonate and water was found to be 0.00023/deg at 25°C. Since high-precision isotope-ratio mass spectrometers permit the measurement of changes in the O^{18}/O^{16} ratio with an accuracy of $\pm 0.01\%$, it becomes possible to determine carbonate paleo temperatures of ocean-water phenomena to $\pm 0.5°C$.

The same principles can be applied to other exchange reactions involving oxygen, such as phosphate-water, sulfate-water, and silica-water. However, the carbonate-water system is the most promising. In this case the hydration-dehydration of CO_2 provides a mechanism for isotopic exchange at low temperatures, and it is reasonable to assume that carbonates are deposited in isotopic equilibrium with the water.

In paleo-temperature determinations, it is important to know to what extent the isotopic content of the carbonaceous material is preserved during geological time. The subject for the first paleo-temperature determination was a Jurassic belemnite (Epstein [165]), an extinct marine animal which flourished in the Mesozoic era, about 120 million years ago. To determine the oxygen isotopic composition of the original calcium carbonate over this long period of time, successive samples were ground off parallel to the growth rings. The carbon dioxide extracted was analyzed mass-spectrometrically. Figure 10.11 shows the variation of the temperature (as calculated from the temperature-δO^{18} relationship) with distance from the center of the shell disk. This variation is taken to be due to seasonal variations in temperature as the shell was deposited in layers and indicates the preservation of the original calcium carbonate. The seasonal variations demonstrated correspond to a maximum temperature variation of from 18 to 24°C.

Finally, there is the serious question as to the constancy of the O^{18} content of the oceans throughout geological time. Since the O^{18} content of the carbonate deposits reflects the O^{18} level of the sea water as well as the temperature of the water at the time of deposition, the temperature measurements of the ancient deposits are meaningful only if we know the

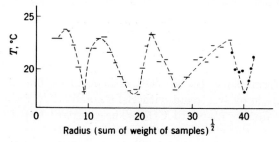

Fig. 10.11. Seasonal temperature variations indicated by the O^{18} content of the growth rings of Jurassic belemnite. (*From Epstein* [165].)

O^{18} content of the water at the time of deposition. It has been assumed that the O^{18} changes in the ocean took place mainly before Cambrian time, and no corrections are usually made for the change in O^{18} concentration with time. Actually, recent studies by Epstein and Mayeda [166] show a variation of 0.3% in the O^{18} content of normal marine waters. Also, since basic igneous rocks are 0.7% enriched in O^{18} over ocean water, and silicates and carbonates are even higher in O^{18} concentration, erosion and sedimentation during geological time must have caused a continual depletion of O^{18} in ocean water. The overall magnitude of this effect is difficult to estimate. However, over short periods of time, this effect must certainly have been small. Carbonates deposited from fresh water can, of course, be distinguished from carbonates deposited from sea water because of the relatively large difference in the O^{18} content of about 1%.

From the investigations to date it seems evident that the ratio of the oxygen isotopes in a fossil shell does reflect the temperature of the water in which the shell grew. This thermometer is now used to study the climate of the past. For example, Emiliani [167] has studied the O^{18} variations in benthonic foraminifera in three deep-sea sediment cores from the equatorial Pacific of Pliocene, Miocene, and Oligocene ages.

The evaluated results indicate a progressive drop in temperature totaling 8°C during 30 million years. The results obtained for the interval from the upper Oligocene to the present are in good agreement with various geological estimates. The deep waters of the ocean basins should reflect the temperatures at the poles, and the uncertainty in the isotopic composition of such waters is believed to be small. Emiliani [168] has also studied the depth habitats of pelagic foraminifera as indicated by data on O^{18} variations in different species and in ocean waters in which they grew. These studies are of great importance in geology and biology, and further progress in isotope thermometry can be expected. A combination of the carbonate-water system with another exchange system involving the oxygen isotopes could eliminate the need to know the O^{18} content of the water during carbonate deposition. For example, Clayton and Epstein [169] have studied the relationship between the O^{18}/O^{16} ratios in coexisting quartz, carbonate, and iron oxides from various geological deposits. Relative temperatures inferred from the isotopic data are in excellent agreement with those estimated from other geological evidence.

Further information than is at present available on the oxygen isotope exchange between silica-water, phosphate-water, and sulfate-water as a function of temperatures would be most useful.

C. Geochemistry of the carbon isotopes

Nier and Gulbransen [147] and Murphy and Nier [148] showed that the C^{13} abundance varied by $\sim 5\%$ in natural materials and that organic carbon from terrestrial plants was depleted in C^{13} as compared with the inorganic carbon of limestone by about 3%. These results have been confirmed and extended by later work (Wickman and Von Ubisch [170]; Baertschi [171]; Mars [172]; Trofimov [173]; Wickman [174]). Craig [175] determined the C^{13}/C^{12} ratio of several hundred samples of carbon from various geological sources with a high-precision isotope-ratio mass spectrometer. The results of his investigations are presented in Fig. 10.12, in which the per-mil deviations ($\delta\%_0$) are plotted by groups. In this work δ is defined as follows:

$$\delta\%_0 = \left(\frac{C^{13}/C^{12} \text{ sample}}{C^{13}/C^{12} \text{ standard}} - 1 \right) 1,000$$

CO_2, prepared from a Cretaceous belemnite from the Peewee Formation in South Carolina, was arbitrarily chosen to give the standard C^{13}/C^{12} ratio. On this scale marine limestone has an average δ value of zero. The total range of variation in the C^{13}/C^{12} ratio is seen to be $\sim 4.5\%$. Subsequently, much larger variations were reported (Thode, Wanless, and Wallauch [176]; Silverman and Epstein [177]). The marine limestones plotted as filled circles in Fig. 10.12 show a range of only 5.7$\%_0$ and have an average δ value very close to zero ($-0.2\%_0$). Since the seasonal growth rings in the carbonaceous shells of Jurassic belemnite show seasonal variations in O^{18} concentrations and irregular variations in C^{13} concentration [178], it is evident that small variations in the C^{13} content of limestone made up of the calcium carbonate parts of organic debris can be expected.   The high C^{13} content of limestone will be due in part to the carbon isotope exchange between

$$HC^{12}O_3^-(aq) + C^{13}O_2(g) \rightleftarrows HC^{13}O_3^-(aq) + C^{12}O_2(g) \qquad \text{(ii)}$$

for which the equilibrium constant is calculated to be 1.009 at 25°C, favoring C^{13} in the carbonate.   The C^{13} concentration of the deposited carbonate of limestone will, of course, vary to some extent, depending on

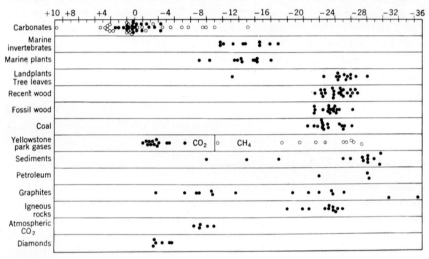

Fig. 10.12. $\delta(\%_0)$ of C^{13}/C^{12} for various carbonaceous samples.   (*From Craig* [175].)

the degree to which equilibration is attained.   Finally, in the precipitation of the carbonate, further enrichment in C^{13} will occur if precipitation is slow and isotopic equilibrium is established between carbonate in solution and solid carbonate.   McCrea [161] showed that calcite precipitated slowly from solution was enriched 3.8$\%_0$ in C^{13} at 25°C.   The exchange process is

$$C^{13}O_3^=(aq) + C^{12}O_3^=(s) \rightleftarrows C^{12}O_3^=(aq) + C^{13}O_3^=(s)$$
where $$K = 1.0038 \text{ at } 25°C$$

The C^{13} concentration of limestone will therefore vary, depending on the degree of attainment of equilibrium during precipitation.

Although there are many features of interest in the distribution of the carbon isotopes in nature, perhaps the main feature is the clear demarcation between limestone carbon, organic carbon of marine origin, and organic carbon of land-plant origin.   Figure 10.12 shows that land and marine plants concentrate C^{12} with respect to atmospheric carbon dioxide, although the latter to a lesser extent. Atmospheric carbon dioxide is assumed to be in isotopic equilibrium with ocean carbonate according to reaction (ii) above and has an average $\delta(C^{13}/C^{12})$ of $\sim -10\%_0$. Marine and land plants are further depleted in C^{13} and have average δ values of $-14\%_0$ and $-26\%_0$, respectively. Wickman [174] discusses these differences and their relationship to the carbon dioxide cycle. He concludes that the preferential assimilation of $C^{12}O_2$ over $C^{13}O_2$ during photosynthesis is influenced by the plants' environment.

The low C^{13} content of the marine and land plants with respect to atmospheric carbon dioxide must certainly be due to kinetic and/or equilibrium isotope effects in the absorption of carbon dioxide and in photosynthesis. That there should be a difference between the magnitude of these isotope effects for marine and land plants is not surprising, since the rate at which CO_2 is supplied and transported across cell walls may be quite different in the two cases. For land plants, the CO_2 diffuses through stomata into substomata cavities along a concentration gradient. In the case of marine plants, carbon dioxide is supplied from solution, and as carbon dioxide is exhausted by photosynthesis, it will be replenished by the shift in the equilibrium reaction

$$CaCO_3 + CO_2 + H_2O \rightarrow Ca^{++} + 2HCO_3^-$$

to the right in a buffered solution. This is the mechanism whereby aquatic plants deposit calcium carbonate during photosynthesis. The magnitude of a kinetic isotope effect in any single step of a multistage process will depend on the extent to which that step is rate-determining. It is clear, therefore, that the rate at which CO_2 is supplied and transported across cell walls may determine the magnitude of a kinetic isotope effect in the overall photosynthetic process. Epstein [179] has studied the fractionation of the carbon isotopes in the absorption of carbon dioxide and in the photosynthesis by plants and has established a mechanism to account for the isotope fractionation that occurs.

Silverman and Epstein [177] studied the carbon isotope ratios in petroleum and related organic materials in marine and nonmarine sediments. Their results indicated a range of C^{13}/C^{12} ratios in petroleum of $\sim 1\%$ as compared with a total range of 4 to 5% in organic carbon. A most interesting feature of these results is the different ranges of $\delta(C^{13}/C^{12})$ values found for petroleum derived from marine and nonmarine sediments. These ranges are -22.5 to -29.4 and -29.9 to -32.5, respectively. The differences between the two ranges are similar to the differences found in the isotopic compositions of present-day nonmarine and marine organisms, although in each case the $\delta(C^{13}/C^{12})$ values are displaced and the petroleums have a lower C^{13} concentration than the organic source material. This means that the C^{13}/C^{12} ratio for petroleum reflects the isotope ratio in the organic source material, although some fractionation occurs in oil-formation processes.

Analyses of chromatographic fractions of petroleum and soluble and insoluble organic constituents of ancient shales indicate that changes in chemical composition do not produce marked changes in isotopic composition. However, hydrocarbon gases occurring in oil-producing areas are considerably lighter in C^{13} than liquid petroleum. This indicates a pronounced isotope fractionation if the gas is derived from petroleum. The relatively narrow range of C^{13}/C^{12} ratios for petroleum suggests that none of the petroleum analyzed to date (Craig [175]; Silverman and Epstein [177]; Wickman [180]) has released large amounts of gas. This implies that the conversion of heavy hydrocarbons into gas and the subsequent loss of gas during geological time is not an important mechanism for destroying oil accumulations. The fact that the petroleums of marine and nonmarine origin are lighter (low C^{13}/C^{12} ratio) than the corresponding organic source materials indicates that not all the biogenic carbon is converted into petroleum. Silverman and Epstein [177] point out that the preservation of the lipids and the loss of the cellulose fractions of plants in petroleum formation would explain the apparent C^{13} depletion in petroleum since the lipids are low in C^{13} as compared with cellulose by about the same amount.

The secondary limestone, the source rock of large sulfur deposits in Texas, which is usually located between layers of anhydrite and gypsum deposits in the cap rock of salt domes, was found to have very low C^{13}/C^{12} ratios (Thode, Wanless, and Wallauch [176]; Feely and Kulp [181]). The $\delta(C^{13}/C^{12})$ values for these secondary limestone deposits range from -25 to -50 and are therefore not like limestone deposited from the sea with a δ value near zero. From these and other sulfur-isotope-fractionation studies reported in the next section, Thode, Wanless, and Wallauch [176] concluded that this limestone, which often contains large amounts of elemental sulfur, is the result of bacterial action on the calcium sulfate cap rock, whereby sulfate is reduced to hydrogen sulfide and sulfur, and at the same time organic material, in this case perhaps petroleum, found around the periphery of salt domes, is oxidized to carbon dioxide and water. The organic matter provides the free-energy source and a source of carbon for the bacteria. The carbon dioxide produced is then the source of carbon for the secondary limestone. Calcium sulfate is therefore changed to calcium carbonate in the overall bacterial process. The fact that this carbon is of organic origin (petroleum found around periphery of salt domes) would account for a δ value for C^{13}/C^{12} of about -25 to -30. However, these secondary limestones are further depleted in C^{13} and have δ values down to -50. Taylor [182] has shown from laboratory experiments that this further depletion can be explained by a kinetic isotope effect in the decarboxylation of pyruvic acid, which occurs in the bacterial process where CO_2 is liberated. In both enzymatic and nonenzymatic decarboxylation, kinetic isotope effects of the order of 1.6% were obtained. The actual C^{13}/C^{12} ratio for CO_2 liberated in nature will therefore vary, depending on the extent of reaction and on the extent to which CO_2 is lost from the system. The range of values obtained is therefore to be expected.

D. Geochemistry of the sulfur isotopes

The sulfur isotopes are of particular interest in geochemistry because of the many forms of sulfur and its wide distribution in the earth's crust. Thode, Macnamara, and Collins [150] showed that the S^{34}/S^{32} and S^{33}/S^{32} ratios for sulfur varied by ~ 8 and 4%, respectively, depending on its origin. Since isotopic fractionation depends on the percentage mass difference, a corresponding variation in S^{36} of 16% is indicated. In general, it was found that sulfates were enriched and the sedimentary sulfides depleted in the heavy-sulfur isotopes. Also, Trofimov [183] reported that sea-water sulfate was enriched in S^{34} as compared with volcanic and meteoritic sulfur. However, in contrast to terrestrial sulfur, Macnamara and Thode [184] found that the S^{34}/S^{32} ratios for meteoritic samples are constant within their limits of error. Also, the value of this ratio coincided approximately with the average value found for terrestrial sulfur. These results led Macnamara and Thode [184] to suggest that the S^{34}/S^{32} ratio for meteorites be used as a standard, since it is constant and probably represents the ratio for terrestrial sulfur at the time the earth was formed (Fig. 10.13). The remarkably constant isotopic ratios found for sulfur in meteorites, however, indicate that this sulfur has not been subjected to the same fractionation processes as sulfur in the earth's crust. A number of investigators have studied the sulfur isotope distribution in sulfur of igneous origin. Since this sulfur is less likely to be altered isotopically following the formation of the earth's crust, it should yield an S^{34}/S^{32} ratio close to the primordial value. Macnamara, Fleming, Szabo, and Thode [185] found that the S^{34}/S^{32} ratio of sulfur in igneous rocks overlapped the value for meteoritic sulfur and showed a spread of only 0.2%.

Although some sulfides of igneous origin show an appreciable enrichment in S^{34} as compared with meteorites, many have S^{34}/S^{32} values within 0.1% of the meteoritic value (Macnamara et al. [185]; Vinogradov et al. [186]; Kulp et al. [187]; Sakai [188]). Also, Rafter [189] and Sakai and Nagasawa [190] have studied the gases flowing from fumaroles and volcanoes. It is found that, when all the sulfur-containing compounds rising from a volcanic vent are collected and analyzed, the average S^{34}/S^{32} ratio is very close to the meteoritic value. From all the evidence to date it seems reasonable to assume that the sulfur in meteorites does give the primordial S^{34}/S^{32} ratio for terrestrial sulfur. Certainly, meteoritic sulfur provides the best standard of comparison for the measurement of S^{34}/S^{32} ratios because of the constancy of this ratio in meteorites.

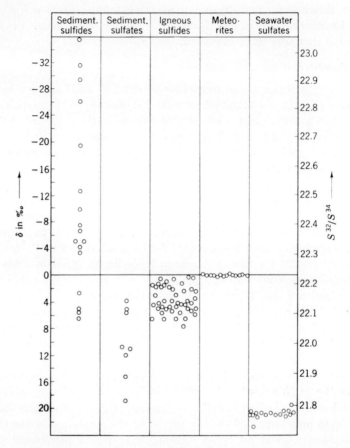

Fig. 10.13. Distribution of the sulfur isotopes in nature.

It is therefore convenient to define

$$\delta(\%_0) = \left(\frac{S^{34}/S^{32} \text{ sample}}{S^{34}/S^{32} \text{ meteorites}} - 1 \right) 1,000$$

The S^{34}/S^{32} ratio for sea-water sulfate has been determined by a number of investigators (Trofimov [183]; Szabo, Tudge, Macnamara, and Thode [192]; Kulp [187]; Sakai [188]). The results of Thode, Monster, and Dunford [191] (Fig. 10.13) show a remarkably constant δ value of $+20.1 \pm 0.5$ for some 50 samples of sea-water sulfate from the Atlantic, Pacific, and Arctic Oceans. This means that the oceans, which are 2% enriched in S^{34} as compared with meteorites, provide a very large reservoir of sulfur at a relatively constant S^{34}/S^{32} level. Since many processes go on in the sea, the fractionation of isotopes that occurs may be reckoned from this base level.

The question arises as to how the distribution of the sulfur isotopes in nature, illustrated in Fig. 10.13, has been brought about. Tudge and Thode [95] calculated exchange constants for isotopic reactions involving

NATIVE SULFUR

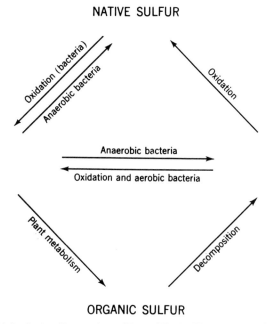

ORGANIC SULFUR

Fig. 10.14. Biological sulfur cycle. (*From Thode, Wanless, and Wallauch* [176].)

the sulfur isotopes. For example, the equilibrium constant for the isotopic exchange reaction

$$S^{32}O_4^{=} + H_2S^{34} \rightleftarrows S^{34}O_4^{=} + H_2S^{32} \tag{iii}$$

was calculated to be 1.072 at 25°C. This means that S^{34} is favored in the sulfate and that there should be 7.2% more S^{34} in the sulfate than in the hydrogen sulfide when isotopic equilibrium is established. It is interesting to note that this is about the spread found in nature between sulfates and sulfides. Szabo, Tudge, Macnamara, and Thode [192] pointed out that isotopic exchange, according to the above reaction, might take place through the well-known sulfur cycle in the sea, illustrated in Fig. 10.14. In this cycle sulfate is continually being reduced either by bacterial action or by plant metabolism. It was suggested that the overall effect of these oxidation-reduction processes would tend to bring about the most favored distribution of the sulfur isotopes according to reaction (iii).

Several stages in the biological sulfur cycle have been investigated in the laboratory from the point of view of sulfur isotope fractionation. Thode, Kleerekoper, and McElcheran [193] found that sulfate-reducing bacteria, *Vibrio desulphuricans*, produce hydrogen sulfide depleted in S^{34} by about 1% as compared with the sulfate medium. Also, Macnamara and Thode [194] showed that free sulfur produced in lake bottoms by bacterial action contained 3.2% less S^{34} than the associated sulfate from which it was produced. Further results of Wallauch [195] showed that this isotope fractionation increased rapidly with the lowering of the temperature, reaching 2% at 10°C. This temperature coefficient is much too large to be explained on the basis of a simple kinetic or equilibrium isotope effect. Jones, Starkey, Feely, and Kulp [196] and Jones and Starkey [197] reported a similar large temperature coefficient and suggested that the size of the isotope effect was affected by factors which influence the rate of growth.

Harrison and Thode [198] studied the fractionation of the sulfur isotopes in the reduction of sulfate to sulfide by bacteria desulphuricans over a wide range of metabolite concentrations, temperature, and conditions of growth. The isotope effect was found to vary from 0 to 2.5%, with the $S^{32}O_4^=$ reacting faster, depending on the rate of the reduction reaction. The magnitude and direction of the isotope effect found can be explained by two consecutive steps as follows:

$$\begin{array}{c} \text{I} \\ K_1 \\ SO_4^= \text{ (sol)} + \text{enzyme} \underset{K_2}{\rightleftharpoons} SO_4^= \cdot \text{enzyme complex} \\ \text{II} \left| \begin{array}{l} K_3 \quad \text{S—O bond-breaking step} \end{array} \right. \\ \downarrow \\ SO_3^= \\ \downarrow \\ \text{rapid reduction relative to I and II} \\ \downarrow \\ H_2S \end{array}$$

where step I involves little or no isotope effect, and step II, a sulfur-oxygen bond-breaking step, involves a large isotope effect similar to the 2.3% kinetic isotope effect reported in the chemical reduction of sulfate (Harrison and Thode [199]). The isotope effect for the process will therefore vary, depending on which step is rate-controlling. Intermediate isotope effects will be obtained when both steps simultaneously influence the rate of sulfate reduction.

Ishii [200] found no evidence of isotope fractionation in the plant metabolism of sulfate in which sulfate is reduced. This was difficult to explain in view of the isotope effect found in the bacterial reduction of sulfate since it is unlikely that nature provides two enzyme systems for sulfate reduction. However, the postulate of Harrison and Thode [198] could explain the difference in the two cases. In the case of plant metabolism of sulfate, step I, the diffusion of sulfate and its transport across a cell membrane by an enzyme mechanism, might always be rate-determining, resulting in a very small or zero isotope effect.

These studies indicate that in nature we can expect kinetic isotope effects in the reduction of sulfate by bacterial action, the magnitude of the effects depending on environmental conditions. However, regardless of the magnitude of the effect produced, fractionation of the sulfur isotopes does occur, with S^{34} enrichment in sulfate. In the bottom sed ments of lakes and shallow seas bacterial action abounds. Under these conditions hydrogen sulfide is produced and either lost from the lake or precipitated as a sulfide. This means that the residual dissolved sulfate will become even richer in S^{34}, like a batch-distillation process, where the distillate becomes richer in the heavier component. The biological sulfur cycle therefore does provide a mechanism for the fractionation of the sulfur isotopes with the enrichment and depletion of S^{34} in sulfate and sulfide, respectively.

If biological processes are largely responsible for the fractionation of the sulfur isotopes between sulfates and sulfides in the sedimentary rocks, then the extent of this fractionation in rocks of a given formation should reflect the biological activity at the time of deposition. With this in mind, Thode, Macnamara, and Fleming [201] investigated sedimentary sulfides (largely pyrite) and some sulfates from sedimentary rocks of different geological age. Their results showed some correlation between the S^{34}/S^{32} ratios in sulfides and the age of the sediments. However, more recent results obtained for large numbers of samples show that there are wide fluctuations in the isotope ratio for a given geological period. For example, the S^{32}/S^{34} ratios obtained by Vinogradov et al. [202] for sedimentary sulfides in samples of different geological age showed a much greater scatter of points in the isotope-ratio-age plot. Although the results indicate a trend toward greater isotope fractionation in the younger sediments, it is clear that the situation is a most complex one. Preliminary results obtained for $CaSO_4$ evaporites of different geological periods suggest a possible cyclic change in the sulfur isotope ratios of sea water with time. Since the sulfides in the marine sediments are formed by the bacterial reduction of sea water sulfate, it is reasonable to suppose that their sulfur isotope ratios will reflect, to some extent, those of the sea water sulfate.

E. Origin of native-sulfur deposits

In the sulfur wells of Louisiana and Texas, native sulfur is found in a matrix of calcitic limestone which usually lies between a layer of gypsum and anhydrite, all of which forms the cap rock of salt domes (Fig. 10.15). Since the native sulfur is always associated with calcium sulfate (gypsum and anhydrite), it was usually assumed that the sulfur was formed from sulfate by some reduction process, although H_2S gas had also been considered a possible source.   Two possibilities exist: either the sulfate was reduced by organic matter at high temperatures (oil found around periphery of salt domes) or the sulfate was reduced by sulfur bacteria, the petroleum providing a source of carbon and free energy for life. Since the isotope fractionation that occurs in the reduction should depend on the mechanism of the process, Thode, Wanless, and Wallauch [176] studied the sulfur and carbon isotope distribution in the sulfur-well core samples ($SO_4^=$, S^0, $S^=$, and $CO_3^=$) in an effort to establish the nature of the process.

The S^{34}/S^{32} ratios for $SO_4^=$, S^0, and $S^=$ are seen to vary considerably about 2% from one sulfur well to another (Fig. 10.16).   In spite of this wide variation in isotopic level from one well to another, the enrichment

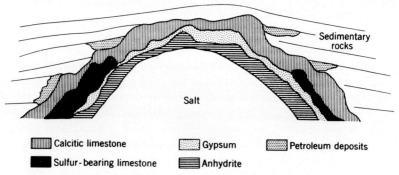

Calcitic limestone   Gypsum   Petroleum deposits
Sulfur-bearing limestone   Anhydrite

Fig. 10.15.   A typical salt dome.   (*From Thode, Wanless, and Wallauch* [176].)

factor between sulfate and sulfur is remarkably constant for the sulfur-well cores of the different salt domes.   It seems clear from these results that the S^0 was in each case derived from the associated sulfate by a process which resulted in a large fractionation of the sulfur isotopes of about 3.9%.   It is difficult to see how a high-temperature hydrocarbon reduction of sulfate could lead to a large isotope fractionation.   The bacterial process, on the other hand, would be expected to do so.   Actually, the average enrichment factor of 1.039 compares with 1.032 found [123] in the case of sulfur-producing lakes of Africa, where bacterial reduction of sulfate is known to take place.   This is convincing evidence that the large sulfur deposits of Louisiana and Texas have been formed by living organisms.

In the reduction of calcium sulfate by *Vibrio desulphuricans*, some organic material is needed to provide a source of free energy for life and a source of carbon.   Thus, as sulfate is reduced, the organic matter is oxidized to carbon dioxide and water.   In this way calcium sulfate would be changed to a porous calcitic limestone with elemental sulfur imbedded in it, in other words, the type of formation present in the salt domes today. The fact that the calcitic limestone matrix of the sulfur wells was found

to be depleted by from 2.5 to 5% in C^{13} as compared with normal limestone deposited in the sea is convincing proof that this calcitic limestone
is of organic origin (Fig. 10.12). Since petroleum, a probable source of
carbon in the salt domes, is depleted by only about 3% in C^{13}, some
further fractionation of the carbon isotopes must have occurred. Actually, some further carbon isotope fractionation does occur in the bacterial
reduction of sulfate, and the higher C^{13} depletions can be explained
(Taylor [182]). The magnitude of this latter fractionation effect would
depend on the extent of reaction and possible losses of CO_2, and this
would account for the wide range of values found for the C^{13}/C^{12} ratios

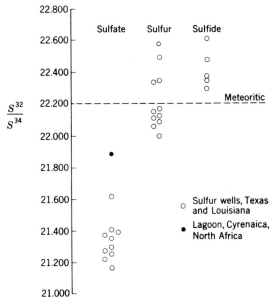

Fig. 10.16. S^{34} variation found in sulfur, sulfate, and sulfide from sulfur wells
of Louisiana and Texas. (*From Thode, Wanless, and Wallauch* [176].)

in the calcitic limestone of salt domes. The sulfur and carbon isotope
data taken together provide overwhelming evidence that the native
sulfur was formed by these living organisms.

Thode, Monster, and Dunford [203] carried out a study of the sulfur
isotope abundances in petroleum and associated sulfates and sulfides in
the reservoir rocks to determine, if possible, the source of petroleum
sulfur and what fractionation, if any, occurs in the formation and maturation of petroleum. Natural crude oil contains sulfur in an organically
bound form, its content varying from 0.1 to over 10%. It is generally
agreed that petroleum is formed from organic matter in a reducing environment in lake and sea-bottom sediments. Sulfur is, of course, present in
relatively large quantities in these sediments in the form of sulfates,

sulfur, and sulfides, with anaerobic- and aerobic-bacteria-reducing sulfates and oxidizing sulfides in the well-known biological sulfur cycle.

Many petroleum samples from Canada and the United States were investigated. Figure 10.17 shows the sulfur isotope abundances in petroleum samples of western Canada of Cretaceous, Devonian, and Mississippian ages. The S^{34}/S^{32} ratio for petroleum was found to vary by 4.5% as compared with a spread of $\sim$9% for terrestrial sulfur. It is seen that a single large oil pool, the Leduc D-2/3, south of Edmonton,

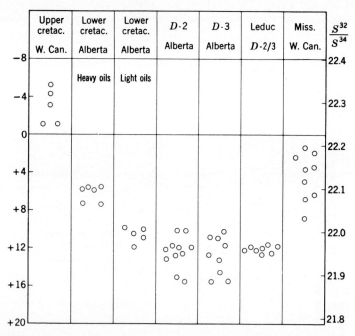

Fig. 10.17. Sulfur isotope abundances in petroleum samples of western Canada, Cretaceous, Devonian, and Mississippian ages. (*From Thode, Monster, and Dunford* [203].)

which covers 100 square miles in area, has a remarkably constant S^{34}/S^{32} ratio. Also, oils from widely separated locations in Alberta found in the same reservoir rock have similar sulfur isotope ratios, but this ratio may vary from one horizon to another. For example, the upper Devonian, lower Cretaceous (heavy oils), and upper Cretaceous oils have sulfur isotope δ values of 12, 8, and -2.5, respectively. Because these oils found in the different horizons have different levels of the sulfur isotope ratio, it is possible to say something about oil migration. It is generally held that oil does migrate horizontally within the reservoir rock, and in this way traps in the reservoir rock (anticlines, faults, etc.) become filled with oil. On the other hand, there is much less likelihood of vertical migration. The sulfur isotope studies, however, support the previous suggestion that the light oils of the lower Cretaceous and the oils of the D-2 and D-3 reefs in Alberta of the middle Devonian are of common origin (Fig. 10.17). These oils are found either just above or just below the unconformity between the lower Cretaceous and the Devonian rocks and usually lie just above or adjacent to a Devonian oil pool (Fig. 10.18). Although oils from widely distributed pools in the same reservoir rocks, e.g., Devonian, have very nearly the same S^{34} content (Table 10.3), they

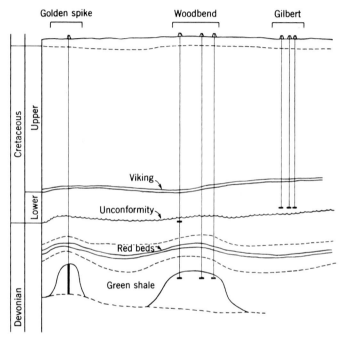

Fig. 10.18. Spatial relation of Gilbert Pools, Leduc Woodben Pools, and Devonian D-3 Pools. (*From E. W. Shaw and G. C. Wells, Western Canada Sedimentary Basin Symposium*, 1954.)

differ greatly in their total sulfur content. This result indicates that there is little or no fractionation of the sulfur isotopes in the maturation of oil in which sulfur is lost as H_2S. The fact that H_2S and its associated petroleum were found to have similar sulfur isotope content is further evidence of this view.

Thode, Monster, and Dunford [203] conclude, therefore, that the sulfur isotope ratio for petroleum does not change materially with time and that it is indicative of the nature of the source sulfur and environment in which the petroleum was formed.

An extensive study has been made of the sulfur isotope distribution in the petroleum-related materials in the sediments of the Uinta Basin,

TABLE 10.3.   Sulfur Isotope Abundances in Petroleum from Devonian
Reservoir Rocks, Alberta D-3

Sample No.	Field or area	Per cent total sulfur	$S^{34}/S^{32}(\delta\%_0)$	
			H_2S	Petroleum
1097–52	Stettler	1.77	+13.7	+10.8
16	Stettler	1.78	+12.6	+11.0
1096–52	Big Valley	0.63	+13.1	+11.7
1419–51	Bashaw	0.62	+13.5	+13.2
7872	Redwater	0.49	......	+10.2
8590	Leduc	0.27	......	+12.3
7005	Leduc	0.26	......	+12.8
1511–51	Wizard Lake	0.29	......	+15.5
3	Golden Spike	0.19	......	+15.5
8587	Golden Spike	0.19	......	+14.6

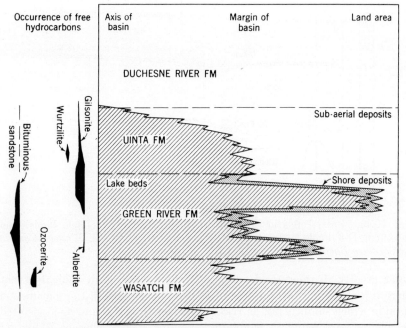

Fig. 10.19.   Subdivision of Tertiary in Uinta Basin.   (*From Hunt, Stewart, and Dickey* [205].)

Utah (Harrison and Thode [204]).   A major problem in any study of the occurrence or origin of petroleums is the identification of the source of a particular hydrocarbon accumulation.   The Uinta Basin is of particular interest because positive identification of the source rock for the various petroleum deposits has been made (Hunt, Stewart, and Dickey [205]).

The Uinta Basin is of interest also because the hydrocarbons are classified as nonmarine in origin. During Cretaceous time, the area was covered by the main epicontinental sea, but at roughly the beginning of the Eocene period the Basin became isolated, forming an inland sea which, as time went on, gradually became more saline. Three deposits of major interest laid down during this period are from the youngest to the oldest, the Wasatch, Green River, and Uinta formations (Fig. 10.19). The major hydrocarbon groups found in the formation are ozocerite in the Wasatch, albertite in the Green River, gilsonite in the late Green River, and wurtzilite in the Uinta formation. In each case the hydrocarbons have been related to their source rocks with reasonable certainty.

The sulfur isotope abundances in the free asphalts, asphalts extracted from the supposed source rocks, and inorganic sulfur compounds of the source rocks are given in Tables 10.4 and 10.5. The results show that the isotope ratios of the organically bound sulfur are very closely related to the isotope ratios of the inorganic sulfur compounds in the source rock. They also show an increase in S^{34} content with decreasing age and

TABLE 10.4.   Sulfur Isotope Abundances in Free Asphalt and Asphalt Extracted from Source Rocks (Uinta Basin, Utah)

Formation	Asphalt (in order of decreasing age)	$S^{34}/S^{32}(\delta\%_0)$	
		Free asphalt	Extracted asphalt
Upper Wasatch........	Ozocerite	$+12.1, +13.0$	$+5.7, +6.8, +4.6$
Lower Green River...	Albertite	$+14.6, +15.0$	$+12.6$
Upper Green River...	Gilsonite	$+23.3, +22.1, +21.9$	$+27.9$
Uinta..............	Wurtzilite	$+30.6, +31.0$	$+28.4, +26.4$

TABLE 10.5.   Sulfur Isotope Abundances in Hydrocarbon Source Rocks (Uinta Basin, Utah)

Formation	Source rock (in order of decreasing age)	$S^{34}/S^{32}(\delta\%_0)$			
		Extracted asphalt	SO_4^-	Pyrite	Total S
Nonmarine, upper Wasatch	No. 1245 Ozocerite	$+5.7, +6.8, +4.6$	$-10.7, -10.0$	$+3.0$	-7.5 -7.2
Lower Green River	No. 2335 Albertite	$+12.6$	$+14.8$	$+14.1$	$+15.0$
Upper Green River	No. 1301 Gilsonite	$+27.9$	$+29.0$	$+30.8, +29.0$	$+29.0$
Uinta	No. 1517 Wurtzilite	$+28.4, +26.4$	$+26.1, +27.2$	$+23.7, +22.8$	$+26.8, +27.2, +26.8$

decreasing depth of the deposits, the youngest sediments being enriched 31% as compared with 20% for sea-water sulfate. This latter comparison confirms beyond a doubt the nonmarine character of these deposits. Further, the one-to-one correspondence between S^{34}/S^{32} ratio and age and depth of deposits is, in view of the changes which occurred in the nature of the controlling environment, strong evidence in favor of a biological theory of S^{34} enrichment in inland lakes and seas (Thode et al. [203]). According to this theory, the sulfur isotopes are fractionated in the bacterial reduction of sulfate, and the two isotopic species $S^{34}O_4^=$ and $S^{32}O_4^=$ are reduced at different rates such that the H_2S produced is depleted in S^{34} whereas the residual $SO_4^=$ becomes enriched in this isotope (Harrison and Thode [204]). The process is like a distillation, the residue becoming richer in the constituent. In view of the large reservoir of sulfate in the oceans, its S^{34}/S^{32} ratio would change very slowly because of such a process. However, in an enclosed lake or sea, as in the case of the Uinta Basin in Eocene time, the total sulfate content would be less and the extent of bacterial action might be great. Under these conditions the residue sulfate in the lake would become richer in S^{34} with time, and the deposits in the bottom of the lake would reflect these changes. This, then, would account for the variation in S^{34} content with depth of sediments in the Uinta Basin, the most recent sediments showing the greatest S^{34} enrichment.

F. Geochemistry of lead isotopes

Since isotopes of high mass are not extensively fractionated by physical or chemical processes, and lead minerals rarely contain uranium and thorium in sufficient concentration to alter significantly the isotopic composition of the lead, it might not be expected that the isotopic composition of lead ores would be of interest in the study of geological processes. There are, however, subtle connections between the isotopic composition of various leads and their geological histories. An excellent account of the importance of lead isotopic studies in geology has been given by Russell and Farquhar [207], to which we refer the reader.

REFERENCES

1. J. J. Thomson: *Proc. Roy. Soc. (London)*, **A89**, 1 (1914).
2. F. W. Aston: *Phil. Mag.*, **38**, 709 (1919).
3. A. J. Dempster: *Phys. Rev.*, **11**, 316 (1918).
4. J. Classen: *Jahrb. Hamburg Wiss. Anst.*, 1907.
5. A. O. Nier: *Rev. Sci. Instr.*, **6**, 254 (1935).
6. F. W. Aston: *Proc. Roy. Soc. (London)*, **A132**, 487 (1931).
7. F. W. Aston: *Proc. Roy. Soc. (London)*, **A149**, 396 (1935).
8. Z. Bay: *Nature*, **141**, 284, 1011 (1938); *Rev. Sci. Instr.*, **12**, 127 (1941).
9. J. S. Allen: *Phys. Rev.*, **55**, 966 (1939).
10. A. A. Cohen: *Phys. Rev.*, **63**, 219 (1943).

11. M. G. Inghram and W. A. Chupka: *Rev. Sci. Instr.*, **24**, 518 (1953).
12. F. A. White, T. L. Collins, and F. M. Rourke: *Phys. Rev.*, **101**, 1786 (1956).
13. F. A. White, T. L. Collins, and F. M. Rourke: *Phys. Rev.*, **97**, 566 (1955).
14. L. Kerwin: *Can. J. Phys.*, **32**, 757 (1954).
15. L. Kerwin: *Can. J. Phys.*, **34**, 1080 (1956).
16. L. Kerwin and D. E. McElcheran: *Can. J. Phys.*, **34**, 1497 (1956).
17. L. Kerwin, D. E. McElcheran, and M. Cottin: *Can. J. Phys.*, **35**, 783 (1957).
18. F. Soddy: *Chem. Soc. Ann. Rept.* 285, 1910.
19. A. O. Nier: *Phys. Rev.*, **77**, 789 (1950); *ibid.*, **79**, 450 (1950).
20. B. B. McInteer, L. T. Aldrich, and A. O. Nier: *Phys. Rev.*, **74**, 946 (1948).
21. M. G. Inghram and R. J. Hayden: A Handbook of Mass Spectroscopy, *Natl. Acad. Sci.-Natl. Res. Council, Nucl. Sci. Ser., Rept.* 14, 1954.
22. F. A. White and T. L. Collins: *Appl. Spectry.*, **8**, 169 (1954).
23. W. R. Smythe and A. Hemmendinger: *Phys. Rev.*, **51**, 178 (1937).
24. J. Mattauch: *Naturwiss.*, **25**, 189 (1937).
25. H. Hintenberger, W. Herr, and H. Voshage: *Phys. Rev.*, **95**, 1690 (1954).
26. M. G. Inghram, R. J. Hayden, and D. C. Hess: *Phys. Rev.*, **72**, 349, 967 (1947).
27. R. W. Pringle, S. Standil, and K. I. Roulston: *Phys. Rev.*, **78**, 303 (1950).
28. R. W. Pringle, S. Standil, H. W. Taylor, and G. Fryer: *Phys. Rev.*, **84**, 1066 (1951).
29. G. I. Mulholland and T. P. Kohman: *Phys. Rev.*, **87**, 681 (1952).
30. W. H. Johnson: *Phys. Rev.*, **87**, 166 (1952).
31. T. P. Kohman and N. Saito: *Ann. Rev. Nucl. Sci.*, **4**, 401 (1954).
32. D. C. Hess and M. G. Inghram: *Phys. Rev.*, **76**, 1717 (1949).
33. W. T. Leland: *Phys. Rev.*, **76**, 1722 (1949).
34. A. O. Nier: *Phys. Rev.*, **55**, 153 (1939).
35. M. G. Inghram, D. C. Hess, P. R. Fields, and G. L. Pyle: *Phys. Rev.*, **83**, 1250 (1951).
36. E. F. Westrum, J. C. Hindman, and R. Greenlee: National Nuclear Energy Series, NNES-PPR IV 14B 22-80, McGraw-Hill Book Company, Inc., New York, 1949.
37. H. G. Thode and R. L. Graham: *Can. J. Res.*, **A25**, 1 (1947).
38. R. K. Wanless and H. G. Thode: *Can. J. Phys.*, **31**, 517 (1953).
39. J. Macnamara, C. B. Collins, and H. G. Thode: *Phys. Rev.*, **75**, 532 (1949).
40. A. J. Dempster: *Phys. Rev.*, **71**, 829 (1947).
41. R. E. Lapp, J. R. Van Horn, and A. J. Dempster: *Phys. Rev.*, **71**, 745 (1947).
42. R. J. Hayden, J. H. Reynolds, and M. G. Inghram: *Phys. Rev.*, **75**, 1500 (1949).
43. M. G. Inghram, D. C. Hess, and R. J. Hayden: *Phys. Rev.*, **71**, 561 (1947).
44. D. C. Hess and M. G. Inghram: *Phys. Rev.*, **76**, 300 (1949).
45. W. H. Walker and H. G. Thode: *Phys. Rev.*, **90**, 447 (1953).
46. J. Macnamara and H. G. Thode: *Phys. Rev.*, **80**, 296 (1950).
47. E. A. Melaika, M. J. Parker, J. A. Petruska, and R. H. Tomlinson: *Can. J. Phys.*, **33**, 830 (1955).
48. J. A. Petruska, E. A. Melaika, and R. H. Tomlinson: *Can. J. Phys.*, **33**, 640 (1955).
49. E. J. Hoagland and N. Sugarman: National Nuclear Energy Series, NNES-PPR IV 9 Paper 69, McGraw-Hill Book Company, Inc., New York, 1951.
50. M. G. Mayer: *Phys. Rev.*, **74**, 235 (1948).
51. D. R. Wiles: M.Sc. Thesis, McMaster University, Hamilton, Ontario, Canada, 1950.
52. D. R. Wiles, B. W. Smith, R. Horsley, and H. G. Thode: *Can. J. Phys.*, **31**, 419 (1953).
53. L. E. Glendenin, E. P. Steinberg, M. G. Inghram, and D. C. Hess: *Phys. Rev.*, **84**, 860 (1951).
54. L. E. Glendenin: *MIT Tech. Rept.* 35, 1949.
55. A. C. Pappas: *MIT Tech. Rept.* 63, 1953.

56. J. A. Petruska, H. G. Thode, and R. H. Tomlinson: *Can. J. Phys.*, **33**, 693 (1955).
57. S. Katcoff: *Nucleonics*, **16**, 78 (1958).
58. T. J. Kennett and H. G. Thode: *Phys. Rev.*, **103**, 323 (1956).
59. M. G. Inghram: *Ann. Rev. Nucl. Sci.*, **4**, 81 (1954).
60. K. Fritze, C. C. McMullen, and H. G. Thode: *Proc. 2d Intern. Conf. Peaceful Uses At. Energy*, Geneva, **15**, 436 (1958).
61. D. M. Wiles, and R. H. Tomlinson: *Can. J. Phys.*, **33**, 133 (1955).
62. D. M. Wiles and R. H. Tomlinson: *Phys. Rev.*, **99**, 188 (1955).
63. A. J. Dempster: *J. Am. Phil. Soc.*, **75**, 755 (1935).
64. N. B. Hannay and A. J. Ahearn: *Anal. Chem.*, **26**, 1056 (1954).
65. G. Hevesy and F. A. Paneth: A Manual of Radioactivity, 2d ed., Oxford University Press, London, 1938.
66. D. Rittenberg: *J. Appl. Phys.*, **13**, 561 (1942).
67. M. G. Inghram: *J. Phys. Chem.*, **57**, 809 (1953).
68. G. R. Tilton et al.: *Bull. Geol. Soc. Am.*, 1955.
69. G. Wetherill: *Phys. Rev.*, **92**, 907 (1953).
70. R. E. Honig: *Anal. Chem.*, **25**, 1530 (1953).
71. F. W. Aston: *Proc. Roy. Soc. (London)*, **A126**, 511 (1930).
72. H. A. Straus: *Phys. Rev.*, **59**, 430 (1941).
73. A. O. Nier, E. P. Ney, and M. G. Inghram: *Rev. Sci. Instr.*, **18**, 294 (1947).
74. A. O. Nier: *Rev. Sci. Instr.*, **18**, 398 (1947).
75. C. R. McKinney, J. M. McCrea, S. Epstein, H. H. Allen, and H. C. Urey: *Rev. Sci. Instr.*, **21**, 274 (1950).
76. R. K. Wanless and H. G. Thode: *J. Sci. Instr.*, **30**, 395 (1953).
77. K. Fajans: *Phys. Z.*, **14**, 131, 136 (1913).
78. A. S. King and R. T. Birge: *Nature*, **124**, 127 (1929).
79. W. F. Giauque and H. L. Johnston: *J. Am. Chem. Soc.*, **51**, 1436 (1929).
80. S. M. Naude: *Phys. Rev.*, **36**, 333 (1930).
81. H. C. Urey, F. G. Brickwedde, and G. M. Murphy: *Phys. Rev.*, **39**, 164, 864 (1931).
82. D. N. Lewis and R. E. Cornish: *J. Am. Chem. Soc.*, **55**, 2616 (1933).
83. E. W. Washburn, E. R. Smith, and M. Frandsen: *J. Res. Natl. Bur. Std.*, **11**, 453 (1933).
84. J. R. Huffman and H. C. Urey: *Ind. Eng. Chem.*, **29**, 531 (1937).
85. M. Randall and W. A. Webb: *Ind. Eng. Chem.*, **31**, 227 (1939).
86. H. G. Thode, S. R. Smith, and F. O. Walkling: *Can. J. Res.*, **B22**, 127 (1944).
87. D. Rittenberg and H. C. Urey: *J. Chem. Phys.*, **2**, 106–107 (1934).
88. L. Farkas and A. Farkas: *Trans. Faraday Soc.*, **30**, 1076 (1934).
89. H. C. Urey and G. K. Teal: *Rev. Mod. Phys.*, **7**, 34 (1935).
90. I. Kirshenbaum: "Physical Properties and Analysis of Heavy Water," McGraw-Hill Book Company, Inc., New York, 1951.
91. H. C. Urey and L. J. Greiff: *Am. Chem. Soc.*, **57**, 321 (1935).
92. L. A. Weber, M. H. Wahl, and H. C. Urey: *J. Chem. Phys.*, **3**, 129 (1935)
93. H. C. Urey: *J. Chem. Soc.*, p. 562 (1947).
94. J. Bigeleisen and M. G. Mayer: *J. Chem. Phys.*, **15**, 261 (1947).
95. A. P. Tudge and H. G. Thode: *Can. J. Res.*, **B28**, 567 (1950).
96. H. C. Urey, J. R. Huffman, H. G. Thode, and M. Fox: *J. Chem. Phys.*, **5**, 856 (1937).
97. H. G. Thode and H. C. Urey: *J. Chem. Phys.*, **7**, 34 (1939).
98. C. A. Hutchison, D. W. Stewart, and H. C. Urey: *J. Chem. Phys.*, **8**, 532 (1940).
99. D. W. Stewart and K. Cohen: *J. Chem. Phys.*, **8**, 904 (1940).
100. H. D. Smyth: "Atomic Energy for Military Purposes," p. 168, Princeton University Press, Princeton, N.J., 1945.
101. H. C. Urey: *J. Appl. Phys.*, **12**, 270 (1941).

102. H. C. Urey: *J. Wash. Acad. Sci.*, **30**, 277 (1940).
103. W. Spindel and T. I. Taylor: *J. Chem. Phys.*, **23**, 1318 (1955).
104. H. M. Feder and H. Taube: *J. Chem. Phys.*, **20**, 1335 (1952).
105. A. C. Rutenberg and H. Taube: *J. Chem. Phys.*, **20**, 825 (1952).
106. H. Taube: *J. Phys. Chem.*, **58**, 523 (1954).
107. W. A. Sheppard and A. N. Bourns: *Can. J. Chem.*, **32**, 4 (1954).
108. H. Taube: *Ann. Rev. Nucl. Sci.*, **6**, 227 (1956).
109. E. W. Washburn and H. C. Urey: *Proc. Natl. Acad. Sci. U.S.*, **18**, 496 (1932).
110. H. Eyring: *Proc. Natl. Acad. Sci. U.S.*, **19**, 78 (1933).
111. J. G. Lindsay, D. E. McElcheran, and H. G. Thode: *J. Chem. Phys.*, **17**, 589 (1949).
112. J. Bigeleisen and L. Friedman: *J. Chem. Phys.*, **17**, 998 (1949).
113. J. G. Lindsay, A. N. Bourns, and H. G. Thode: *Can. J. Chem.*, **29**, 192 (1951).
114. J. G. Lindsay, A. N. Bourns, and H. G. Thode: *Can. J. Chem.*, **30**, 163 (1952).
115. P. E. Yankwich and A. L. Promislow: *J. Am. Chem. Soc.*, **76**, 4648 (1954).
116. P. E. Yankwich and E. C. Stivers: *J. Chem. Phys.*, **21**, 61 (1953).
117. A. Marcelin: *Ann. Phys.*, **3**, 158 (1915).
118. M. Polanyi and M. G. Evans: *Trans. Faraday. Soc.*, **31**, 875 (1935).
119. H. Eyring: *J. Chem. Phys.*, **3**, 107 (1935).
120. J. Bigeleisen: *J. Chem. Phys.*, **17**, 425, 675 (1949).
121. J. Bigeleisen and M. Wolfsberg: *Advan. Chem. Phys.*, **1**, 15 (1958).
122. J. Bigeleisen: *J. Phys. Chem.*, **56**, 823 (1952).
123. N. B. Slater: *Trans. Roy. Soc. (London)*, **A246**, 57 (1953).
124. S. Glasstone, K. J. Laidler, and H. Eyring: "The Theory of Rate Processes," McGraw-Hill Book Company, Inc., New York, 1941; see also H. S. Johnson: *Advan. Chem. Phys.*, **3**, 131 (1961).
125. K. B. Wiberg: *Chem. Rev.*, **55**, 713 (1955).
126. L. Melander: *Acta Chem. Scand.*, **3**, 95 (1949).
127. L. Melander: *Nature*, **163**, 599 (1949).
128. L. Melander: *Arkiv Kemi*, **2**, 213 (1950).
129. V. Berglund-Larson and L. S. Melander: *Arkiv Kemi.*, **6**, 219 (1953).
130. G. Hammond: *J. Am. Chem. Soc.*, **77**, 334 (1955).
131. H. C. H. Zollinger: *Helv. Chim. Acta*, **38**, 1597, 1617, 1623 (1955).
132. H. C. H. Zollinger: *Experientia*, **12**, 165 (1956).
133. E. Grovenstein and U. V. Henderson: *J. Am. Chem. Soc.*, **7**, 569 (1956).
134. E. Grovenstein and G. A. Ropp: *J. Am. Chem. Soc.*, **78**, 2560 (1956).
135. W. Hückel, W. Tappe, and G. Lagutke: *Ann.*, **543**, 191 (1940).
136. E. R. Alexander and A. Mudrak: *J. Am. Chem. Soc.*, **72**, 1810, 3194 (1950).
137. E. R. Alexander and A. Mudrak: *J. Am. Chem. Soc.*, **73**, 59 (1951).
138. D. J. Cram: *J. Am. Chem. Soc.*, **71**, 3863 (1949).
139. G. O'Connor and H. Nace: *J. Am. Chem. Soc.*, **74**, 5454 (1952).
140. G. O'Connor and H. Nace: *J. Am. Chem. Soc.*, **75**, 2118 (1953).
141. P. Stevens and J. Richmond: *J. Am. Chem. Soc.*, **63**, 3132 (1941).
142. A. N. Bourns and R. W. F. Bader: Handbook XIV International Congress of Pure and Applied Chemistry, p. 86, 1955.
143. M. Dole: *J. Am. Chem. Soc.*, **57**, 2731 (1935).
144. M. Dole: *J. Chem. Phys.*, **4**, 4268 (1936).
145. M. Dole: *Science*, **109**, 77 (1949).
146. M. Dole: *Chem. Rev.*, **51**, 263 (1952).
147. A. O. Nier and E. A. Gulbransen: *J. Am. Chem. Soc.*, **61**, 697 (1939).
148. B. F. Murphy and A. O. Nier: *Phys. Rev.*, **59**, 771 (1941).
149. H. G. Thode, J. Macnamara, F. P. Lossing, and C. B. Collins: *J. Am. Chem. Soc.*, **70**, 3008 (1948).

150. H. G. Thode, J. Macnamara, and C. B. Collins: *Can. J. Res.*, **B27**, 361 (1949).
151. H. G. Thode: *Research*, **2**, 154 (1949).
152. E. Ingerson: *Bull. Geol. Soc. Am.*, **64**, 301 (1953).
153. K. Rankama: "Isotope Geology," Pergamon Press, New York, 1954.
154. E. H. Riesenfeld and T. L. Chang: *Z. Physik. Chem.*, **B33**, 127 (1936).
155. C. H. Greene and R. J. Voshuyl: *J. Am. Chem. Soc.*, **58**, 693 (1936).
156. I. Friedman: *Geochim. Cosmochim. Acta*, **4**, 89 (1953).
157. G. Boato and H. Craig: *Ann. Rev. Phys. Chem.*, 1955.
158. S. Epstein: private communication, 1958.
159. H. C. Urey: *Science*, **108**, 489 (1948).
160. H. C. Urey, S. Epstein, C. McKinney, and J. M. McCrea: *Bull. Geol. Soc. Am.*, **59**, 1359 (1948).
161. J. M. McCrea: *J. Chem. Phys.*, **18**, 849 (1950).
162. S. Epstein, R. Buchsbaum, H. Lowenstam, and H. C. Urey: *Bull. Geol. Soc. Am.*, **62**, 417 (1953).
163. R. N. Clayton: *J. Chem. Phys.*, **30**, 1246 (1959).
164. W. Compston, and S. Epstein: Abstract, American Geophysical Union Meeting, May, 1958.
165. S. Epstein: Mass Spectroscopy in Physics Research, *Natl. Bur. Std. (U.S.) Circ.* 522, p. 133, 1953.
166. S. Epstein and T. Mayeda: *Geochim. Cosmochim. Acta*, **4**, 213 (1953).
167. C. Emiliani: *Science*, **119**, 853 (1954).
168. C. Emiliani: *Am. J. Sci.*, **252**, 149 (1954).
169. R. N. Clayton and S. Epstein: *J. Geol.*, **66**, 352 (1958).
170. F. E. Wickman and H. von Ubisch: *Geochim. Cosmochim. Acta*, **1**, 119 (1951).
171. P. Baertschi: *Nature*, **168**, 288 (1951).
172. K. E. Mars: *J. Geol.*, **59**, 131 (1951).
173. A. Trofimov: *Dokl. Akad. Nauk S.S.S.R.*, **85**, 169 (1952).
174. F. E. Wickman: *Geochim. Cosmochim. Acta*, **2**, 243 (1952); *Nature*, **169**, 1051 (1952).
175. H. Craig: *Geochim. Cosmochim. Acta*, **3**, 53 (1953).
176. H. G. Thode, R. K. Wanless, and R. Wallauch: *Geochim. Cosmochim. Acta*, **5**, 286 (1954).
177. S. R. Silverman and S. Epstein: *Bull. Am. Assoc. Petrol. Geol.*, **42**, 998 (1958).
178. H. C. Urey, H. A. Lowenstam, S. Epstein, and C. R. McKinney: *Bull. Geol. Soc. Am.*, **62**, 399 (1951).
179. S. Epstein: *Natl. Acad. Sci.-Natl. Res. Council, Bull.* 400, p. 20, 1956; private communication, 1959.
180. F. E. Wickman: *Geochim. Cosmochim. Acta*, **9**, 136 (1956).
181. H. W. Feely and J. L. Kulp: *Bull. Am. Assoc. Petrol. Geol.*, **41**, 1802 (1957).
182. E. W. Taylor: M.Sc. Thesis, McMaster University, Hamilton, Ontario, Canada, 1955.
183. A. Trofimov: *Dokl. Akad. Nauk S.S.S.R.*, **66**, 181 (1949).
184. J. Macnamara and H. G. Thode: *Phys. Rev.*, **78**, 307 (1950).
185. J. Macnamara, W. H. Fleming, A. Szabo, and H. G. Thode: *Can. J. Chem.*, **30**, 73 (1952).
186. A. P. Vinogradov, M. S. Chupakhin, and V. A. Grinenko: *Geokhimua*, vol. 4 (1957).
187. J. L. Kulp, W. V. Ault, and H. W. Feely: *Econ. Geol.*, **51**, 139 (1956).
188. H. Sakai: *Geochim. Cosmochim. Acta*, **12**, 150 (1957).
189. J. A. Rafter: *New Zealand J. Sci.*, **2**, 154 (1958); **1**, 103 (1958).
190. H. Sakai and H. Nagasawa: *Geochim. Cosmochim. Acta*, **15**, 32 (1958).

191. H. G. Thode, J. Monster, and H. B. Dunford: *Geochim. Cosmochim. Acta*, **25**, 150 (1961).
192. A. Szabo, A. P. Tudge, J. Macnamara, and H. G. Thode: *Science*, **111**, 464 (1950).
193. H. G. Thode, H. Kleerekoper, and D. E. McElcheran: *Research*, **4**, 581 (1951).
194. J. Macnamara and H. G. Thode: *Research*, **4**, 582 (1951).
195. R. Wallauch: unpublished work, McMaster University, Hamilton, Ontario, Canada, 1953.
196. G. E. Jones, R. L. Starkey, H. W. Feely, and J. L. Kulp: *Science*, **123**, 1124 (1956).
197. G. E. Jones and R. L. Starkey: *Appl. Microbiol.*, **5**, 111 (1957).
198. A. G. Harrison and H. G. Thode: *Trans. Faraday Soc.*, **54**, 84 (1958).
199. A. G. Harrison and H. G. Thode: *Trans. Faraday Soc.*, **53**, 1648 (1957).
200. M. Ishii: M.Sc. Thesis, McMaster University, Hamilton, Ontario, Canada, 1953.
201. H. G. Thode, J. Macnamara, and W. H. Fleming: *Geochim. Cosmochim. Acta*, **3**, 253 (1953).
202. A. P. Vinogradov, M. S. Chupakhim, V. A. Grinenko, and A. Trofimov: *Geochimua*, **1**, 96 (1956).
203. H. G. Thode, J. Monster, and H. B. Dunford: *Bull. Am. Assoc. Petrol. Geol.*, **42**, 2619 (1958).
204. A. G. Harrison and H. G. Thode: *Bull. Am. Assoc. Petrol. Geol.*, **42**, 2642 (1958).
205. J. M. Hunt, F. Stewart, and P. A. Dickey: *Bull. Am. Assoc. Petrol. Geol.*, **38**, 1671 (1954).
206. L. Melander: "Isotope Effects on Reaction Rates," The Ronald Press Company, New York, 1960.
207. R. D. Russell and R. M. Farquhar: "Lead Isotopes in Geology," Interscience Publishers, Inc., New York, 1960.
208. A. O. Nier: *Advan. Mass Spectrometry Proc.*, 1959, p. 507.
209. J. C. D. Milton and J. S. Fraser: *Can. J. Phys.*, **40**, 1626 (1962).
210. H. Farrar and R. H. Tomlinson: *Can. J. Phys.*, **40**, 943 (1962).

11

Mass Spectrometry of Free Radicals

F. P. Lossing

Division of Pure Chemistry, National Research Council
Ottawa, Ontario

11.1. Introduction

The realization, in the years immediately following 1925, of the widespread occurrence of free radicals in the gas-phase reactions of organic compounds provided an immense impetus to the study of kinetics. The work of Taylor [1], of Paneth and Hofeditz [2], of Rice [3], and of many others led to a rapid development in this field. However, the experimental difficulties involved in detecting free radicals, identifying them, and measuring their concentrations have proved to be very great, and our knowledge of the properties of free radicals has consequently developed rather slowly. Although the kinetic evidence for the participation of free radicals in a given reaction is frequently quite unambiguous, the kineticist has generally no means available for detecting and identifying the radicals. He is therefore obliged to identify a radical by inference from the reaction products and to measure the rate at which it undergoes a particular reaction from observations of the overall kinetics. In this way a large body of data has been accumulated, from which, on the basis of intercomparisons, the kineticist attempts to build a self-consistent structure describing the properties and reactivities of free radicals.

So far, no completely satisfactory method has been developed for making measurements on free radicals at the low steady-state concentrations at which they occur in ordinary gas-phase thermal and photochemical reactions. All the methods of observation, whether by absorption spectroscopy, by electron-spin resonance, by mass spectrometry, or by other means, either require special conditions of radical concentration or are applicable only to certain radicals with special properties.

As will be seen later, the detection of radicals by mass spectrometry is not limited by the specific nature of the radical, and in fact more than 75 different free radicals have been detected to date. Since total ionization cross sections of molecules are to a first approximation a constitutive property of the atoms they contain, it seems fairly safe to state that any radical present in the ionizing electron beam to the extent of a few per cent or greater can be detected, provided only that the radical ion is stable. This property gives the method a special value for kinetic studies in that the failure to detect a certain radical among the products of a reaction, when the experimental conditions are such that its detection should be possible, can be considered an observation of some significance. This is not always the case in experiments in which the means of detection depends on some property of the radical, such as its absorption spectrum or its reactivity in special reactions, which is less universal than the ability to form a stable ion. On the other hand, for radicals of not too great complexity, absorption spectroscopy can give detailed structural information which cannot be obtained from mass-spectrometric studies.

The use of mass spectrometry as a means of studying the role played by free radicals in reactions stems mainly from the pioneering work of G. C. Eltenton [4–7]. In this work, which he began in the laboratories of N. N. Semenov and V. N. Krondatiev in Leningrad, prior to 1939, Eltenton succeeded in overcoming the many formidable difficulties involved in designing reactors suitable for the study of thermal-decomposition reactions at low and intermediate pressures and combustion reactions at pressures from 30 to 140 mm. This work will be discussed in more detail later.

The first measurements of ionization potentials of free radicals by mass spectrometry were carried out by Hipple and Stevenson in 1943 [8]. They succeeded in introducing into the ionization chamber of a mass spectrometer a stream of methyl and ethyl radicals of sufficient concentration to permit the measurement of the ionization efficiency curves. These ionization potentials, together with appearance potentials of methyl and ethyl ions from various compounds, gave values for the heats of formation of methyl and ethyl radicals which were in excellent agreement with values from kinetic data.

Present-day work in free-radical mass spectrometry has, in general, followed along the lines laid down in these two pioneering studies.

In addition to studies in which free radicals have been investigated, a number of workers have employed mass spectrometers to investigate molecular and atomic species which are not found under ordinary conditions, but which are not free radicals in the familiar (but not rigorously defined) sense. One important field, the investigation of the atomic and molecular species formed in the vaporization of solids, is discussed elsewhere in this volume (Chaps. 3 and 12). Mass spectrometers have been used to follow the concentrations of non-free-radical intermediates. In 1942 Leifer and Urey [9] investigated the thermal decomposition of dimethyl ether and of acetaldehyde, using a mass spectrometer for continuous sampling of the reaction products and nonradical intermediates. A similar study was made by Zemany and Burton, who monitored the concentration of normal and deuterated methanes during the pyrolysis and photolysis of acetaldehyde and deuteroacetaldehyde [10]. The pyrolysis of polystyrene was followed by a similar means, and a number of dimers, trimers, etc., were identified by Bradt et al. [11]. Although these are interesting applications of mass spectrometry, which led to the present use of process-monitor instruments, they will not be discussed here in detail since they lie outside the scope of this chapter.

A number of review articles have been published which deal wholly or in part with the free-radical aspects of mass spectrometry [12–19].

11.2. Principles of the Method

A. Detection of radicals using low-energy electrons

The principles involved in the detection of a free radical formed in a reaction can be illustrated by reference to Fig. 11.1, which shows a

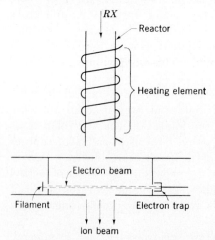

Fig. 11.1. Schematic diagram of reactor and ionization chamber.

schematic drawing of an ionization chamber into which a stream of reaction products from a thermal reactor is directed. With the reactor at a temperature at which no thermal decomposition of the molecule RX can occur, the ions formed by electron impact on RX molecules will be only those occurring in the customary mass spectrum of RX, namely:

$$RX + e \rightarrow RX^+ + 2e \qquad V_1 = I(RX) \qquad (11.1)$$
$$RX + e \rightarrow R^+ + X + 2e \qquad V_2 = A(R^+) \geq I(R) + D(R{-}X) \qquad (11.2)$$
$$RX + e \rightarrow R + X^+ + 2e \qquad V_2 = A(X^+) \geq I(X) + D(R{-}X) \qquad (11.3)$$

The energy thresholds for the formation of the ionic species RX^+, R^+, and X^+ are the appearance potentials V_1, V_2, and V_3, respectively. If the reactor is now heated to a temperature at which a small fraction of the RX molecules is thermally dissociated into the radicals or atoms R and X, two new ionic processes can occur:

$$R + e \rightarrow R^+ + 2e \qquad V_4 = I(R) \qquad (11.4)$$
$$X + e \rightarrow X^+ + 2e \qquad V_5 = I(X) \qquad (11.5)$$

The new threshold V_4 for the formation of R^+ ions will be lower than the previous one, V_2, by an energy of the order of $D(R—X)$, that is, about 2 to 4 volts for most compounds. If, with the reactor cold, the ionizing voltage had been initially set at a value less than V_2 but greater than V_4, the ion current of R^+ would initially have been zero. If the temperature of the reactor were then raised to a value at which a fraction of RX

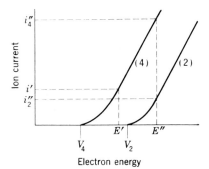

Fig. 11.2. Foot of ionization efficiency curve showing method of detecting free radicals.

molecules dissociated thermally into R and X radicals, a current of R^+ ions would be obtained, resulting from ionization of the free radical R. A similar experiment would reveal the presence of the radical X.

The situation with regard to the best choice of the ionizing voltage for detection can be illustrated by reference to Fig. 11.2. Curves 2 and 4 represent the ionization efficiency curves for processes (11.2) and (11.4) near their respective thresholds V_2 and V_4. It is obvious that, although an electron energy E' will give an ion current i' arising only from radicals R produced in the thermal reaction, a considerably higher ion current for radicals may be obtained at an electron energy E''. However, at this electron energy E'', the observed current $i''(= i_4'' + i_2'')$ includes a contribution from process (11.2). The contribution i_2'' decreases as the amount of RX dissociated in the reactor is increased, and suitable corrections for the decrease must be made if the maximum degree of sensitivity is desired. It is evident that the sensitivity of detection of a given radical becomes larger as $V_2–V_4$ increases, or as the curvature at the feet of the ionization efficiency curves is decreased. From a consideration of the factors governing the shape of the ionization efficiency curves, and the energy distribution of the electrons, Robertson [12] has made an analysis of the sensitivity of detection for free radicals. Assuming the limit to be a condition where $i_4 = i_2$, Robertson derived the expression

$$\frac{C_M}{C_R} = \frac{2k^3T^3 + (E - I)k^2T^2}{2k^3T^3 + (V - E)k^2T^2} \exp\left(\frac{V - E}{kT}\right) \tag{11.6}$$

where C_R = concentration of radicals R in a concentration C_M of molecules which can give rise to R^+ ions

I, V = ionization potential of R and appearance potential of R^+ from the molecule M

E = electron energy used for detection

k = Boltzmann constant

T = temperature, °K

Since the exponential term is the most important one, this expression shows that the energy distribution of the electrons governs the sensitivity of detection and that consequently a low temperature for the electron-emitting filament is to be preferred. Robertson gives two examples for the limit of sensitivity of detection for methyl radicals in the presence of methane, using an electron-emitting filament temperature of 2000°K. Taking $V(R^+) = 14.4$ volts, $I(R) = 9.95$ volts, and using $E = 14.0$ volts, the limit of detection is 1 radical in 10^2 molecules. Using $E = 13.0$ volts, about 1 radical in 10^4 molecules could be detected. It is possible that a considerable improvement in this ratio could be brought about by the use of an ion source of the retarding-potential-difference design [20], in which the energy spread of the ionizing electrons is greatly reduced and the curvature at the foot of the ionization efficiency curves correspondingly decreased.

A number of interfering factors which are encountered in practice are listed below.

(1) **Formation of other products giving rise to R^+.** Since free radicals are extremely reactive, R and X may give rise to the formation of a number of other compounds such as RR, RX', etc., some of which may be formed in relatively large amounts and for which the appearance potential of R^+ may be appreciably lower than V_2.

(2) **Formation of R on the electron-emitting filament.** The compound RX or some product such as RR may pyrolyze on the electron-emitting filament, thereby bringing about an apparent lowering of the appearance potential V_2. Alternatively, the compound RX may pyrolyze on the filament to produce products other than R and X, the detection of which causes the wrong conclusions to be drawn as to the course of the reaction being examined.

(3) **Formation of an interfering substance having the same mass as R.** It might be thought, at first sight, that identification of a radical by the mass number would be unique, at least for species of low molecular weight. It is remarkable, however, how many examples of

coincidence in mass between radicals and between molecules and radicals do occur. For example, CH_3 and NH, NH_2 and CH_4, OH and NH_3, CN and C_2H_2, HCN and CH_2CH, C_2H_5 and CHO, CH_3O and HNO are pairs which are difficult to distinguish in instruments with the usual resolving power. The development of modern high-resolving instruments for the study of organic molecules provides a way to eliminate this difficulty [21]. A further difficulty is the interference with a radical peak by an isotopic peak of a substance whose ionization potential is not greatly different from that of the radical. An example of this was pointed out by Eltenton [5]. The difference between the appearance potential of $C_2H_5^+$ ion from ethane and the ionization potential of ethyl radical is about 4 volts, and one might consequently expect the sensitivity for detection of ethyl radicals thermally produced from ethane to be correspondingly high. However, when ethyl is formed from ethane, the temperature is high enough to decompose an appreciable fraction of the radicals to ethylene, which has an isotopic peak $C^{12}C^{13}H_4$ at mass 29. Consequently, in order to avoid interference from ethylene, the electron energy used for detecting ethyl must be below 10.6 volts, the ionization potential of ethylene, and the available energy range extends to only 2 volts, instead of 4 volts, above the ionization potential of ethyl. For this reason it was necessary, in the measurement of the ionization potential of the ethyl radical, to correct the height of the mass 29 peak at low energies for the $C^{12}C^{13}H_4$ contribution [8].

(4) **Interference from vibrationally excited molecules.** At temperatures of the reactor below which RX can dissociate, it is possible that an appreciable fraction of the RX molecules could have sufficient thermal excitation to lower V_2 by a detectable amount:

$$RX^* + e \rightarrow R^+ + X + 2e \qquad V'(R^+) = I(R) + D(R—H) - E^* \qquad (11.7)$$

If this were so, one might expect the observed ionization potential of RX to be similarly reduced. Experimentally, this effect is small, even when using very narrow energy distributions in the bombarding electrons or photons, and it must be assumed that such excited molecules are rapidly deactivated by collisions with the walls of the ionization chamber.

Momigny [22] has suggested that, with certain ion-source designs, successive impacts with electrons having less than the energy required for ionization can produce ions. An examination of the tails of the ionization efficiency curves obtained with a given instrument should reveal whether this difficulty will be encountered.

(5) **Reaction of R or RX on the walls of the ionization chamber.** It has been suggested that the detection of some radical species can be invalidated by reaction of the radicals with the walls of the ionization chamber [12, 23]. The kinetics of the wall reaction has been considered

in some detail by LeGoff [24, 25]. He found the yield of recombination of CH_3 radicals on the walls of the reactor and ion source to be less than 1 in 10^3. In a later paper, LeGoff and Letort [26] examined the reactions of CH_3 radicals with reactor walls which were assumed to be covered with an adsorbed layer of reaction products and reactant. With CH_3I as a reactant, the yields of CH_4 and C_2H_6 were found to be of the order of 1×10^{-4} to 6×10^{-4}. In the decomposition of $Pb(CH_3)_4$, the yields of CH_4 and C_2H_6 were significantly higher, of the order of 4×10^{-3} to 20×10^{-3}, showing that the nature of the adsorbed layer was important for such reactions. Fabian and Robertson [27] have examined the reaction of propyl radicals, formed by the pyrolysis of propane on a platinum filament, on the interior surfaces of the mass spectrometer. The yield of propyl radicals was compared at two pumping speeds differing by a factor of about 11. When the total area available for reaction, about 300 cm^2, was considered, the probability of reaction for propyl at a single collision was found to be about 4×10^{-5}. However, if only the area of the catalytic platinum filament was considered, a value of 0.022 was obtained, which was nearly the same as that found for propane. This work demonstrates quite clearly that propyl radicals, at least, are quite insensitive to collisions with surfaces at ordinary temperatures. Such a conclusion is qualitatively supported by the observations of Lossing and coworkers [28] that, in the pyrolysis of suitable compounds, many hydrocarbon radicals can be detected in high yields. A mass spectrometer described by Blanchard and LeGoff [29, 30] should provide further information regarding the reaction of radicals at surfaces. In this instrument the ionization chamber can be cooled with liquid nitrogen. At very low temperatures, where almost every wall collision results in condensation, the molecules which are ionized should be only those that have not collided with the wall. A comparison of the mass spectrum of a given radical under such conditions with the spectrum at normal ionization-chamber temperatures should reveal the extent to which radicals have been decomposed at the wall. This instrument is discussed more fully in Sec. 11.3A.

In some cases, a radical may react with residual traces of substances adsorbed on the walls of the ionization chamber. The formation of methane in small yields when a stream of methyl radicals is directed into the ionization chamber has been observed [31]. When CD_3 radicals were substituted for CH_3 radicals, the methane produced contained appreciable quantities of CD_3H. After a period of 1 hr, in which CD_3COCD_3 was passed into the source, further introduction of a beam of CH_3 radicals resulted in the formation of CH_3D at first, which was rapidly substituted by CH_4 [32]. The source of abstractable hydrogen was evidently present in the ionization chamber and was presumably water which was adsorbed on the plates of the ion source and which could be deuterated by exchange with the deuteroacetone.

Although it appears that ion sources of a conventional nature are suitable for the detection of most hydrocarbon radicals, it seems fairly certain that some other species, particularly atoms and oxygenated radicals, are more susceptible to wall reactions. Ion sources in which the radicals enter as a molecular beam directed through the ionizing region into a pumping lead [12] are to be preferred for studies of radicals of this type.

An alternative means of decreasing interference by reactions in the ion source was devised by Foner and Hudson for investigations of radicals and ions in flames [33]. The beam of radicals, which was collimated by a number of orifices, was interrupted by a vibrating beam chopper driven magnetically at about 200 cps, and phase detection of the ion beam was employed. This provided a means of eliminating the background resulting from products formed by reactions on the filament and on the plates of the ion source. A discrimination factor of about 10^4 was obtained by this means. This large factor undoubtedly contributed to their success in detecting the OH and HO_2 radicals.

A study of the characteristics of molecular beams collimated by a triple-aperture system and of simple-aperture sampling systems for mass spectrometers has been made by Nutt et al. [34, 35]. They found that a simple aperture in a metal foil followed theoretical prediction in that a change from molecular effusion to adiabatic expansion occurred when the upstream pressure was such that the mean free path became smaller than the aperture dimensions.

(6) Radicals formed in the ionization chamber by electron impact. In dissociative ionization processes, such as process (11.2) or (11.3), ion production is accompanied by formation of one or more neutral fragments which may be free radicals. There is a small but finite possibility that one of these fragments can be ionized by collision with a second electron, giving rise to a small current of radical ions. This could lead to a spurious detection of radicals. The problem has been considered by Robertson [36], who concluded that the upper concentration limit of radicals produced by such an effect would be about 1 in 10^4. This concentration is just at the limit of detection. See, however, Sec. 11.3F.

(7) Radical ions formed by ion-pair processes. The threshold V_2 for the appearance of R^+ ions would be appreciably lowered if an ion-pair process

$$RX + e \rightarrow R^+ + X^- + e \qquad (11.8)$$

were to occur. The cross sections for such processes are generally small by comparison with processes like (11.2) and (11.3), but in certain compounds this effect could materially lower the sensitivity of detection of radicals by limiting the effective energy interval V_2-V_4.

B. Detection of radicals using charge-exchange ionization

The possibility of detecting free radicals by mass spectrometry following charge-exchange ionization has been suggested by Tal'roze [37]. To detect a free radical, the recombination energy of the bombarding ion must be equal to, or greater than, the ionization potential of the free radical, but less than the appearance potential of the radical ion in dissociative ionization of the bombarded molecule. Since there are a large number of excitation levels in polyatomic ions, the charge-exchange cross section is not limited by the resonance requirement. This method has an inherent advantage over low-energy electron impact in that, for energies in the region of the ionization potential of the radical, the cross section for charge exchange is much larger than for ionization by electron impact. A mass spectrometer for free-radical studies which was designed to permit the use of charge-exchange ionization as well as the use of low-energy electrons has been described by Tal'roze et al. [38]. In this instrument a beam of ions from an auxiliary mass spectrometer was directed into the ionization chamber of the main instrument. This ionization chamber also contained provision for electron bombardment. With this instrument the authors examined the thermal decomposition of hydrazine and detected the presence of the N_2H_3 radical. A relative increase in the ion currents at masses 30, 29, 16, and 15 arose partly from ionization of the neutral species N_2H_2, N_2H, NH_2, and NH and partly from unknown contributions from dissociative charge exchange of N_2H_3 and other radicals produced in the thermal reaction. A separation of these contributions was not made in this preliminary work. One difficulty to be encountered in the application of this method was pointed out by the authors. The charge-exchange spectrum of hydrazine with NH_3^+ included the fragment ions $N_2H_3^+$ and NH_2^+, for which the appearance potentials are 0.6 volt and 3.5 volts, respectively, above the NH_3 ionization potential. Evidently, the recombination energy of NH_3^+ included contributions from excited meta-stable NH_3^+ ions present in the bombarding ion beam. The consequent absence of a sharp high-energy limit for the charge-exchange process may present some difficulties in fully realizing the inherent advantage of this method of radical detection.

C. Measurement of free-radical concentrations

An alternative method of distinguishing R^+ ions formed by reactions (11.2) and (11.4) can be used, provided that the radicals constitute an appreciable fraction of the reaction products. In this technique a fixed electron energy of 50 to 75 volts is used, and the proportional contributions from all stable products to the ion current for R^+ are subtracted as in the customary method of gas analysis. Provided that all the stable products of reaction can be identified and that interference at mass R^+ by other radicals can be shown to be absent, the ion current remaining after such contributions have been subtracted can be attributed to the radical. Since the mass-spectral patterns of the stable compounds are not entirely constant over long periods, this method is not generally useful for radical concentrations less than a few per cent of the total products. However, it has certain advantages for quantitative measurements of radical concentrations, since with 50-volt electrons the ionization cross section for a radical is many times larger than with low-energy electrons and is much less dependent on fluctuations in filament temperature and contact potential. Although this mode of operation is less sensitive and convenient than the low-energy method for the detection of radicals, it is much to be preferred for quantitative measurements of radical concentrations.

The usual method of determining the sensitivity coefficient for stable compounds, that of introducing over the leak a known pressure of a pure compound, must of course be somewhat modified in the case of extremely reactive species such as free radicals or atoms. Provided, however, that such a calibration can be made for a free radical, the ability so gained for measuring the partial pressures of the radical in a reacting system can provide essential information about the nature of a reaction in a most direct way. This information can then be used to supplement the information about reaction mechanisms obtained by the more conventional methods.

A sensitivity determination of this kind was carried out by Lossing and Tickner [39] for the methyl radical. The apparatus used was essentially the same as that shown in Fig. 11.3, which is a later modification. A stream of helium at 6 to 20 mm containing a small partial pressure of mercury dimethyl (8 to 14 μ Hg) was passed through a heated quartz tube into which projected a quartz cone containing an orifice of about 30 μ in diameter. The walls of the top of the cone in the region of the orifice were about 15 μ thick. The residence time in the heated portion of the tube was about 0.001 sec. A sample of the reacting-gas stream was in this way allowed to flow continuously into the ionization chamber by an essentially collision-free path. Observations of the mass spectrum of the reaction products of mercury dimethyl showed that only methyl radicals, mercury, ethane, and a trace of methane were formed. Sensitivity calibrations for mercury dimethyl, ethane, and methane were made, with suitable corrections for temperature effects, so that the partial pressures of these substances in the reaction stream could be measured at any temperature. At various temperatures the "net" peak at mass 15, after subtraction of contributions from mercury dimethyl, ethane, and methane, was determined. On the assumption of a 100% carbon balance, the sensitivity coefficient for the mass 15 peak of the methyl radical could then be calculated from the data at the higher temperatures at which CH_3 radicals were most abundant. Using this value, a total carbon balance was calculated for each temperature.

As can be seen from Table 11.1, the carbon balance obtained was close to 100%, even though the ratio of methyl to other products varied over a wide range, showing that the derived sensitivity coefficient for methyl was approximately correct. Measurements were also made at flow rates so low that all methyl radicals had sufficient time to combine. Using the sensitivities determined for $Hg(CH_3)_2$, CH_4, and C_2H_6, carbon

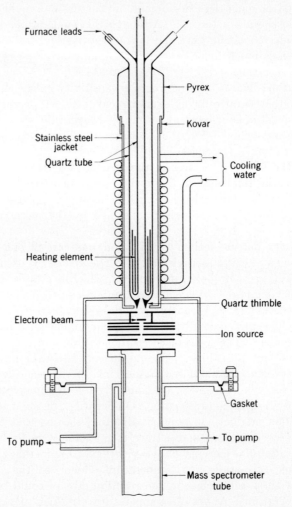

Fig. 11.3. Thermal reactor and ion source of mass spectrometer for study of homogeneous thermal decompositions. (*From Lossing* [16].)

balances were calculated at different temperatures on the assumption that CH_4 and C_2H_6 were the only products. It can be seen from the agreement obtained (Table 11.2) that this assumption and the calibrations were approximately correct.

Since the actual sensitivity of a component depends on instrumental factors, the sensitivity coefficient for methyl radical is best expressed

TABLE 11.1.   Carbon Balance in Two Experiments on Thermal Decomposition of Mercury Dimethyl in a Fast-flow Reactor [39]

Furnace temperature, °C	Hg(CH₃)₂ press, μ		% decomposition	Pressure of products, μ			Sum of C₂H₆ + ½[CH₄ + CH₃] + Hg(CH₃)₂ remaining	Carbon balance, %
	Initial	Remaining		C₂H₆	CH₄	CH₃		
617	8.15	7.86	3.6	0.055	0.032	0.071	7.96	97.7
650	8.12	7.37	9.2	0.172	0.107	0.70	7.99	98.4
692	8.10	6.39	21.1	0.579	0.272	1.87	8.05	99.4
850	8.06	0.35	95.8	3.46	1.13	7.59	8.18	101.5
805	8.03	1.10	86.2	3.17	0.979	6.41	7.97	99.3
672	8.01	6.95	13.2	0.335	0.204	1.42	8.10	101.1
634	7.99	7.61	4.7	0.112	0.081	0.49	8.01	100.3
611	14.23	13.98	1.8	0.097	0.049	0.18	14.19	99.7
650	14.18	13.17	7.1	0.407	0.185	1.05	14.20	100.1
696	14.11	10.96	22.3	1.59	0.504	2.80	14.20	100.6
850	14.04	0.60	95.7	8.15	1.83	8.70	14.02	100.0
810	13.99	1.71	87.8	7.38	1.62	8.15	13.98	100.0
677	13.89	11.80	15.1	0.833	0.330	2.06	13.83	99.6
630	13.81	13.08	5.3	0.186	0.089	0.71	13.66	98.9

TABLE 11.2.   Carbon Balance as in Table 11.1, but at Slow Flow Rates, with All Methyl Radicals Recombined [39]

T, °C	% decomposition of mercury dimethyl	% recovery of carbon
175	0.0	100
630	22.9	105
672	48.1	105
731	78.2	103
896	94.4	94
816	92.7	97
763	86.6	100
696	65.1	99
650	40.3	96
602	15.2	101
175	0.0	101

as a ratio of that of methane.   In this work the ratio S_{CH_3}/S_{CH_4} was found to be 0.47.   This ratio has subsequently been measured by other workers. LeGoff [26] obtained $S_{CH_3}/S_{CH_4} = 0.15$ in the course of a study of the pyrolysis of methyl iodide on heated filaments.   Osberghaus and Taubert [40], from an examination of the radicals formed on the heated walls of an ionization chamber, found the ratio to be 1.0.   The discrepancies between these values are large enough to cause some concern [23, 26, 28].

It should be emphasized, however, that this ratio is subject to a number of experimental factors which may cause it to be quite different from the actual ratio of the ionization cross sections for parent-peak formation for CH_3 and CH_4. In the experiments of Lossing and Tickner, the radicals were formed on the high-pressure side of the leak and then passed through the leak separately. Provided that the flow characteristics of the leak were approximately those of molecular flow, the $M^{1/2}$ term involved in the flow is balanced by a corresponding $M^{1/2}$ term for the rate of pumping in the ion source. The ratio of the partial pressures of CH_3 and CH_4 in the source should then be (to a first approximation) the same as in the reactor. On the other hand, in LeGoff's apparatus, methyl iodide passed through the leak and was dissociated in a low-pressure region. In this case, the CH_3 and CH_3I were subject to pumping rates differing by a factor of about $(15/142)^{1/2}$, or about $1/3$. It is interesting, but possibly fortuitous, that the value for the ratio S_{CH_3}/S_{CH_4} obtained in the two instruments differed by approximately this factor [26]. Using the method of Otvos and Stevenson [41], the total cross section for ionization for CH_3 can be calculated to be approximately 0.88 that for CH_4. Taking an average of the mass spectra obtained for CH_3 by LeGoff and Letort [26] and by Osberghaus and Taubert [40], the intensity of the parent peak corresponds to about 0.64 of the total ionization. For CH_4 the parent peak accounts for about 0.47 of the ions formed. The relative cross sections for the production of the parent ions CH_3^+ and CH_4^+ should consequently be given by

$$\frac{\sigma_{15} \text{ (in } CH_3)}{\sigma_{16} \text{ (in } CH_4)} = \frac{0.64 \times 0.88}{0.47} \approx 1.2 \tag{11.9}$$

This is remarkably close to the value of 1.0 given as a lower limit by Osberghaus and Taubert. On the other hand, if a value of 1.2 were applied to the data of Lossing and Tickner in Table 11.1, it would result in a carbon balance greater than 220% at the temperature where methyl was most abundant. It is evident that this calculated value does not correspond to the conditions of these experiments nor to those of LeGoff and Letort. At present it is not possible to provide a complete explanation of the differences found for the ratio S_{CH_3}/S_{CH_4} under different experimental conditions, and it is obvious that the best procedure is to use a value which has been determined by a suitable calibration in the particular instrument rather than to attempt to apply to a calculated value suitable corrections for pumping rates and other instrumental parameters.

The possibility that certain radicals may not be stable to wall collisions, as discussed earlier, must be taken into account when attempting to measure radical concentrations by this method. LeGoff [23] has given an expression for the loss of radicals by such means:

$$\frac{R}{R_0} = \frac{1}{1 + bS/a} \tag{11.10}$$

where R_0 = concentration of radicals in stream before entering ionization chamber

R = observed concentration

S/a = ratio of areas of walls to openings in ionization chamber

b = probability per collision that a radical will disappear at the wall

For a Nier-type ionization chamber, in which S/a is about 50, LeGoff concluded that, for a 50% reduction in radical concentration, b would have to be only 2×10^{-2}. An error of this magnitude, if not recognized, would of course be quite serious. It is clear, therefore, that before attempting to derive a sensitivity coefficient for a radical, one should, if possible, produce the radical by a reaction in which it is known to be formed in high yields and examine its behavior in the ion source. The possibility of decomposition (or combination) at the wall can then be evaluated from the mass spectrum of the radical. Provided that the disappearance of radicals in the ionization chamber leads to products with an identifiable mass spectrum, suitable corrections can be made. For example, Ingold and Lossing [31] found that part of the methane, reported by Lossing and Tickner [39] as formed by an unidentified reaction in the pyrolysis of mercury dimethyl, was actually produced by a reaction in the ionization chamber. The amount so formed was found to be a constant proportion of the methyl concentration, and a correction was made to the methyl and methane yields. The implied assumption that the proportion of CH_3 radicals appearing as CH_4 was independent of the partial pressure of other hydrocarbons present is not necessarily valid for large concentration changes.

With due regard for such effects, this method of calibration for the measurement of concentrations of free radicals is in principle perfectly general and can be applied to any radical provided that a substance can be found from which the radical can be produced in a reasonably high state of purity. The smallest relative concentration of radicals which can be measured with any degree of accuracy is of the order of a few per cent. This limit is, of course, much higher than the concentrations of radicals existing under the conditions found in conventional kinetic studies. This means that, at the present state of development of the method, special reaction conditions such as fast-flow systems at relatively high temperatures, irradiation by very intense illumination, and so on, must be employed. The method is therefore more appropriate to the study of a given elementary step under such special conditions than to the measurement of the concentrations of radicals in conventional kinetic experiments.

11.3. Applications

Although the application of mass spectrometry to free-radical studies has been carried out in relatively few laboratories, a fairly wide range of problems has been attacked. Reactions in which the participating atoms and radicals have been identified include the following: heterogeneous and homogeneous thermal-decomposition reactions, bimolecular reactions, combustion flames, electric discharges, reactions with excited mercury atoms, direct photolytic reactions, and radicals produced by dissociative ionization. For convenient reference, a list of radicals and atoms detected by mass spectrometry is given in Table 11.3 under these headings, together with the reactant used and the literature reference.

A. Heterogeneous thermal reactions

A number of interesting and productive studies have been made by A. J. B. Robertson at King's College, London, and by P. LeGoff at the University of Nancy, which deal with the formation of free radicals on heated surfaces of various metals. For such work, which must necessarily be carried out at very low pressures ($\sim$10^{-5} mm), the mass-spectrometric method is without rival. The detection and identification of free radicals which are produced within a mean free path of the ionizing electron beam is relatively free from the complicating factors present in earlier studies of such reactions using mirrors as a means of detection. In such reactions even the stable products are formed in quantities far too minute to be analyzed by chemical means. Moreover, the changes in the nature and proportions of the products which accompany the often rapid changes in the condition of the heated surfaces can be followed. Such transient phenomena were noted by Eltenton and have been examined in more detail by the later workers.

One of the most interesting of Eltenton's findings was the demonstration [5] that the pyrolysis of methane on carbon filaments yielded only methyl radicals, and not methylene radicals, as had been concluded from earlier experiments with iodine mirrors. That the absence of methylene was not attributable to a failure of the method of detection was demonstrated in experiments in which he found methylene in considerable abundance when diazomethane was pyrolyzed in the same apparatus. A more complete examination of the methane pyrolysis was made by Robertson [42], who found that the reaction on a hot platinum wire also yielded only methyl, but no methylene, radicals. He concluded that the production of methylene iodide in the iodine mirror experiments must have resulted from secondary reactions at or near the mirror.

TABLE 11.3.   Radicals Detected by Mass Spectrometry

Radical	Reactant	Reference
\multicolumn{3}{c}{Produced by Thermal-decomposition Reactions}		
OH	Water	100
N_2H_3	Hydrazine	38
CF CCl CBr	Halogenated methanes	102
CS	Carbon disulfide	47
CH_2	Diazomethane	5, 6, 97, 98
CF_2	Fluoromethane	102
CCl_2	Carbon tetrachloride	47
	Chloromethane	102
CBr_2	Bromomethane	102
CHO	Glyoxal, formic acid	101
CH_3	Tetramethyl lead	5, 6, 8, 23, 24, 40, 98
	Dimethyl mercury	39, 52, 53, 98, 104, 105
	Dimethyl disulfide	98
	Methyl iodide	23, 24, 28
	Methane	5, 6, 42
	Ethane	5, 6
	Di-t-butyl peroxide	23, 24, 39
	Azomethane	28
	1-Butene	104
	Ethylene oxide	48
	Propylene oxide	48
	Dimethyl ether	48
	Methyl ethyl ketone	28
	Propane	28
	Anisole	50
	n-Propyl radical	28, 106
	n-Butane	40
	Isobutane	40
	Tetramethylethylene	40
CD_3	Di(trideuteromethyl) mercury	104
CF_3	Perfluoroazomethane	111
	Fluoromethane	102
	Trifluoroethyl nitrate	113
CCl_3	Trichlorobromomethane	111
	Chloropicrin	111
CH_2F	Di(monofluoromethyl) mercury	113
CH_2Cl	Di(monochloromethyl) mercury	113
	Chloroethyl nitrite	113
	Chloromethane	103
CH_2Br	Bromoethyl nitrite	113
	Bromomethane	103

Table 11.3.   Radicals Detected by Mass Spectrometry (Continued)

Radical	Reactant	Reference
CHF_2	Difluoroethyl nitrate	113
$CHCl_2$	Dichloroethyl nitrite	113
	Chloromethane	103
$CHBr_2$	Bromomethane	103
CH_3S	Methyl benzyl sulfide	28
C_2H_3	Divinyl ether	28
	Methyl vinyl mercury	112
C_2H_5	Tetraethyl lead	8
	Diethyl mercury	28, 106
	Ethane	28, 43
	Butane	42
	Azoethane	28, 106
CH_3CO	Acetic anhydride	101
CH_2CN	Cyanoethyl nitrate	115
$CH\equiv CCH_2$	Propargyl iodide	106
Cyclopropyl	Cyclopropylmethyl nitrite	116
Allyl	Allyl iodide, 1-butene, 1,5-hexadiene	104
n-Propyl	n-Butyl nitrite	114
sec-Propyl	Azoisopropane	106
	Isobutyl nitrite	114
	Propane	27
CH_3CHCN	β-Cyanopropyl nitrate	115
CH_2CH_2CN	γ-Cyanopropyl nitrate	115
γ-Methallyl	γ-Methallyl iodide	107
β-Methallyl	β-Methallyl iodide	107
Cyclobutyl	Cyclobutylmethyl nitrite	116
n-Butyl	n-Pentyl nitrite	114
Isobutyl	3-Methyl butyl nitrite	114
sec-Butyl	2-Methyl butyl nitrite	114
t-Butyl	Neopentyl nitrite	114
n-Butyloxy	n-Butyl nitrite	114
$(CH_3)_2CCN$	Azobisisobutyronitrile	115
Cyclopentadienyl	Anisole	121
Cyclopentyl	Cyclopentylmethyl nitrite	116
n-Pentyloxy	n-Pentyl nitrite	114
Phenyl	Anisole, phenyl ether	50
Phenyloxy	Anisole, phenyl ether	50
Cyclohexyl	Cyclohexylmethyl nitrite	116
Benzyl	Benzyl ether, toluene	50
	Benzyl methyl sulfide	28
	Benzyl iodide	104
	Benzylamine	108
Cycloheptatrienyl	Dihydroheptafulvalene	121
Benzoyl	Benzyl ether, benzaldehyde	50
m-NO_2 benzyl	m-NO_2 benzyl bromide	117
m-F benzyl	m-F benzyl bromide	117
p-F benzyl	p-F benzyl bromide	117

TABLE 11.3.   Radicals Detected by Mass Spectrometry (Continued)

Radical	Reactant	Reference
p-Cl benzyl	p-Cl benzyl iodide	117
o-Xylyl	α-Bromo-1,2-dimethyl benzene	109
p-Xylyl	α-Bromo-1,4-dimethyl benzene	109
m-Xylyl	α-Bromo-1,3-dimethyl benzene	109
p-CN benzyl	p-CN benzyl bromide	117
m-CN benzyl	m-CN benzyl bromide	117
p-OCH₃ benzyl	β-(p-Methoxyphenyl) ethyl nitrite	117
p-Isopropylbenzyl	p-Isopropylbenzyl iodide	117
α-Naphthylmethyl⎫ β-Naphthylmethyl⎭	Bromides	118
Diphenylmethyl	Diphenylmethylamine	118

Produced in Bimolecular Reactions

OH	$H_2 + O_2$, $CH_3 + O_2$	53
HO₂	$H + O_2$	58, 59
	$H_2 + O_2$, $CH_3 + O_2$	53
CH₂	$CH_3 + O_2$	53
CHO (?)	$CH_3 + O_2$	53
CH₃O	$CH_3 + O_2$	53
CH₃O₂	$CH_3 + O_2$	53
C₂H₅	$CH_3 + C_2H_6$	5, 6
Allyl	CH_3 + propylene	5, 6

Produced in Combustion Flames

H	$H_2 + O_2$	33
O	$H_2 + O_2$	33
OH	$H_2 + O_2$	33
HO₂(?)	$CH_4 + O_2$	5, 6
CHO (?)	$CH_4 + O_2$	5, 6

Produced in Electric Discharges

H	H_2O	76
N(⁴S)	N_2	65–69
O	O_2, H_2O	57, 76
$O_2(^1\Delta_g)$	O_2	73, 75
O₃ (?)	O_2	72
O₄ (?)	O_2	72
OH	H_2O, H_2O_2	58, 60, 62, 76
HO₂	$O_2 + H$	58–60
NH₂	Hydrazine	64
N₂H₃	Hydrazine	64
CH₃	CH_4	77

TABLE 11.3.   Radicals Detected by Mass Spectrometry (Continued)

Radical	Reactant	Reference
	Produced by Collision with Excited Mercury Atoms	
CH₃	Acetone	32, 79, 90
	1-Butene	81
	β-Methallyl radical	81
	Acetaldehyde	32
	1,3-Butadiene	83
	1,2-Butadiene	83
	Acetylacetone	84
	Methyl formate	80
	Biacetyl	84
	Acetic acid	80
	Methanol	86
	Dimethyl ether	86
CHO	Acetaldehyde	32
	Methyl formate	80
CH₃O	Methyl formate	80
CH₃CO	Acetone	32, 79, 90
	Acetylacetone	84
CH₃OCO	Methyl formate	80
CH₃COCH₂	Acetylacetone	84
Propargyl	Allene, 1,2-butadiene ⎱	
	1,3-Butadiene ⎰	83
Allyl	Propylene, 1-butene	81
Methallyl	Isobutene, 1-butene, 2-butene	81
	Produced by Direct Photolysis	
CHO	Formaldehyde	88
CH₃CO	Acetone	90
CH₃	Acetone	90
NH₂	NH₃, hydrazine ⎱	
	Benzylamine ⎰	89
N₂H₃	Hydrazine	89
C₆H₅CH₂	Benzylamine	89
C₆H₅CH₂NH	Benzylamine	89
	Produced by Dissociative Ionization	
H	Propane	96
CH₃	Propane ⎱	
	n-Hexane ⎰	96
C₂H₃	n-Hexane	96
C₂H₅	Propane ⎱	
	n-Hexane ⎰	96
C₃H₅	Propane ⎱	
	n-Hexane ⎰	96
C₃H₇	n-Hexane	96

The pyrolysis of ethane and n-butane on a platinum filament was also examined by Robertson [42]. The main products from the pyrolysis of n-butane at 1050°C were ethane and ethylene and a significant quantity of ethyl radicals. The pyrolysis of ethane, on the other hand, proceeded at 950°C mainly by dehydrogenation to ethylene. In later work [43, 44] Robertson demonstrated clearly the importance of the nature of the surface. Using a platinum surface which was freshly cleaned, the pyrolysis products of ethane were found to be ethyl radicals and hydrogen. As the filament became carburized, the catalytic activity decreased and only ethylene was formed. Further carburization reduced the activity markedly until no decomposition was observed. The addition of small amounts of water to the ethane resulted again in the formation of ethylene, but substances such as O_2, H_2, and CH_3OH were not effective in this respect. A poisoned filament could be reactivated by heating in oxygen, after which ethylene was again produced. After a short period, attributed by the authors to the time required for removal of an oxide layer, the formation of ethyl radicals was observed. These observations and the measurement of the probability of reaction of ethane on collision with the filament under various conditions of surface, and at different temperatures, enabled the authors to draw certain conclusions regarding the nature of the transition state and its mobility with regard to the surface. The formation of O atoms from O_2 on the platinum filament at around 1500° was observed.

In a series of papers, LeGoff and his associates have reported some investigations of the nature of reactions on hot metal surfaces. A drawing of the apparatus used, which is similar in many respects to that of Robertson, is given in Fig. 11.4. In some preliminary experiments, LeGoff [24] found that $Pb(CH_3)_4$, CH_3I, and di-t-butyl peroxide were decomposed at about the same rate on platinum, nickel, and uncarburized tungsten, and the reaction surface of the metals was found to be of the same magnitude as the geometric surface. Methyl radicals were formed in the reaction in each case. In later work, LeGoff and Letort examined in detail the effect of carburization on the activity of tungsten surfaces toward a number of organic compounds [45, 25, 26]. The yield of CH_3 radicals from $Pb(CH_4)_3$, CH_3I, and di-t-butyl peroxide was found to pass through a maximum as the filament temperature was raised. They concluded that the carbon in the surface layer of the filament was removed at the higher temperatures by diffusion of the carbon into the bulk of the filament. A number of their findings concerning the rate of reaction of CH_3 radicals and H and I atoms with the interior surfaces of the instrument are described in Sec. 11.2A (5). LeGoff and Letort [46] also studied the decomposition of acetaldehyde on a tungsten filament. The only products were CO, H_2, and free carbon, and no free radicals were formed. The mechanism of action of the surface layer was explained as

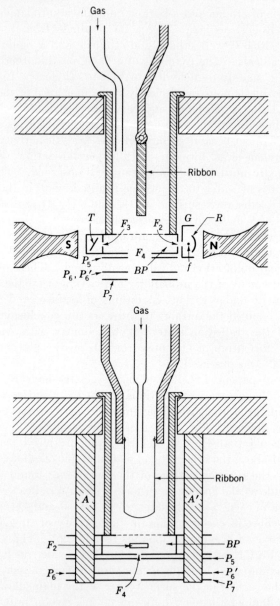

Fig. 11.4.   Reactor and ion source of mass spectrometer for study of reactions on metallic surfaces.   (*From LeGoff* [25].)

follows.   Each molecule of CH_3CHO decomposing in the filament leaves carbon, which hinders the later decomposition.   The carbon, however, simultaneously diffuses into the bulk of the metal, so that the fraction of the surface covered by carbon at any moment depends on the competition between these two processes.   In O_2-CH_3CHO mixtures, simultaneous oxidation and decomposition reactions were observed, the former presumably caused by the action of a layer of O atoms on the tungsten.

The reactions of CS_2, H_2S, SO_2, and CCl_4 on pure- and carburized-tungsten surfaces were also examined [47].   The decomposition of CS_2 on both pure and carburized tungsten was found to result in the formation of CS radicals and S_2.   The following reactions evidently occurred:

$$CS_2 \xrightarrow{\text{ribbon}} CS + S \qquad (11.11)$$

$$2S \xrightarrow{\text{walls}} S_2 \qquad (11.12)$$

A decomposition into C and S_2 directly was ruled out by the observation that no CO was formed on subsequently treating the hot filament with oxygen.   The yield of CS was greater when a carburized filament was used.   This is consistent with a decarburization reaction of the filament:

$$CS_2 + C_{(W)} \rightarrow 2CS \qquad (11.13)$$

The decomposition of H_2S on pure tungsten yielded H_2 and S_2 as products, and the HS radical was not detected.   The reactions occurring were evidently

$$H_2S \xrightarrow{\text{ribbon}} H_2 + S \qquad (11.14)$$

$$2S \xrightarrow{\text{walls}} S_2 \qquad (11.15)$$

On a carburized filament the decomposition of H_2S was faster than on pure tungsten, and the additional products CS and CS_2 were found:

$$H_2S + C_{(W)} \rightarrow H_2 + S + C_{(W)} \qquad (11.16)$$
$$S + C_{(W)} \rightarrow CS \qquad (11.17)$$
$$2CS \rightleftharpoons CS_2 + C \qquad (11.18)$$

In the decomposition of SO_2 on pure tungsten, SO and O_2 were detected, and the following reactions were postulated:

$$SO_2 \rightarrow SO + \tfrac{1}{2} O_2 \text{ (below } 1700°K) \qquad (11.19)$$
$$SO_2 \rightarrow S + O_2 \text{ (above } 1700°K) \qquad (11.20)$$

On carburized tungsten the products were CO and CS.

The decomposition of CCl_4 was found to give CCl_2 radicals and chlorine on both pure and carburized tungsten, and the possible reaction

$$CCl_4 + C_{(W)} \rightarrow 2CCl_2 \qquad (11.21)$$

was definitely excluded.   An interesting observation was that CS_2, H_2S, and SO_2 were not decomposed at all on pure-carbon filaments even at

2100°K.   Evidently the carbon present on a carburized tungsten filament is in a different form, possibly as tungsten carbide.

An interesting study of the free radicals produced in heterogeneous reactions was made by Osberghaus and Taubert [40], using a quite different experimental technique.   In their apparatus the ionization chamber itself could be heated to 1200°C.   This arrangement presents certain advantages in that all the surfaces with which the reactants and products come into contact are at the same temperature, but is less convenient for the study of the reactivity of surfaces of different specific nature.   Figure 11.5 shows a diagrammatic representation of their combined reactor and ionization chamber and a part of the mass-spectrometer tube.   The yield of methyl radicals from methane, ethane, and propane on the chrome-nickel steel walls of the ionization chamber was found to be less than about 0.2% at 1200°C, but the yields from n- and isobutane and tetramethyl methane were considerably higher.   A calibration of the sensitivity for CH_3 radicals, made using tetramethyl lead at about 800°C, allowed the yield of radicals from the other compounds to be measured quantitatively.   The mass spectrum and the ionization potential of CH_3 radical were also measured, as discussed in Sec. 11.2C.   In experiments with methyl chloride the production of HCl was observed, but the corresponding CH_2 radical was absent.   The temperature dependence of HCl production corresponded to an activation energy of about 26 kcal/mole.

Blanchard and LeGoff [29, 30] constructed a mass spectrometer for the study of the reactions of radicals at metallic surfaces.   A schematic diagram of the ion source of this instrument is shown in Fig. 11.6.   Among other ingenious features, the construction was such that the ionization chamber could be cooled with liquid nitrogen while a stream of radicals was directed into it from a heated quartz reactor.   The effect of wall temperature on the sensitivity of radicals to wall collision could then be examined under controlled conditions.   In preliminary investigations, some calculations were made on the proportions of molecules in the electron beam which had not collided with the wall, which had collided with the wall, and which had entered the ionization chamber from the surrounding space.   In experiments with iodine vapor at low temperatures where the condensation probability was unity, the intensity of the I_2^+ peak was found to be 6% of its value at 30°C.   This value corresponds to the concentration of molecules which are in the molecular jet and which have not collided with the walls.   The agreement with a calculated value of 14.4% was satisfactory, taking into account the approximations made in the calculations and the disturbing effects of condensed layers on the ion-source efficiency.   The effect of wall temperature on the recombination of I atoms in a stream of I_2 and I issuing from the heated reactor was also examined.   The recombination probability rose very slowly from 1.7×10^{-3} at room temperature to 4×10^{-3} at $-25°C$.

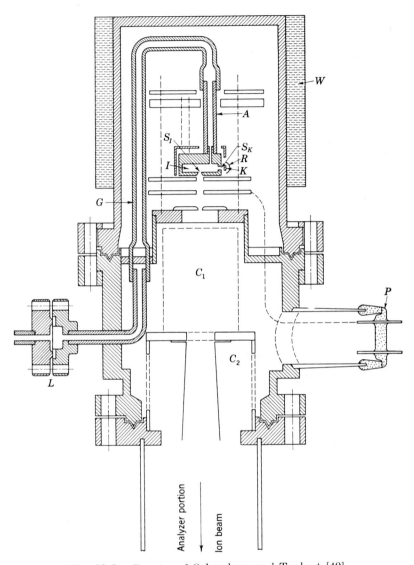

Fig. 11.5.   Reactor of Osberghaus and Taubert [40].

$C_{1,2}$ ion lens W water-cooling
I ionization region R electron reflector
G gas-inlet tube P press seal for high-tension leads
K cathode S_I ion exit slit
L leak S_K cathode slit
A Al_2O_3 insulation tube

Below $-25°C$ both I_2 molecules and I atoms were condensed. No I atoms could be detected on warming the walls.

Further studies with this instrument should throw considerable light on the reactions of radicals in condensed films and their interactions with metallic surfaces.

A number of other investigations in which radicals were produced by a heterogeneous reaction, but in which the primary interest lay not in the

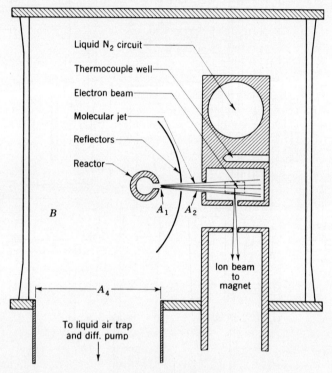

Labels on figure:
Liquid N₂ circuit
Thermocouple well
Electron beam
Molecular jet
Reflectors
Reactor
B
A_1 A_2
Ion beam to magnet
A_4
To liquid air trap and diff. pump

Fig. 11.6. Ion source for low-temperature studies. (*From Blanchard and LeGoff* [29].)

reaction but in obtaining radicals for ionization-potential measurements, will be discussed below.

B. Homogeneous thermal reactions

For the study of reactions in the intermediate pressure region (1 to 100 mm), it is necessary, as for heterogeneous reactions, that the reaction products pass from the reaction zone to the ionizing electron beam by an essentially collision-free path. A further requirement is that the sample of reaction products which are being analyzed must be withdrawn from a portion of the reaction zone which is representative of the whole. Since the concentration of radicals existing under the steady-state conditions of the conventional thermal decomposition is far too low to be detected, it is evident that the method is useful only under conditions in which the reaction is proceeding very rapidly. In order to obtain a representative sample and scan its spectrum in a time which is short compared with the reaction time, special techniques must be employed. Two methods of operation are possible:

1. The reaction is carried out in a static system under conditions of high reaction velocity, and the mass spectrum of a continuously sampled portion is scanned periodically at a speed which is fast compared with the duration of the reaction.

2. The reaction is carried out continuously in a portion of a fast flow system, and the mass spectrum of a continuously sampled portion is scanned at leisure. By this arrangement the "time" scale is converted into a "distance scale." If provision is made for moving the locus of the reaction with respect to the sampling orifice, the concentration of products representative of the different pseudo-steady-state conditions along the reaction zone can be examined.

Until quite recently the first alternative could not be used because of the lack of high-speed amplifiers and recording devices. The development of rapid-scanning mass spectrometers and the combination of time-of-flight instruments with secondary electron multipliers have opened up great possibilities for this method. Some of the applications of these techniques to the study of cool flames and of flash photolysis are discussed in Secs. 11.3C and 11.3E.

The second alternative, that of a flow system, has particular advantages for thermal reactions and has been more widely used. One of the reactors used by Eltenton fulfilled many of the requirements for this method, and a second reactor (Sec. 11.3C) was even more effective. In these reactors a stream of reactant at a small partial pressure was carried in a stream of inert gas through a tubular furnace and over an orifice in a thin diaphragm. Through this orifice a small fraction of the products entered the ionization chamber. The remainder was removed through a concentric tube by a fast pump. The maximum useful pressure which can be maintained in a reactor of this type depends on the size of the orifice and the pumping speed on the low-pressure side of the orifice. The upper limit of pressure for quantitative work is that at which the peak-height vs. partial-pressure relation becomes seriously nonlinear. For qualitative work the limit is considerably higher, being that at which this relation approaches the reversal point. Eltenton was able to employ pressures up to 100 mm for qualitative detection in the reactors of his design. He demonstrated the formation of methyl radicals from methane, ethane, and tetramethyl lead and of methylene radicals from diazomethane. The formation of ethyl radicals by the reaction

$$CH_3 + C_2H_6 \rightarrow C_2H_5 + CH_4 \qquad (11.22)$$

and of allyl radicals by the reaction

$$CH_3 + C_3H_6 \rightarrow CH_2 : CHCH_2 + CH_4 \qquad (11.23)$$

was also observed [5].

A modification of Eltenton's reactor used by Lossing and coworkers and shown in Fig. 11.3 had the advantage that all the interior surfaces were of quartz. Consequently, the possibility of disturbing the concentration of radicals and other products by catalytic reactions was substantially reduced. A further advantage was that the orifice was in the tip of a quartz cone projecting into the heated zone. The stream of reaction products entering the ion source was consequently more representative of the composition in the hottest part of the reaction zone. Whether or not the sample is still truly representative after the molecules have undergone a number of collisions with the walls of the ionization chamber depends on the nature of the radicals. As discussed above, it appears that many hydrocarbon radicals are not seriously affected by wall collisions.

Using this reactor, Lossing and coworkers examined the thermal decomposition of a large number of compounds, with particular attention to the presence and identity of free radicals [28, 48, 49]. In one case, that of di-t-butyl peroxide [39], measurements were made of the activation energy for the decomposition reaction in a temperature region where $k = 10$ to 300 sec^{-1} The values of E and the preexponential factor obtained in this abnormally high temperature region were in good agreement with those obtained by conventional kinetic methods at lower temperatures. Although such a method for measuring activation energies appeared at first to be rapid and convenient, it was found that serious errors were involved at temperatures above 600°C. Under these conditions the errors undoubtedly resulted from the nonlinearity of the temperature profile of the reactor. With suitable changes in design, the effective temperature range might be extended considerably.

Using this apparatus, the thermal decomposition of ethylene oxide was shown to yield at least 0.6 methyl radical per molecule [48]. An interesting feature of this reaction was the failure to detect the expected intermediate, acetaldehyde. The conclusion that the decomposition did not proceed by the intermediate formation of a normal acetaldehyde molecule was further supported by the observation that, in a separate experiment, acetaldehyde itself was not decomposed at the temperatures employed. The thermal decomposition of propylene oxide also yielded methyl in large amounts, together with ethylene, acetylene, carbon monoxide, hydrogen, and other products. In agreement with Eltenton [5] methyl radicals were also found in abundance in the decomposition of dimethyl ether. The corresponding methoxy radical CH_3O was not detected, presumably as a result of its instability at high temperatures.

The decomposition of dioxane was briefly examined. No significant amounts of CH_3 or CH_2 radicals were found.

The decompositions of a number of benzene derivatives at temperatures up to 1450°C were investigated qualitatively by Ingold and Lossing [50]. The products of decomposition of benzene, toluene, benzaldehyde, anisole, and diphenyl and dibenzyl ethers were generally complex mixtures, but radicals such as phenyl, benzyl, phenoxy, and benzoyl were detected in small quantities in some of the reactions.

The detection of methyl, ethyl, vinyl, allyl, sec-propyl, benzyl, and CH_3S radicals, produced in the thermal decomposition of various derivatives, has been reported [28, 49]. The difference in the stabilities of the n-propyl and sec-propyl radicals and the corresponding difference in their modes of dissociation was apparent from the products of decomposition of azo-n-propane and azoisopropane. At 660°C the sec-propyl radical was obtained in moderate yields, together with its thermal-decomposition product, propylene. On the other hand, the n-propyl radical was almost completely dissociated at 660°C into ethylene and methyl radicals. The radical itself could not be detected at this temperature.

Since the concentrations of free radicals in a stream of carrier gas can be measured directly using the technique described earlier, it is possible to measure radical combination rates by this method. Rates of combination are of great importance in kinetic experiments since the rates of other radical reactions, such as the abstraction of H atoms, are usually obtained in the form of a ratio, in which the denominator is the combination rate. An estimate of the rate of dimerization of methyl radicals was made by Lossing and Tickner [39] from the radical concentrations found in a stream of decomposing $Hg(CH_3)_2$. However, this estimate was very approximate, since the design of the reactor was not suitable for such measurements. The concentration gradients of reactant and products in this reactor were quite steep, and the contact time available for dimerization could not be calculated with any degree of certainty.

In later work by Ingold and Lossing [31], this difficulty was largely overcome by the use of a furnace which could be moved with respect to the orifice. A diagram of this arrangement is given in Fig. 11.7. By means of a rack and pinion, the movable furnace could be placed at various distances from the orifice. A stream of helium at 5 to 18 mm carrying dimethyl mercury at a partial pressure of a few microns was pumped through the reactor at a high flow rate. With the movable furnace at the position shown at the left in Fig. 11.7, some 80 to 90% of the theoretical yield of methyl radicals was present at the orifice, the remainder being ethane and a little methane. After the furnace had been retracted 0.5 cm, as shown at the right in Fig. 11.7, the concentration of products at the orifice was again measured. The concentration of methyl was found to have decreased, and the amount of ethane had increased proportionately. This amount of reaction could be assigned to the new 0.5-cm reaction zone appearing just above the orifice as a result of the retraction of the furnace. Since the rate of flow could be measured, the residence time for this reaction zone could be calculated with reasonable accuracy. The number of methyl radicals reacting in

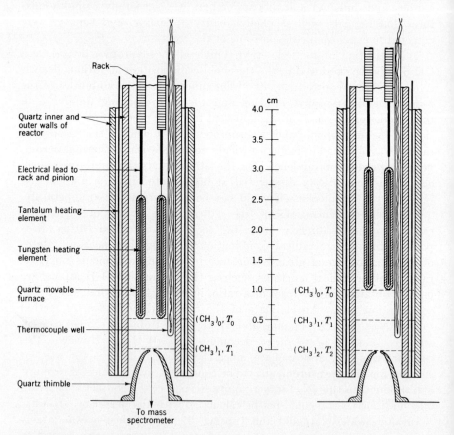

Fig. 11.7. Schematic diagram of movable furnace for study of combination of methyl radicals. (*From Ingold and Lossing* [31].)

this time could then be determined. By means of successive retractions of the furnace, a set of data similar to that given in Table 11.4 was obtained. As a result of the temperature change at the orifice as the distance to the orifice increased, and the corresponding changes in sensitivity and contact time, a number of correcting factors were required. After allowance had been made for these factors, as measured in suitable calibration experiments, it was found that the rate of combination of methyl radicals had a small negative temperature coefficient, the collision yield ranging from 0.018 at 1087°K to 0.11 at 434°K. This value is

TABLE 11.4. Composition of Products from Thermal Decomposition of Mercury Dimethyl (Pressures in μ)

Column 1 gives the distance from the end of the movable furnace to the leak. Columns 3 and 5 show how the CH_3 partial pressure decreased and the C_2H_6 partial pressure increased as this distance (and consequently the time) was increased.

Initial Hg(CH$_3$)$_2$	(1) Position of furnace, cm	(2) Mean temp., °K	(3) CH$_3$	(4) CH$_4$	(5) C$_2$H$_6$	(6) ΔCH$_3$	(7) 2 ΔC$_2$H$_6$	(8) ΣCH$_3$ + CH$_4$ + 2C$_2$H$_6$
2.2	0.2	....	1.69	0.065	0.0222	....	.....	1.80
	0.5	1087	1.53	0.065	0.0423	0.16	0.04	1.68
	1.0	872	1.43	0.065	0.0971	0.10	0.11	1.69
	1.5	713	1.27	0.065	0.190	0.16	0.19	1.69
	2.0	617	1.10	0.065	0.312	0.17	0.24	1.79
	2.5	551	0.882	0.065	0.457	0.21	0.29	1.86
	3.0	502	0.732	0.065	0.571	0.15	0.23	1.94
	3.5	466	0.580	0.065	0.682	0.15	0.22	2.01
	4.0	434	0.480	0.065	0.747	0.10	0.13	2.04
3.6	0.2	....	2.75	0.10	0.113	....	.....	3.08
	0.5	1087	2.68	0.10	0.156	0.07	0.086	3.09
	1.0	872	2.36	0.10	0.269	0.32	0.23	3.00
	1.5	713	1.94	0.10	0.511	0.42	0.48	3.06
	2.0	617	1.53	0.10	0.726	0.41	0.43	3.09
	2.5	551	1.16	0.10	0.894	0.37	0.34	3.05
	3.0	502	0.903	0.10	1.05	0.26	0.31	3.10
	3.5	466	0.703	0.10	1.18	0.20	0.26	3.16
	4.0	434	0.539	0.10	1.25	0.16	0.14	3.14
7.5	0.2	....	5.69	0.275	0.306	....	.....	6.57
	0.5	1087	5.30	0.275	0.430	0.39	0.25	6.43
	1.0	872	4.34	0.275	0.847	0.96	0.83	6.31
	1.5	713	3.32	0.275	1.49	1.02	1.28	6.57
	2.0	617	2.34	0.275	2.01	0.98	1.04	6.64
	2.5	551	1.66	0.275	2.33	0.68	0.64	6.59
	3.0	502	1.22	0.275	2.58	0.44	0.50	6.65
	3.5	466	0.904	0.275	2.75	0.32	0.26	6.68
	4.0	434	0.656	0.275	2.93	0.25	0.36	6.79
25.3	0.2	....	17.07	1.12	2.33	....	.....	22.85
	0.5	1087	14.41	1.12	3.36	2.66	2.06	22.25
	1.0	872	9.29	1.12	5.41	5.12	4.10	21.23
	1.5	713	5.76	1.12	7.13	3.53	3.44	21.14
	2.0	617	3.38	1.12	8.55	2.38	2.84	21.60
	2.5	551	1.90	1.12	9.12	1.48	1.14	21.26
	3.0	502	1.22	1.12	9.80	0.68	1.36	21.94

about one-third of that obtained by sector photolysis experiments. In later work [51] a dependence of the rate on the helium pressure was found, and it became evident that if allowance were made for the difference in third-body effectiveness in the two cases, the rates of combination obtained by sector photolysis and by mass spectrometry were in satisfactory agreement. The agreement between the two methods was particularly gratifying in view of the very different nature of the errors and assumptions involved.

The reaction of methyl radicals with NO is of importance in kinetic studies, and the rate of this reaction has been the subject of some controversy. A brief examination of this reaction was made using mass spectrometry [48] in which NO was added to a stream of methyl radicals at a known partial pressure. An approximate value of 2×10^{-4} for the collision efficiency for removal of CH_3 radicals was obtained. A more detailed study of this reaction was made by Bryce and Ingold [52]. A number of products were identified: NH_3, H_2O, HCN, CO, N_2, CH_3CN, and a substance of mass 45. The last, probably CH_3NO, became the predominant one at short contact times. A collision efficiency of 3×10^{-4} to 4×10^{-4} for the $CH_3 + NO$ reaction at low pressures was obtained.

The same authors identified a number of products resulting from the reaction of methyl radicals and molecular oxygen [53]. Using a movable furnace, dimethyl mercury was pyrolyzed, and the resulting methyl radicals were mixed with oxygen just below the furnace. From an analysis of the products, which included the free radicals CH_3O_2, CH_3O, CH_2, OH, and HO_2, it was evident that the reaction was complex. Three modes of reaction were proposed:

$$CH_3 + O_2 \rightarrow CH_3O_2 \tag{11.24}$$
$$CH_3 + O_2 \rightarrow CH_2O + OH \tag{11.25}$$
$$CH_3 + O_2 \rightarrow CH_2 + HO_2 \tag{11.26}$$

The products of the reactions (11.25) and (11.26) may, alternatively, represent two modes of dissociation of the CH_3O_2 radical, which would be expected to be unstable at the temperature of the reaction stream. The total collision yield of the reaction

$$CH_3 + O_2 \rightarrow products \tag{11.27}$$

was found to be in the region 10^{-3} to 10^{-4}. Attempts to detect the CHO radical in this reaction were inconclusive. Using the same technique, the H_2-O_2 reaction was investigated at elevated temperatures, but at pressures much too low for a flame to be produced. The presence of OH and HO_2 radicals in this reaction was clearly demonstrated. The nonradical products were H_2O and H_2O_2. No H or O atoms were detected, but the sensitivity of detection for these was probably rather low as a result of reactions in the ionization chamber.

C. Studies on combustion flames

One of the applications of the mass spectrometer to chemical kinetics investigated originally by Eltenton was the possibility of detecting radicals and other short-lived intermediates in combustion flames. A diagram of the central portion of the reactor which he used, shown in Fig. 11.8, presents a number of ingenious features. In this reactor a

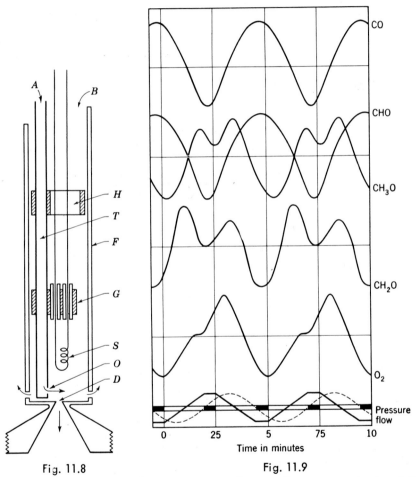

Fig. 11.8   Fig. 11.9

Fig. 11.8. Reactor for study of low-pressure flames. Gas components A and B enter as shown, and the flame extends from the side hole O across the diaphragm D. The distance OD can be varied by oscillating the tube T in a vertical direction, lateral movement being prevented by the spring guide H. A retractable Pt-Ir coil S, moving in guides G, serves to ignite the mixture. (*From Eltenton* [5].)

Fig. 11.9. Summarized phase curves for intermediates measured with a pulsating oxygen-in-methane flame. (*From Eltenton* [5].)

stationary flame was propagated in a flow system.   The distance from the orifice to the point of origin of the flame could be varied so that the composition at different parts of the flame could be examined.   Although the mixture of reaction products was rather complex, this feature allowed the phase relations between the intensities of various ions to be observed, and consequently provided a means of determining whether a particular ion was a fragment or a parent ion.   An automatic device compensated for the pressure changes following ignition of the flame and, in addition, allowed the flame to be pulsated at a regular rate.   The combustion flames of methane, propane, and carbon monoxide were examined.

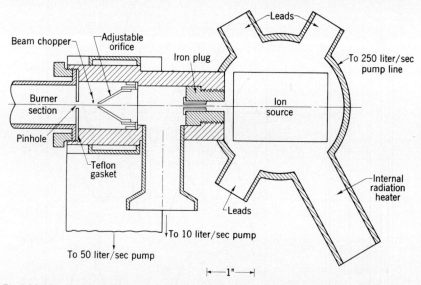

Fig. 11.10.   Cross section of mass-spectrometer gas-inlet system employed in flame studies.   (*From Foner and Hudson* [33].)

Although Eltenton emphasized that his intention was to explore the possibilities of the method rather than to study exhaustively a given reaction, nevertheless some interesting conclusions were reached.   The phase curves which he obtained for various ions (not necessarily parent ions) in a pulsating oxygen-methane flame are reproduced in Fig. 11.9. Metallic surfaces in the reactor were found to render the detection of oxygenated radicals rather uncertain.   The presence of the CH_3 radical in the methane-oxygen flame was demonstrated clearly, however, and its phase relation showed that it was not produced by thermal dissociation of methane, but originated in a zone between the oxygen- and methane-rich zones.

A considerable advance in instrumental design for studying flames was made by Foner and Hudson [33].   The arrangement of the sampling system and the ionization chamber which they used is shown in Fig. 11.10.

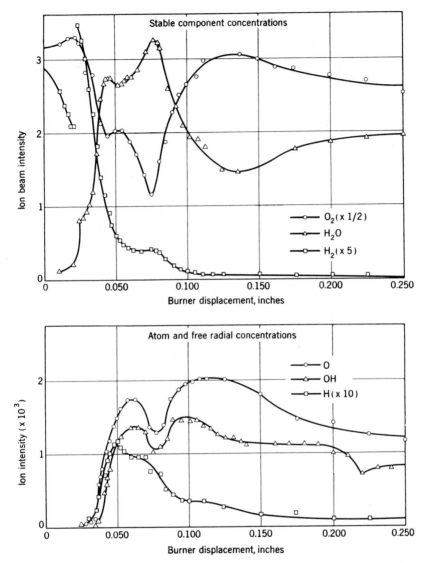

Fig. 11.11.   Ion-intensity measurements of stable components and intermediates in the low-pressure hydrogen-oxygen flame.   The abscissa is the relative displacement of the central burner tube from the pinhole.   (*From Foner and Hudson* [33].)

A series of apertures were employed to ensure that the stream of reaction products which entered the ionization chamber coaxial with the ionizing electron beam formed a molecular beam.   A magnetically driven beam chopper which interrupted the molecular beam at 200 cps provided a useful method of discriminating against background from products formed on the ionizing filament or on the walls of the ionization chamber, as referred to earlier. An examination of the hydrogen-oxygen flame showed the presence of H and O atoms and OH radicals in considerable abundance. The concentration profiles of stable and free-radical intermediates which they obtained are shown in Fig. 11.11. Foner and Hudson found that the interpretation of these curves was complicated by diffusion effects, turbulent mixing of the components of the mixture, and particularly by changes in the configuration of the flame as it was moved with respect to the diaphragm. The CH_4-O_2 flame was found to yield an extremely complex mixture of products, among which were identified CO, CO_2, H_2O, C_2H_2, H_2, and C_4H_2. Because of the complexity of the mass spectrum, the only radical which could be identified unambiguously was CH_3.

A rapid-scanning mass spectrometer for the study of intermediates in cool-flame oxidation reactions has been constructed by Ouellet and a number of coworkers. With this instrument [54] spectra could be obtained at a repetition rate of about 60 cps and were recorded by an oscilloscope and a high-speed motion-picture camera. The cool flames of ethyl ether [55] and acetaldehyde [56] were examined, the latter in some detail. The spectra were quite complex, and no free radicals could be identified. In the slow oxidation of acetaldehyde, however, the build-up and decay of the peracetic acid concentration and its behavior at the onset of the cool flame could be followed.

D. Radicals and atoms produced by the action of electrical discharges

The applications of mass spectrometry to the study of radicals and atoms formed by the action of electrical discharges have been reviewed by Schiff [57]. The formation of ions in such discharges has been studied also by mass spectrometry, but will not be discussed here.

The detection of HO_2 radicals in the reaction between molecular oxygen and hydrogen atoms from a Wood's discharge tube was carried out by Foner and Hudson [58] and by Robertson [59]. In the experiments of Robertson, oxygen was added to a stream of atomic hydrogen at a total pressure of 0.1 to 0.5 mm. The sampling orifice opening to the mass spectrometer was located in a short side tube leading from the tube which carried the main reaction stream. The presence of the "dead" volume thus created, through which the radical products had to diffuse, may have provided some disturbance of the concentrations of products. Nevertheless, Robertson was able to make an unambiguous detection of the HO_2 radical. Hydrogen peroxide was found under all the experimental conditions and was attributed to the addition of a second H atom to a HO_2 radical already formed. A discussion of the probable third-body restrictions was given. The OH radical was not detected, and it was concluded that its concentration must have been less than 10% of the HO_2 concentration. No ozone, O atoms, or species such as H_2O_4 were found.

The detection of HO_2 in the same reaction by Foner and Hudson [58] was carried out at much higher pressures ($\sim$30 mm), using a modification of their sampling system described above. An inert gas, helium, was added to the flow system. They detected the HO_2 and OH radicals and proposed the following modes of formation:

$$H + O_2 + M \rightarrow HO_2 + M \tag{11.28}$$
$$HO_2 + H_2 \rightarrow OH + H_2O \tag{11.29}$$

It was estimated that about 1% of the H atoms ended up as HO_2. In a further publication, Foner and Hudson [60] obtained increased yields of HO_2 by reacting H_2O_2 with a stream of H_2O (or H_2O_2) through which an electrical discharge was passed. Under these conditions an additional mode of formation was found to occur:

$$OH + H_2O_2 \rightarrow HO_2 + H_2O \tag{11.30}$$

The ionization potential of the HO_2 radical was measured, and from the appearance potential of HO_2^+ from H_2O_2, a value of

$$D(H\text{---}HO_2) = 89.5 \text{ kcal/mole}$$

was derived. From thermochemical data this leads to

$$D(H\text{---}O_2) = 47 \pm 2 \text{ kcal/mole}$$

These authors also made an investigation of the effect of surface on the formation of H_2O_2 and OH in the $H + O_2$ reaction [61]. For this purpose, a reactor, the walls of which could be cooled by liquid nitrogen, was constructed of Pyrex glass. The walls were coated with phosphoric acid. At room temperature and 1 mm pressure they found water to be the chief product, and, contrary to the results of Robertson, no H_2O_2 was formed. OH radicals were present, even at very low pressures, where formation by a bimolecular reaction in the gas phase would be excluded by third-body restrictions. They consequently concluded that OH radicals must be formed on the walls rather than by the reaction they had proposed earlier. Cooling the walls of the reactor with liquid nitrogen reduced the OH concentration to zero. On warming the reactor the (integrated) evolution of H_2O_2 indicated that the rate of production was about 150 times the room-temperature rate.

Foner and Hudson [62] also measured the ionization potential of OH radicals, produced by a discharge in water and in hydrogen peroxide. From this value (Table 11.5) and the appearance potential of OH^+ ion from water, they derived $D(H\text{---}OH) = 116 \pm 5$ kcal/mole, and consequently $D(O\text{---}H) = 103 \pm 5$ kcal/mole. These values are in good agreement with other measurements for this dissociation energy.

Foner and Hudson also examined the products formed by an electrodeless discharge in hydrazoic acid and in hydrazine [63, 64]. In this work they identified, for the first time, a number of unusual hydronitrogen compounds whose existence had been postulated: diimide N_2H_2, triazene N_3H_3, and tetrazene N_4H_4. The free radicals NH_2 and N_2H_3 were also detected. The following ionization potentials were measured: NH_2

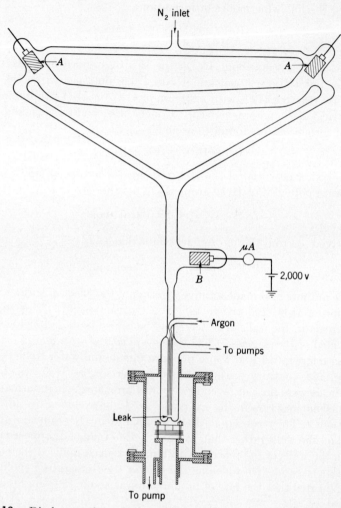

Fig. 11.12. Discharge tube and mass-spectrometer ion source. Discharge occurs between electrodes A; electrode B acts as an electron trap. (*From Jackson and Schiff* [66].)

radical, 11.4 volts; N_2H_3 radical, 7.88 volts; diimide, 9.85 volts; triazene, 9.6 volts. From appearance potentials a number of bond-dissociation energies were evaluated: $D(NH_2\!-\!H) = 106 \pm 3$ kcal/mole,

$$D(H_2N\!-\!NH_2) = 58 \pm 9 \text{ kcal/mole}$$

and $D(H\!-\!N_2H_3) = 76 \pm 5$ kcal/mole. The first two of these values are in good agreement with bond-dissociation energies obtained from kinetic data. The third has not been previously measured.

The mass-spectrometric method has been applied to the study of active nitrogen by Schiff and coworkers at McGill University and Kistiakowsky and coworkers at Harvard University. In the experiments of Jackson and Schiff [65, 66], a stream of active nitrogen from a "condensed"

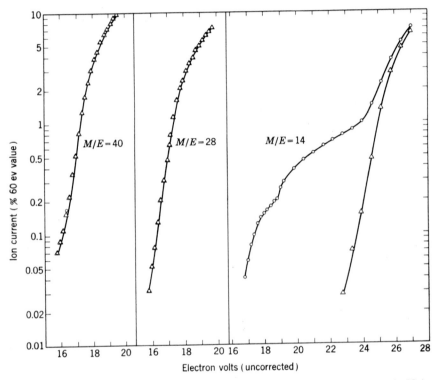

Fig. 11.13. Semilogarithmic ionization efficiency curves for the ions Ar^+, N_2^+, and N^+. The circles represent values obtained with excitation of the discharge; the triangles represent values obtained in the absence of the discharge. (*From Jackson and Schiff* [66].)

discharge tube was pumped rapidly through a glass tube and over a quartz cone containing an orifice leading to the ionization chamber. The experimental arrangement is shown in Fig. 11.12. A small amount of argon was added to the stream, and the height of the mass 40 peak enabled corrections to be made for pressure changes resulting from the operation of the discharge. Appearance potential curves for the ions of mass 14 and mass 28 were measured with and without the discharge. The appearance potential curves of Fig. 11.13 show that a new species at mass 14 was formed by the discharge. An appearance potential of 14.7 ± 0.3 volts was obtained. This was in good agreement with the accepted value of 14.54 volts for the ionization potential of the ground state ^4S nitrogen atom, demonstrating clearly the presence of N atoms in the active nitrogen. Some evidence was found for a second appearance potential for N^+ ions, corresponding to a dissociative electron impact on excited N_2 molecules or on N_3 molecules, but no positive identification of these species could be made. From measurements of the decrease in the N_2^+ peak height and the increase in the N^+ peak under carefully controlled conditions, the authors were able to measure the sensitivity of the instrument to N atoms, and consequently the concentrations could be measured. Depending on the extent to which the walls of the reactor were poisoned by meta-phosphoric acid, the concentration of N atoms at the orifice ranged from 0.1 to 1.0%. This is in good agreement with estimates of 0.3 to 2.0% measured by calorimetric and spectroscopic methods by other workers.

Active nitrogen was also investigated by Berkowitz, Chupka, and Kistiakowsky [67], using a similar reaction system, but with the additional provision for simultaneous measurement of the optical intensity of the afterglow. These authors confirmed the observations of Jackson and Schiff that ^4S nitrogen atoms were present in the active nitrogen stream. They concluded that the appearance potential curve for mass 14, obtained with the discharge on, showed no evidence for the existence of a second source of N^+ ions, and that the curve for N_2^+ ions also showed no irregularities attributable to a meta-stable N_2 species. The concentration of N atoms in the stream was measured by means of a calculation based on the changes in the mass 28 and mass 14 peaks when the discharge was turned on. An attempt was made to allow for the combination of atoms on surfaces in the neighborhood of the source, and a correction factor of about 2.6 was obtained, based on the change in the mass 14 peak brought about by the addition of a glass insert between the orifice and the ionization chamber. The derivation of this factor from the observed change depends on what assumptions are made about the shape of the molecular "beam" issuing from the orifice. This in turn depends on whether the orifice behaves as an ideal orifice or as a short tube at the pressures used. From his own experience the present writer feels that the assumptions made should be considered as representing a limiting case, and the derived correction factor might well be considerably smaller. In any event, the accuracy of measurement was great enough to show a number of interesting relations. From simultaneous observations of the optical intensity and the net mass 14 peak due to atoms, it was clearly shown that the intensity of afterglow was proportional to the square of the $N(^4S)$ atom concentration under various conditions. Cooling the walls of the flow system to dry-ice temperatures reduced the atom concentration by about 20%. The addition of helium or argon to the stream increased the afterglow intensity, and these gases appeared to be more effective third bodies for recombination than nitrogen itself. They were able also to show that the presence of oxygen is not essential to the formation of the afterglow, although its presence reduced the rate of combination at the walls. The addition of NO to the stream resulted in a rapid disappearance of N atoms, but the mechanism proposed, the direct addition of N atoms to NO to form N_2O (identified by the mass 44 peak), seems improbable on the basis of recent evidence as to the effectiveness of NO in promoting the recombination of N atoms. The ratio of the mass 46 and 44 peaks produced when NO was added was not greatly different from the ratio found in CO_2, which appears in the mass spectrum of NO at high pressures when the electron-emitting filament is carburized. (In later work N_2O was not detected in this reaction.) They concluded that the afterglow results from a termolecular reaction involving two ^4S nitrogen atoms, the rate constant being 2×10^{-33} $(cm^3)^2$ mole^{-2} sec^{-1}, and that the reaction probably involves the intermediate formation of the $^5\Sigma_{g^+}$ molecular state, followed by a collision-induced radiationless transfer into the $^3\Pi_g$ state.

In an extension of this work Kistiakowsky and Volpi [68] investigated the reaction of N atoms with NO, N_2O, NO_2, and O_2. An electrodeless discharge operating on 2,400 Mc/sec at 125 watts output was used and was found to yield about the same atomic concentration as the 60-cycle discharge used previously, but with improved stability. The reaction of N atoms with NO was found to proceed as follows:

$$N + NO \rightarrow N_2 + O \qquad (11.31)$$
$$O \,(+ \text{ wall}) \rightarrow \tfrac{1}{2}\, O_2 \,(\text{wall}) \qquad (11.32)$$
$$NO_2{}^* + M \rightarrow NO_2 + M \qquad (11.33)$$
$$NO_2{}^* \rightarrow NO + O \qquad (11.34)$$
$$O + NO_2 \rightarrow NO + O_2 \qquad (11.35)$$

A lower limit of 4×10^{11} cm^3 mole^{-1} sec^{-1} for the rate of the first reaction was estimated, and approximate values for several other rate constants were derived. A value derived for the accommodation coefficient of O atoms on the walls of the reactor agrees well with other estimates. The reaction of N atoms with O_2 was found to conform to the mechanism:

$$N + O_2 \rightarrow NO + O \qquad (11.36)$$
$$N + NO \rightarrow N_2 + O \qquad (11.37)$$
$$O \,(+ \text{ wall}) \rightarrow \tfrac{1}{2}\, O_2 \,(+ \text{ wall}) \qquad (11.38)$$

The rate constant for the first of these reactions was calculated to be 2×10^{12} exp $(-6,200/RT)$ cm^3 mole^{-1} sec^{-1}. The reaction of N atoms with N_2O was found to be very slow, probably less than 2.5×10^6 cm^3 mole^{-1} sec^{-1}. The reaction of N atoms with equimolar mixtures of NO and N_2O, in which O atoms would be formed, enabled the authors to conclude that the reaction of O atoms with N_2O is also very slow, k being less than 2×10^8 cm^3 mole^{-1} sec^{-1} at room temperature. The reaction of

N atoms with NO_2 was found to exhibit complex behavior, and only a tentative mechanism could be put forward:

$$N + NO_2 \rightarrow N_2O + O \tag{11.39}$$
$$O + NO_2 \rightarrow NO + O_2 \tag{11.40}$$
$$N + NO \rightarrow N_2 + 0 \tag{11.41}$$
$$O\,(+\text{ wall}) \rightarrow \tfrac{1}{2}\,O_2\,(+\text{ wall}) \tag{11.42}$$

The possibility of two additional primary steps existed:

$$N + NO_2 \rightarrow 2NO \tag{11.43}$$
$$N + NO_2 \rightarrow N_2 + O_2 \tag{11.44}$$

Kistiakowsky and Volpi [69] also found that the rates of reaction of nitrogen atoms with H_2, CO, and NH_3 at room temperature and at 250°C were too slow to be detected. The reaction of N atoms with NO and NO_2 were reinvestigated, using a modified reaction system. Even with this system, the rates were too fast to measure, although a lower limit of 5×10^{13} cm^3 mole^{-1} sec^{-1} could be set for the reaction with NO.

The kinetics of N atom recombination was examined by Herron et al. [70] by titrating N atoms with NO and measuring the NO consumed using a mass spectrometer. They found the reaction above 3 mm Hg to be mainly homogeneous and third-order, with a rate constant of 5.7×10^{15} (cm^3)^2 mole^{-2} sec^{-1}. On a glass wall partly poisoned by water, a pseudo-first-order recombination occurred with a coefficient of 1.6×10^{-5} sec^{-1}. The third-order combination was found to be independent of temperature in the range 195 to 450°K. The effectiveness of helium and argon as third bodies was also investigated.

Herron and Dibeler [71] examined the possibility of detecting N atoms by evaporation of condensates from an electrical discharge in nitrogen and nitrogen-rare gas mixtures. They obtained qualitative evidence for the presence of N atoms and set an upper limit for the recovery of N atoms of $10^{-3}\%$ of the atoms originally condensed.

The reactions of N atoms with acetylene, ethylene, and propylene were also investigated [71a]. Although the flow rate was not high enough for radicals to be detected directly, some interesting stable products were observed. The formation of cyanoacetylene from acetylene, and possibly of cyanoethylene from ethylene, suggests that the CN radical can undergo a replacement reaction with a hydrogen atom. The main product in all cases was HCN, together with smaller amounts of cyanogen. The formation of ammonia was also observed.

The products formed by an electrical discharge through oxygen have also been studied. Grundland [72] reported the production of O_3 and O_4 in an ozonizer at atmospheric pressure. Foner and Hudson [73] detected O_2^+ ions in the products of a discharge through O_2 at an appearance potential 0.93 volt below the first ionization potential of the O_2

molecule. This energy difference is in good agreement with the separation of $^1\Delta_g$ and $^3\Sigma_g$ states of O_2, and they concluded that 10 to 20% of the O_2 molecules were in the $^1\Delta_g$ state. In some preliminary experiments, Herron and Schiff [74, 75] found an increase in the ratio O^+/O_2^+ of about 50% corresponding to about a 3% change in composition of the gas stream, when a discharge was passed through oxygen. The additional source of O^+ ion was very probably the O atoms formed in the discharge, although some interference from O^+ ions formed by the dissociative ionization of the excited O_2 molecules, reported by Foner and Hudson, would be expected. A search for ozone showed that, at the pressures used, the concentration was less than 0.02%. Ozone was, however, formed in a trap at liquid-nitrogen temperature further along the gas stream, evidently by a surface reaction.

In a continuation of this work [75] they found that ozone was not present in the gas stream, following the discharge, in amounts greater than 0.02% at pressures up to 2 mm. In agreement with Foner and Hudson [73] they found that excited oxygen molecules in the $^1\Delta_g$ state were present at a concentration of about 10%. The concentration of O atoms was measured by adding NO_2 to the stream and observing the amount of NO_2 remaining after 0.5 sec as a function of the initial concentration. The concentration was found to range from 4 to 12%, being enhanced by the addition of traces of water vapor. From a comparison of measurements of the O atom concentration in the gas stream by NO_2 addition and in the ionization chamber by mass-spectral comparisons, they concluded that only 1 oxygen atom in 21 is ionized before recombining on the wall. The rate of combination of O atoms on the Pyrex walls of the tube leading from the discharge was found to be 1.1×10^{-4}, in good agreement with earlier work. The reactions of O atoms with N_2O, NO, and NO_2 were also investigated. No reaction with the former could be detected, and an upper limit of 1×10^8 cm^3 mole^{-1} sec^{-1} for the bimolecular rate constant was estimated. In the reaction of O atoms with NO and NO_2, no NO_3 was detected, and the only product in each case was the other oxide.

Tal'roze et al. [76] have examined the products of an electrical discharge through water vapor. Their apparatus was similar in some respects to that of Foner and Hudson [58] in that they used a beam chopper to reduce background resulting from wall reactions in the ion source. In their apparatus, however, the molecular beam from the discharge entered the ionization coaxially with the ion beam rather than at 90°. H and OH radicals were found in abundance when the discharge pressure was 0.5 to 1.5 mm Hg, and O atoms when the discharge pressure was raised to 3 mm. The formation of O atoms was attributed to reactions between OH^+ and H_2O and OH and H_2O^+, which became important at the higher pressures. A calibration for the H and O atom sensitivities was carried out by measuring the electrical equivalent of the heat developed by atom recombination on a platinum wire located near the orifice.

Dogramadze and Zmbov [77] examined products from a 10 Mc/sec electrodeless discharge through methane in a fast flow system at 0.2 to 0.8 mm Hg. A careful comparison of the mass 15/16 ratio with and without the discharge showed a net increase of 1 to 2.6% with the discharge operating. Under conditions of high residence time no net change in the 15/16 ratio was observed, although other products such as C_2H_2, C_2H_4, and C_2H_6 were formed. This experiment showed that the increase in the mass 15/16 ratio under fast-flow conditions resulted from the formation of CH_3 radicals in the discharge, and not from instrumental factors. On account of interference at mass 14 by CH_3^+ fragmentation and background, a small net increase for CH_2^+ was not taken as confirmation of the presence of CH_2 radicals in the discharge.

Herron et al. [78] examined the behavior of hydrazoic acid in an electrical discharge. Although condensation of the reaction products showed a blue coloration, they found no trace of radicals or excited products in the product stream. They concluded that the concentrations of such species must be exceedingly low.

E. Radicals produced in photochemical reactions

The possibilities for the use of mass spectrometry as a means of detecting and identifying the free radicals and other intermediates produced in photochemical reactions appear promising. Possibly the greatest difficulty in such an application is the relatively low steady-state concentration of radicals present when such reactions are carried out under the conditions employed in conventional photolytic experiments. As we have seen, the corresponding difficulty in thermal reactions can be generally overcome by increasing the reaction temperature and employing fast flow systems. However, increasing the number of quanta absorbed per unit reaction volume is a problem of a different nature. The number of effective quanta entering the reaction volume per second is limited by the intrinsic intensity per unit area of the surface of the lamp. Their current efficiency not being very high, most ultraviolet lamps are usually more efficient as sources of heat than of light, and the intensity of ultraviolet available is effectively limited by the rate at which heat can be dissipated by the electrodes or removed from the body of the lamp. In addition, the absorption coefficients of most organic compounds for radiation of kinetically effective wavelengths are so small that very little energy can be transferred per second to a unit volume of reactant.

One way out of this difficulty is to take advantage of the large absorption coefficient of mercury vapor for the 2,537-A mercury resonance line. It is well known that this absorption coefficient is so large that a partial pressure of a few microns of mercury vapor will provide essentially complete absorption of the resonance radiation in a path of a few millimeters. The exchange of energy between the electronically excited (6^3P_1) mercury atoms so formed and various organic compounds, which by themselves may not absorb in the ultraviolet, results in a "sensitized" reaction of the latter, usually a dissociation into free radicals or atoms. In some preliminary experiments using mercury photosensitization, Farmer, Lossing, Marsden, and Steacie [79] found that acetone could be decomposed in a flow system to the extent of some 80% in an exposure time of 0.002 sec. They detected CH_3 and CH_3CO radicals, as well as CO, $CH_2=CO$, and biacetyl, as products of the reaction. A later modification of this reaction system is shown in Fig. 11.14. The reactant at a partial pressure of a few microns was carried in a stream of helium at about 8 mm through the mercury saturator and through a tubular quartz reaction cell which was illuminated strongly by a low-pressure mercury lamp of cylindrical form. Just beyond the illuminated zone the gas stream passed over a quartz cone containing an orifice, of about 28 μ diameter, which opened into the ionization chamber of the mass spectrometer. The length of the illuminated zone could be varied by means of a cylindrical shutter. The linear rate of flow through the reactor could be varied from 10 to 60 m/sec, although in practice the time spent in the illuminated zone was most conveniently controlled by the shutter. Using this experimental arrangement, Lossing, Marsden, and Farmer [81] investigated the mercury-photosensitized reactions of the C_2—C_4 olefins. They found that the decomposition reaction of ethylene proceeded almost exclusively by a nonradical split into acetylene and molecular hydrogen. The alternative mode of decomposition to form vinyl radicals, if it occurred at all, was estimated to be less than 3% of the total reaction at the temperature used (60°C). The decomposition of propylene, on the other hand, led to the formation of allyl radicals and a hydrogen atom. Allene was also formed, but since the ratio of allene to allyl produced increased by a factor of over 6 for a 3.5-fold increase in the time of illumination, it was evident that most of the allene was not produced in a primary dissociation step. Neither could the allene be formed by a spontaneous dissociation of a "hot" allyl radical formed in the primary act. Of the 112 kcal/mole imparted to a propylene molecule by a $^3P_1$ Hg atom, about 77 are required to rupture the allyl-H bond, and the resulting allyl radical could have a maximum excitational energy of 35 kcal. From thermochemical data the authors concluded that D(allene-H) in allyl is about 65 kcal/mole and that consequently allene was not formed by this means. An alternative method of production, made possible by the unusually high concentration of radicals and of $^3P_1$ Hg atoms, was suggested:

$$\text{Allyl} + \text{Hg}(^3P_1) \rightarrow \text{allene} + \text{H} + \text{Hg}(^1S_0) \tag{11.45}$$

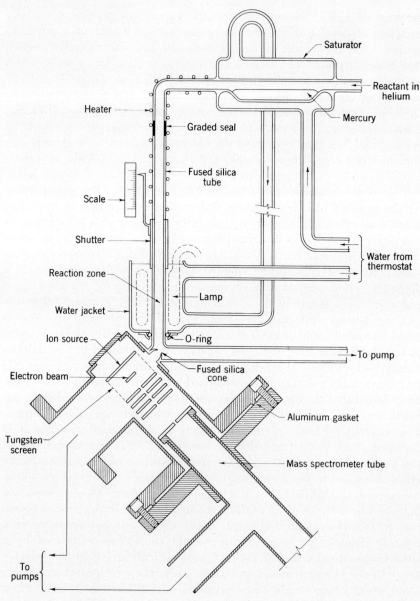

Fig. 11.14. Reactor for study of mercury-photosensitized reactions. (*From Kebarle and Lossing* [80].)

On the assumption that all the allene was formed by this means, the authors estimated that, on this basis, the allyl radical would have a cross section for reaction with excited mercury atoms about five times larger than that of propylene. A larger quenching cross section, or a larger quantum yield, or a combination of both, could be responsible. The mercury-photosensitized decompositions of isobutene and of 2-butene proceeded mainly by rupture of the allylic C—H bond to form a methallyl radical and a H atom:

$$CH_2 : C(CH_3)CH_3 + Hg(^3P_1) \rightarrow CH_2 : C(CH_3)CH_2 + H + Hg(^1S_0) \quad (11.46)$$

and

$$CH_3CH : CHCH_3 + Hg(^3P_1) \rightarrow CH_3CH : CHCH_2 + H + Hg(^1S_0) \quad (11.47)$$

In the dissociation of 1-butene, methyl, allyl, and methallyl radicals were found, and it was concluded that two primary modes of dissociation were occurring:

$$CH_2 : CHCH_2CH_3 + Hg(^3P_1) \rightarrow C_3H_5 + CH_3 + Hg(^1S_0) \quad (11.48)$$

$$CH_2 : CHCH_2CH_3 + Hg(^3P_1) \rightarrow C_4H_7 + H + Hg(^1S_0) \quad (11.49)$$

The first of these two modes predominated. Since $D(C_3H_5—CH_3)$ is about 64 kcal/mole and $D(C_4H_7—H)$ is about 77 kcal/mole, the relative frequency of the two modes of bond rupture appeared to be governed by energetic considerations.

The primary dissociation steps in the mercury-photosensitized reactions of acetaldehyde and acetone were also examined by this method [32] in an attempt to determine the importance of dissociation steps involving rearrangement into two molecules as compared with a dissociation into two radicals. The primary dissociation of acetone was found to proceed almost exclusively by the formation of CH_3 and CH_3CO radicals. The amount of methane produced was negligibly small, and hence dissociation by the rearrangement reaction

$$CH_3COCH_3 + Hg^* \rightarrow CH_4 + CH_2{=}CO + Hg \quad (11.50)$$

evidently occurred to a very small extent or not at all. Ketene was a product, but its formation could be explained only by a reaction between CH_3CO radicals and excited mercury atoms, comparable with the reaction of allyl radicals mentioned above.

In the decomposition of acetaldehyde, the products were CO, CH_3 radicals and appreciable quantities of CH_4, C_2H_6, and H_2. The detection of CHO radical at low electron energies was not unambiguous, although at higher aldehyde pressures a peak at mass 58, assumed to be the parent of glyoxal (the dimer of CHO) was clearly evident. The fact that the methane did not arise from a rearrangement of the acetaldehyde molecule to CH_4 and CO was demonstrated by experiments in which mixtures of CH_3CHO and CD_3CDO were simultaneously decomposed. With pure

CD_3CDO, the ratio of CD_4 formed per CD_3CDO molecule was determined. Addition of CH_3CHO to the reaction stream, where it was decomposed simultaneously with the CD_3CDO, caused a proportional decrease in the yield of CD_4, as shown by the relation given in Fig. 11.15. If CD_4 had been formed by a molecular rearrangement of CD_3CDO, addition of CH_3CHO would not have affected the ratio. Extrapolation of the line in Fig. 11.15, although somewhat uncertain, showed that the

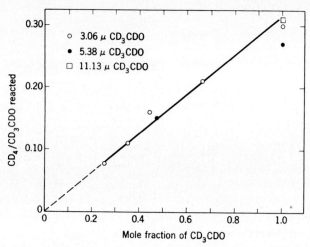

Fig. 11.15. CD_4 production from mercury-photosensitized decomposition of CD_3CDO-CH_3CHO mixtures. (*From Lossing* [32].)

molecular rearrangement constituted less than 5% of the total decomposition. The methane was therefore considered to arise by the reaction

$$CH_3 + HCO \rightarrow CH_4 + CO \qquad (11.51)$$

which would have to be very fast at 55°C. The virtual absence of the mixed aldehydes CD_3CHO and CH_3CDO suggested that the combination of CH_3 and CHO must be quite slow at the low pressures used in the experiment.

Formaldehyde, under the same conditions, was found to dissociate in part by an intramolecular elimination of H_2 [82]. A plot, similar to that of Fig. 11.15, of H_2/H_2CO reacted against the mole fraction of H_2CO in H_2CO-D_2CO mixtures gave again a straight line, but with an intercept at H_2/H_2CO reacted = 0.4. The authors therefore concluded that about 40% of the formaldehyde dissociated by intramolecular elimination of H_2, and the remainder, by a free-radical process. Using the same apparatus, Collin and Lossing [83] examined the mercury-photosensitized decomposition of allene and the butadienes. The only products found in the allene reaction were hydrogen, a radical of mass 39 (C_3H_3), and a substance of mass 78, evidently the dimer of the radical. Two electronic configurations for the C_3H_3 radical were considered possible, $H_2C=C=\dot{C}H$ (allenyl) and $H_2\dot{C}—C\equiv CH$ (propargyl). Provided that resonance could occur, the actual structure of the radical would be intermediate between these two forms. Although the dimer could not be unambiguously identified as dipropargyl or diallenyl, the use of methyl radicals added to the reaction stream as a radical "trap" enabled an identification of the nature of the C_3H_3 radical to be made. Assuming that the reaction $C_3H_3 + CH_3$ would proceed most rapidly at the carbon containing the free electron, the two possible forms of C_3H_3 would give, respectively,

$$CH_3 + \dot{C}H=C=CH_2 \rightarrow \text{1,2-butadiene} \tag{11.52}$$
$$CH_3 + \dot{C}H_2—C\equiv CH \rightarrow \text{1-butyne} \tag{11.53}$$

Since the spectra of these two products differ greatly in the mass 53/ mass 54 peak ratio, it was possible to identify the product as 1-butyne, no 1,2-butadiene being formed. Since only one product was found, it was concluded that the reactivity of the C_3H_3 radical was confined to one end only and that the radical had a configuration much closer to that of propargyl than to that of allenyl. The decomposition of 1,2-butadiene was found to yield mainly CH_3 and C_2H_3 radicals, but C_4H_4 and H_2 were also formed. A rather surprising observation was that 1,3-butadiene also yielded mainly CH_3 and C_3H_3 radicals, rather than two vinyl radicals, as might be expected from the structure of 1,3-butadiene. A mechanism involving a 3,1-hydrogen atom migration in the excited butadiene molecule was proposed to account for the experimental results. Some tentative conclusions were drawn with respect to the two available values for the heat of formation of the vinyl radical.

The authors also pointed out that the reactions observed using this technique may in some cases appear to be quite different from those observed in mercury-photosensitized reactions carried out at the higher reactant pressures used in the conventional methods. At the higher pressures the excited molecules formed by collision with Hg^* may not have time to dissociate before colliding with a second reactant molecule, and thereby initiating polymerization. Moreover, the addition of the free radicals to butadiene under the higher-pressure conditions would also tend to promote polymerization. Consequently, an identification of the modes of dissociation of the primary excited species is greatly facilitated by the unusually low partial pressures used in the mass-spectrometric studies.

Using the improved reactor shown in Fig. 11.14, Kebarle and Lossing [80] investigated the Hg-photosensitized decomposition of formic acid, acetic acid, and methyl formate. They concluded that the primary decomposition of formic acid proceeded by way of two intramolecular rearrangement reactions, to give CO_2 and H_2, and CO and H_2O in the ratio $3:7$. No free radicals were detected. In the methyl formate reaction, CH_3, CHO, CH_3O, and CH_3OCO radicals were detected. Addition of excess CH_3 radicals [from $Hg(CH_3)_2$ added to the reaction stream] produced large yields of dimethyl ether and methyl acetate, by combination of CH_3 with CH_3O and CH_3OCO radicals. In the reaction of acetic acid, only CH_3 radicals could be detected directly in the mass spectrometer. The formation of acetone, however, which increased greatly in amount when excess CH_3 radicals were added to the reaction stream, showed that CH_3CO radicals were present. The low sensitivity for detection of CH_3CO, CHO, and probably other oxygenated radicals is one of the disadvantages of this apparatus. It is not clear whether this arises from loss of these radicals on collision with metal surfaces or whether radicals of this type have inherently low parent-ion sensitivities. On this account, Harrison and Lossing, in further work on ketones and aldehydes, supplemented the direct detection of radicals by the addition of CH_3 or CD_3 radicals to the reaction stream and the detection and identification of the combination products of methyl with other radicals produced in the primary dissociation steps. This was done by adding $Hg(CH_3)_2$ or $Hg(CD_3)_2$ to the reaction stream. The mercury-photosensitized decomposition of the dimethyl mercury yielded methyl radicals in abundance. Using this technique, the mercury-photosensitized decomposition of biacetyl, acetylacetone, and acetonyl acetone [84] and benzaldehyde, acrolein, and crotonaldehyde [85] were investigated.

In the sensitized photodissociation of CD_3OH [86], addition of CD_3 radicals resulted in the formation of CD_3OCD_3, and not CD_3CD_2OH. It was concluded, therefore, that CH_3OH dissociated by formation of CH_3O radicals rather than CH_2OH radicals. In the same investigation, the dissociation of dimethyl ether was shown to proceed by two modes: $CH_3OCH_2 + H$ and $CH_3O + CH_3$.

An extension of this method to direct photolytic reactions would be of great interest, and a number of drawbacks inherent in the use of mercury photosensitization could be avoided. For this purpose, the instrument constructed by Kistiakowsky and Kydd [87] had some important potential advantages. The mass spectrometer was of the Wiley time-of-flight design and was used in conjunction with a flash-photolysis reaction system. The main advantage was the high repetition frequency, up to 2 kc/sec, with which mass spectra could be obtained. Such a repetition frequency requires a very high pumping speed in the source, and this was facilitated by the open construction possible in this type of instrument. A further advantage of this instrument, that the ion source is at approximately ground potential, and that consequently the gas inlet can be located very close to the electron beam, can also be had using the conventional type of mass spectrometer, but less conveniently. Kistiakowsky and Kydd examined the flash photolysis of ketene and of nitrogen dioxide, using a flash of 500 joules intensity with a duration of about 10 μsec. Ketene, at a pressure of 0.3 to 1.2 mm, diluted with neon in varying amounts, was found to be decomposed to the extent of about 10% per flash. Methylene radicals were not detected, presumably on account of the relatively high partial pressure of ketene. CO and ethylene were formed, evidently by the rapid reaction

$$CH_2 + CH_2 = CO \rightarrow C_2H_4 + CO \qquad (11.54)$$

The presence of a product of mass 56 suggested that this reaction may proceed by addition to give an intermediate such as cyclopropanone, which can either decompose to C_2H_4 and CO or be stabilized.

In the flash photolysis of NO_2, O atoms and NO_3 were not detected, but the O_2 peak reached its greatest height only after 100 μsec. It was concluded that the probability of the reaction

$$NO_2 + O \rightarrow NO + O_2 \qquad (11.55)$$

is 10^{-2} or larger. A small amount of N_2O was also found. This novel method of studying flash-photolysis reactions looks to be most promising.

In the course of an electron-impact study by Reed [88] on formaldehyde, some experiments were performed in which formaldehyde vapor in the inlet system of a mass spectrometer was irradiated by a mercury vapor lamp, with the intention of producing sufficient CHO radicals to allow a direct measurement of the ionization potential to be made. It was found that insufficient radicals were generated to permit accurate measurements on the parent peak of the radical, and the ionization potential was derived by the indirect method.

The dissociation of a number of compounds into free radicals by direct photolysis has been observed by Terenin and coworkers [89]. Their mass spectrometer had an ionization chamber through which passed an electron beam and also a beam of photons from a hydrogen lamp. With the lamp off, the electron energy was lowered until all fragment ions disappeared and only the parent ion was formed. When the lamp was turned on, peaks were observed resulting from photodissociation of the original compound, followed by ionization of the resulting free radicals by the low-energy electrons. In this way the authors detected NH_2 radicals in the photolysis of NH_3; and N_2H_2, N_2H_3, and NH_2 in the photolysis of hydrazine. With benzylamine they detected $C_6H_5CH_2NH$ radicals and also benzyl and NH_2 radicals. Further work along this line should be of great interest.

The use of a field-emission ion source for photochemical studies has been investigated by Beckey and Groth [90]. In this ion source, which was developed by Inghram and Gomer [91–93], essentially only parent ions are formed. Conditions in this respect are favorable for the detection of free-radical fragments. One difficulty is found, however. Ions having an unpaired electron show a strong tendency to pick up a H atom from substances adsorbed on the tungsten point [94, 95]. Beckey and Groth [90], using an instrument with a focused field-emission source, examined the photolysis of acetone with 2,537 A radiation in the presence and absence of mercury vapor. Irradiation in the absence of mercury gave a fourfold increase in the CH_3CO^+ ion current. In the presence of mercury vapor there was a further tenfold increase, and CH_3 radicals were also detected. Ketene and acetaldehyde were produced, and also a peak at mass 45, presumably resulting from H-atom abstraction by CH_3CHO^+. The authors concluded that, before the field-emission source can be useful for quantitative work, further improvements must be made in its characteristics.

F. Radicals produced by dissociative ionization

An interesting study of free radicals has been described by Beck and Osberghaus [96], who examined the radicals and other neutral fragments resulting from the dissociative ionization of propane and hexane. Their mass spectrometer had two electron beams, one of high intensity for producing a large amount of ionization and a second for ionizing a sample of the gas leaving the first beam. The gas entered first an "impact chamber," through which passed an electron beam consisting of 32-cps square-wave pulses of 40 ma at 190 volts. This was sufficient to ionize about 10% of the gas. Using suitable field gradients, the charged particles were drawn out and rejected, while the unionized gas and neutral dissociation fragments passed into a second ionization chamber of the usual kind. The mass spectrum produced in this second chamber consisted of a normal mass spectrum of the original gas plus the superposed spectra of uncharged reaction products. The a-c component of the ion current at a given mass consisted of two parts: a "modulation" of gas pressure of about 1%, resulting from the ionization in the impact chamber, and a further modulation, opposite in sign, resulting from the arrival of pulses of neutral reaction products from the impact chamber. The extent of the gas-modulation signal could be determined by choice of a suitable peak which was free from interference by neutral fragments, for instance, the parent molecule ion. The gas-modulation part of the signal being the same proportion for all peaks in the spectrum of the gas, this part of the signal could be eliminated by calculation, and the intensity resulting from ionization of neutral fragments could be measured. In propane, for example, the neutral products identified were propylene, propyne (or allene), ethylene, acetylene, methane, hydrogen, and the following radicals: allyl, ethyl, and methyl. From n-hexane they found propyl, allyl, ethyl, vinyl, and methyl radicals. By varying the electron accelerating potential in the second ionization chamber, they measured the ionization potentials of allyl, ethyl, and vinyl radicals, obtaining results in excellent agreement with other work (Table 11.5). Further development of this technique should be of value, both for free-radical studies and for the identification of neutral fragments formed in various dissociative ionization processes.

G. Ionization potentials of free radicals

The ionization potential of a free radical is a quantity of considerable importance to our understanding of the electronic structure of radicals and radical ions. In addition, it has been proposed that the ionization potential of a radical is an important factor in determining the rates of certain organic reactions. The extent to which the ionization potential plays a significant role in such reactions can be established only as more values become available. Moreover, a knowledge of the ionization potential of a radical, together with ionic heats of formation obtained from the appearance potentials of fragment ions, permits the estimation of heats of formation of radicals, and consequently of bond-dissociation energies in derivatives. As is well known, bond-dissociation energies measured in this way can be quite unreliable, as a result of excitation energy given to the ion, either in the direct ionization of the radical or in the dissociative ionization process. Although translational energy imparted to a fragment ion in dissociative ionization can be measured, there is at present no way of taking account of other forms of excitation.

Following the success of Stevenson et al. in measuring the ionization potentials of methyl, ethyl [8, 98], and methylene [97, 98], a large number of other radicals have been investigated. In Table 11.5 are given the ionization potentials of 60 free radicals measured by direct electron impact on radicals produced by various methods. Ionization potentials measured by electron impact are usually considered to correspond to vertical transitions and may therefore be higher than the adiabatic values by a few tenths of a volt. Although the adiabatic ionization potentials, as derived from Rydberg series, are more precisely defined, the only ionization potentials of free radicals measured so far by spectroscopic methods have been for CH and CH_3. Moreover, the experimental difficulties are so great that it appears doubtful that spectroscopic values will be available for more than a few radicals in the near future. Consequently, electron- and photon-impact measurements, although leading to results that are less well defined energetically, provide the only source of information at the present time.

Some questions may be raised as to the reproducibility of radical-ionization-potential measurements on different mass spectrometers, and also as to the related problems involved in placing such measurements on an absolute energy scale. As to the first point, it may be seen from Table 11.5 that the total spread among six measurements of $I(CH_3)$ on different instruments and using several different methods of evaluating

Table 11.5.   Vertical Ionization Potentials of Free Radicals Measured by
Electron Impact

Radical	Ionization potential, volts	Reference
	Inorganic Radicals	
OH	12.7	100
	13.17 ± 0.1	62
HO$_2$	11.53 ± 0.1	60
NH$_2$	11.4 ± 0.1	64
N$_2$H$_3$	7.88 ± 0.2	64
	Hydrocarbon Radicals	
CH$_2$	11.9 ± 0.2	97
	11.9 ± 0.1	98
	11.82 ± 0.05	132
CH$_3$	10.07 ± 0.1	8
	9.85 ± 0.2	24
	10.11 ± 0.1	105
	9.95 ± 0.03	104
	9.85 ± 0.1	98
	9.88 ± 0.1	40
	9.84 (spectroscopic)	99
CD$_3$	9.95 ± 0.03	104
C$_2$H$_3$	9.45 ± 0.05	112
	9.35 ± 0.15	96
C$_2$H$_5$	8.67 ± 0.1	8
	8.78 ± 0.05	106
	8.80 ± 0.05	96
CH≡CCH$_2$	8.25 ± 0.08	106
CH$_2$=CHCH$_2$	8.16 ± 0.03	104
	8.15 ± 0.1	96
CH$_3$CH$_2$CH$_2$	8.69 ± 0.05	114
(CH$_3$)$_2$CH	7.90 ± 0.05	106
CH$_3$CH=CHCH$_2$	7.71 ± 0.05	107
CH$_2$=C(CH$_3$)CH$_2$	8.03 ± 0.05	107
CH$_3$CH$_2$CH$_2$CH$_2$	8.64 ± 0.05	114
(CH$_3$)$_2$CHCH$_2$	8.35 ± 0.05	114
CH$_3$CH$_2$CHCH$_3$	7.93 ± 0.05	114
(CH$_3$)$_3$C	7.42 ± 0.07	114
	Cyclic Radicals	
Cyclopropyl	8.05 ± 0.1	116
Cyclobutyl	7.88 ± 0.05	116
Cyclopentyl	7.79 ± 0.03	116
Cyclohexyl	7.66 ± 0.05	116
Cyclopentadienyl	8.69 ± 0.1	121
Benzyl	7.76 (average)	104, 108
Cycloheptatrienyl	6.60 ± 0.1	121

TABLE 11.5.   Vertical Ionization Potentials of Free Radicals Measured by
Electron Impact (Continued)

Radical	Ionization potential, volts	Reference
m-Xylyl	7.65 ± 0.03	109
o-Xylyl	7.61 ± 0.05	109
p-Xylyl	7.46 ± 0.03	109
m-CN benzyl	8.58 ± 0.1	117
m-NO₂ benzyl	8.56 ± 0.1	117
p-CN benzyl	8.36 ± 0.1	117
m-F benzyl	8.18 ± 0.06	117
p-Cl-benzyl	7.95 ± 0.1	117
p-F benzyl	7.78 ± 0.1	117
β-Naphthylmethyl	7.56 ± 0.05	117, 118
p-iso-C₃H₇ benzyl	7.42 ± 0.1	117
α-Naphthylmethyl	7.35 ± 0.1	118
p-CH₃O benzyl	6.82 ± 0.1	117
Diphenylmethyl	7.32 ± 0.1	118
Halogenated Radicals		
CF	13.81 ± 0.12	102
CCl	12.9 ± 0.10	102
CBr	10.43 ± 0.02	102
CF₂	13.30 ± 0.12	102
CCl₂	13.2 ± 0.2	47
	13.10 ± 0.12	102
CBr₂	10.11 ± 0.09	102
CF₃	10.10 ± 0.05	111
	10.15 ± 0.06	102
	10.2	113
CCl₃	8.78 ± 0.05	111
CH₂F	9.37	113
CH₂Cl	9.32	113
	9.70	103
CH₂Br	9.30	113
	8.34	103
CHF₂	9.45	113
CHCl₂	9.30	113
	9.54	103
CHBr₂	8.13	103
Miscellaneous Radicals		
CS	11.8 ± 0.2	47
CHO	9.88 ± 0.05	101
CH₃CO	8.05 ± 0.17	101, 102
CH₂CN	10.87 ± 0.1	115
CH₂CH₂CN	9.85 ± 0.1	115
CH₃CHCN	9.76 ± 0.1	115
(CH₃)₂CCN	9.15 ± 0.1	115

the ionization efficiency curves is 0.26 volt. In addition, the average of these values, assuming they are all of equal reliability, is only 0.1 volt higher than the adiabatic ionization potential of CH_3 recently obtained from a Rydberg series by Herzberg and Shoosmith [99]. This is the only radical for which a comparison can be made at the present time between the spectroscopic and the electron-impact values. Although the agreement in this case is quite good, it is possible that, for more complex radicals, larger differences between the vertical and adiabatic ionization potentials may exist.

The method of measurement is essentially the same as that used for obtaining the ionization potential of a stable molecule by electron impact. The radicals, usually generated by a thermal-decomposition reaction, are allowed to flow into the ionization chamber together with a reference gas of known ionization potential. The ionization efficiency curves for the radical and for the standard gas are then obtained in the usual way. A number of methods have been employed to normalize the two curves, and in many cases corrections must be applied to remove contributions from nonradical species. It is usually necessary to make a more or less complete analysis of the spectrum of products to ensure that the ion current attributed to the radical is not in fact arising from a dissociative ionization process. The most serious difficulty encountered is that of finding a reaction which will produce the radical in amounts that are sufficiently large relative to the background resulting from other substances present. This is quite a simple matter for some radicals, particularly those having a high degree of thermal stability. For example, it can be seen from Table 11.3 that the methyl radical can be produced by the thermal decomposition of a wide variety of derivatives. The lower the activation energy for rupture of the radical-X bond in the derivative compared with the activation energy necessary for the decomposition of the radical, the greater, of course, is the thermal stability of the radical at the temperature required for the pyrolysis of the derivative. For radicals of low stability, for example, the higher alkyl radicals, the activation energy for the thermal decomposition of the radical may be considerably smaller than for even such easily dissociated compounds as the mercury dialkyls or the azo derivatives. Consequently, in a homogeneous thermal decomposition, the radicals are destroyed almost as soon as they are formed and the yield of detectable radicals may be small or even zero. In such cases some advantage may be had from the use of pyrolytic reactions at low pressures on a heated filament. In such circumstances the radical once formed will, on the average, leave the zone of reaction without further collisions with the heated surface or with other molecules. The ionization potentials of a number of radicals produced by pyrolysis on heated filaments have been reported: CH_3 by LeGoff [24], CS and CCl_2 by Blanchard and LeGoff [47], an approximate value for OH by Tsuchiya

[100], the CHO and CH_3CO radicals by Reed and Brand [101], and the CF, CCl, CBr, CF_2, CCl_2, CBr_2, and CF_3 radicals by Reed and Snedden [102]. Because of the dependence of modes of reaction on the specific nature of heated metal surfaces, it may not be possible to obtain a satisfactory yield of radicals under "steady-state" conditions. For example, Reed and Brand found the production of CHO and CH_3CO radicals on a heated wire to be a transitory effect during heating and cooling of the wire. They measured the ionization potentials of these radicals by carrying out a heating and cooling cycle at different electron energies and observing the maximum current for the radical ion at each cycle. The success of such an expedient depends on the reproducibility of changes in the surface conditions of the wire, and the degree of precision attainable would consequently be expected to be inferior to that reached under "steady-state" conditions. Nevertheless, they were able to obtain for the ionization potential of the CHO radical, as produced from two different compounds, values which agree to within 0.2 volt.

In later work Reed and Snedden decomposed halogenated methanes on a heated filament and measured the ionization potentials of CH_2Cl, $CHCl_2$, CH_2Br, and $CHBr_2$ radicals [103]. From these and appearance potential data they derived bond-dissociation energies in some derivatives. As can be seen from Table 11.5, there are some unresolved differences among the experimentally measured ionization potentials of halogenated methyl radicals.

The use of electrical discharges for the production of radicals for ionization-potential measurements has been a most fruitful method for di- and triatomic inorganic free radicals. As discussed above, Foner and Hudson were by this means able to measure the ionization potentials of HO_2 [60], OH [62], and NH_2 and N_2H_3 radicals, as well as the unusual compounds diimide and triazine [64]. From these ionization potentials and the appearance potentials of radical ions in dissociative ionization processes, they have obtained values for bond-dissociation energies in a number of derivatives. Where values for the dissociation energies obtained by other methods were available for comparison, the agreement was quite satisfactory.

A number of measurements have been made on radicals produced by the pyrolysis of iodides and mercury alkyls. The ionization potentials of methyl [104, 105] and ethyl [106], prepared from the mercury alkyls, were found to be in good agreement with the earlier measurements [8]. The isopropyl radical [106], produced from the thermal decomposition of azoisopropane at 655°C, was found to have a vertical ionization potential of 7.90 volts, appreciably higher than values derived from the appearance potentials of fragment ions. The propargyl radical $CH_2C{\equiv}CH$, prepared from propargyl iodide, was also investigated [106]. Allyl [104] and methyl-substituted allyl [107] were produced from the iodides by thermal decomposition. In both the allyl and benzyl radicals substitution of H by CH_3 lowered the ionization potential by amounts ranging from 0.1 to 0.5 volt, depending on the position of substitution. From the appearance potentials of allyl and methylallyl ions from the halide derivatives, the authors derived heats of formation of these radicals which are in reasonable agreement with estimates from kinetic data on thermal-decomposition reactions. A brief investigation was made of the relative stability of p-, m-, and o-xylyl radicals [110]. The p- and o-xylyl radicals dissociated by loss of a H atom to form the corresponding quinodimethanes. In agreement with structural considerations, the m-xylyl radical was found to have a much greater stability toward thermal dissociation than the other isomers.

The vertical ionization potentials of CF_3 and CCl_3 radicals, produced from the thermal decomposition of CF_3NNCF_3 and CCl_3NO_2, have been measured by Farmer et al. [111]. From these, and from appearance potentials, heats of formation of CF_3 and CCl_3 of -117 ± 2 and $+13.4 \pm 3$ kcal/mole were obtained. For CCl_3, this heat of formation was in agreement with estimates from kinetic data. For CF_3, some discrepancies were found in the appearance potential data, and the value for $\Delta H_f(CF_3)$ must be regarded as being less reliable. The ionization potential measured by Reed and Snedden [102] for CF_3 was in good agreement with that found in this work.

The thermal decomposition of methyl vinyl mercury provided a means for obtaining a sufficient yield of vinyl radicals for ionization potential measurements [112]. Earlier attempts to produce this radical in significant yields had been unsuccessful [5, 28]. The ionization potential measured, 9.45 volts, combined with appearance potential data, gave 105 kcal/mole for the C—H bond-dissociation energy in ethylene. When combined with the standard heats of formation of acetylene and ethylene, this leads to a C—H bond-dissociation energy in the vinyl radical of 41 kcal/mole. Vinyl is consequently rather unstable with respect to loss of H atom, and the difficulties encountered in earlier work in detecting it at elevated temperatures [5, 28] can be easily understood on this basis.

The measurement of the ionization potentials of allyl, ethyl, and vinyl radicals [96] formed as neutral products of dissociative ionization has been described in Sec. 11.3F.

For the preparation of a number of relatively unstable radicals, such as the n-propyl, n-butyl, isobutyl, and sec-butyl radicals, a convenient method is provided by the thermal decomposition of the appropriate alkyl nitrites [113, 114]. It has been known for some time that the thermal decomposition of alkyl nitrites proceeds according to the following reactions:

$$RCH_2ONO \rightarrow RCH_2O + NO \tag{11.56}$$
$$RCH_2O \rightarrow R + CH_2O \tag{11.57}$$
$$R \rightarrow R_1 + R_2 \tag{11.58}$$

The activation energy for reaction 56 is about 37 kcal/mole, appreciably lower than for the decomposition of mercury dialkyls or azo derivatives. The second step, reaction 57, is very fast, the activation energy being of the order of 12 kcal/mole or less. Since the activation energy for the dissociation of n-propyl and n-butyl radicals is probably greater than 25 kcal/mole, conditions for the production of relatively high radical concentrations are not too adverse. Experimentally, it was found that a good yield of propyl and butyl radicals could be produced in this way. The ionization potentials of the alkyl radicals were found to decrease in a

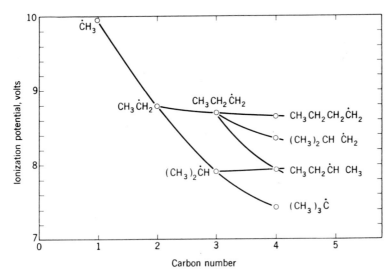

Fig. 11.16. Vertical ionization potentials of alkyl radicals, showing effect of substitution by methyl groups. (*From Lossing and De Sousa* [114].)

systematic manner with increasing methyl substitution, as shown in Fig. 11.16. Along the series methyl, ethyl, *sec*-propyl, and *t*-butyl, in which substitution is at the carbon containing the free electron, the ionization potential decreases along a smooth curve. The decrease from n-propyl to *sec*-butyl, in which the substitution is also on the carbon containing the free electron, is of the same order of magnitude. On the other hand, substitution by methyl of a hydrogen on a carbon adjacent to or once removed from the carbon containing the free electron brings about only a small decrease in ionization potential. It was found that these relative changes could be closely reproduced by a set of ionization potentials calculated by a modified molecular-orbital method.

Substitution of H by CN in methyl, ethyl, and isopropyl, as in the CH_2CN, CH_3CHCN, and $(CH_3)_2CCN$ radicals, brought about an increase in the ionization potential of about 1 volt [115]. In this work the bond-dissociation energy $D(H—CH_2CN)$ was found to be ≤ 79 kcal/mole.

The weakness of this bond agreed with other evidence for the resonance stabilization of the CH_2CN radical and its homologs. The ionization potentials of C_3—C_6 cycloalkyl radicals have been measured in the same way by Pottie et al. [116]. The radicals were produced by thermal decomposition of the appropriate cycloalkyl nitrites. The ionization potentials were in reasonable agreement with values calculated by the modified molecular-orbital method. At temperatures above 800°C, the cyclopropyl radical was isomerized to allyl.

The effect of para and meta substitution, by various characteristic groups, on the ionization potential of benzyl radical has been examined [117]. The substituents, in order of decreasing ionization potential of the radical, were m-CN, m-NO$_2$, p-CN, m-F, p-Cl, p-F, p-i-Pr, βC_4H_4, and p-CH$_3$O (Table 11.5). This order is in agreement with the known electron-releasing or electron-attracting powers of the substituents, as derived from the rates of solvolysis of the corresponding halides in solution. A plot of ionization potential against σ^+ values from the literature showed a linear relationship between these quantities. Since the effect of substituents on the neutral radicals is very small, these results permit a consideration of the stabilization of the ions under conditions where solvation effects do not occur.

The ionization potentials of the conjugated hydrocarbon radicals α- and β-naphthylmethyl and diphenylmethyl were measured by Harrison and Lossing [118]. The results were in excellent agreement with ionization potentials calculated by Hush and Pople [119], using a self-consistent molecular-orbital method, and by Streitwieser, using the omega technique [120].

Measurements of the ionization potentials of cyclopentadienyl and cycloheptatrienyl radicals presented some interesting features [121]. The cyclopentadienyl radical was produced by the thermal decomposition of anisole at 950°C. This reaction followed a rather unusual course:

$$\langle\rangle{-}O{-}CH_3 \xrightarrow{\,950°\,} CH_3 + \left[\langle\rangle{-}\dot{O} \leftrightarrow \langle\rangle{=}O\right] \longrightarrow \langle\rangle + CO \quad (11.59)$$

The ionization potential of the C_5H_5 radical produced in this reaction was in good agreement with the ionization potential of C_5H_5 radicals formed by the thermal decomposition of dicyclopentadienyl nickel in a separate experiment. Cycloheptatrienyl radical, produced by the pyrolysis of bitropenyl (dihydroheptafulvalene), was found to have a low ionization potential (6.60 volts), in agreement with the large degree of stabilization predicted for cycloheptatrienyl ion by molecular-orbital theory.

In Table 11.6 the experimentally measured ionization potentials of some free radicals are compared with the values calculated by Hush and Pople [119] and Streitwieser [120]. With a few exceptions, the agreement is satisfactory, and in many cases it is within the reproducibility of measurement.

TABLE 11.6.   Comparison of Experimental Vertical Ionization Potentials of Radicals with Values Calculated from Molecular-orbital Theory

Radical	Ionization potential, volts	
	Observed	Calculated
Allyl	8.16[96,104]	8.23[119]
β-Methylallyl	8.03[107]	8.04[120]
α-Methylallyl	7.71[107]	7.75[120]
Cyclopentadienyl	8.69[121]	8.82[120]
Cycloheptatrienyl	6.60[121]	6.41[120]
Benzyl	7.76[104,108]	7.78[119]
α-Naphthylmethyl	7.35[117,118]	7.35[119]
β-Naphthylmethyl	7.56[118]	7.57[118]
Diphenylmethyl	7.32[118]	7.26[119], 7.42[120]
m-Xylyl	7.65[109]	7.65[120]
p-Xylyl	7.46[109]	7.56[120]
Propargyl	8.25[106]	8.34[120]
Ethyl	8.67[8], 8.78[106], 8.80[96]	8.67[120]
sec-Propyl	7.90[106]	7.78[120]
t-Butyl	7.42[114]	7.07[120]
CH_2Cl	9.32[113], 9.70[103]	9.42[120]
$CHCl_2$	9.30[113], 9.54[103]	9.02[120]
CCl_3	8.78[111]	8.64[120]

11.4. Conclusion

In conclusion, attention might be drawn to a number of recent developments whose application to mass-spectrometric studies of free radicals may be valuable.   Mention has already been made of the advantages of the RPD method of Fox et al. [20].   Although a decrease in absolute intensity would result, the elimination of the curvature at the foot of the ionization efficiency curves would greatly increase the ability to discriminate between radicals and fragment ions.

In recent years photoionization studies have proved to be an effective method for measuring molecular ionization potentials [122, 123]. Although a simple mass analyzer was employed many years ago to examine the production of ion pairs by photoionization [124], it is only recently that interest has been shown in the application of photoionization to mass spectrometry.   In an exploratory examination of the possibilities, Lossing and Tanaka [125], using the 1,263- and 1,165-A lines from a krypton lamp, were able to produce resolved ion currents of 10^{-12} amp by photoionization of several organic compounds.   Methyl radicals produced by pyrolysis of mercury dimethyl were easily detected by photoionization. A number of advantages of photoionization for free-radical studies were pointed out: the steep rise of photoionization efficiency curves above the threshold, the absence of contact potential, and the consequent freedom from drift in energy with time and from the pyrolytic effects associated with heated filaments.

In a further study of photoionization, Herzog and Marmo [126] used a radio-frequency mass spectrometer of the Bennett type, with a hydrogen-discharge lamp as a source of ultraviolet. They were able to identify and eliminate the source of a "secondary" spectrum observed by Lossing and Tanaka and obtained single peaks for a number of compounds.

A realization of the experimental system necessary for free-radical studies and for ionization-potential measurements was made by Hurzeler, Inghram, and Morrison [127], who combined a vacuum ultraviolet monochromator with a mass spectrometer having a mass resolution of 1 in 300. With this instrument they were able to use a narrow energy range of ionizing radiation and to vary its wavelength continuously—both necessary conditions for useful free-radical and ionization-potential work. With this instrument they carried out rather extensive photo-ionization studies and were able to provide an interpretation of the structure observed near the threshold of the photoionization efficiency curves. An instrument of a similar type has been described by Weissler et al. [128] and Schönheit [129]. Although a review of this important work is beyond the scope of this chapter, it should be pointed out that an instrument of this type should provide a considerable advance over present methods of studying the production and the ionization potentials of radicals.

A high-speed cycloidal-focusing mass spectrometer with possible applications to free-radical studies has been developed [130] by the Consolidated Electrodynamics Corporation. This instrument appears to possess a number of advantages over conventional instruments for this type of work. The Bendix time-of-flight mass spectrometer [131] used by Kistiakowsky and Kydd in their flash-photolysis work has rather similar advantages and will undoubtedly be used more widely in free-radical work in the next few years.

REFERENCES

1. H. S. Taylor: *Trans. Faraday Soc.*, **21**, 560 (1925).
2. F. A. Paneth and W. Hofeditz: *Ber.*, **B62**, 1335 (1929).
3. F. O. Rice and K. K. Rice: "The Aliphatic Free Radicals," The Johns Hopkins Press, Baltimore, 1935.
4. G. C. Eltenton : *J. Chem. Phys.*, **10**, 403 (1942).
5. G. C. Eltenton: *J. Chem. Phys.*, **15**, 455 (1947).
6. G. C. Eltenton: *J. Phys. Coll. Chem.*, **52**, 463 (1948).

7. G. C. Eltenton: *Rev. Inst. Franc. Petrole Ann. Combust. Liquides*, **4**, 468 (1949).
8. J. A. Hipple and D. P. Stevenson: *Phys. Rev.*, **63**, 121 (1943).
9. E. Leifer and H. C. Urey: *J. Am. Chem. Soc.*, **64**, 994 (1942).
10. P. D. Zemany and M. Burton: *J. Phys. Coll. Chem.*, **55**, 949 (1951).
11. P. Bradt, V. H. Dibeler, and F. L. Mohler: *J. Res. Natl. Bur. Std.*, **50**, 201 (1953).
12. A. J. B. Robertson: "Mass Spectrometry," Methuen & Co., Ltd., London, 1954.
13. W. J. Dunning: *Quart. Rev. (London)*, **9**, 23 (1955).
14. J. D. Craggs and C. A. McDowell: *Rept. Progr. Phys.*, **18**, 375 (1955).
15. C. Ouellet: *Experientia, suppl.*, **7**, 144 (1957).
16. F. P. Lossing: *Ann. N.Y. Acad. Sci.*, **67**, 499 (1957).
17. H. D. Beckey: *Angew. Chem.*, **70**, 327 (1958).
18. J. Collin: *Ind. Chim. Belge*, **24**, 25 (1959).
19. J. Cuthbert: *Quart. Rev. (London)*, **13**, 215 (1959).
20 R. E. Fox, W. M. Hickam, D. J. Grove, and T. Kjeldaas, Jr.: *Rev. Sci. Instr.*, **26**, 1101 (1955).
21. J. H. Beynon: "Mass Spectrometry and Its Application to Organic Chemistry," American Elsevier Publishing Company, New York, 1960.
22. J. Momigny: *Bull. Soc. Chim. Belges*, **66**, 33 (1957).
23. P. LeGoff: "Applied Mass Spectrometry," pp. 120, 122, 126, The Institute of Petroleum, London, 1954.
24. P. LeGoff: *J. Chim. Phys.*, **50**, 423 (1953).
25. P. LeGoff, *J. Chim. Phys.*, **53**, 269 (1956).
26. P. LeGoff and M. Letort, *J. Chim. Phys.*, **53**, 480 (1956).
27. D. J. Fabian and A. J. B. Robertson: *Trans. Faraday Soc.*, **53**, 363 (1957).
28. F. P. Lossing, K. U. Ingold, and I. H. S. Henderson: "Applied Mass Spectrometry," p. 102, The Institute of Petroleum, London, 1954.
29. L. P. Blanchard and P. LeGoff: "Advances in Mass Spectrometry" (J. Waldron, ed.), Pergamon Press, New York, 1959, p. 570.
30. L. P. Blanchard and P. LeGoff: *Can. J. Chem.*, **37**, 515 (1959).
31. K. U. Ingold and F. P. Lossing: *J. Chem. Phys.*, **21**, 1135 (1954).
32. F. P. Lossing: *Can. J. Chem.*, **35**, 305 (1957).
33. S. N. Foner and R. L. Hudson: *J. Chem. Phys.*, **21**, 1374 (1953).
34. C. W. Nutt, J. S. M. Botterill, G. Thorpe, and G. W. Penmore: *Trans. Faraday Soc.*, **55**, 1500 (1959).
35. C. W. Nutt, G. W. Penmore, and A. J. Biddlestone: *Trans. Faraday Soc.*, **55**, 1516 (1959).
36. A. J. B. Robertson: "Mass Spectrometry," p. 47, The Institute of Petroleum, London, 1952.
37. V. L. Tal'roze: *Pribory i Tekhn. Eksperim.*, no. 5, p. 116 (1957).
38. V. L. Tal'roze et al.: *Pribory i Tekhn. Eksperim.*, no. 6, p. 78 (1960).
39. F. P. Lossing and A. W. Tickner: *J. Chem. Phys.*, **20**, 907 (1952).
40. D. Osberghaus and R. Taubert: *Z. Physik. Chem.*, **4**, 264 (1955).
41. J. W. Otvos and D. P. Stevenson: *J. Am. Chem. Soc.*, **78**, 546 (1956).
42. A. J. B. Robertson: *Proc. Roy. Soc. (London)*, **A199**, 394 (1949).
43. D. J. Fabian and A. J. B. Robertson: *Proc. Roy. Soc. (London)*, **A273**, 1 (1956).
44. A. J. B. Robertson: "Advances in Mass Spectrometry" (J. Waldron, ed.), Pergamon Press, New York, 1959, p. 559.
45. P. LeGoff and M. Letort: *Compt. Rend.*, **239**, 970 (1954).
46. P. LeGoff and M. Letort: *J. Chim. Phys.*, **54**, 3 (1957).
47. L. P. Blanchard and P. LeGoff: *Can. J. Chem.*, **35**, 89 (1957).
48. F. P. Lossing, K. U. Ingold, and A. W. Tickner: *Discussions Faraday Soc.*, **14**, 34 (1953).
49. F. P. Lossing: *Ind. Chim. Belge*, **19**, 613 (1954).

50. K. U. Ingold and F. P. Lossing: *Can. J.Chem.*, **31**, 30 (1953).
51. K. U. Ingold, I. H. S. Henderson, and F. P. Lossing: *J. Chem. Phys.*, **21**, 2239 (1953).
52. W. A. Bryce and K. U. Ingold: *J. Chem. Phys.*, **24**, 1968 (1956).
53. K. U. Ingold and W. A. Bryce: *J. Chem. Phys.*, **24**, 360 (1956).
54. E. G. Leger: *Can. J. Phys.*, **33**, 74 (1955).
55. E. G. Leger and C. Ouellet: *J. Chem. Phys.*, **21**, 1310 (1953).
56. L. P. Blanchard, J. B. Farmer, and C. Ouellet: *Can. J. Chem.*, **35**, 115 (1957).
57. H. I. Schiff: *Ann. N.Y. Acad. Sci.*, **67**, 518 (1957).
58. S. N. Foner and R. L. Hudson: *J. Chem. Phys.*, **21**, 1608 (1953).
59. A. J. B. Robertson: "Applied Mass Spectrometry," pp. 112, 122, The Institute of Petroleum, London, 1954; *Chem. Ind. (London)*, 1485 (1954).
60. S. N. Foner and R. L. Hudson: *J. Chem. Phys.*, **23**, 1364 (1955).
61. S. N. Foner and R. L. Hudson: *J. Chem. Phys.*, **23**, 1974 (1955).
62. S. N. Foner and R. L. Hudson: *J. Chem. Phys.*, **25**, 602 (1956).
63. S. N. Foner and R. L. Hudson: *J. Chem. Phys.*, **28**, 719 (1958).
64. S. N. Foner and R. L. Hudson: *J. Chem. Phys.*, **29**, 442 (1958).
65. D. S. Jackson and H. I. Schiff: *J. Chem. Phys.*, **21**, 2233 (1953).
66. D. S. Jackson and H. I. Schiff: *J. Chem. Phys.*, **23**, 2333 (1955).
67. J. Berkowitz, W. A. Chupka, and G. B. Kistiakowsky: *J. Chem. Phys.*, **25**, 457 (1956).
68. G. B. Kistiakowsky and G. G. Volpi: *J. Chem. Phys.*, **27**, 1141 (1957).
69. G. B. Kistiakowsky and G. G. Volpi: *J. Chem. Phys.*, **28**, 665 (1958).
70. J. T. Herron, J. L. Franklin, P. Bradt, and V. H. Dibeler: *J. Chem. Phys.*, **29**, 230 (1958); **30**, 879 (1959).
71. J. T. Herron and V. H. Dibeler: *J. Chem. Phys.*, **31**, 1662 (1959).
71a. J. T. Herron, J. L. Franklin, and P. Bradt: *Can. J. Chem.*, **37**, 579 (1959).
72. I. Grundland: *Compt. Rend.*, **236**, 476 (1953).
73. S. N. Foner and R. L. Hudson: *J. Chem. Phys.*, **25**, 601 (1956).
74. J. T. Herron and H. I. Schiff: *J. Chem. Phys.*, **24**, 1266 (1956).
75. J. T. Herron and H. I. Schiff: *Can. J. Chem.*, **36**, 1159 (1958).
76. A. K. Lavrovskaya, V. E. Skurat, V. L. Tal'roze, and G. D. Tantsyrev: *Dokl. Acad. Nauk S.S.S.R.*, **117**, 641 (1957).
77. N. N. Dogramadze and K. F. Zmbov: *Bull. Inst. Nucl. Sci. Boris Kidrich*, **9**, 104 (1959).
78. J. T. Herron, J. L. Franklin, P. Bradt and V. H. Dibeler, *J. Am. Chem. Soc.*, **80**, 6188 (1958).
79. J. B. Farmer, F. P. Lossing, D. G. H. Marsden, and E. W. R. Steacie: *J. Chem. Phys.*, **23**, 1169 (1955).
80. P. Kebarle and F. P. Lossing: *Can. J. Chem*, **37**, 389 (1959).
81. F. P. Lossing, D. G. H. Marsden, and J. B. Farmer: *Can. J. Chem.*, **34**, 701 (1956).
82. A. G. Harrison and F. P. Lossing: *Can. J. Chem.*, **38**, 544 (1960).
83. J. Collin and F. P. Lossing: *Can. J. Chem.*, **35**, 778 (1957).
84. A. G. Harrison and F. P. Lossing: *Can. J. Chem.*, **37**, 1478 (1959).
85. A. G. Harrison and F. P. Lossing: *Can. J. Chem.*, **37**, 1696 (1959).
86. R. F. Pottie and F. P. Lossing: *Can. J. Chem.*, **39**, 102 (1961).
87. G. B. Kistiakowsky and P. H. Kydd: *J. Am. Chem. Soc.*, **79**, 4825 (1957).
88. R. I. Reed: *Trans. Faraday Soc.*, **52**, 1195 (1956).
89. F. I. Vilesov, B. L. Kurbatov, and A. N. Terenin: *Dokl. Acad. Nauk S.S.S.R.*, **122**, 94 (1958); *Tr. Chim. i Chim. Tekhn.*, **1**, 181 (1961).
90. H. D. Beckey and W. Groth: *Z. Phys. Chem. (Frankfurt)*, **20**, 307 (1959).
91. M. G. Inghram and R. Gomer: *J. Chem. Phys.*, **22**, 1279 (1954).

MASS SPECTROMETRY OF FREE RADICALS   505

92. M. G. Inghram and R. Gomer: Z. *Naturforsch.*, **10a**, 863 (1955).
93. R. Gomer and M. G. Inghram: J. *Am. Chem. Soc.*, **77**, 500 (1955).
94. H. D. Beckey: *Naturw.*, **45**, 259 (1958).
95. H. D. Beckey: Z. *Naturforsch.*, **14a**, 712 (1959).
96. D. Beck and O. Osberghaus: Z. *Physik*, **160**, 406 (1960).
97. A. Langer and J. A. Hipple: *Phys. Rev.*, **69**, 691 (1946).
98. A. Langer, J. A. Hipple, and D. P. Stevenson: J. *Chem. Phys.*, **22**, 1836 (1954).
99. G. Herzberg and J. Shoosmith: *Can. J. Phys.*, **34**, 523 (1956).
100. T. Tsuchiya: J. *Chem. Phys.*, **22**, 1784 (1954).
101. R. I. Reed and J. C. D. Brand: *Trans. Faraday Soc.*, **54**, 478 (1958).
102. R. I. Reed and W. Snedden: *Trans. Faraday Soc.*, **54**, 301 (1958).
103. R. I. Reed and W. Snedden: *Trans. Faraday Soc.*, **55**, 876 (1959).
104. F. P. Lossing, K. U. Ingold, and I. H. S. Henderson: J. *Chem. Phys.*, **22**, 621 (1954).
105. J. Waldron, *Trans. Faraday Soc.*, **50**, 102 (1954).
106. J. B. Farmer and F. P. Lossing: *Can. J. Chem.*, **33**, 861 (1955).
107. C. A. McDowell, F. P. Lossing, I. H. S. Henderson, and J. B. Farmer: *Can. J. Chem.*, **34**, 345 (1956).
108. J. B. Farmer, I. H. S. Henderson, C. A. McDowell, and F. P. Lossing: J. *Chem. Phys.*, **22**, 1948 (1954).
109. J. B. Farmer, F. P. Lossing, D. G. H. Marsden, and C. A. McDowell: J. *Chem. Phys.*, **24**, 52 (1956).
110. J. B. Farmer, D. G. H. Marsden, and F. P. Lossing: J. *Chem. Phys.*, **23**, 403 (1955).
111. J. B. Farmer, I. H. S. Henderson, F. P. Lossing, and D. G. H. Marsden: J. *Chem. Phys.*, **24**, 348 (1956).
112. A. G. Harrison and F. P. Lossing: J. *Am. Chem. Soc.*, **82**, 519 (1960).
113. F. P. Lossing, P. Kebarle, and J. B. de Sousa: "Advances in Mass Spectrometry" (J. Waldron, ed.), Pergamon Press, New York, 1959, p. 431.
114. F. P. Lossing and J. B. de Sousa: J. *Am. Chem. Soc.*, **81**, 281 (1959).
115. R. F. Pottie and F. P. Lossnig: J. *Am. Chem. Soc.*, **83**, 4737 (1961).
116. R. F. Pottie, A. G. Harrison, and F. P. Lossing: J. *Am. Chem. Soc.*, **83**, 3204 (1961).
117. A. G. Harrison, P. Kebarle, and F. P. Lossing: J. *Am. Chem. Soc.*, **83**, 777 (1960).
118. A. G. Harrison and F. P. Lossing: J. *Am. Chem. Soc.*, **82**, 1052 (1960).
119. N. S. Hush and J. A. Pople: *Trans. Faraday Soc.*, **51**, 600 (1955).
120. A. Streitwieser, Jr.: J. *Am. Chem. Soc.*, **82**, 4123 (1960).
121. A. G. Harrison, L. R. Honnen, H. J. Dauben, Jr., and F. P. Lossing: J. *Am. Chem. Soc.*, **82**, 5593 (1960).
122. K. Watanabe, F. F. Marmo, and E. C. Y. Inn: *Phys. Rev.*, **91**, 1155 (1953).
123. N. Wainfan, W. C. Walker, and G. L. Weissler: *Phys. Rev.*, **99**, 542, (1955).
124. A. Terenin and B. Popov, *Phys. Z. Sowjetunion*, **2**, 299 (1932).
125. F. P. Lossing and Ikuzo Tanaka: J. *Chem. Phys.*, **25**, 1031 (1956).
126. R. F. Herzog and F. F. Marmo: J. *Chem. Phys.*, **27**, 1201 (1957).
127. H. Hurzeler, M. G. Inghram, and J. D. Morrison: J. *Chem. Phys.*, **27**, 313 (1957); **28**, 76 (1958).
128. G. L. Weissler, J. A. R. Samson, M. Ogawa, and G. R. Cook: J. *Opt. Soc. Am.*, **49**, 338 (1959).
129. E. Schönheit: Z. *Physik*, **149**, 153 (1957); Z. *Naturforsch.*, **15a**, 125 (1960).
130. L. G. Hall, C. K. Hines, and J. E. Slay: "Advances in Mass Spectrometry" (J. Waldron, ed.), Pergamon Press, New York, 1959, p. 266.
131. W. C. Wiley and I. H. McLaren: *Rev. Sci. Instr.*, **26**, 1150 (1955).
132. E. W. C. Clark and C. A. McDowell: *Proc. Chem. Soc.*, **1960**, 69.

12

The Ionization and Dissociation of Molecules

C. A. McDowell

Department of Chemistry, The University of British Columbia
Vancouver, B.C.

12.1. Introduction

In this chapter we shall be concerned with the results of studies of the ionization and dissociation of atoms and molecules when these are subjected to electronic impact. A vast literature has been built up in this field during the past twenty years. Many of the results are extremely difficult to interpret unambiguously, but considerable progress has nevertheless been made. The main areas of interest which we shall discuss in some detail include the measurement and interpretation of appearance potentials and ionization potentials and the use of such data in the evaluation of bond-dissociation energies, heats of formation of molecular ions, and the determination of electron affinities. We shall also be interested in the attempts, both theoretical and experimental, which have been made to understand the nature of the threshold ionization laws for different types of ionization processes. In addition, we shall be much interested in the light which these studies throw on problems concerned with the electronic structures of molecules and ions. As electron impact is the general means by which ions are produced for mass analysis, we shall also be interested in the mode of formation of ions and shall discuss such matters as meta-stable ion formation and the use of mass-spectral data in the study of the structures of complex organic and inorganic molecules.

It will be realized, of course, that we cannot in one chapter survey the whole of this vast field of knowledge in very great detail. One treatment will aim at discussing critically the various main developments which have occurred in the field. We shall do this with a view to enabling the

506

reader to form a balanced view of the validity of some of the approaches. Also, we shall be much concerned to indicate, where possible, what might be fruitful lines for future research.

12.2. Ionization Potentials and Appearance Potentials

The ionization potential of a molecule is the energy required to remove an electron from the highest occupied molecular orbital. Such data are of importance as an aid to the understanding of the electronic structures of molecules. Electrons with sufficient energy can ionize molecules by the following process:

$$XY + e = XY^+ + 2e \qquad (12.1)$$

Ionization can also be caused by the interaction of quanta of sufficiently high energy to cause the process

$$XY + h\upsilon = XY^+ + e \qquad (12.2)$$

this process being called photoionization.

The potential which has to be applied to accelerate the ionizing electrons in Eq. (12.1) so that they have just sufficient energy to cause a particular ion to appear in the mass spectrum of a compound under study is known as the appearance potential. It will be denoted by the symbol $V(XY^+)$ or $V(X^+)$. For the process here discussed, where we are simply concerned with the formation of a parent molecular ion, this quantity is the molecular ionization potential. Likewise, the energy of the quantum which causes the photoionization process (12.2) to occur is also called the ionization energy, or potential.

Ionization potentials of molecules can also be determined by studying that portion of the far-ultraviolet spectrum which is caused by Rydberg transitions. Once the Rydberg transitions have been identified, they can be fitted to a series of the type

$$\upsilon = \upsilon_\infty - \frac{a}{(n + b)^2} \qquad (12.3)$$

where a and b are constants for a particular molecule and n takes on integral values representing the different Rydberg bands. The ionization potential can readily be calculated once υ_∞ is known. This method has been used to obtain accurate values for many molecular ionization potentials. There are, however, often difficulties in the interpretation of the spectra, and an unambiguous assignment of the Rydberg transitions is not then possible. It is, however, almost always possible to obtain a value for the ionization potential of a molecule from electron-impact studies. In certain cases, such as carbon tetrafluoride and silicon hexafluoride where the parent ions CF_4^+ and SiF_6^+ are so unstable, it is difficult to detect these in sufficient abundance in the mass spectrometer to enable a satisfactory determination of their appearance potentials.

12.3. The Franck-Condon Principle and the Ionization and Dissociation of Molecules by Electron Impact

The absorption of energy by a molecule as the result of electron impact can be described in a satisfactory manner in terms of the Franck-Condon principle. For the case of a diatomic molecule, the various possible

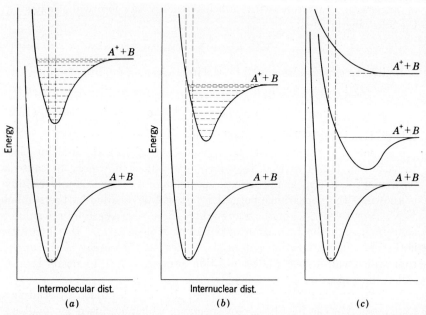

Fig. 12.1.   Possible Franck-Condon transitions for the diatomic-molecule case as the result of electron impact.

Franck-Condon transitions are shown in Fig. 12.1a to c.   Diagram a of Fig. 12.1 depicts the vertical transitions leading to the formation of XY^+ molecules by electronic impact.   These transitions lead to the formation of the XY^+ ion in various vibrationally excited states, the lowest transition giving, of course, the vertical ionization potential of the molecule.   The probability of a transition induced by electron impact is a function of the energy of the transition E_t, the energy of the bombarding electron E_e, and the overlap integral.   For a transition between vibrational levels v'' and v' of the electronic states e'' and e', this probability can be written as

$$P \propto [M^{e'',v''}_{e',v'}]^2 \tag{12.4}$$

where
$$M^{e'',v''}_{e',v'} = \int_0^\infty \psi_i^{v''} X_{e'',e'} \psi_j^{v'} \, d\tau \tag{12.5}$$

and $\psi_i^{v''}$ = eigenfunction of upper electronic state e'' in its vibration level v''

$\psi_j^{v'}$ = eigenfunction of lower electronic state e' in its vibrational level v'

$X_{e'',e'}$ = matrix element of perturbing function between the two vibronic states

The value of the probability of the transition can be found by graphical methods used by Hagstrum and Tate [1]. This method gives one an idea of the type of ionization efficiency curve to be expected, the ionization efficiency curve being the graphical representation of the variation of the intensity of the ion current studied as a function of the energy of the ionizing electrons. For the case of transitions producing the

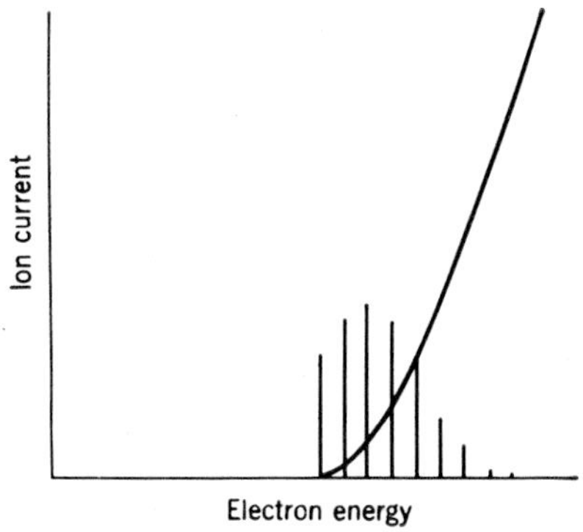

Fig. 12.2. The ionization efficiency curve expected for electron-impact-induced Franck-Condon transitions depicted in Fig. 12.1a. The vertical lines represent transition probabilities to various vibrational levels of the molecular ion.

parent molecular ion (Fig. 12.1), this method indicates that different types of ionizing efficiency curves may be expected [2]. In the case where the interatomic distance is the same for the molecular ion and the neutral molecule, then the appearance potential will be the same as the adiabatic ionization potential. Figure 12.2 shows the type of ionization efficiency curve to be expected in this case. The vertical lines represent peaks at the various vibrational levels, the total area under the peaks being proportional to the electronic transition area for that state [Eq. (12.4)].

When the interatomic distances for the two states differ appreciably, as depicted in Fig. 12.1b and c, it is to be expected that transitions will occur to higher vibrational levels of the ionic state. There will, of course, be a finite probability that ionization to the lowest vibrational level of the ionic state can also occur. High sensitivity is needed, however, to detect the states with the lowest amount of vibrational excitation. If the difference in the internuclear distances for the two states is quite large, it is possible that transitions will also occur to a repulsive ionic state as indicated in Fig. 12.1c. Such transitions result in the dissociation of the molecular ion, and so here the situation is the same as that shown in Fig. 12.1b. For these latter cases the application of the above method, based on the Franck-Condon principle, leads to a means of estimating and determining the amount of kinetic energy with which fragments resulting from the dissociation process (12.6) may be endowed.

$$XY + e = X^+ + Y + 2e + K.E. + E.E. \qquad (12.6)$$

where E.E. = excitation energy (electronic, vibrational, or rotational) which ion or neutral fragment may possess

K.E. = excess kinetic energy in the process

Though the measurement of the amount of excess kinetic energy possessed by ions will be discussed later, here we may indicate the application of the Franck-Condon principle to show that the C^+ ions produced from carbon monoxide can be formed with a wide range of kinetic energies. The process to be discussed is that represented by Eq. (12.7).

$$CO + e = C^+ + O + 2e + K.E. \qquad (12.7)$$

Hagstrum and Tate [1] have considered this case, and Fig. 12.3 illustrates the Franck-Condon transitions which lead to the production of C^+ ions through process (12.7). Curve 1 shows the lower part of the ground $^1\Sigma^+$ electronic state of the CO molecule; curve 2 represents the square of the eigenfunction of the lowest vibrational state ($v' = 0$); curve 3 is the portion of the upper potential energy curve of the CO^+ ion to which transitions will yield curve 4, which is the distribution of the kinetic energy of the C^+ ions resulting from process (12.7).

It will, of course, be realized that the foregoing considerations are applicable only to diatomic molecules. In the case of polyatomic molecules, the potential curves are replaced by multidimensional potential energy surfaces. Ionization will probably occur so that there is produced that ion which can most readily be formed without any nuclear displacements and with the lowest expenditure of energy. At slightly higher energies the molecular ion may be produced with more than one vibrational mode excited. If many vibrational modes are excited or a few are excited to high quantum numbers, the resulting vibrationally excited molecular ion may, by internal energy transitions, find itself in such a condition that it has enough vibrational quanta so that, if these are all mainly excited in mode, dissociation may occur at a particularly weak bond. Later we shall discuss these processes in more detail, as they are of considerable interest and importance in understanding meta-stable ion transitions. These processes are also of fundamental importance in understanding the quasi-equilibrium theory of the decomposition of molecular ions.

For large polyatomic molecules it is impractical to attempt fully to interpret processes occurring in electron impact in terms of detailed potential hypersurfaces for the various states of the molecule and its molecular ions. As mentioned earlier, however, the molecular ion

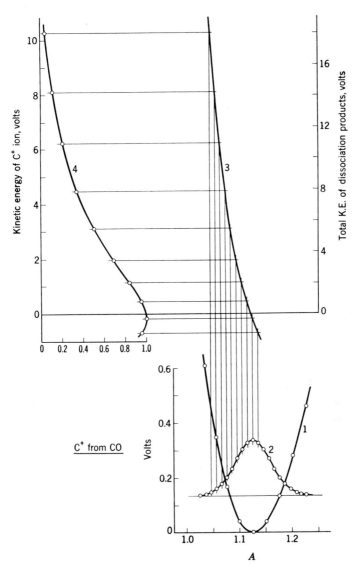

Fig. 12.3.   The Franck-Condon transitions resulting in the production of C⁺ ions from CO by the process shown in Eq. (12.7).   (*From Hagstrum and Tate* [1].)

possessing a certain amount of electronic and vibrational energy may not dissociate within the time of one vibration, but instead it can rapidly and randomly distribute this energy among the various vibrational modes. It may undergo unimolecular decomposition along energetically available paths if it can attain certain configurations with sufficient vibrational energy concentrated in the proper modes.

The rate of decay of the parent polyatomic molecular ion for a particular mode of dissociation would be expected to be a function of the excess energy above that necessary to cause the dissociation. In most mass spectrometers the products of dissociation which are analyzable are those which occur in appreciable amounts in times of several microseconds. It therefore becomes important to consider the amount of excess energy which a molecular ion must possess so that its dissociation may be detectable by the mass spectrometer. Rosenstock, Wallenstein, Wahrhaftig, and Eyring [3] have developed a statistical theory which indicates that the excess energy necessary to reduce the lifetime the order of a microsecond increases with the number of internal degrees of freedom of the molecular ion. As a consequence of these considerations, Chupka [4] has pointed out that the ionization efficiency curve for a fragment ion should approach the energy axis with curvature, quite apart from any effect resulting from energy spread in the electron beam, and the ion intensity may become vanishingly small at energies appreciably above the theoretical appearance potential.

It has already been mentioned that transitions to the dissociation limit of a molecular-ionic potential energy curve (or hypersurface) or to a repulsive state (Fig. 12.1b) lead to the appearance of fragment ions in the mass spectrum of the compound being studied. The measurement of the appearance potentials of such ions leads to much useful information of a chemical-physical nature. Obviously, transitions can occur which lead to the production of ions with and without excess kinetic energy. The products may also be produced either in their ground states or in excited electronic states; and they may be produced in vibrationally excited states. These various possibilities are envisaged in the way Eq. (12.6) is written.

Later we shall discuss in considerable detail the experimental methods available for measuring the appearance potentials of ions. We shall also consider the theoretical interpretation and implications of such measurements. Little more need therefore be said about these matters at this point.

Several workers have attempted to test the detailed applicability of the Franck-Condon principle to electron-induced dissociations. Schaeffer and Hastings [5] determined the relative yields of atomic and molecular ions, that is, X^+/X_2^+ ratios, for hydrogen, deuterium, and tritium. After allowing for possible effects due to mass discrimination, the experimentally determined ratios, as determined at 4,000 volts, were found to be in good agreement with those calculated from the theory of the Franck-Condon principle.

More recently, however, Stevenson [6] has claimed that neither the average appearance potentials of H_2^+ and D_2^+ nor the distribution in kinetic energies of the protons or deuterons are those expected from the conventional application of the Franck-Condon principle to the potential-energy diagrams of H_2 and H_2^+. To account for the experimental observations, it is suggested by Stevenson that one must assume the existence of a transient complex D_2^- or H_2^- with a distribution of lifetimes of 10^{-14} to 10^{-13} sec. Attention is drawn, however, to the footnote at the end of Ref. 6.

12.4. Experimental Methods for Determining Appearance Potentials

Though we have only indicated the formation of positive ions as the result of electron impact in our earlier discussion, it will of course be realized that negative ions can also be formed. In what follows we shall first describe methods which are generally applicable to appearance potentials, though frequently only to positive ions. Later we shall discuss the particular problems which arise in measuring the appearance potential of negative ions.

The various methods available for the determination of the appearance potentials of ions depend on the accurate measurement of the ionization efficiency curve for the ion under study. Normally, the compound in its gaseous state is introduced to the mass spectrometer at a suitable pressure, and it is ionized by an electron beam, the energy of which can be varied in known amounts. The problem of evaluating the appearance potential is mainly one of interpreting the meaning of the ionization curve.

It is obviously of considerable importance to have the electron energy scale properly calculated. This is necessary because the true electron energy can differ considerably from that indicated on a meter on the instrument because of contact potentials in the electron gun and other effects in the ion chamber of the mass spectrometer. Sometimes ion chambers are gold-plated to reduce contact potential and to prevent surface reactions which may lead to variations in contact potentials and other undesirable effects. The inert gases Ar, Kr, etc., are frequently used in calibrating standards to fix the electron energy scale, for the ionization potentials of these elements are known with high accuracy [7]. The normal procedure is to admit the calibrating gas along with the compound being studied. The partial pressures of the two substances are adjusted to give ion beams of comparable magnitudes. The ionization efficiency curves of the calibrating gas and the compound under investigation are determined, and from the resulting graphs representing the ionization efficiency curves, the difference between the ionization potential and the appearance potential of the ion being studied can be evaluated. Figure 12.4 shows typical ionization efficiency curves obtained by this method.

In evaluating the difference between the ionization potential of the calibrating gas and the ion under study, many different methods have been used. It will be noticed that the beginning portions of the ionization efficiency curves in Fig. 12.4 are curved. The vanishing current method of evaluating ionization potentials introduced by Smyth [8] assumed that the appearance potential of an ion was given by the electron voltage at which the ion current just vanished. A modification of this method, introduced by Warren and McDowell [9], is easier to apply. The ordinates of the ionization efficiency curves are chosen to make the linear portions of the curves parallel. The differences in the electron

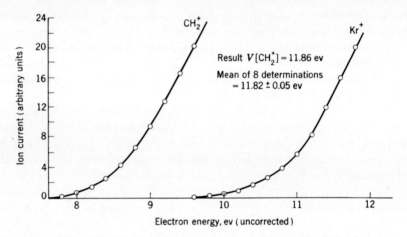

Fig. 12.4. A typical ionization efficiency curve produced by the normal simple electron-impact method.

voltage ΔV corresponding to various values of the ion current I are measured, and a graph ΔV against I drawn and extrapolated to zero ion current. The extrapolated value ΔV is taken as the difference between the ionization potential of the calibrating gas and the appearance potential of the ion being studied. Either of the above methods gives more reliable and reproducible values than those given by the earlier linear extrapolation method due to Smith 10], which is not now used.

Among other methods used to evaluate appearance potentials are several which depend on logarithmic plots of ionization efficiency curves. All these stem from the method introduced by Honig [11] and arise from considerations of the energy spread of the electrons emitted by the filament. If we assume that the electrons have a Maxwellian energy distribution, we may write, for the electron distribution function,

$$dN_e(U) = \frac{4\pi mA}{h^3} \, U \exp\left(\frac{-\phi + U}{kT}\right) dU \qquad (12.8)$$

where $dN_e(U)$ = number of electrons with energies between U and
$U + dU$ leaving the filament per second
U = thermal energy of electrons
m = mass of electron
h, k = Planck's and Boltzmann's constants, respectively
T = absolute temperature of the filament metal

On leaving the filament, the electrons are accelerated by potential difference V, so that the ionizing electrons have energy

$$E = U + V \qquad (12.9)$$

The total number of ions $N_i(V)$ produced per second with an electron accelerating voltage V is given by

$$N_i(V) = \int_{U=0}^{U=\infty} Ne(U)p(E)\, dU \qquad (12.10)$$

where $p(E)$ is the probability that an electron of total energy E produces, an ion which reaches the collector. The exact form of $p(E)$ is not known but, on the assumption that it is proportional to the square of the electron energy in excess of the energy V_e for the production of ion, i.e.,

$$
\begin{aligned}
p(E) &= 0 & \text{for } E \le V_c \\
p(E) &= C_1(E - V_e)^2 & \text{for } E \ge V_c
\end{aligned}
\qquad (12.11)
$$

then Eq. (12.10) can be solved by substituting Eqs. (12.8) and (12.11). This yields the result

$$N_i(V) = 2C_2kT^3[(N_c + V) + 3kT]\exp\left(\frac{-\phi + V_c - V}{kT}\right)$$
$$\text{for } V \le V_c \quad (12.12)$$

and

$$N_i(V) = C_2T^2[6k^2T^2 + 4kT(V - V_c) + (V + V_c)^2]\exp\left(\frac{-\phi}{kT}\right)$$
$$\text{for } V \ge V_c \quad (12.13)$$

If $\ln N_i$ is plotted as a function of V, the curve approaches an exponential straight line below V_c with a final slope

$$\frac{d}{dV}(\ln N_i)_F = \frac{1}{kT} \qquad (12.14)$$

For the critical energy $V = V_c$, the critical slope

$$\frac{d}{dV}(\ln N_i)_c = \frac{2}{3kT} \qquad (12.15)$$

Honig tested these ideas experimentally and found that, for argon, there was excellent agreement between the experimental results and the theoretical predictions. This method enabled him to obtain experimental values for the ionization potentials of hydrocarbons which seemed to be quite accurate.

Several other workers have utilized logarithmical plots of ionization current against energy obviously based on simple modifications of Honig's work. Lossing et al. [12] in their method adjusted the relative pressures of the gas to be studied and the calibrating substance until the peak heights were approximately equal at an ionization potential of 50 volts. The peak heights of both substances were then measured at different values of the ionizing electron energy. A semilogarithmic plot of the peak heights, as a percentage of the peak heights at 50 volts, against the electron ionizing energy yields curves which have parallel linear regions near the onset of ionization. These workers take the difference between the linear parts of the semilogarithmic plots of the calibrating gas and the substance being studied at an arbitrary peak height of 1% of the peak height at 50 volts as being the difference in the appearance potentials of the ions. The values obtained in this way are reproducible and of quite good accuracy. This method has been used by Lossing and his collaborators in nearly all their work on the measurement of the ionization potentials of free radicals (Chap. 11).

The semilogarithmic method used by Morrison [13] was a straightforward adaptation of the earlier and more detailed method due to Honig which we have described above. In this approach Morrison plots the logarithm of the ion current against the electron energy, and the point at which the semilogarithmic curve ceases to be a straight line is taken to be the appearance potential. This somewhat arbitrary choice is based on the assumption that the ionization probability rises very abruptly above the critical energy. Though this method would appear to be difficult to justify theoretically and might be expected to lead to difficulties in choosing just the correct value of the electron energy at which the semilogarithmic plot deviates from linearity, nevertheless, Morrison's application of it leads to values for the appearance potentials of a number of compounds which are in good agreement with those obtained by Honig by the application of the somewhat less arbitrary critical-slope method. Further application of this method by Morrison and Nicholson [14] has enabled them to provide values for the ionization potentials of a large number of hydrocarbons and their derivatives. Their values for methane, n-hexane, and benzene (13.16, 10.54, and 9.52 ev) are in fair agreement with the results obtained by Honig (13.04, 10.48, and 9.45 ev). Table 12.1 lists the values obtained by Morrison and Nicholson.

Another method of interpreting ionization efficiency curves has been introduced by Morrison [15, 16]. This technique involves the evaluation of the differential curves $\Delta V/\Delta I$, where I is the ion current and V represents the energy of the ionizing electrons. The method aims at

TABLE 12.1. The Ionization Potentials of Some Organic Compounds [14]

	H	CH_3	C_2H_5	nC_3H_7	nC_4H_9	$CH_2{=}CH$	C_6H_5
—CN	13.91	12.39	11.85	†	†	10.75	9.95
—Cl	12.78	11.46	11.18	10.96	.....		9.42
—OH	12.76	10.95	10.60	10.46	†	.....	9.03
—Br	11.69	10.73	10.49	10.29	10.12	.....	9.41
—COOH	11.51	10.70	10.47	10.22	†	10.90	†
—CHO		10.28	10.06	10.01	.....	10.34	9.82
—CMeO	10.28	9.92	9.76	9.59	9.58	9.91	9.77
—I	10.48	9.67	9.47	9.41	9.32	.....	9.10
—NH₂	10.52	9.41	9.32	9.17	9.19	.....	†
—C_6H_5	9.52	9.23	9.12	9.14	9.14	8.86	†

† Molecular ion not abundant enough for its appearance potential to be measured.

detecting structure on the ionization efficiency curves due to the formation of ions in excited states. It is, of course, well known that changes in curvature are more readily exhibited by plotting either the first or second derivative of the generating parameter. Morrison's theoretical analysis is reproduced below.

It is assumed that the ionization probability P for the excitation of a molecule in its ground state to a single electronic level of the ion can be denoted by

$$P = 0 \qquad \text{for } E < V_c$$
$$P = p(E) \qquad \text{for } E \geq V_c$$

where E = ionizing electron energy

V_c = ionization potential

The resulting ion current produced at any nominal electron energy V by an electron beam with a Maxwellian energy spread denoted by $N_e(U)$ is proportional to

$$I = \int_{V_c}^{\infty} N_e(U)p(E) \, dU \qquad (12.16)$$

The total electron energy $E = U + V$, where U is the thermal energy of the electrons. Hence we may write

$$I = \int_{V_c}^{\infty} N_e(E - V)p(E) \, dE \qquad (12.17)$$

Differentiating Eq. (12.17) with respect to the nominal energy V, we obtain

$$\frac{dI}{dV} = \int_{V_c}^{\infty} \frac{\partial N_e(E - V)}{\partial V} p(E) \, dE \qquad (12.18)$$

since

$$\frac{\partial N_e(E - V)}{\partial V} = - \frac{\partial N_e(E - V)}{\partial E}$$

Eq. (12.18) can be integrated by parts to give

$$\frac{dI}{dV} = -N_e(E - V)p(E)\Big]_{V_c}^{\infty} + \int_{V_c}^{\infty} N_e(E - V)\frac{dp(E)}{dE}\,dE \quad (12.19)$$

If $p(V_c) = 0$, the integrated term vanishes. Provided that $p(E)$ and all its derivatives up to and including order $n - 1$ vanish at $E = V_c$, the process can be repeated and leads to the formula

$$\frac{d^nI}{dV^{n+1}} = N_e(V_c - V)p_k(V_c) + \int_{V_c}^{\infty} N_e(E - V)\frac{d^{k+1}p(E)}{dE^{k+1}}\,dE \quad (12.20)$$

If the first nonvanishing derivative is of order k, Eq. (12.20) holds for $n \leq k$, but for the $(k + 1)$st derivative,

$$\frac{d^{k+1}I}{dV^{k+1}} = N_e(V_c - V)p_k(V_c) + \int_{V_c}^{\infty} N_e(E - V)\frac{d^{k+1}}{dE^{k+1}}p(E)\,dE \quad (12.21)$$

where $p(V_c)$ is the kth derivative of $p(E)$ evaluated at $E = V_c$. If $p(E)$ is proportional to $(E - V_c)$, then the $(k + 1)$st derivative vanishes and

$$\frac{d^{k+1}I}{dV^{k+1}} \propto N_e(V_c - V) \quad (12.22)$$

This shows that, if $p(E)$ is a polynomial of kth degree, the $(k + 1)$st derivative of the ionization efficiency curve reduces to the electron energy distribution, reversed with respect to the energy scale, and with the cutoff on the high-energy side at V_c.

Morrison first evaluated dI/dV as $\Delta I/\Delta V$ by subtraction of the data obtained as ionization efficiency curves. Later, experimental details were given of instruments which evaluated the differential ionization efficiency curves $\Delta I/\Delta V$ and the second differentials $\Delta^2I/\Delta V^2$ by special electronic circuits [17]. Hercus and Morrison [18] have also described an instrument for the rapid determination of ionization efficiency curves in which the time taken for a complete measurement is of the order of a few minutes.

Plotting the first and second derivatives of the ionization efficiency curves for the rare gases, Morrison showed that the technique revealed the presence of many structural details in the data. When the method was applied to molecules, the second differential curve clearly indicated [19] that the structure revealed was due to the formation of the ions in excited states. The agreement with known excited states of the various ions, however, was not very good.

Perhaps the most convincing example of the second-differential ionization-efficiency-curve plotting is that given by Morrison [20] comparing the results of electron and photon impact for the ground state of the NO^+ ion. Figure 12.5 shows these curves. The electron-impact curve refers to a plot of $\Delta^2 I/\Delta V^2$, whereas the first derivative of the photon-impact data is plotted. There is no doubt that the second-derivative representation does disclose the presence of details in the ionization-efficiency data which can rightly be attributed to the formation of the NO$^+$ ion in various vibrational states. The lack of quantitative agreement is undoubtedly caused by the spread in energy of the ionizing electrons, as was clearly understood by Morrison, who makes some comments on it in his paper.

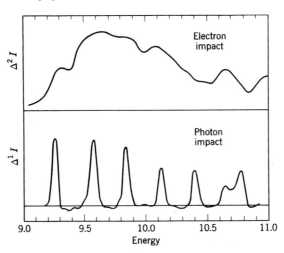

Fig. 12.5. Comparison of the results of electron and photon impact for the lowest state of NO$^+$. The upper curve shows the second differential ionization efficiency $\Delta^2 I/\Delta V$ for electron impact; the lower curve represents the first differential ionization efficiency $\Delta^1 I/\Delta V$ for photon impact. (*From Morrison* [20].)

One of Morrison's close collaborators, Nicholson [21], has shown that this second-differential ionization-curve method yields good results when applied to the determination of the difference in the ionization potentials of Kr and Ar. Obviously, much more detailed work, particularly with a wide range of molecules of different types, is needed before the accuracy of this method can be regarded as being as well established as those discussed earlier. It does, however, seem to be a promising approach, and its wide use by other workers in the field will be awaited with considerable interest. It has already yielded interesting results in estimating relative electronic transition probabilities [218].

Dibeler and his colleagues, who have made a large number of appearance potential measurements on a great variety of ions, use a variation of the critical-slope method. They [22] evaluate ionization and appearance potentials by measuring the voltage interval between two parallel straight lines plotted on the same normalized scale. One of the lines corresponds to the ion of unknown appearance potential, and the other to an ion of known ionization potential. Their method is thus similar to that used by Lossing and his collaborators, except that Dibeler normalizes the two curves by adjusting the partial pressures of the gases so as to give approximately equal ion current for the two ions at ionizing electron energies of 75 volts.

From the previous discussion it will be readily apparent that the spread of energy of the ionizing electrons plays an important part in determining the accuracy with which appearance and ionization potentials may be measured. This factor is also, as has been seen, of considerable importance when we seek to obtain information about the excited states of molecular ions. We have already drawn attention to the fact that, whereas the method of plotting the second-differential ionization efficiency curves does display the presence of excited states of molecular ions, the spread in the energy of the ionizing electrons does not permit accurate estimates to be made of the excitation energy.

It is emphasized that all the methods discussed above use an electron beam obtained by thermal emission from a filament. A tungsten filament is the one most frequently used, and the energy spread of the emitted electrons can be quite considerable. Some workers have attempted to use filaments which give a somewhat narrow energy spread to the electrons. Thus Stevenson and Hipple [23] and others [24] used a tungsten ribbon coated with BaO and SrO and claimed a half-width of the electron energy as about 0.3 ev. Morrison and Nicholson [25] later used either a filament of thoriated iridium or an oxide cathode of BaO/SrO on platinum and claim that the energy spread of the resulting electrons has a half-width of the order of 0.15 ev.

Tungsten and other filaments are often attacked by the compounds being studied in a mass spectrometer. This is particularly the case with corrosive substances like Cl_2 and compounds containing oxygen and the halogens (F_2, Br_2, I_2). Furthermore, many hydrocarbons react with tungsten filaments and form a film of tungsten carbide. When such a film has formed, the thermionic emission, and presumably the energy spread, of the electrons remains fairly constant. In analytical applications of mass spectrometry, many different carbonizing techniques are used. A favored technique is the conditioning of a new tungsten filament by exposing it to butene. The type of carbide formed and the extent of carbonization have a marked effect on the work function of the tungsten.

For the foregoing reasons, there has been a tendency toward the use of filament materials other than tungsten. Thoriated iridium has been found to be satisfactory [26]. The main advantages of this filament material are (1) lower operating temperature, with reduced chance of pyrolysis of organic compounds; (2) no preconditioning required; (3) chemical inertness toward Cl_2, O_2, etc.; (4) constant-emission characteristics. Rhenium, and also carbon, have also found use as filament materials. Rhenium has the advantage that it is not so brittle as tungsten at high temperatures and its carbides are less stable than those of tungsten. It is very useful in surface-ionization sources.

Carbon filaments are used frequently when organic compounds are being studied. Such filaments have also been used in mass-spectrometric studies of organic free radicals [27]. The electron emission of a carbon filament follows approximately the same law as that found for tungsten filament. It is not much affected by exposure to oxygen, the oxidation rate being about 500 to 1,000 times less than tungsten. Another advantage is that most organic compounds do not decompose readily on a carbon filament.

12.5. Monoenergetic-electron-impact Methods

All the experimental methods described above suffer from the defect that the ionizing electrons used have a more or less wide energy spread. This means that fine details in the ionization efficiency curves are not completely or accurately shown despite the use of various techniques such as plotting the second-derivative ionization efficiency curve. One needs, ideally, a monoenergetic electron source to measure ionization and appearance potentials accurately. This requirement is also, of course, of prime importance if the excited states of molecular ions are to be investigated accurately.

Several such sources have been described in the literature. Early types such as that described by Nottingham [28], based on velocity selection of electrons, gave interesting and promising results. In Nottingham's studies much fine detail was evident in the ionization efficiency curve which he obtained for the Hg^+ ion. These simple methods suffered from the defect that the electron emission for a narrow energy spread of almost 0.1 ev was quite low, and this led to considerable difficulties in the detection of the resulting ions. The development of electron multipliers has, of course, made it possible to measure small ion beams with considerable accuracy, and counting techniques can now also be used. It is therefore not surprising that considerable interest has developed in the further use of monoenergetic electron guns in the ionization chambers of mass spectrometers. One of the most successful reports is that by Clarke [29], who studied the ionization efficiency curves of N_2^+, N^+, Xe^{2+} for the ions produced by electrons from a simple electron energy selector.

Clarke's apparatus used a 127° electrostatic velocity selector of the type which Hughes and McMillan [30] showed would produce a beam of nearly monoenergetic electrons. In this case it was generally found that the energy spread of the electron beam was almost 0.2 to 0.3 ev. The selector was followed by a grid to accelerate the electron beam to energies sufficient to ionize the molecules being studied. The electron energy selector used by Clarke is shown diagrammatically in Fig. 12.6.

The electron selector was constructed of stainless-steel str ps set into grooves cut in plates of Mycalex. Electrons from a tungsten filament F were accelerated by 3 volts between the filament and the first 0.7- by 4-mm slit of the selector. The inside curved deflection plate P_1, of ion radius, was maintained at $+4.5$ vo!ts, while the outer curved deflection plate P_2 of 1.5 cm radius could be varied in potential from 0 to $+3$ volts,

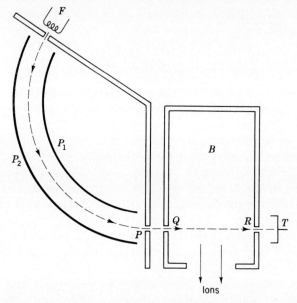

Fig. 12.6. The electron energy selector used by Clarke [29]. F represents a tungsten filament. P_1 and P_2 are deflection plates forming the electrostatic selector whose exit slit is at P. Q and R are the entrance and exit slits of the ionization chamber B. The trap T measures the electron current.

the usual value for optimum work being about $+1.3$ volts. The additional energy necessary to cause ionization was given to the electrons by the potential difference between the exit slit of the selector P and the entrance slit Q of the ionization chamber B. A trap T was provided to measure the electron current and, when not in use, was grounded to the case B.

The mass spectrometer with which the above energy selector was used was equipped with a Be-Ca multiplier which could provide a gain of almost 10^6. Adjusting the voltage on the outer electrode of the selector enabled different parts of the Maxwellian distribution of the energy of the electrons to be selected. Generally, this outer plate was operated with a voltage of 1.3 ev, the trap current being about 10^{-8} amp and the energy distribution about 0.3 ev.   Distributions of as low as 0.2 ev were also obtainable.

Using this selector, Clarke was able to detect fine structure on the N_2^+, the N^+, and the Xe^{2+} ionization efficiency curves, as has been mentioned earlier.   He detected ionization leading to the $^2\Pi_u$ state of N_2^+ at 1.0 $\pm$ 0.2 ev above the initial ionization potential of N_2.   He also detected the production of N^+ ions from N_2 by electron impact, with the atomic ions produced in their $^2D^\circ$ and $^2P^\circ$ excited states.   In the case of Xe^{2+}, he measured the second ionization potential to be 21.09 $\pm$ 0.05 ev.   The

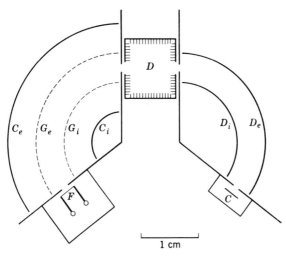

Fig. 12.7.   Energy selector used by Marmet and Kerwin [31, 32]. F is the filament; C_e and C_i are collector electrodes; G_i and G_e are the main selector electrodes made of 90% transparent tungsten mesh. D is the ionization chamber; D_i and D_e are the analyzer electrodes; and C is the collector plate.

ionization efficiency curve for the Xe^{2+} ion was found to obey a square law rather than a nearly linear one.

An important advance in the design of electron energy selectors has been made by Marmet and Kerwin [31, 32], who have described the construction and performance of an electrostatic electron selector which provides an 0.1 = μamp beam of electrons with an energy spread of only 0.04 ev.   The construction of their selector is shown in Fig. 12.7.   The main selector electrodes were made of 90% transparent tungsten mesh. Electrons approaching these electrodes passed through, for the main part, and were prevented from returning by a positive potential maintained on the outer collectors.   The formation of a space change was prevented, even at beam energies as low as 0.5 ev, with a consequent reduction in energy spread.   When the beam of electrons was directed into the ionization chamber D, reflections from the walls were prevented by the use of a unique method of construction incorporating a special material called *electron velvet*.

Preliminary studies, using this electron energy selector in a simple vacuum jar, indicated that the doublets $^2P_{3/2}$ and $^2P_{1/2}$ of the ground state of the Ar^+ ion could be resolved.   These are separated by only 0.18 ev. Later studies on hydrogen showed that structure was clearly present on the ionization efficiency curves for the H_2^+ ion which was attributed to the formation of the ion in its various excited vibrational states.   Similar results were observed for the N_2^+ ion.   It is obvious that the incorporation of this improved type of electron energy selector into a mass spectrometer will lead to many important and significant results.   Such studies are at present under way [33].

A somewhat different method for obtaining what are essentially monoenergetic electrons has been introduced by Fox and his colleagues [34, 35].

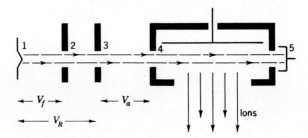

Fig. 12.8.   Diagrams of the RDP source of Fox et al. [34, 35].   The filament (1) with electrons which pass through the slits in electrodes (2 and 3).   They enter the ionization chamber (4) and are affected by the electrode (5).

This they have called the retarding-potential-difference, or RPD, method. A specially designed electron gun is used.   Figure 12.8 shows a simplified representation of the source.   The filament 1 emits electrons which are caused to pass through the slits in electrodes 2 and 3.   They enter the ionization chamber 4 and are collected by electrode 5.   If the initial thermal energies and contact potentials are neglected, it will be seen that the energy of electrons which enter the ionization chamber is determined solely by the potential difference V between the cathode and electrode 4, regardless of the potential of electrode 2.   Any initial thermal energy which the electrons may have possessed upon emission from the cathode must, of course, be added to that acquired in traversing the potential difference V.   Since electrode 3 has no effect on the energy of those electrons which reach the ionization chamber, this electrode then may be made negative in potential with respect to the cathode by an amount V_R such that all electrons which leave the cathode with energies less than V_R will not pass through the slit in electrode 3.   Those electrons with initial energy just equal to V_R will leave the slit in electrode 3 with essentially zero velocity, and their kinetic energy on reaching the slit in electrode 4 will be determined by the difference in potential between electrodes 3 and 4, indicated as V_a in Fig. 12.8. Electrons possessing initial energies greater than V_R will have their kinetic energy in the ionization chamber unaffected by the potential of electrode 2. If the value of V_a is such that the electrons have sufficient energy to produce ionization, this will be produced by an electron beam with an energy distribution characterized by a sharp lower limit determined by V_R. If V_R is now changed by a small amount ΔV_R, the number of electrons entering the ionization chamber will change accordingly. The energy of the electrons will, however, be unaltered. In this case, the ionization will be caused by an electron beam whose energy distribution is characterized by a sharp lower limit determined by $V_R + \Delta V_R$. The difference in ionization for these two cases will be due to those electrons lying within the energy band ΔV_R. By varying V, the difference in ion current can be measured as a function of the measured potential V_a. In this way one can obtain the ionization efficiency curve for a process produced by essentially monoenergetic electrons. It has generally been possible to arrange for ΔV_R to be of the order of 0.1 to 0.2 ev. When the value of ΔI (the difference in the ion current for a change of ΔV_R in the electron energy) is plotted against the energy of the electrons, the resulting ionization efficiency curves normally exhibit fine structure. This structure can be correlated with the production of the ions in excited states.

A modified RPD method has been introduced by Cloutier and Schiff [36]. These authors use a retarding potential method incorporating a space-change-limited diode. The design makes use of the potential minimum existing in the region between the filament and anode of space-change-limited diode to produce the retarding potential for the electrons leaving the cathode. The slit arrangement and the voltage distribution are shown schematically in Fig. 12.9. It can be shown that a fine control of the change in retarding potential V_m is obtained by altering the anode voltage V_a.

Using this type of source, Cloutier and Schiff were able to detect the first excited state of the N_2^+ ion. They were not able, however, to observe the higher excited states of N_2^+ detected by Hickam and coworkers [37] and by Frost and McDowell [38]. The method will doubtless prove of greater value when further developed, and its use by other workers will be awaited with interest.

It should perhaps be mentioned here that the RPD method of Fox et al. has proved to be of considerable use in studying the formation of atomic and molecular ions in their excited states. These studies will be discussed in detail later in this chapter. Here we may note that, in nearly all cases studied, it has been possible to correlate the experimental results directly with the excited state of molecular ions to be expected from molecular orbital theory.   It has also been possible to check and assess many theoretical predictions for which no other experimental method can at present provide the necessary data.

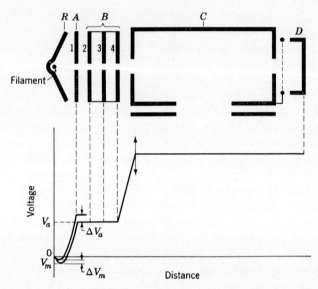

Fig. 12.9.   Schematic diagram of the slit arrangement and the voltage distribution of the electron gun in the method used by Cloutier and Schiff.   (*From Cloutier and Schiff* [36].)

12.6. Appearance Potentials of Negative Ions

The negative ions frequently observed in the mass spectra of compounds arise from three main processes.   First, an electron-capture process may occur such as that indicated in Eq. (12.23).   It must be emphasized that stable molecular negative ions of the type there indicated are rare.

$$XY + e = XY^- \qquad (12.23)$$

However, the SF_6^- ion is so stable that the process

$$SF_6 + e = SF_6^- \qquad (12.24)$$

can be used to calibrate the electron-energy voltage scale at low energies near zero (see below).

An electron-capture process of more frequent occurrence is that indicated in Eq. (12.25).

$$XY + e = X + Y^- \qquad (12.25)$$

In that process dissociation occurs presumably after the formation of the unstable XY^- molecular negative ion with the formation of either X^- or Y^-, that is, an atomic negative ion such as Cl^- or a negative ion of the type CN^-.

Both the processes represented by Eqs. (12.23) and (12.25) usually occur in the low-energy region of the electron energy scale. This can readily be seen when it is recalled that the appearance potential of the negative Y^- ion formed by the process represented by Eq. (12.25) is given by the relation

$$V(Y^-) = D(XY) - EA(Y) + \text{K.E.} + \text{E.E.} \tag{12.26}$$

where K.E., E.E. = excess kinetic and excitation energy if any
$D(XY)$ = dissociation energy of XY band
$EA(Y)$ = electron affinity of Y atom

Since the dissociation energies of most chemical bonds lie in the range 3 to 5 ev and the electron affinities of most atoms are in the range of 1.5 to 3.5 ev, it follows that the values of $V(Y^-)$ to be expected will lie in the range 0.5 to 4 ev.

Negative ions can, of course, also arise from an ion-pair process such as that shown in the following equation:

$$XY + e = X^+ + Y^- \tag{12.27}$$

The appearance potential for the Y^- ion formed by the process indicated in Eq. (12.27) will be given by the expression

$$V(Y^-) = D(XY) + I(X) - EA(Y) + \text{K.E.} + \text{E.E.} \tag{12.28}$$

where $I(X)$ represents the ionization potential of the X atom, and the other symbols have the same meaning as given in (12.26). Using the values of Eq. (12.26) and keeping in mind that the first ionization potentials of most atoms lie between 6 and 15 ev, we see that the appearance potentials of negative ions arising by an ion-pair process as indicated in Eq. (12.27) will generally be in the range of 6 to 13 ev.

Figure 12.10 shows the formation of the SF_6^- ion from SF_6 by an electron-capture process of the type indicated in Eq. (12.23). In this case the RPD method was used so that this figure indicates the type of molecular electron resonance-capture process to be expected when essentially monoenergetic electrons are used. The very narrow width of the resonance-capture peak is noteworthy. Figure 12.11 shows the formation of negative ions by both of the processes represented by Eqs. (12.23) and (12.25). These data were taken using a normal electron-impact source where the electrons have a quasi-Maxwellian distribution of energies. It is apparent from a comparison of the resonance-capture processes in both Figs. 12.10 and 12.11 that the main contribution to the width of the resonance-capture peak in Fig. 12.11 must come from the more or less wide spread in the energies of the ionizing electrons. Furthermore, it is evident, from the ionization efficiency curve shown in Fig. 12.11, that all the difficulties discussed above with regard to the accurate determination of the appearance potentials of positive ions are even

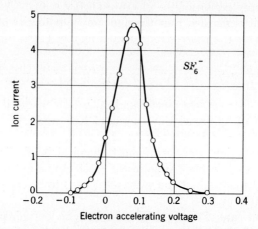

Fig. 12.10. Electron-capture process leading to the formation of the SF_6^- molecular negative ion as indicated in Eq. (12.23).

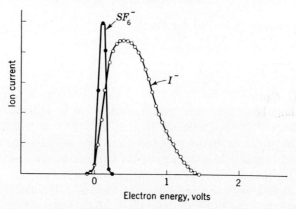

Fig. 12.11. Electron-capture process for the formation of the I^- ion from I_2 by a dissociative process as represented by Eq. (12.25). (*From Frost and McDowell* [42].)

greater for negative ions. Nevertheless, practically all the methods discussed have, at one time or another, been used to determine the appearance potentials of negative ions.

Another major problem arises in the case of negative-ion appearance potentials. The calibration of the electron energy scale is particularly difficult at low energies, for there are no spectroscopic standards available in that region. At somewhat higher energies, i.e., in the region where ion-pair processes usually occur, the problem is less acute. One method applied in this case is to calibrate the scale using the appearance potential of the positive ion produced in the ion-pair process depicted in Eq. (12.27). This is, however, not always a reliable method. Furthermore, it is difficult to be certain that the extrapolations to very low electron-energy values which are often required to observe the negative ions produced by electron-capture processes such as those shown in Eqs. (12.23) and (12.25) are at all reliable.

These problems have been discussed in some detail by Hagstrum [39] in his interesting work on CO, NO, and O_2. They have also been considered carefully by Craggs, McDowell, and Warren [40]. Many of the problems associated with the earlier methods of calibrating the electron energy scale for negative-ion studies have been overcome by the work of Hickam and Fox [41], who first showed that the retarding-potential-difference (RPD) technique could readily be applied to the study of negative ions. As mentioned earlier, these authors were able to observe the formation of the SF_6^- ion from SF_6 by an electron-capture process. This SF_6^- peak was observed to occur at an electron energy of 0.03 ev. It thus provides an excellent means of establishing the electron energy scale for negative-ion work. An interesting example of its use in this way is provided by the study of the formation of the I^- ion from I_2 by electron impact [42], shown in Fig. 12.11. It may perhaps be mentioned that another calibrating point for negative ions is now available. It has been established that the ion-pair process which occurs [43] when I_2 is bombarded by electrons, i.e.,

$$I_2 + e = I^+(^3P_2) + I^-(^1S) \qquad (12.29)$$

produces the ions with a maximum of 0.2-ev excess kinetic energy [44]. This means that the appearance potential of the I^+ ion, which is 8.84 ev, also fixes this point on the electron energy scale for negative ions.

It will of course be realized that the mass spectrometer itself is a unique instrument for detecting the nature of an ionization process, and it can readily be established whether or not an ion-pair process such as that shown in Eq. (12.27) is occurring. Morrison [20] has indicated that the ionization probability for such processes may be regarded as step functions. In the case of certain studies this argument has been used [44] to infer that an ion-pair process like that indicated in Eq. (12.29) was taking place. This procedure is somewhat unnecessary when the mass spectrometer itself can uniquely show the occurrence of processes such as (12.29), and thus provide a definitive answer and one not subject to any ambiguity of interpretation.

It will be appreciated that extreme care must be taken in the use of the RPD method in establishing the energy scale for negative-ion formation at near-zero energy. The method has, however, been used successfully by several different groups of research workers and has produced many new and interesting results. Furthermore, it will be understood that it is essential to use monoenergetic electrons for the measurement of absolute cross sections when electron capture takes place over an extremely small energy range, and this is usually the case in practice. There can, however, be considerable differences in the capture cross sections for negative-ion formation. This is particularly the case when we compare the cross sections for ions which are produced by a truly electron-capture process as indicated in Eq. (12.23) with a dissociative capture process such as that shown in Eq. (12.25). These differences are clearly indicated for the cases of the formation of SF_6^- and I^- in Fig. 12.11. The RPD method for studying the occurrence of electron-capture processes has been applied by Frost and McDowell to the hydrogen halides [45], oxygen [46], and NO [47].

Equation (12.26) shows that, when negative ions are formed, there is the possibility that the ions may be produced with excess kinetic energy. When this happens, the determination of the appearance potential alone, even though this may be done accurately with an energy scale correctly calibrated, provides but little fundamental information. It is therefore necessary to have an experimental method for determining the excess kinetic energy with which negative ions may be produced. One of the most successful methods for studying the kinetic energy of ions formed in ionic dissociative processes is that introduced by Lozier [48]. This method has been applied with much success to the study of negative ions. Lozier's apparatus does not, however, have any provision for mass analysis, and though recently it has been used in some interesting studies of the ionization and dissociation of simple molecules [49, 50], the results are of somewhat limited value, for it is frequently not possible to identify unambiguously the dissociation or ionization processes involved. Concurrent mass-spectrometric work is necessary to provide this information.

The work which Craggs and his colleagues carried out with the Lozier apparatus, to which reference has been made, illustrates the points mentioned above. They studied the attachment of electrons in carbon dioxide [50], carbon monoxide [51], and oxygen [52]. In the case of CO and O_2, earlier work by Hagstrum [39] and others had shown clearly the nature of the dissociative processes which occurred on electron capture. However, similarly detailed mass-spectral data were not available for carbon dioxide. Craggs and Tozer [50] found that electron attachment takes place only for electrons with energies between 6.7 ev and about 10 ev. This is characteristic of dissociative attachment of the type indicated by Eq. (12.25). Direct electron attachment such as indicated by Eq. (12.23) is more or less ruled out, for the electron affinity of carbon dioxide would be expected to be much less than 6.7 ev. There are therefore only three processes really possible:

$$CO_2 + e = CO + O^- \tag{12.30}$$
$$CO_2 + e = C^- + 2O \tag{12.31}$$
$$CO_2 + e = CO^- + O \tag{12.32}$$

Craggs and Tozer consider that the process $CO_2 + e = C^+ + O_2$ is precluded as a primary process by the structure of the carbon dioxide molecule, but this is not a very compelling argument. Independent mass-spectrometric measurements were necessary which showed that, in their laboratory, the only negative ions identified were O^- ions. This would establish the process represented by Eq. (12.30) as being the only one operative.

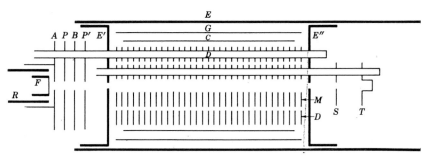

Fig. 12.12. Lozier ionization apparatus. A is the anode; P and P' are space-change control grids; B, the electron retarding electrode; E, the electron accelerating electrode; M, the uniform field electrode; R, the electron refiller electrode; D, the ion discriminator; C, the ion collector; G, a guard cylinder; S, the secondary-electron suppressor grid; T, the electron trap; F, the filament.

In the familiar form of the Lozier apparatus, a normal electron gun is used to provide the ionizing beam of electrons. Above, it has been pointed out that the cross sections for the formation of negative ions by different attachment processes can vary considerably. In their work on O_2, Craggs et al. [52] made calculations to determine the effect of electron energy spread on the shape of the attachment peak, for ions of all energies. It was concluded that the energy spread of the electrons had little effect on the shape of the peak. That conclusion is true, however, only when the width of the attachment peak is considerably greater than the electron energy spread, and consequently it is to be expected that a narrow peak may be considerably affected. The distortion caused by nonmonoenergetic electrons on the shape of a very narrow attachment peak is well illustrated by Hickam and Fox [41] for the case of electron attachment to form SF_6^- from SF_6.

Figure 12.12 illustrates the essential parts of the Lozier apparatus. The electron-impact process occurs in the ionization chamber I, and the ions formed move (essentially perpendicular to the ionizing electron beam)

through the two sets of concentric vanes M and D so that they are collimated. They are collected on the collector C, and the ion current can
thus be measured. By applying retarding potentials to the ion discriminator plates D it is possible to measure the appearance potentials for ions
formed with differing amounts of excess kinetic energy. In an ionic
dissociative process, the total kinetic energy is divided between the
fragments according to the law of conservation of momentum. The
kinetic energy of the (Y^-) ion, formed according to Eq. (12.25), is given by

$$\text{K.E.}(Y^-) = \frac{m(Y^-)}{m(X) + m(Y^-)} \, [\text{K.E.}(X) + \text{K.E.}(Y^-)] \quad (12.33)$$

Hence the appearance potential of the Y^- ion formed with kinetic
energy is

$$V(Y^-) = V_0(Y^-) + \frac{m(X) + m(Y^-)}{m(Y^-)} \, \text{K.E.}(Y^-) \quad (12.34)$$

where $V_0(Y^-)$ = appearance potential of Y^- ion formed with zero
kinetic energy

ion (X), $m(Y^-)$ = masses of ions indicated

K.E.(Y^-) = kinetic energy of Y^- ion of which appearance potential
is $V(Y^-)$

Plotting $V(Y^-)$ against the value of K.E.(Y^-) obtained from the retarding
potentials applied to the ion discriminator plates in the Lozier apparatus

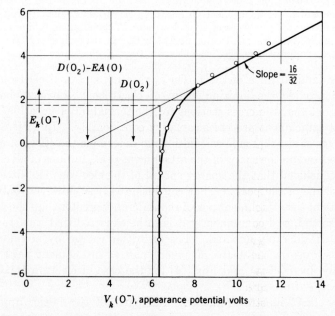

Fig. 12.13. Typical plot of Eq. (12.34), here shown for the case of the formation
of the O^- ion from O_2 by electron impact.

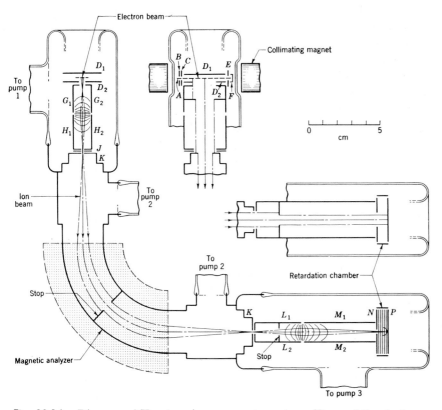

Fig. 12.14. Diagram of Hagstrum's mass spectrometer. Views of the electron-beam and retardation-chamber ends of the apparatus at 90° from the principal view are shown to the right of these portions and above them, respectively. Equipotential lines are indicated in the G-H and L-M ion lenses and in the retardation chamber. Dimensions of the slit in electrode B are 0.08×1.0 cm; C, 0.03×1.0 cm; E, 0.1×1.2 cm; D_2, 0.4×2.0 cm; G_1G_2, 0.4 cm wide; H_1H_2, 0.05 cm wide; J, 0.1×2.0 cm; K (entrance), 0.02×1.0 cm; K (exit), 0.03×1.0 cm; L_1L_2, 0.05 cm wide; stop in L_1L_2, 0.1×0.8 cm; N, 0.025×3.0 cm. Other dimensions may be estimated from the scale of the drawing. The shaded area indicates the extent of the analyzer magnetic field. Typical electrode potentials for $E_k(e^-) = 100$ ev follow: A, -100 volts; B, -80 volts; C, D_1, D_2, and E, zero; F, $+45$ volts; G_1G_2, -15 volts; H_1H_2, J, and K, -100 volts; L_1L_2, -150 volts; M_1M_2 and N, -30 volts; P, $-2 \rightarrow +10$ volts.

should yield a straight line of slope $[m(\mathrm{X}) + M(\mathrm{Y}^-)]/m(\mathrm{Y}^-)$. The intercept on the $V(\mathrm{Y}^-)$ axis at K.E.$(\mathrm{Y}^-) = 0$ gives $V_0(\mathrm{Y}^-)$. Figure 12.13 shows a typical graph of Eq. (12.34). Once $V_0(\mathrm{Y}^-)$ has been determined, one can proceed to evaluate the other parameters in Eq. (12.26), for generally negative ions are formed only in their ground states. In particular, if the dissociation energy $D(\mathrm{XY})$ is known accurately, the electron affinity $EA(\mathrm{Y}^-)$ of the Y atom can be evaluated. Much of the work carried out with the Lozier apparatus [49–52] has had the evaluation of electron affinities of atoms as one of the aims, and we shall discuss such matters in detail later in this chapter.

Before passing on to other matters, it should be mentioned that a Lozier apparatus incorporating an RPD electron gun has been successfully used in the study of negative-ion formation in carbon monoxide [53].

A mass-spectrometric method for determining the kinetic energy of fragment ions which has been applied most successfully to negative ions

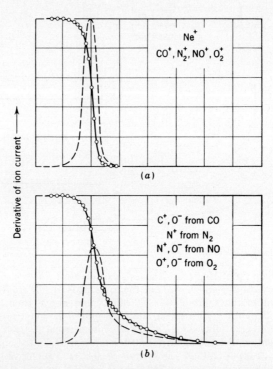

Fig. 12.15. Retarding-potential and ion-derivative curves for ions formed (a) without kinetic energy and (b) with kinetic energy. (*From Hagstrum* [39].)

has been described by Hagstrum [39]. This apparatus, which is illustrated in Fig. 12.14, consists of a specially designed mass-spectrometer tube having a retarding field between the electrodes N and P. It is possible to apply various retarding potentials to these electrodes so that the ion current I_p can be determined as a function of the applied retarding potential V_p. Ion current I_p, versus retarding potential V_p, curves, for a number of ions, are shown in Fig. 12.15a. The derivative curves for dI_p/dV_p plotted against V_p are also given. The curves there shown are typical for ions formed without excess kinetic energy. When ions are produced with excess kinetic energy, the plot of dI_p/dV_p is no longer symmetrical about the point $V_p = 0$, as shown in Fig. 12.15b, where the curves for ions formed with excess kinetic energy are presented. Frequently, cases are found where extra maxima occur in the dI_p/dV_p versus V_p curve, thus indicating the presence of ions with different amounts of kinetic energy.

To determine the appearance potentials of negative ions referred to zero kinetic energy, Hagstrum used the same procedure outlined above in the discussion of the Lozier apparatus. This method again yielded satisfactory results, and Hagstrum data will be used later in discussions concerning the energetics of dissociation processes.

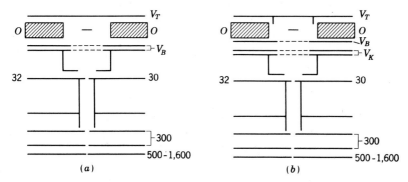

Fig. 12.16. Ion source used by Randolph and Geballe [54] to measure the kinetic energy of fragment ions. The dotted lines in the electrodes V_B and V_K indicate finely woven gold-plated tungsten gauze. The kinetic energies of ions are determined by accelerating the ionic species out of the ionization region by applying appropriate potentials to the electrodes V_T and V_B and then retarding the ions by potentials applied to electrode V_K.

Another mass-spectrometric method which has been successfully used to determine the kinetic energy with which O^- ions are formed from O_2 is that described by Randolph and Geballe [54]. These authors designed a special ion source in which retarding plates were inserted between the ion chamber and the first accelerating and focusing slits. A diagram of their source is shown in Fig. 12.16. V is the electrode used to measure the kinetic energy with which the ions are formed. The dotted lines in electrodes V_B and V_K represent a finely woven gold-plated tungsten gauze. For kinetic-energy measurements, the ions are accelerated out of the ionizing region by the electrodes V_T and V_B and then retarded by potentials applied to V_K. After the appearance potentials of ions with different amounts of kinetic energy have been determined, the same method is used to estimate the appearance potentials of the ions with zero kinetic energy $V_0(Y^-)$ as was described in the case of the Lozier method.

12.7. Determination of Ionization and Appearance Potentials by Photoionization Method

Because of the difficulties encountered in the accurate determination of ionization and appearance potentials by electron-impact methods which we have discussed above, there is considerable interest in the study of photoionization methods. Most of the work has been carried out using a monochromator to provide a beam of photons with a known narrow energy spread. These photons are then caused to interact with gas molecules in a special chamber, and the resulting photoions are detected by measuring the ionization current produced [55–57]. Very accurate values have been obtained by this method for the ionization potentials of a large number of molecules. The method has, however, some serious limitations. Because of the lack of mass analysis, one has to be sure that the sample being studied is quite free of impurities with lower ionization potentials. It thus suffers from the same sort of disadvantages as the Lozier electron-impact method.

These difficulties can be overcome by combining the photoionization source with a mass spectrometer. The application of mass analysis to photoionization studies was first carried out by Terenin and Popov [58]. They were able to demonstrate the formation of ion pairs according to Eq. (12.35):

$$XY + h\upsilon = X^+ + Y^- \tag{12.35}$$

in the photoionization of the thallium halides. More recently, Lossing and Tanaka [59] described some preliminary, but most promising, work on the use of a photoionization source for mass spectrometry. The type of apparatus used by Lossing and Tanaka is shown in Fig. 12.17. The ultraviolet-light source was a krypton discharge lamp with a lithium fluoride window. The output consisted mainly of the 1,236- and 1,165-A lines in the ultraviolet; these wavelengths correspond to 10.03 and 10.64 ev. The radiation from this lamp entered the ionization chamber of a mass spectrometer through a slit 4 by 6 mm, replacing the usual filament slit.

The mass spectra of butadiene obtained by photoionization alone and by electron impact alone are shown in Fig. 12.18, the simple pressure and the sensitivity of recording being the same for both spectra. The radiation from the lamp had sufficient energy to ionize butadiene (ionization potential 9.07 ± 0.01 ev), but insufficient to form ionic fragments. The ion of mass 54 was produced with a high intensity. The peaks at mass 39, 27, and 26 are probably formed by photoelectrons which have been accelerated to higher energies by strong electric fields. Lossing and Tanaka were also able to detect the free methyl radical from the pyrolysis of mercury dimethyl by observing the ion formed by a photoionization process.

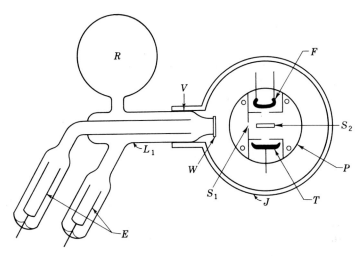

Fig. 12.17.   Diagram of the arrangement of lamp and ionization chamber. The body of the lamp L was of Pyrex, and the electrodes E were of tantalum sheet. A reservoir R contained krypton.   Light from the lamp passed through the lithium fluoride window W, through the slit S_1, and over the ion exit slit S_2, in the plate P.   A filament F and trap T provided a means for obtaining electron-impact spectra.   The top plate of the ionization chamber is not shown.   The lamp was cemented to an opening V in the vacuum jacket J of the mass spectrometer.   (*From Lossing and Tanaka* [59].)

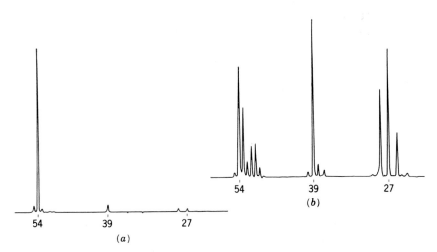

Fig. 12.18.   (*a*) Mass spectrum from butadiene as the result of photoionization using the apparatus depicted in Fig. 12.17; (*b*) mass spectrum of butadiene as the result of electron impact. (*From Lossing and Tanaka* [59].)

These workers did not make any appearance potentials or ionization potentials on the ions found by photon impact, but they discussed, briefly, the possible applications of the technique to the measurement of such quantities. Shortly after this work, a number of investigations reported on the mass-spectrometric analysis of ion fragments produced by photon impact using undispersed radiation [60–62].

Inghram and his coworkers [63] were able successfully to use the combination of a vacuum monochromator and a mass spectrometer in a detailed study of the formation of ions by photon impact. In these experiments the residual gases in the monochromator were isolated from those to be studied by photoionization by using a lithium fluorite window which cut off radiation below 1,050 A. These workers used an extremely convenient design of monochromator due to Seya [64] and Namioka [65]. An improved apparatus of similar design but without the disadvantage of the lithium fluorite window is shown in Fig. 12.19. This is the apparatus used by Frost, Mak, and McDowell [216] in their work on NO_2. Slit widths of 0.020 to 0.040 in., which give an energy spread of about 0.05 ev at a photon energy of about 10 ev, made possible the determination of the ionization potentials with high accuracy. Furthermore, many fine details of considerable theoretical interest and importance were shown up in the ionization efficiency curves for the formation of ions by photon impact. Pertinent details of these researches will be discussed later.

About the same time Vilesov et al. [66] reported on similar mass-spectrometric photoionization studies, in which they investigated the photoionization and photodissociation of various amines such as ammonia, hydrazine, benzylamine, and aniline. They did not, however, make any determinations of ionization or appearance potentials. Almost concurrently, Weissler and his associates [67] similarly reported their very extensive studies on the photoionization of Ar, He, Ne, O_2, N_2, CO, NO, CO_2, N_2O, and NO_2, in an apparatus using a mass spectrometer. Their apparatus had the advantage that, by careful design and the extensive use of differential pumping, they were able to dispense with windows between the monochromator and the ionization chamber, and thus use photons of quite high energy. These workers made use also of the Seya-Namioka design for their monochromator. They obtained a great amount of most valuable experimental data, not only on ionization potentials, but also on dissociative photoionization processes.

In the photoionization studies whose aim is the determination of appearance potentials, the ionization efficiency curves are obtained by plotting as ordinates the number of ions per photon in arbitrary units. This represents the ratio of the ion current measured at the collector end of the mass spectrometer to the output amplitude of the sodium salicylate–coated photomultiplier (or other photocathode) which monitors the light beam. The abscissa is, of course, the energy of the photon beam.

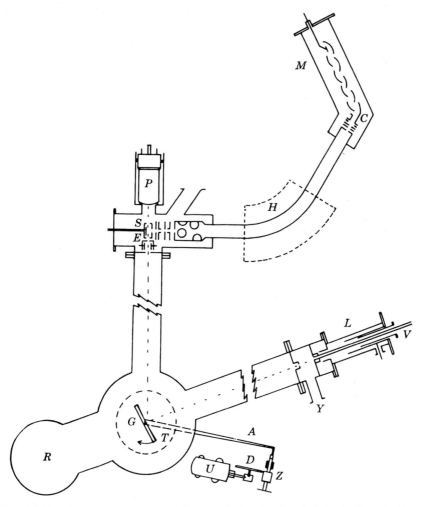

Fig. 12.19. Mass spectrometer with monochromator for photoionization studies. *G* is the grating; *T*, the grating mount; *R*, the vacuum pump; *A*, a lever to rotate the grating. Motor *U* with associated drive mechanisms *D* and *Z* controls the rate of scanning the radiation emitted by the light source *L*. *Y* is a pumping lead; *V*, a quartz capillary. The ion source of the mass spectrometer is at *S*. *E* is the filament; *P*, an electron multiplier phototube. *H* is the magnet; *C*, the collector. *M* is an ion-electron multiplier to measure the ion current.

Weissler et al. [67] used a light source which consisted of repetitive low-pressure sparks through a ceramic capillary. A mixture of air and helium at a total pressure of about 200 μ provided a satisfactory distribution of intense emission lines, and occasionally argon was used in those cases where more lines were desirable below 700 A. This source thus provided photons with energies as high as about 30 ev. These workers were able to discern much detailed structure on their ionization efficiency curves, which could readily be ascribed to the production of ions in excited molecular states or to atomic ions in various electronic states arising from many high-energy dissociation processes. The nature of their light source prevented certain fine details from being observed because of the gaps in the energy distribution of the ionizing photons. Nevertheless, the results reported are of considerable importance and marked a major advance in this field. For the same reasons, many of the appearance potentials determined are less accurate than one would desire, but many new data became available for the first time.

Hurzeler, Inghram, and Morrison [63] used a form of Lyman tube as their light source. Further as noted, they used a lithium fluoride window between their monochromator and the photon collision chamber. Hence their work was restricted to ions produced by impact with photons with energies less than 11.5 ev. These authors found that a more closely spaced series of bands could be obtained by using an equimolecular mixture of hydrogen and deuterium, rather than hydrogen alone. They plot their data in the same manner as Weissler et al. [67], but interpret it differently by following the differential ionization efficiency curve method used earlier by Morrison [15, 16], discussed in Sec. 12.4. It is assumed that the photoionization probability $p(E)$ near the threshold for ionization can be approximated by a step, or Heaviside, function. If the ionizing photon beam has an energy distribution given by $m(E - V)$ around its center V, the resulting ion current will be

$$I(V) = \int_{V_e}^{\infty} m(E - V)p(E)\, dE \qquad (12.36)$$

Since
$$\frac{\partial m(E - V)}{\partial V} = -\frac{\partial m(E - V)}{\partial E}$$

and assuming $p(E)$ to be a Heaviside function,

$$\frac{dI(V)}{dV} \propto m(V_e - V)p(E_e) \qquad (12.37)$$

where $p(E_e)$ represents the photoionization probability function at the critical potential V_c.

Thus, for a step function, the shape of the first derivative will give the energy distribution of the photon beam with the maximum value at $V = V_c$ and the amplitude is proportional to $p(E_c)$, that is, the transition probability.

Using the above technique for treating the data, Hurzeler, Inghram, and Morrison were able to determine the appearance potentials for parent molecular ions, as well as fragment ions formed as the result of photoionization. In the plots of the first-differential photoionization curves, much fine structure was frequently revealed. We have already made reference to the detection of vibrational fine structure in the curve representing the formation of NO^+ from NO by photoionization.

An interesting study of the photoionization efficiencies for Br_2, I_2, HI, and CH_3I has been carried out by Morrison, Hurzeler, Inghram, and Stanton [44]. These authors point out that it is not possible to account for all the features of photoionization efficiency curves in terms of step functions. In particular, some of the results obtained by Weissler and his colleagues [67] show clearly that peaks occur in the ionization efficiency curves. Similar peaks were observed by Morrison et al. [44] in their work, and these are attributed to processes arising from autoionization. Such processes, it is suggested, would be more readily expected to be represented by a delta function. Figure 12.20 shows the photoionization

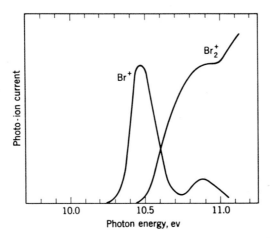

Fig. 12.20. The photoionization efficiency curves for the formation of Br^+ ions from bromine. (*From Morrison and Stanton* [110].)

efficiency curves which were obtained [44] for the Br_2^+ and Br^+ ions formed from bromine. The first Br^+ peak is attributed to the formation of this ion in our ion-pair process. The energy necessary for a vertical transition to the excited state of Br_2 giving rise to Br^+ and Br^- is estimated, from the position of the peak maximum, to be 10.48 ± 0.02 ev. The ion was shown to be produced with a kinetic energy of 0.15 ev. The second peak in the Br^+ curve was interpreted as being due to a second excitation-dissociation process, with a vertical ionization potential of 10.88 ± 0.05 ev. This is consistent with the energy difference (0.39 ev) between the $^3P_2$ and $^3P_1$ levels of the Br^+ ions.

In the ionization efficiency curves observed for the photoionization of HI and CH_3I, there was evidence of fine structure which could be identified with known Rydberg levels. Some of the observed curves are thought to be produced by the superposition of sharp peaks, due to the excitation-autoionization processes, on the normal type of threshold curve, due to the production of the molecular ions in their lowest states.

12.8. Appearance Potentials of Free Radicals

The determination of the appearance potentials of free radicals presents many unique and difficult experimental problems. One major difficulty is that a large number of free radicals have an extremely short lifetime, and so special techniques are necessary to ensure that a sufficiently large concentration of the radicals can be built up in the ionization chamber of the mass spectrometer. Most of the techniques used stem from the initial work of Eltenton [68] and of Hipple and Stevenson [69]. Since this pioneer work the most important experimental developments have been due to Lossing and Tickner [70].

The compound from which the free radicals are to be formed is usually decomposed thermally by the process

$$R - R = 2R \qquad (12.38)$$

in a quartz furnace, and the radicals in the form of a crude molecular beam are introduced to the ionization chamber in a stream of inert gas. Using this type of apparatus, Lossing and his collaborators, in an important series of papers, have described the determination of the ionization and appearance potentials of a large number of free radicals. These data have been most important in the calculation of the bond-dissociation energies. A complete account of the application of mass-spectrometric methods to the study of free radicals is given in Chap. 11.

All the methods for determining the appearance potentials of ions have, in principle, been used in the study of free radicals. We have already mentioned that Lossing and Tanaka [59] were able to detect the CH_3^+ ion formed from the methyl radical by photoionization. Recently the ionization potentials of alkyl radicals have been determined by photoionization methods [217].

12.9. Molecular Ionization Potentials

The ionization potential of a molecule is the energy required to remove an electron from the lowest occupied molecular orbital. These data are therefore of considerable importance in understanding the arrangements of electrons in molecular orbitals. A knowledge of the first ionization potential alone often enables many important chemical and physical deductions to be made [71]. It will, of course, be realized that a molecule has as many ionization potentials as it has orbitals of different energies. One very important use of electron-impact molecular-ion appearance potentials or ionization potentials is in checking the correct assignments and interpretations of the far-ultraviolet Rydberg-series spectra of molecules and free radicals [72]. The data are also useful in checking the results of photoionization studies, particularly those which may have been performed without mass analysis. In this regard attention is called to the determination of the ionization potentials of a large number of molecules by the photoionization method, with mass analysis, which have been reported by Watanabe [73]. These data refer almost wholly to the determination of the first ionization potentials of the molecules, though in many cases inner ionization potentials have been obtained also from the photoionization measurement.

Of very great fundamental importance is the determination of as many of the inner ionization potentials as possible for a wide variety of molecules. These can generally be obtained only from studies using the methods described earlier, utilizing monoenergetic electron techniques. These more advanced methods are necessary because the spread in the energy of the electrons in the ordinary electron-impact methods is usually much greater than the separation between the different ionization potentials of a molecule. It is for this reason that photoionization mass-spectrometric methods are regarded as so promising, for the spread in the energy of the ionizing photons can generally be made as little as 0.02 to 0.05 ev.

Furthermore, though a large amount of information is available on the ionization potentials of unsaturated compounds from Rydberg-series studies, similar data are not readily available for saturated compounds, for extensive optical-spectroscopic research of this type has not been carried out. For this sort of molecule the electron-impact studies are of particular importance. It must also be realized that there are very few data in the literature on the absorption spectra of molecular ions, and the mass-spectrometric techniques discussed above can be applied to obtain values for the excited states of a wide variety of polyatomic molecular ions. The methods described in Sec. 12.5 have been success-fully applied to obtain energy values for many of the excited states of the molecular ions N_2^+, O_2^+, HF^+, HCl^+, HBr^+, H_2O^+, H_2S^+, NH_3^+, and others. These results have been used to interpret many details of the electronic structures of the molecules mentioned, and some account of these studies will now be outlined.

Probably one of the first examples of an excited state of a molecular ion being correctly identified by electron impact was provided by Tate and Smith's [74] work on the O_2^+ ion. Smith found a break in the ionization efficiency curve ~ 4.1 ev above the appearance potential of the ground $^2\Pi_g$ state of O_2^+. Mulliken [75] identified this excited state of the O_2^+ ion as being the $^4\Pi_u$ state. In later work, using the RPD method, Frost and McDowell [46] were able to observe several excited states for the O_2^+ ion. All these, except the highest, can be identified with

states of the O_2^+ molecule known from optical-spectroscopic data.
Table 12.2 shows the various states of the O_2^+ ion which were identified.

TABLE 12.2. Adiabatic and Vertical Ionization Potentials and Interatomic
Distances for O_2 and O_2^+

Electronic state of O_2^+ ion	Spectroscopic ionization potential of O_2, ev	Electron-impact ionization potential of O_2, ev	Interatomic distance, A
$^2\Pi_g$	12.16	12.21 ± 0.04	1.1227
$^4\Pi_u$	16.11[78]	16.30 ± 0.03	1.3813
$^2\Pi_u$	16.97	17.18 ± 0.02	1.4089
$^4\Sigma_g$	18.16[78,79]	18.42 ± 0.02	1.2795
?	.....	21.34 ± 0.02	.....
			$1.207(O_2)$

We have already mentioned that certain excited states of the N_2^+ ion
were detected by Clarke [29], using his electron selector source. The
RPD method was first shown by Fox and Hickam [80] and Frost and
McDowell [38] as being useful for the detection of excited states of
molecular ions. Fox and Hickam were able to detect the formation of
the $^2\Pi$ and $^2\Sigma_u^+$ states of N_2^+ and CO^+, as well as, of course, the $^2\Sigma_g^+$
ground states of these ions. They also found evidence for the formation
of the propylene and benzene molecular ions in excited states. Frost
and McDowell were able to detect and measure accurately the appearance
potentials for four states of the N_2^+ ion. The values obtained are given
in Table 12.3.

TABLE 12.3. Adiabatic and Vertical Ionization Potentials of Nitrogen

Electronic state of ion	Spectroscopic, ev	Electron impact, ev
X $^2\Sigma^+$	15.576	15.63
A $^2\Pi_u$	16.71	16.84
B $^2\Sigma_u^+$	18.748	18.76
C $^3\Sigma_u^+$	23.581	23.53

The value given for the first spectroscopic ionization potential depends
on the value chosen for $D(O_2^+)$ in its $^2\Pi_g$ state. A short extrapolation of
the $^2\Pi_g$ state gives $D(O_2^+)^2\Pi_g$ as 6.57[76,77] ev. Hence, from the cycle

$$D(O_2) + I(O) = D(O_2^+) + I(O_2)$$

we get $I(O_2) = 12.16$ ev. The true value for $I(O_2)$ is likely to be nearer
the value of 12.075 ev obtained by Watanabe [73]. This gives $D(O_2^+)^2\Pi_g$

as 6.65 ev. Assuming Watanabe's value for $I(O_2)$ to be correct, then the ionization potential of oxygen referring to the formation of the O_2^+ ion in its $^2\Pi_u$ state is 16.89 ev.

Cloutier and Schiff [36] also detected the presence of N_2^+ ions excited to their $^2\Pi_u$ state and measured the appearance potential of the ion. Dorman, Morrison, and Nicholson [81], by plotting differential ionization efficiency curves, claim to be able to identify the presence of excited states of the O_2^+ and N_2^+ ions. Certainly there are signs of structure on their curves, but the assignments are somewhat forced.

Among other compounds studied by the RPD method, it has been found possible, in the cases of methane and the methyl halides, to detect and measure the appearance potentials of the molecular ions in most of their excited electronic states [83]. It is interesting to note that, in the cases of the molecular ions from methyl bromide and iodide, it has been possible to resolve the components of the ground ^2E states of these ions due to spin-orbital coupling [84]. These spin-orbital splittings were known from the spectroscopic studies of Price [85], and the agreement between these data and those obtained from the electron-impact studies was quite good.

Excited states of the molecular ions of HF, HI, H_2O, H_2S, and NH_3 likewise were detected by the application of the RPD method [86]. All the states detected could readily be interpreted in terms of molecular-orbital theories of the electronic structures of the compounds studied. The case of ammonia is of some interest. The first ionization potential of NH_3 occurs at 10.4 ev, which refers to the production of the NH_3^+ ion in its $^2A_1$ state. A higher ionization potential was found at 15.31 $\pm$ 0.04 ev [87] due to the formation of the NH_3^+ ion in its first excited ^2E electronic state. Earlier, Walker and Weissler [88] has found that the photoionization cross section for the formation of ions from NH_3 rose sharply just above 15 ev. They attributed this increase almost wholly to the process

$$NH_3 + e = NH_2^+ + H + 2e \qquad (12.39)$$

which occurs at about 15.7 ev. However, as we have already mentioned, the occurrence of a second ionization potential for NH_3 at 15.31 ev will lead to an increase in the cross section for the production of ions, and so process (12.39) is obviously not the sole cause of the observed increase.

Electron-impact studies have also led to the discovery of excited states of the aromatic hydrocarbons benzene, naphthalene, anthracene, and phenanthrene [89, 90]. These studies are of importance in that they have enabled quantitative assessments to be made of certain molecular-orbital theories of the electronic structures of these molecules. These data are valuable because, though some spectroscopic values for the first ionization potentials of certain of these compounds are available, a better and more complete test of the theory is possible if the inner ionization potentials of the molecules are known. The comparison between the theoretical and experimental values shows that there is good agreement between these two sets of data [90].

Another interesting feature is that these results came at a time when Lennard-Jones and his colleagues [91] were developing a new orbital theory of the electronic structures of unsaturated compounds. Hall [91] has discussed the case of the ionization potentials of saturated hydro-carbons, but the theoretical treatment of these molecules is largely retarded, because, in general, only the first ionization potentials are known. The application of this theory to unsaturated hydrocarbons can be briefly outlined, and we shall later indicate how it has been extended and applied to molecular structural problems by the use of electron-impact data. In this theory the electronic structure of the molecule is described in terms of orbitals, each of which describes the behavior of one electron. In order that these orbitals may be the best possible ones, the state function must satisfy equations which can be written as

$$(H + V + A)\psi_n = \sum_m E_{mn}\psi_m \qquad (12.40)$$

where H = Hamiltonian for an electron in the field of a bare nucleus
  V, A = coulombic and exchange operators representing the effect of
           remaining electrons
and E_{mn} is defined as

$$E_{mn} = \int \psi_m^*(H + V + A)\psi_n \, d\tau \qquad (12.41)$$

The behavior of the electrons can equally well be described by several different types of orbitals, including what are called *equivalent* orbitals. These orbitals are identical with each other, except for their position and spatial distribution. For equivalent orbitals the quantities E_{mn} are replaced by the terms e_{mn}, which are defined as

$$e_{mn} = \int \chi_m(H + V + A)\chi_n \, d\tau \qquad (12.42)$$

where χ_m and χ_n are any two orbitals from an equivalent set.

In applying this theory to unsaturated hydrocarbons, it is necessary to consider, instead of the ground state, an excited state—"the standard excited state"—in which the π electrons are in singly occupied orbitals with parallel spins. These singly occupied π orbitals can be transformed among themselves into equivalent orbitals localized around a single atom, so that, from the normal equations, the ionization potentials can now be calculated. If the internuclear distances are the same, the ionization potentials and molecular orbitals of the ground state are equal, to a good approximation, to those of the lower members of the set corresponding to the excited state; thus the ionization potentials of the ground state can be found. The application of this form of the theory to unsaturated hydrocarbons leads to a matrix which is of the form

$$
\begin{vmatrix}
e - E & f & g & \cdot & \cdot & \cdot \\
f & e - E & f & g & \cdot & \cdot \\
g & f & e - E & \cdot & \cdot & \cdot \\
\cdot & \cdot & \cdot & & \cdot & \cdot & f \\
\cdot & \cdot & \cdot & & \cdot & f & e - E
\end{vmatrix} = 0 \qquad (12.43)
$$

where $e = e_{mn}$, $f = e_{m,n\pm1}$, $g = e_{m,n\pm2}$, etc.

This method forms the basis of the work which Franklin [92] has described, in which he calculated the ionization potentials of a large number of organic compounds such as paraffins, olefins, and alkyl bromides. The agreement with experimental results is generally quite good. Wacks and Dibeler [90] have applied the above methods and also molecular-orbital calculations to obtain data which they have compared with their experimental results on second-ionization potentials of benzene, naphthalene, anthracene, and phenanthrene, and again good agreement between the two sets of data is found. The equivalent orbital method has also been applied with success to the alkyl hydrocarbon radicals by Stevenson [93] and to the various isomeric propyl and butyl radicals by Lossing, Kebarle, and De Sousa [94]. In view of the rather drastic simplifying assumptions implicit in the method of calculation, among which is the assumption of planarity in the molecule and the ion, the degree of agreement is remarkably good.

12.10. Threshold Laws for the Probability of Ionization by Electron and Photon Impact

In many of the earlier sections of this chapter, particularly in Sec. 12.4 dealing with the experimental methods for determining appearance potentials, and elsewhere, we have commented on the problem of determining the nature of the threshold law for the probability of ionization by both electron and photon impact. A detailed discussion of the theory of ionization processes is given by Massey and Burhop (Ref. 95, chap. 3). Bates, Fundaminsky, Leech, and Massey [96] have given a lengthy and elegant analysis of the theory of excitation and ionization processes and have shown that the Born approximation leads to a threshold-ionization-efficiency law which varies with the first power of the excess energy of the ionizing electron. Another and somewhat different treatment has been given by Wigner [97]. Wigner, in discussing two-particle collisions in general, showed that the nature of the transitions need not be considered and that the probability of a given process near the threshold depends only on the dissociation of the collision complex. Wannier [98] has used these ideas to treat the case of direct ionization of an atom or ion by electron impact. In this treatment, where the dissociating particles separate in a coulombic field, the way in which the excess energy is carried off determines the threshold law. This approach of Wannier yields the result that the probability of ionization just above the threshold is proportional to $\epsilon^{\frac{1}{2}\mu - \frac{1}{4}}$, where ϵ is the excess energy and

$$\mu = \frac{1}{2}\left(\frac{100z - 9}{4z - 1}\right)^{\frac{1}{2}}$$

where z is the charge on the ion. For atoms or molecules when $z = 1$, this law leads to the result that the ionization probability above the threshold is proportional to $\epsilon^{1.127}$. In terms of our earlier discussion, we thus have $p(E) \propto (V - V_c)$.

Geltman [98] has used a modified form of the Born approximation which is a development of the treatment of Wigner [97]. The generalization of Geltman's results shows that the excess-energy dependence of the threshold law for multiple (n-fold) ionization is found to yield the forms $p(E) \propto (V - V_e)^n$ for ionization by electron impact and $p(E) \propto (V - V_c)^{n-1}$ for photoionization. At the time of publication, Geltman compared his theory with the then available experimental results, which, though meager, did in the main substantiate the theory. Later, Geltman [99] derived the energy dependence of the photodetachment cross section near the threshold for diatomic negative ions. The threshold law for such an ion may be written

$$p(E) = V(V - V_c)^{\frac{3}{2}}[A_o + A_1(V - V_c) + A_2(V - V_c)^2 + \cdots] \quad (12.44)$$

where we have written, for the sake of consistency, V for the photon energy, V_c the energy at the onset of ionization, and A_0, A_1, and A_2, the values of special constants. At sufficiently small values of excess energy, only the first term in (12.44) is important. The more general form for $p(E)$ is $pE \propto V(V - V_c)^{x/2}$, where $x = 1$ for atoms and heteronuclear molecules, and 3 and 1 for homonuclear molecules, according to whether they have *gerade* or *ungerade* symmetry. As has been pointed out [81], in its simplest form this expression is similar to the result obtained by Wigner for the separation of two particles with no forces between them. For these he predicted the result that $p(E) \propto (V - V_c)^{\frac{1}{2}}$.

Several workers have studied the nature of the probability for ionization for single and multiple ionization processes. There are, however, few experimental points lying in the small region near the threshold in which the calculations are applicable. Thus there is no really adequate experimental verification of the exact shape of the ionization probability curve in this region. Hickam, Fox, and Kjildaas [100], using the RPD method, obtained a linear ionization probability for the He$^+$ ion down to about 0.2 ev above the threshold.   They also obtained a linear-ionization-probability law for Xe^{2+}, whereas Clarke [29], as we have already mentioned, earlier obtained a quadratic curve initially for Xe^{2+}, which is in agreement with the requirements of Geltman's theory [98].   Fox [100a] has reported studies with He^3 which have shown that, for the formation of the doubly charged ion, the ionization probability is proportional to the square of the excess energy above the threshold.   Studies with xenon were also carried out in which multiply charged ions, possessing up to seven charges, were shown to be quite complex.   All the xenon ions, however, showed an apparent linearity near the initial onset of ionization. The ionization efficiency curves for double ionization by electron impact of Ne, Ar, and Xe have been studied by Morrison and Nicholson [25]. These authors found that, near the threshold, the probability of double ionization varied as the square of the excess electron energy.   In many cases upper states of the ions were detected and the appearance potentials measured.   Later work by Dorman, Morrison, and Nicholson [101], on the ionization efficiency curves for the threefold ionization of Ar and the three-, four-, five-, and sixfold ionization of Xe, indicated that the probability for n-fold ionization varies as the nth power of the excess energy for $n = 3$ and $n = 4$ and possibly also for the cases of $n = 5$ and $n = 6$. Further extensive work by Dorman and Morrison [101a] confirmed and expanded these earlier conclusions.   Of particular interest are their studies on multiple ionization processes in molecules.

12.11. Molecular-structural Aspects of Ionization and Dissociation Processes

Frequent reference has been made throughout this chapter to the need for a more precise understanding of the nature of the ionization processes which occur as the result of electron impact.   In the previous section we discussed the various attempts which have been made to provide an understanding of the physical processes and rearrangements which occur on ionization.   The various theories discussed, with the possible exception of the work of Geltman [99], do not explicitly take account of any factors of molecular or atomic structure.   The same is true of the various experimental studies which have been carried out to determine the nature of the threshold laws for the probabilities of excitation and ionization by electron impact.   Morrison [15–17] has, it is true, made some attempt at indicating the presence of excited molecular ionic species and, in a discussion of threshold laws, attempted to interpret some of the observed data on the hypothesis that many ionization and dissociation processes may involve the prior excitation to a state from which autoionization occurs to yield the ions observed [81].   The main argument is based on an interpretation of the observed shapes of the differential ionization efficiency curves. Though the ideas [97–99] on which Morrison builds are undoubtedly sound, it is still an open question as to whether the experimental evidence [81] is sufficiently impressive as to be wholly convincing. Some of the types of processes to be expected can of course be discerned at certain parts of the experimental curves. Other identifications suggested are perhaps not as well established as would be necessary for their general acceptance. There can, however, be no doubt but that this approach is quite well worthwhile and can surely be further developed to yield much useful information. What has already been published is impressive.

Another approach is concerned with understanding what might be called the spectroscopy of the ionization or dissociation process. This work arises out of attempts to apply spectroscopic and quantum-theoretical ideas about the behavior of electrons, to understand the structure of ions produced by electron (and photon) impact, and to try to use the laws of quantum chemistry to predict the electronic notes and configurations of the ions which may be produced by the dissociation of the parent ion.

The nature of this theoretical treatment can easily be understood by considering a simple case. When a distinct molecular ion is considered, it is often possible to say into which electronic states of the products it may dissociate, for the adiabatic correlation rules for the formation of diatomic molecules from different electronic states of atoms have been given by Wigner and Witmer [102]. These correlation rules can be extended immediately to linear polyatomic molecules and ions. The case of nonlinear polyatomic molecules is more difficult, but group-theoretical methods have been used by Mulliken [103] to classify and correlate the electronic states of polyatomic molecules belonging to a large variety of symmetry groups. These adiabatic correlation rules are readily found by investigating how a state which has the symmetry of the irreducible representation of a particular symmetry group transforms under the operations of another group with different symmetry. For example, in the discussion below, it will be important to know that, when the tetrahedral symmetry (group T_d) of methane is distorted by pulling a hydrogen atom out of position to yield a molecule with symmetry C_{3v}, the molecular orbital with the symmetry of the irreducible representation T_2 of group T_d correlates with two new orbitals of symmetry classes A_1 and E of the group C_{3v}.

In making the correlations between the state of a polyatomic ion and its dissociation products, it is necessary to consider not only the electronic states, but the electronic-vibrational, or vibronic, states.

We shall indicate [105] how this method is applied to give detailed information on the electron-induced dissociation of methane shown in Eq. (12.45).

$$CH_4 + e = CH_3^+ + H + 2e \qquad (12.45)$$

The electronic structure of methane as given by the method of molecular orbitals has been discussed many times [104]. For present purposes it is sufficient to say that the molecule has tetrahedral symmetry, group T_d, and the 10 electrons occupy 5 orbitals. An adequate description of the electronic structure of methane for the present purpose is that given in the equation

$$(1s_c)^2[sa_1]^2[pt_2]^6 \quad {}^1A_1 \qquad (12.46)$$

Calculation shows that the triply degenerate orbit has the lowest ionization potential. The vertical ionization potential of methane is known from electron-impact studies to be 13.1 ev, and this refers to the removal of an electron from the $[pt_2]$ orbital to leave a $CH_4{}^+$ ion with the electronic structure $(1s_c)^2[sa_1]^2[pt_2]^5$, ${}^2T_2$ if the tetrahedral symmetry of the parent molecule is maintained in this ion. This was the assumption made by McDowell in the original theoretical treatment [105]. Later, Liehr [106] pointed out that, because of the degeneracy of the ground state of the $CH_4{}^+$ ion, if it has tetrahedral T_d symmetry, it is necessary to take into account the effect of the Jahn-Teller theorem. This theorem indicates that the ${}^2T_2$ state cannot describe a stable configuration. To determine the equilibrium configuration of the $CH_4{}^+$ ion, it is necessary to consider which nuclear displacements will destroy the assumed tetrahedral symmetry. Liehr shows that the normal vibration $v_3(t_2)$ is most likely to produce the greatest perturbing force.

$CH_3{}^+$, into which $CH_4{}^+$ dissociates, has a nondegenerate electronic state for both C_{3v} and D_{3h} symmetry. It is therefore likely that the parent $CH_4{}^+$ ion will also have C_{3v} geometry in its equilibrium configuration. Thus there is obtained the correlation

$$ {}^2T_2(T_d) \xrightarrow{v_3(a_1)} {}^2A_1(C_{3v}) + {}^2E(C_{3v}) \qquad (12.47)$$

The ${}^2E$ is not a stable state in C_{3v} symmetry and must undergo further Jahn-Teller displacements, finally forming an ion with C_{2v} geometry. Therefore we get the ground state of the $CH_4{}^+$ ion to be $(1sa_1)^2[\pi_e]^4(\sigma a_1)^1$, ${}^2A_1(C_{3v})$.

With a C_{3v} geometry one of the C—H bonds of $CH_4{}^+$ differs from the others. If the molecule is sufficiently excited vibrationally by the $v_3(a_1)$, component of the $v_3(t_2)$ model, this nonequivalent bond will dissociate. This dissociation process leads to a methyl ion of symmetry C_{3v} or D_{3h}. One cannot predict in which of these two geometries the $CH_3{}^+$ ion will be produced. One can only say that the $CH_3{}^+$ ion is formed predominantly in a ${}^1A_1(C_{3v})$ or ${}^1A_1'(D_{3h})$ vibronic state [105].

Similarly, it has been possible to discuss in detail the adiabatic electron correlations for the formation of methyl ions from methyl iodide and methyl cyanide [105]. Laidler [107] has applied similar methods to elucidate many of the details of the electron-induced decomposition of the water molecule. McDowell [108] also applied these methods to simple diatomic molecules such as H_2, N_2, and O_2, as well as to NH_3, HCN, etc.

One important and interesting aspect of this work is the fact that it suggests what excited states of the various molecular ions might be observed by detailed studies of ionization efficiency curves, particularly when monoenergetic sources or photoionization methods are employed. We have already made reference to the detection of the first excited state of the NH_3^+ ion [87]. This state is likely to be the electronic state if the ion is assumed to have symmetry C_{3v}. This is an electronically degenerate state, and so the Jahn-Teller effect will destroy the symmetry by nuclear displacements of symmetry $\alpha_1 + \alpha_2 + \epsilon$. Only nuclear displacements of symmetry ϵ may destroy the equivalence of the NH bonds [106]. Hence we have the correlation

$$^2E(C_{3v}) \xrightarrow{v_2(\alpha_1)} [^2E'(D_{3h})] \xrightarrow{v_3(\alpha_1)} {}^2A_1(C_{2v}) + {}^2B_2(C_{2v}) \tag{12.48}$$

because the NH_2^+ ion into which NH_3^+ dissociates has symmetry C_{2v}. In the C_{2v} geometry, one of the NH bonds in NH_3^+ differs from the other two. If the molecular ion is sufficiently excited vibrationally by the component of the $v_3(\epsilon')$ mode, this nonequivalent bond will dissociate to yield a NH_2^+ ion of symmetry C_{2v}.

In the case of methane, it follows, from Eq. (12.46), that the second ionization potential will refer to the removal of an electron from the $[sa_1]$ orbital, and so must involve the production of a CH_4^+ ion in the excited $^2A_1$ electronic state. This state was detected at 19.42 ev by Frost and McDowell [83]. Now the C^+ ion which is formed from methane by bombardment with electrons with 26.2-ev energy must obviously arise from this $^2A_1$ excited electronic state of the CH_4^+ ion, assuming that it has symmetry T_d. This ion is not orbitally degenerate, and so the Jahn-Teller theorem does not apply. As has already been indicated by McDowell [108], from this state there can arise the following vibronic states: $A \times a_1 = A_1$; $A_1 \times e = E$; $A_1 \times t_2 = T_2$. The ground state of the C^+ ion is the ^2P_u state. It is readily shown [108] that this ^2P_u state of the C^+ ion correlates only with the T_2 vibronic state of CH_4^+. If, therefore, the CH_4^+ ion in the $^2A_1$ electronic, or $^2T_2$ vibronic, state is vibrationally excited in either the $v_3(t_2)$ or $v_4(t_2)$ modes, it can dissociate adiabatically to form the ^2P_u state of C^+. In either of these modes the carbon atom has a definite motion [109]; hence the conclusion of Morrison and Stanton [110], based on a somewhat naïve picture of the dissociation process, is erroneous. It of course follows that their other conclusions (which will be discussed later) concerning the energetics of the process

$$CH_4 + e = C^+(^2P_u) + 4H(^2S_{1/2}) + 2e \tag{12.49}$$

are likewise vitiated.

12.12. Bond-dissociation Energies

The bond-dissociation energy of a molecule is the energy required to dissociate it into known fragments in known states. As we have indicated earlier, the dissociation energy of the molecule XY is usually denoted by the symbol $D(XY)$. Electron-impact studies can lead to values for dissociation energies which often cannot be readily evaluated by other methods. If, for example, an ion X^+ is known to arise by the process

$$XY + e = X^+ + Y \tag{12.50}$$

the appearance potential of the X^+ ion is given by the equation

$$V(X^+) = D(XY) + I(X) + \text{K.E.} + \text{E.E.} \tag{12.51}$$

where $I(X)$ is the ionization potential of the atom X, and K.E. is the excess kinetic energy, and E.E. the excitation energy, of the ion or of the neutral atom Y. Thus, having determined the appearance potential of the ion X by one of the methods outlined in Sec. 12.4, the dissociation energy of the XY bond, that is, $D(XY)$, can be calculated provided both K.E. and E.E. are known or can be measured.

The method used by Hagstrum [39] and described earlier can be used to determine the kinetic energy with which a particular ion is formed. It can also be used to determine the appearance potential which refers to the ion being formed with zero kinetic energy (Sec. 12.6). Other methods are available such as those described by Washburn and Berry [111] and a more recent development used successfully by Taubert [112]. This method is, however, only qualitative, but can be made semiquantitative by calibrating the instrument with ions of known kinetic energy. A simple method which indicates the presence of excess kinetic energy was demonstrated by McDowell and Warren [113], though this is not quantitative. Several authors have, however, used this method to detect the presence of kinetic energy in the fragment ions [114]. Morrison and Stanton [110] have employed a method due to M. G. Inghram, in which ions are subjected to energy analysis in a cylindrical condenser before entering the mass analyzer. These authors claim that an error of only 0.05 ev in kinetic energy can be detected. Fuller details of the methods will be awaited with interest, and provided this preliminary claim is not too extravagant, the method should be most useful. Though many ions formed in electron-induced dissociation processes possess excess kinetic energy, it has been found in a large number of cases that ions are produced with practically near-zero excess kinetic energy. In this connection we may mention an empirical observation known as Stevenson's rule, to which, of course, exceptions are known. It is that, in a dissociation process of the type indicated in Eq. (12.50), the ion which has the smaller ionization potential is formed without excess kinetic energy [119].

For dissociation processes in which the ions are formed without excess kinetic energy, one can use the ionization potentials of free radicals (Chap. 11) to derive reliable values for the dissociation energies of a large number of bonds; that is, always provided that the ion and the neutral fragments are formed in their ground states, which again generally happens to be the case. There are, however, instances where this is known not to be so, and we shall discuss a few of these more difficult problems later. Generally, there are two methods available for using electron-impact data. These have been called the direct and indirect methods. The direct one has been mentioned earlier and is most easily applied when the ion is produced without excess kinetic energy in a known dissociation process; e.g.,

$$CH_4 + e = CH_3^+ + H + 2e \tag{12.52}$$

The appearance potential $V(CH_3^+)$ is known to be 14.39 ev [113], and the ionization potential of the methyl radical, 9.56 ev [115–118]; hence the dissociation energy of the CH_3—H bond in methane $D(CH_3$—H) is $14.39 - 9.86 = 4.53$ ev.

The indirect method uses thermochemical data in conjunction with electron-impact data to derive dissociation energies. This is necessary when there is no value available for the ionization potential of the radical from which the ion is formed. Thus, if it is desired to determine the dissociation energy $D(R_3$-$R_4)$, one first determines the appearance potential of the R_1^+ ion from, say, the molecule R_1R_2 and also from R_1R_3 and assumes that this ion is produced by the following processes:

$$R_1R_2 + e = R_1^+ + R_2 + 2e + V_1 \tag{12.53}$$
$$R_1R_3 + e = R_1^+ + R_3 + 2e + V_2 \tag{12.54}$$

The assumption is also made that *no* excess *kinetic energy* is included in the value of the appearance potentials V_1 and V_2. One can then deduce the relation

$$R_1R_2 - R_1R_3 = R_2 - R_3 + V_1 - V_2 \tag{12.55}$$

If the heats of formation of the compounds R_1R_2, R_3R_4, R_1R_3, and R_2R_4 are known and also the dissociation energy $D(R_2$-$R_4)$, then the values of the heat of reaction (12.56) can be calculated:

$$R_1R_2 + R_3R_4 = R_1R_3 + R_2 + R_4 + \Delta H \tag{12.56}$$

On subtracting Eqs. (12.55) and (12.56), it is easily seen that the process R-R = R + R, that is, $D(R$-$R)$ is given by the equation

$$D(R_3$-$R_4) = \Delta H + V_1 - V_2 \tag{12.57}$$

where

$$\Delta H = \Delta H°(R_1R_3) + \Delta H°(R_2) + \Delta H°(R_4) - \Delta H°(R_1R_2) - \Delta H°(R_3R_4)$$

Here $\Delta H°(XY)$ is the heat of formation of the compound XY in its standard state.

Compounds can always be chosen so that R_2R_4 is H_2, or CH_4, or some other compound for which there is available a reliable value for $D(R_2\text{-}R_4)$. This method has been extensively used by Stevenson [119] to deduce values for the bond-dissociation energies of various C—H and C—C bonds in hydrocarbons. The results he obtained are in good agreement with those estimated in other ways. Now that a large number of values are becoming available for the ionization potentials of free radicals, this method is being less used.

Table 12.3 lists values for some bond-dissociation energies which have been evaluated from electron-impact data. Most of the data have been calculated by the direct method, using the ionization potentials of the free radicals. In all cases the main ion and the other dissociation products are apparently formed in their ground states and without excess kinetic energy.

Several examples are of course known where electron-induced dissociations lead to ions being produced with excess kinetic energy, and also where the ion or the other dissociation products are produced in excited states. Perhaps the best example is the electron-induced dissociation of oxygen. Careful measurements by Hagstrum [39] established that many of the processes leading to the production of O^+ and O^- ions do occur with excess kinetic energy, and also some produce either the O^+ or the O^- ion, or even both, in excited states. Recently the RPD method has been used by Frost and McDowell [46] to demonstrate directly, for the first time, the formation of at least five distinct processes leading to the production of O^+ from molecular oxygen when this is bombarded with electrons with energies between 17 and 23 ev. Subsequent work by Morrison et al. [101] has confirmed the earlier findings of Frost and McDowell [46]. Table 12.4, which is taken from this last reference, indicates the complexities of the dissociation processes occurring in oxygen. It is readily

TABLE 12.4.   Minimum Energies for Ionic Dissociation Processes in Oxygen [46]

Eq.	Process	Products	Minimum energy, ev
1	$O_2(^3\Sigma_g^-) \rightarrow$	$O(^3P_g) + O^-(^2P_u)$	3.67
2	$O_2(^3\Sigma_g^-) \rightarrow$	$O^+(^4S_u) + O^-(^2P_u)$	17.28
3	$O_2(^3\Sigma_g^-) \rightarrow$	$O^+(^4S_u) + O(^3P_g)$	18.73
4	$O_2(^3\Sigma_g^-) \rightarrow$	$O^+(^2D_u) + O^-(^2P_u)$	20.60
5	$O_2(^3\Sigma_g^-) \rightarrow$	$O^+(^4S_u) + O(^1D_g)$	20.70
6	$O_2(^3\Sigma_g^-) \rightarrow$	$O^+(^2D_u) + O(^3P_g)$	22.50
7	$O_2(^3\Sigma_g^-) \rightarrow$	$O^+(^2P_u) + O^-(^2P_u)$	22.30
8	$O_2(^3\Sigma_g^-) \rightarrow$	$O^+(^4S_u) + O(^1S_g)$	22.90

apparent that an extremely refined electron-impact technique would be necessary to distinguish the occurrence of all the eight possible processes indicated as theoretically possible in this table.

The study of the electron-induced dissociation of nitrogen provides another interesting example where a knowledge of the excited states of the products leads to the deduction of the correct value for the dissociation energy of the molecule. Clarke [29] had found evidence that there were several processes leading to the production of N^+ ions. At 2.36 $\pm$ 0.07 ev above the initial appearance potential, a break in the ionization efficiency curve was found. A further break was found at 1.4 $\pm$ 0.1 ev higher. The difference in energy between the $^4S°$ and the $^2D°$ levels of the N atom is 2.383 ev, and the $^2D°$ and $^2P°$ levels are separated by 1.191 ev. Thus it seems that Clarke's work indicates that the N^+ ion is produced in its ground state with the N atom in successive excited states. These results lead to the conclusion that $D(N_2)$ must equal 9.756 ev.

Shortly after Clarke's work, Burns [120] reported the results of an RPD study of the ionic dissociation of N_2. These new results indicated breaks in the ionization efficiency curve at 1.9 and 2.4 ev about the first appearance potential for N^+ ions. The first value was interpreted as forming the nitrogen atom in its ground state $^4S°$ and the nitrogen ion, N^+, in its ^1D state. At the second break, 2.4 ev above the onset, is the process yielding $N(^2D°) + N^+(^3P)$. The detailed interpretation of these results is not wholly satisfactory. Later studies by Frost and McDowell [121], also using the RPD method, showed that there are three distinct processes leading to the production of N^+ ions. These occur at 24.32, 26.66, and 27.93 ev, and it was shown that the processes by which atomic nitrogen ions are produced are those in which the accompanying nitrogen atom is formed in its three states, $^4S°$, $^2D°$, and $^2P°$, the atomic nitrogen being always found in its ground (^3P) state. Since it was known from Hagstrum's [39] work that no nitrogen negative ions are formed and that the atomic nitrogen ions are formed without excess kinetic energy, the above results led to the conclusion that $D(N_2) = 9.756$ ev.

The RPD method has also been used to study the ionization and dissociation of the halogen molecules Cl_2, Br_2, I_2, ICl, and IBr. Measurements of the appearance potentials of both the positive and negative ions enable previous estimates of the bond-dissociation energies and the electron affinities of the halogen atom to be assessed. In these cases it is not possible to evaluate both a dissociation energy and an electron affinity from the same data. It is, however, possible to show that certain values are in agreement with the whole range of electron-impact data, particularly when both electron-capture processes and ion-pair processes are studied [43].

In some cases it is extremely difficult to decide whether or not one of the dissociation products is produced in an excited state.   An example of this kind is provided by the electron-impact studies on cyanogen and its compounds.   Stevenson [119] in his work assumed that, in the dissociation process

$$C_2N_2 + e = CN^+ + CN + 2e \qquad (12.58)$$

the cyanogen radical was produced in its $A\,^2\Pi i$ excited state.   Measurements on the CN^+ ion by the beam-half-width method, however, led McDowell and Warren [113] to conclude that this ion was produced with excess kinetic energy and that both it and the CN radical were formed in their ground states.   On this basis, they decided that the value for $D(CN—CN)$ could be obtained only from other data, as the total energetics of process (12.58) were not known.   By studying the electron-induced dissociation of methyl cyanide according to the equation

$$CH_3CN + e = CH_3^+ + CN + 2e \qquad (12.59)$$

it was possible, from the appearance potential of the CH_3^+ ion, to deduce a value for the heat of formation ΔH_f° of the CN radical.   This datum, together with other thermochemical data, enabled a value of 4.99 ev to be calculated for $D(CN—CN)$.   As a corollary to this value for $D(CN—CN)$, it followed that the ionization potential of the CN radical must be ≤ 15.6 ev.   Kandel [122] showed that the CH^+ ion produced from C_2N_2 was endowed with 0.57 ev excess kinetic energy and, from this, determined the ionization potential of the CN radical as 15.13 ev.   Herron and Dibeler [123] have studied the ionization of $CNBr$, $CNCl$, and CNI. They acknowledge that the experimental data do not make it possible to decide between the two interpretations mentioned above.   Nevertheless, they chose to follow the ideas suggested by Stevenson and obtained the value of 89 ± 2 kcal/mole as the heat of formation ΔH_f° of the CN radical.   This value is in good agreement with that obtained by McDowell and Warren [113] by studying the ionization of methyl cyanide and also by Brewer, Templeton, and Jenkins [124] from their thermochemical studies.   It is obvious from this discussion that these compounds should be restudied and that accurate measurements of the kinetic energies of the ions should be made.   Such data should enable the correct mechanism for the ionic dissociation of cyanogen to be elucidated.

Attention is called to a study of related compounds, namely, hydrozoic acid and methyl azide reported by Franklin, Dibeler, Reese, and Krauss [125].   These authors give a detailed discussion of the various possible dissociation processes and, from their measurements of the appearance potentials of various ions, calculate values for the dissociation energies of several bonds.   In view of the complexity of these dissociation processes, it is difficult to be certain just how accurate are the derived values for dissociation energies given in their paper.

We have just emphasized how difficult it can be to derive values for bond-dissociation energies in an electron-impact process which involves a complex dissociation of the molecular ion. Unfortunately, all too many attempts have been made, and there are numerous examples in the literature—too many to be recorded here—where this sort of thing has been done. Generally, there is no way of knowing if the various processes assumed are, in fact, the correct dissociation processes. Accordingly, all bond-dissociation energies derived from complex dissociation processes should be regarded with reserve unless the authors have given a reasonably detailed discussion of the physical reasons justifying their choice of dissociation mechanism.

An excellent example of the difficulties involved is provided by considering the electron-induced dissociation of methane. The process leading to the production of the CH_3^+ ion is clear, and all workers [119, 113] are agreed that it is

$$CH_4 + e = CH_3^+ + H + 2e \qquad (12.52)$$

It is also apparent that the CH_3^+ is produced in its ground state with little or no excess kinetic energy, and so, as has been pointed out earlier, a reliable value can be deduced for the dissociation energy $D(CH_3—H)$ from the equation

$$V(CH_3^+) = I(CH_3) + D(CH_3—H) \qquad (12.60)$$

since the ionization potential of the methyl radical is accurately known.

At slightly higher electron energies than that required to form the CH_3^+ ion, there appears the CH_2^+ ion. This ion could be formed by either of the processes

$$CH_4 + e = CH_2^+ + H_2 + 2e \qquad (12.61)$$
$$CH_4 + e = CH_2^+ + 2H + 2e \qquad (12.62)$$

Evidence was found [113] for both these processes, and accordingly, the first appearance potential for the CH_2^+ ion at 15.3 ev was assigned to process (12.61). From the available data it was deduced that $D(CH_2\text{-}H)$ $\leq 3.45 \pm 0.2$ ev. The uncertainty here is caused by the lack of knowledge of any excess kinetic energy which the CH_2^+ ion may possess, though the beam-half-width method indicated that the ion possessed little kinetic energy.

The production of the CH^+ ion is possible from either of the two processes

$$CH_4 + e = CH^+ + 3H + 2e \qquad (12.63)$$
$$CH_4 + e = CH^+ + H_2 + H + 2e \qquad (12.64)$$

So far no experimental data are available which would enable a decision to be made between these alternatives. Furthermore, though there is evidence [113] that the CH^+ ion is produced with very little excess kinetic energy, the fact that other lighter fragments occur in the dissociation makes it difficult to say that appearance potential value does not include also a measure of some kinetic energy contribution. On the reasonable assumption that (12.63) represents the dissociation process, it is possible to derive the value of $D(CH - H) \leq 3.4 \pm 0.3$ ev. This is, however, based on a value for the ionization potential of the CH radical of 11.1 ± 0.2 ev [126], which may not be highly accurate.

Though the above processes indicate how difficult is the detailed interpretation of some multiple-dissociation processes, the most difficult, and in some ways the most important, processes to consider in the electron-induced dissociation of methane are those leading to the production of the C^+ ion.

It is apparent that the C^+ ion can be formed in the following ways as the result of the electron bombardment of methane:

$$CH_4 + e = C^+ + 4H + 2e \qquad (12.65)$$
$$CH_4 + e = C^+ + 2H_2 + 2e \qquad (12.66)$$
$$CH_4 + e = C^+ + H_2 + 2H + 2e \qquad (12.67)$$

The ionization efficiency curves do not allow a decision to be made as to which of these processes is the one leading to the production of C^+ ions at the lowest observed appearance potential of 26.2 ev. Most workers have assumed that the process represented by Eq. (12.65) is the one taking place at 26.2 ev [113, 127]. It is then possible to derive a value of $D(C—H)$ of ≤ 3.6 ev.

The importance of assigning correctly the mechanism for the production of C^+ ions is that this of course could lead to a knowledge of the heat of atomization of methane, and from this one can calculate the latent heat of vaporization of carbon. This is an important fundamental datum, about which there has been much discussion [113, 128–131]. The relationships between the latent heat of vaporization of carbon and the dissociation energies for the various bonds in methane are best seen as follows. If we denote by a, b, c, and d the respective dissociation energies $D(CH_3—H)$, $D(CH_2—H)$, $D(CH—H)$, and $D(C—H)$, then the following relation holds between these quantities of the latent heat of vaporization of carbon L_1 at 25°C:

$$L_1 = a + b + c + d - 226.1 \quad \text{kcal} \qquad (12.68)$$

When the above-mentioned values for a, b, c, and d were used in Eq. (12.68), it followed *on the assumptions* that (1) no excess-kinetic-energy terms were neglected, and (2) the assumed dissociation mechanisms were correct, the value of $L_1 \sim 120$ kcal/mole. This was obviously a *minimum* value.   Direct experiment had indicated, however, that $L_1 \sim 170$ kcal mole [132, 133].   Later and more accurate studies [134, 135] of the vaporization of carbon have established the higher value without any shadow of doubt.   In some recent electron-impact studies it has been claimed that they lead to this value of $L_1 = 7.386$ ev.   However, as these involve studies [136] on the electron-induced dissociations of CF_4, CCl_4, and CBr_4, the same comments made about the ionic dissociation of methane apply.   All CX_4 molecules will be subjected to the same ambiguities of interpretation.   Furthermore, though Reed and Snedden discuss the various possible dissociation mechanisms, they arbitrarily assume that the C^+ ion is produced by the same type of mechanism as outlined in Eq. (12.67), namely,

$$CX_4 + e = C^+ + X_2 + 2X + e \qquad (12.69)$$

Similarly, little compelling evidence is produced for their choice of mechanisms for a corresponding study on the electron-induced dissociation of the compounds CH_3X, where X = H, Br, Cl, CN, I, and OH and also CH_2Cl_2 and $CHCl_3$[137].   Their work is certainly of interest, but as we have already pointed out, little reliance can be placed on the assumed mechanism of such complex dissociation processes unless the choices made are supported by much reliable corroborative evidence. Though they can, undoubtedly, interpret their results in a satisfactory manner by assuming that $L_1 = 7.386$ ev, their work by no means provides *proof* that this value is correct, though this is known to be the case from other work [134, 135], as we have mentioned above.

The situation regarding the electron-induced dissociation of methane and all compounds of the CX_4 type in so far as studies on the appearance potential of the C^+ ion are concerned may not be very relevant to the problem of evaluating the latent heat of vaporization of carbon, L_1, for a number of important reasons.   We have already shown, in Sec. 12.11, that it is possible to obtain information on the detailed molecular mechanisms of various electron-induced ionization and dissociation processes by the application of adiabatic-correlation theories.   In the particular case of the production of the C^+ ion from methane, it has already been pointed out that this ion, in its ground state ^2P_u, correlates with the ^2T vibronic state of the excited ^2A state of the CH_4^+ ion if it is assumed that the parent ion has tetrahedral, or T_d, symmetry.   We have also indicated that the vibrationally excited CH_4^+ ion may decompose into, say, 4H atoms by exciting one of the fundamental vibrations of t_2 symmetry, namely, v_3 or v_4.   Should it vibrate with the v_3 mode excited, the C^+ ion formed will undoubtedly have a component of the excess kinetic energy liberated in the process.   That this is so has been found by Morrison and Stanton [110] by study of the C^+ ions formed by means of an energy selector.   Figure 12.21 shows the results they obtained.

Fig. 12.21 shows the second-derivative ionization efficiency curves for the Ne^+ and the C^+ ions. It is at once apparent that C^+ is formed with excess kinetic energy, and Morrison and Stanton measure the amount of excess kinetic energy to be 0.15 ev. Because they thus determine that this C^+ ion is produced with excess kinetic energy, it is assumed by Morrison and Stanton that what they call the "symmetrical" dissociation of methane,

$$CH_4 + e = C^+ + 4H + 2e \qquad (12.65)$$

cannot occur. As we have already pointed out, in Sec. 12.11, the vibration need not necessarily be one leaving the C^+ at rest! Furthermore, there is no reason to suppose that the CH_4^+ parent ion is in fact tetrahedral with symmetry T_d. It equally well may have C_{3v} symmetry,

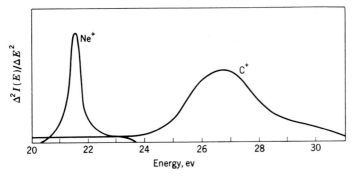

Fig. 12.21. Curves of $\Delta^2 I/\Delta V^2$, the second derivative of the ionization efficiency, as a function of the electron energy for Ne^+ and the C^+ ion from methane. (*From Morrison and Stanton* [110].)

and in this case, undoubtedly, the CH_4^+ ion could dissociate to produce a C^+ ion with excess kinetic energy. It must also be realized that, in a complex dissociation process of the type indicated in Eq. (12.65), if the C^+ ion is produced with kinetic energy, this is only a *component* of the total kinetic energy liberated in the whole process. Not only will the lighter fragments carry off most of the excess kinetic energy, but it will never be possible to compute the total kinetic energy liberated in the dissociation unless a complete knowledge is available of the kinetic energy and direction of motion of *all* the fragments liberated in the dissociative act. No satisfactory methods are at present available for the determination of the kinetic energies and directions of motion of neutral fragments released in a dissociation of the type indicated by Eq. (12.65), and so at present this problem seems insoluble. It follows, therefore, that process (12.65) may, in fact, be the one leading to the production of C^+ ions from methane when it is bombarded with electrons with 26.2 ev of energy. The C^+ ion is, undoubtedly, produced with excess kinetic energy [110], and the *total* kinetic energy involved in the process could be large enough to make the electron-impact results lead to the correct value of L_1, namely, 7.386 ev.

Because of the great difficulties thus encountered in deriving bond-dissociation energies from electron-impact data on processes involving complex dissociations, we shall not discuss these in any detail. Our aim is to outline the principles of the method and to discuss a few of the more significant results. In a later section we shall follow the practice introduced by Field and Franklin [130] and discuss heats of formation of molecular ions rather than bond-dissociation energies as such. It must, however, be pointed out that, where values for heats of dissociation of molecular ions are derived from complex dissociation processes, the same criticisms as have been made about bond-dissociation energies also apply.

In Table 12.5 we list values for some bond-dissociation energies which have come from electron-impact studies where the dissociation process is known with a high degree of certainty. Usually, the processes from which these data arise are relatively simple ones, and in all cases considerable care has been taken to establish the mechanism. Furthermore,

TABLE 12.5.   Some Recent Electron-impact Values for Bond-dissociation Energies

Bond	Dissociation energy, ev	Reference
N_2	9.756	121
NO	6.34	138
CH_3—H	4.42	
C_2H_5—H	4.25	139
sec-C_3H_7—H	4.22	139
C_2H_3—H	4.55	140
C_2H_3—CH_3	4.01	140
C_2H_3—C_2H_3	4.51	140
C_2H_2—H	1.78	140
C_3H_3—H (allene)	3.52	141
C_3H_3—H (propyne)	3.59	141
$CH_3 \cdot C_3H_3$ (1,2-butadiene)	2.96	141
$CH_3 \cdot C_3H_3$ (1-butyne)	2.93	141
CH_3—I	2.28	142
CH_3CN	4.52	113
sec-C_3H_7—I	1.84	139
sec-C_3H_7—Cl	3.18	139
sec-C_3H_7—Br	2.55	139
$CH \equiv C \cdot CH_2$—Br	2.51	139
$CH \equiv C \cdot CH_2$—I	1.98	139
C_2H_3—Cl	3.71	140
C_2H_3—Br	3.18	140

many of the values have been calculated by the direct method and use known values for the ionization potentials of the cognate free radical or use values that have been determined specifically in the researches quoted. It is further pointed out that this list is by no means all-inclusive; it represents only a selection of the recent data.

12.13. Heats of Formation of Atomic and Molecular Ions

As we have stressed above, the dissociation energy of the bond R_1R_2 can be calculated if the appearance potential of the R_1^+ ion in the process is available,

$$R_1R_2 + e = R_1^+ + R_2 + 2e \qquad (12.70)$$

and either it is known that the ion is formed without excess kinetic energy or the amount of kinetic energy is known. The appearance potential of the R_1 ion $V(R_1^+)$ for process (12.71) is related to the dissociation energy of the R_1R_2 bond by the equation

$$V(R_1^+) = D(R_1\text{-}R_2) + I(R_1) \qquad (12.71)$$

and also, following (12.57), the appearance potential of the R_1^+ ion $V(R_1^+)$ is related to the heats of formation of the compound R_1R_2, etc., by the equation

$$V(R_1^+) = \Delta H_f^\circ(R_1^+) + \Delta H_f^\circ(R_2) - \Delta H_f^\circ(R_1R_2) = \Delta H \text{ reaction} \quad (12.72)$$

In this equation $\Delta H_f^\circ(R_1^+)$ is the standard heat of formation of the R_1^+ ion, $\Delta H_f^\circ(R_2)$ is the heat of formation of the radical R_2, and $\Delta H_f^\circ(R_1R_2)$ is the standard heat of formation of the compound R_1R_2. The heat of formation of the radical R_2 can be calculated from the equation

$$D(R_1 - R_2) = \Delta H_f^\circ(R_1) + \Delta H_f^\circ(R_2) - \Delta H_f^\circ(R_1R_2) \qquad (12.73)$$

and the heat of formation of the ion R_1^+, $\Delta H_f^\circ(R_1^+)$, can be calculated from the equation

$$H(R_1^+) = I(R_1) - \Delta H_f^\circ(R_1) \qquad (12.74)$$

The compilation of tables of heats of formation of various ions by Field and Franklin [130] has been useful in that electron-impact and thermochemical data in that convenient form can often be used directly in calculations of chemical interest. Further, they have shown that, from the data, certain relationships became apparent which are not readily seen in other ways and that simple methods of predicting energy values can be devised. Field and Franklin show that results from a large number of methyl derivatives lead to the conclusion that

$$\Delta H_f^\circ(CH_3^+) = 265 \text{ kcal/mole}$$

and $\Delta H_f^\circ(C_2H_5^+)$, evaluated from data on ethyl derivatives, is about 225 kcal/mole. In other cases, as for example with the $C_3H_7^+$ ion, it soon becomes apparent that this ion is formed in different isomeric structures from different compounds, as the heat of formation is not constant. Field and Franklin show also how values of the heats of formation of molecular ions may be used to evaluate activation energies for decomposition processes whereby the ions yield simpler ionic fragments. This is done by calculating the endothermicity of the reactions from the equation

$$\Delta H_R = \Delta H_f^\circ(Y^+) + \Delta H_f^\circ(Z) - \Delta H_f^\circ(X^+) \qquad (12.75)$$

for the process

$$X^+ = Y^+ + Z \qquad (12.76)$$

The activation energy $\Delta H^\ddagger$ is given by

$$\Delta H^\ddagger = \Delta H_{obs} - \Delta H_R \qquad (12.77)$$

It is found that, even for complex ionic decomposition reactions involving the simultaneous breaking and making of several bonds, in many cases little or no activation energy, calculated as indicated above, is involved.

Bernecker and Long [143] have more recently prepared an extensive tabulation of the heats of formation of many organic positive ions and their parent radicals and molecules. Such data are of great value in the interpretation of the mass spectra resulting from electron-induced dissociations of complex molecules. They are also necessary in applying reaction-rate theories to the data [3, 4].

12.14. Electron Affinities by Mass-spectrometric Methods

The electron affinity of an atom is defined as the energy difference between the negative ion and the neutral atom in their ground states. When negative ions are formed in electron-induced dissociation processes, e.g.,

$$XY + e = X^+ + Y^- + e \qquad (12.27)$$
$$XY + e = Y^- + X \qquad (12.25)$$

the appearance potentials for the negative ion (Y^-) formed by these two processes are given by either of the two equations

$$V(Y^-) = D(XY) + I(X) - EA(Y) + \text{K.E.} + \text{E.E.} \qquad (12.28)$$
$$V(Y^-) = D(XY) - EA(Y) + \text{K.E.} + \text{E.E.} \qquad (12.26)$$

In these equations the items have the same meanings as given earlier.

Therefore, if in an experiment one can determine the appearance potential of the negative ion $V(Y^-)$ and also measure the kinetic energy and any excitation energies of the ion or the products, one can compute the electron affinity of the atom Y. It is to be noted that there are theoretical reasons for believing that negative ions are always formed in their ground states. Furthermore, it is generally also the case that the other products are also formed in their ground states. This is, of course, an indirect method for determining electron affinities, but we shall describe it first, and later we shall discuss a direct mass-spectrometric method.

The indirect method for determining electron affinities has been developed most highly by Hagstrum [39]. An account of his work was given in Sec. 12.6. Hagstrum's main efforts were directed toward determining the electron affinity of the oxygen atom, but the value he obtained, namely, 2.2 ev, was too high, mainly because of the lack of a sufficiently accurate method for calibrating the negative-ion electron energy scale. It must be mentioned, however, that there were other errors in interpretation which also contributed to this erroneous result. The method he developed is, however, sound in principle, and with the use of sulfur hexafluoride to calibrate the electron energy scale and an RPD ion source, it should be possible to obtain quite accurate values for the electron affinities of a number of atoms. Using an apparatus with modifications of this type, Randolph and Geballe [54] have found a value of 1.48 ± 0.05 ev for the electron affinity of the oxygen atom. This is in very good agreement with the results of photodetachment measurements by Branscomb, Burch, Smith, and Geltman [144], who found $EA(O)$ equal to 1.465 ev. Mention has already been made of the use of a Lozier apparatus by Craggs and his coworkers [50–52]. Though the earlier results obtained by these authors for the electron affinity of oxygen were in agreement with the widely accepted value of Branscomb et al., their later studies [52] are in better agreement with this very precise work.

It may be mentioned here that this value of 1.45 ev for the electron affinity of the oxygen atom is the only one which is consistent with recent electron-impact studies on the formation of O^- from NO by an electron resonance-capture process as in

$$NO + e = N + O^- \qquad (12.78)$$

The appearance potential for the formation of the O^- ion given by

$$V(O^-) = D(NO) - EA(O) + K.E. + E.E. \qquad (12.79)$$

was found experimentally to be 6.99 ev. Hagstrum [39] had earlier shown that these fragments are formed with 2.1 ev total excess kinetic energy. Substituting these data in Eq. (12.79) with the now accepted value for $D(NO)$ indicates that the Branscomb, Smith, and Burch value for $EA(O)$ must be correct.

An interesting example of the application of the indirect method of determining electron affinities is found in the results of Frost and McDowell [43] on the study of the formation of ion-pair processes in bromine and iodine, to which reference has already been made (Sec. 12.6). It was found by these authors that the process

$$I_2 + e = I^+ + I^-$$ (12.29)

occurs at 8.62 ev. It was assumed that the I^+ ion is formed in its ground ($^3P_2$) state and the I^- ion in its ground (^1S) state. Hence, since

$$D(I_2) = 1.542 \text{ ev}$$

$I(I) = 10.45$, there results a value for the electron affinity of the iodine atom of 3.37 ev, assuming that the products in process (12.29) are formed without excess kinetic energy. More recently, Morrison, Hurzeler, Inghram, and Stanton [44], in studying the photoionization of I_2 in a mass spectrometer, observed the formation of I^+ ions at a photon energy corresponding to 8.95 ev. They assumed that these ions were produced by the process

$$I_2 + h\nu = I^+ + I^-$$ (12.80)

and state that the products are produced with 0.2 ev excess kinetic energy. From these data they deduced a value of 3.14 ev for the electron affinity of the iodine atom. Both these values are in good agreement with those derived from other sources. It should also be pointed out that similarly good agreement is found for the values of the electron affinity of the bromine atom deduced from the experimental results of these two sets of workers.

The most satisfactory way to determine electron affinities is by the *direct surface-ionization method*. This method employs a mass spectrometer to measure the currents of ions emitted from a hot tungsten filament placed in a diffuse beam of alkali halide molecules. Essentially, all the alkali atoms leave the filament, as positive-ion current gives the neutral X atom current. The partial pressures P_i can be calculated from the numbers Z_i of the species emitted from the hot filament at temperature $T°K$, using the equation

$$P_i = Z_i(2\pi m_i kT)^{1/2}$$ (12.81)

Since the electron affinity of an atom X is defined as $-\Delta E°$, at $0°K$, for the reaction

$$X + e^- = X^-$$ (12.82)

it may be determined by measuring the equilibrium constant K for the reaction (12.82). The equilibrium constant K_p is related to the change in free energy in reaction (12.82) by Eq. (12.83).

$$\ln K_p = \ln \frac{P_{X^-}}{P_X P_{e^-}} = -\frac{\Delta F°}{RT_f}$$ (12.83)

and $$RT_f \ln K_p = -\Delta E° + F_{e^-}^{(0)} + F_X^{(0)} - F_{X^-}^{(0)}$$ (12.84) where $F_{e^-}^{(0)}$, $F_X^{(0)}$, and $F_{X^-}^{(0)}$ are the free energies of the electron, the X atom, and the X^- ion, chosen so that $F_i^{(0)} = 0$ at $T = 0°K$ and $P = 1$ atm.

As it is easy to measure the number of M^+ ions produced, and the electron current and the X^- ion current can be measured separately, K_P can be calculated for reaction (12.82). This method has been used extensively by Yonov and his collaborators [145, 146] and by Bailey [147]. The theory of the method is quite straightforward, and we shall give an outline of it below.

As the filament temperatures are quite high, $\sim 2000°K$, and as only low gas pressures are used, it is obvious that only atomic ions will be formed on the hot filament. The mass balance

$$Z_{M^+} + Z_M = Z_{X^-} + Z_X \tag{12.85}$$

holds, where Z_i is the number of species i emitted from unit area of the filament in unit time. Under the experimental conditions used, $Z_X \gg Z_{X^-}$. It is also readily shown, from considering the equilibrium $M = M^+ + e^-$, that

$$\frac{Z_{M^+}}{Z_M} = r = \tfrac{1}{2} \exp\left(\frac{\phi - I_M}{kT_f}\right) \tag{12.86}$$

where ϕ = work function of filament metal
I_M = ionization potential of atom M
Richardson's equation gives the electron emission as

$$I(e^-) = 120.1 T^2 \exp\left(-\frac{\phi}{kT}\right) \tag{12.87}$$

and so ϕ and r may be calculated from the electron current, and measurements of the filament temperature by an optical pyrometer. It is here that Yonov and Bailey differ. Yonov assumed that the work function of the surface was that of pure tungsten and that the total positive and ion currents observed were caused solely by M^+ and X^- ions formed on the filament in equilibrium amounts.

Since $Z_X \gg Z_{X^-}$, Eqs. (12.86) and (12.87) give $Z_X = Z_{M^+}(r + 1/r)$. Z_{e^-} is obtained from the electron current $I(e^-)$ in amperes per square centimeter of emitting surface by use of the relation

$$Z_{e^-} = I(e^-) \frac{N_0}{\mathfrak{F}} \tag{12.88}$$

where N_0 = Avogadro's number
$\mathfrak{F}$ = Faraday equivalent
The partial pressure of the electron gas in equilibrium with the number Z_e of electrons emitted is, from Eq. (12.81),

$$P_{e^-} = \frac{I(e^-)N_0}{\mathfrak{F}} (2\pi m_e kT_f)^{1/2} \tag{12.89}$$

Thus the expression for the equilibrium constant K_p becomes

$$K_p = \frac{Z_{X^-}}{Z_X P_{e^-}} = \frac{Z_{X^-}}{Z_{M^+} I(e^-)} \left(\frac{r}{r} + 1 \right) \frac{\mathfrak{F}}{N_0} (2\pi m_e \text{-} kT)^{\frac{1}{2}} \qquad (12.90)$$

$F_{e^-}^{(0)}$, $F_X^{(0)}$, and $F_{X^-}^{(0)}$ are given by

$$F_i^{(0)} = RT_f \left\{ \ln \left[\frac{h^2}{(2\pi m_i k T_f)^{\frac{3}{2}}} \frac{1}{kT_f} \right] - \ln Q_i \right\} \qquad (12.91)$$

where Q_i is the internal partition function of the species i. Substituting from Eqs. (12.91) and (12.90) into (12.83), the expression for the electron affinity $EA(X)$ becomes

$$EA(X) = RT_f \ln \left\{ \left[\frac{Z_{X^-}}{Z_{M^+} I(e^-)} \right] \left(\frac{r}{r} + 1 \right) \frac{Q_e Q_x}{Q_{x^-}} \left(2\pi m_{e^-} \frac{k^2 T_f^2}{N_0 k^3} \right) \right\} \qquad (12.92)$$

Experimentally, Z_{X^-}/Z_{M^+} is measured as the ratios of the corresponding ion currents observed in a mass spectrometer. $I(e^-)$ is determined by direct measurement of the total temperature-limited negative current leaving the filament and the surface area of the filament; and the temperature of the filament is measured, as has been mentioned before, with an optical pyrometer.

Bakalina and Yonov [146] have used a modification of this method with mixed alkali halides, and so derived values for the differences between the electron affinities of various atoms. Because the work function may be different for different areas of the filament on which surface ionization occurs, Yonov has modified the original method. The use of a mixed beam of alkali halides enables the difference in the electron affinities to be determined by measuring the ratio of the negative-ion currents Z_{Y^-}/Z_{X^-}, the filament temperature, and Z_{M^+}, the alkali metal-ion current. In the experiments, crossed molecular beams of the alkali salts are used and the intensities of these are controlled so that the M^+ ion current measured by the mass spectrometer is equal for each beam. Under these conditions $Z_{M^+} = Z_{X^-}$. Periodic checks are made of the constancy of the atomic beams by measuring their M^+ ion currents. The ratio Z_{Y^+}/Z_{X^-} is measured at various temperatures, and the electron affinity difference between atoms Y and X calculated. As this difference is to be expected, this is generally found to be independent of temperature. Bakalina and Yonov [148] determined the difference between the electron affinities of bromine and sulfur to be 1.23 ± 0.05 ev by studying the surface ionization of Br and S atoms in a mixed stream of the vapors of sodium sulfide and sodium bromide. Taking $EA(\text{Br}) = 3.6$ ev, they deduced the electron affinity of sulfur to be 2.37 ev. In later experiments [149] they determined the differences between the electron affinities of all the halogen atoms, and between the sulfur atom and the CN radical. The values obtained for the differences between the electron affinities of the halogens are given in Table 12.6.

Table 12.6.   Differences between the Electron Affinities of the Halogen Atoms [148]

Halogen pair	Electron-affinity difference, ev
Cl—Br	0.25 ± 0.06
Cl—I	0.53 ± 0.03
Cl—F	0.20 ± 0.03
Br—I	0.27 ± 0.02
F—Br	0.02 ± 0.02
F—I	0.29 ± 0.04

Taking the value of 3.56 ev [150] for the electron affinity of the bromine atom as a standard, Bakalina and Yonov give, as the values for the electron affinities of the other halogen elements,

$$EA(\text{F}) = 3.58 \text{ ev}, \qquad EA(\text{Cl}) = 3.81 \text{ ev}, \qquad \text{and} \qquad EA(\text{I}) = 3.29 \text{ ev}$$

For the CN radical they give the value

$$EA(\text{CN}) = 3.7 \text{ ev}$$

These may be compared with the values obtained by Bailey [147], using a similar method but calculated from measurements on the vapor of one halogen salt, i.e., from determinations of the ratio Z_{X^-}/Z_{M^+}. Bailey's values are $EA(\text{F}) = 3.53$ ev (mean of 3.60 and 3.46 ev), $EA(\text{Cl}) = 3.76$ ev, and $EA(\text{Br}) = 3.51$ ev.

A modification of the above method has been used by Bailey [147], which he has called the *electron-affinity-difference method*. This is in some respects similar to the method used by Yonov, but instead of measuring the ratio of the negative-ion currents Z_{Y^-}/Z_{X^-} from two atomic beams of different halide salts to one alkali metal, say, sodium, Bailey used interhalogen compounds such as ClF_3 and IBr. When the hot filament is surrounded with interhalogen gas $X_m Y_n$, the equilibrium between the species emitted from the surface will be

$$X + Y^- = X^- + Y \tag{12.93}$$

For this equilibrium we have

$$-\frac{\Delta F^{(\text{o})}}{RT_f} = \ln K_p = \ln \left(\frac{Z_X \cdot Z_Y}{Z_Y \cdot Z_X} \right) \tag{12.94}$$

and

$$-\Delta F^{(\text{o})} = EA(\text{X}) - EA(\text{Y}) + (F_X^{(\text{o})} - F_{X^-}^{(\text{o})}) - (F_Y^{(\text{o})} - F_{Y^-}^{(\text{o})}) \tag{12.95}$$

Substituting from (12.91) into (12.94) and (12.95), we get the following expression for the difference of the electron affinities of the X and Y atoms:

$$EA(X) - EA(Y) = RT_f \ln \left(\frac{Z_X - Z_Y Q_Y - Q_X}{Z_Y - Z_X Q_Y Q_X -} \right) \qquad (12.96)$$

At sufficiently high temperatures there is complete dissociation of the $X_m Y_n$ molecules on striking the hot filament surface, and the ratio Z_Y/Z_X is simply n/m. Equation (12.96) then becomes

$$EA(X) - EA(Y) = RT_f \ln \left(\frac{n}{m} \frac{Z_X - Q_X Q_Y -}{Z_Y - Q_Y Q_X -} \right) \qquad (12.97)$$

Thus the determination by mass-spectrometric measurements of the ratio Z_X-/Z_Y- of the ions emitted from the filament at temperature T enables the difference in the electron affinities of the atoms X and Y, that is, $EA(X) - EA(Y)$, to be calculated. An underlying assumption is that the accommodation coefficients for electron exchange between the filament surface and the X and Y atoms on or very near the surface are all unity. This is a reasonable assumption which is likely to be correct. Using this method with ClF_3 and IBr, Bailey [147] obtained the results that the difference between the electron affinities of chlorine and fluorine was $EA(Cl) - EA(F) = 0.238$ ev and the difference between the electron affinities of bromine and iodine was $EA(Br) - EA(I) = 0.333$ ev.

12.15. Meta-stable Ions

Our discussion of the molecular ions so far has been restricted wholly to ions of integral mass number. In the mass spectra of large molecules it has been found that frequently small diffuse peaks are observed, usually at nonintegral mass numbers. These were recognized by Hipple, Fox, and Condon [151] as being due to the decomposition of ions in the field-free region between the source and the magnetic analyzer. It is generally found that the intensity of these ions varies linearly with the pressure of the original sample.

The origin of these meta-stable ions can be understood from the following elementary analysis. It is assumed that the ion of mass m_1 passes through a potential difference V_1 before undergoing a meta-stable transition into the ion of mass m_2. This resulting ion of mass m_2 then passes through a region with an accelerating voltage of $V - V_1$, where V is the original total accelerating voltage; and it enters the field-free region between the source and the magnetic analyzer. The meta-stable transition is assumed to be almost thermally neutral, and the unchanged fragment resulting from the dissociation is assumed to move with the same velocity as the ion of mass m_2. This latter ion will enter the magnetic

The structure of the $C_6H_7^+$ ion was shown by labeling experiments not to be the benzenium ion. The indicated sequence of steps leading to the ormation of the $C_6H_5^+$ ion is

$$\langle \bigcirc \rangle\text{-}CH_2\text{·}CH_3 \xrightarrow[-e]{-H} \langle \bigcirc \rangle\text{-}\overset{+}{C}H\text{·}CH_3 \xrightarrow{-C_2H_2} C_6H_7^+ \xrightarrow{-H_2} C_6H_5^+ \qquad (12.106)$$

$$106 \qquad\qquad\qquad\qquad 105$$

i.e., ring rearrangement is an important process.

The $C_8H_7^+$ are not all formed by the loss of a hydrogen atom from the α-carbon. Some are apparently formed by the loss of a hydrogen atom from the benzene ring to form the methyltropylium ion [157], as indicated:

$$\langle \bigcirc \rangle\text{-}CH_3 + e \longrightarrow \langle \bigcirc^+ \rangle\text{-}CH_3 + H + 2e \qquad (12.107)$$

Further work [158] on toluene-α-C^{13} and cycloheptatriene have confirmed that the $C_7H_7^+$ ion from toluene is not benzyl, but tropylium.

The formation of the $C_5H_5^+$ ion from a precursor other than the tropylium $C_7H_7^+$ is shown [158] by the occurrence of a meta-stable peak at 44.5 in the spectrum of toluene-α-d_3. This peak is due to the process

$$C_6H_5\text{-}CD_3^+ = C_5H_5^+ + C_2D_3 \qquad (12.108)$$

$$95 \qquad\quad 65 \qquad 30$$

Ionic rearrangements often occur which do not at the same time lead to a manifestation of meta-stable ions in mass spectra. We shall discuss some of these cases later, as they are of considerable chemical interest. They are also of some importance in molecular-structural interpretations of mass-spectral data, for one has always to keep in mind that the ions observed may not necessarily have the same structural identity or groups as the parent molecule.

It has been observed that certain meta-stable peaks correspond to the formation of an ion of a particular structure, and also to the breaking up of an ion of this empirical formula. Beynon et al. [159] have shown that, in the mass spectrum of anthraquinone ($C_{14}H_8O_2$) of molecular weight 208, meta-stable peaks are observed at masses 155.8 and 128.3, corresponding to the meta-stable transitions $208^+ \rightarrow 180^+ + 28$ and $180^+ \rightarrow 152^+ + 28$. Accurate mass measurements by high-resolution mass spectrometers enable the atomic fragments lost in the transition to be identified. These measurements show that the formulas of the ions of masses 180 and 152 are $C_{13}H_8O$ and $C_{12}H_8$, respectively. Thus the loss of mass 28 corresponds to the elimination of CO. It is assumed that the carbon atoms lost are those attached to the ketonic oxygens, and a two-stage fragmentation would seem necessary to preserve the final ion as a single entity.   This process can be represented as proceeding through a meta-stable ion as follows:

$$(12.109)$$

Anthraquinone$^+$ Fluorenone$^+$ Diphenylene$^+$

It is often important to identify elimination reactions of this type, which frequently occur consequent upon a meta-stable transition.   Of course, it could be that such elimination processes are only part of an overall change, of which the meta-stable ions observed are the external manifestation.   A good example is provided by the work of Meyerson et al. on the formation of the tropylium ion and the $C_5H_5^+$ ion [157] from alkyl benzenes, and also from benzyl chloride and benzyl alcohol [160]. Similar work on labeled p-xylenes has shown that the resulting ions do not have the tolyl structure, and the spectra indicate drastic rearrangement before dissociation [161].   Meyerson and his coworkers have also found [162] that the three C_7H_8 isomers toluene, cycloheptatriene-1,3,5, and spiro-2,4-heptadiene-1,3, dissociate under electron impact to yield a common $C_7H_7^+$ species which they identify as the tropylium ion.   Furthermore, they present evidence which indicates that the data on toluene and cycloheptatriene decompose to this common product from a common excited state, $C_7H_8^+$, of the molecular ion.

Many examples of molecular rearrangements occurring in large organic molecular ions are known from the extensive study of the mass spectra of a vast number of compounds.   These are too numerous to mention here in detail.   Extensive reviews of this aspect of mass-spectral data have been given by McLafferty [163, 164], Beynon [159, 165, 166], and Biemann [219].

12.16. Quasi-equilibrium Theory of Mass Spectra of Large Molecules

In the early sections of this chapter, we discussed in some detail how the ionization and dissociation of small molecules (mainly those with fewer than five atoms) could be understood in terms of the known energy levels of such molecular systems.   When the number of atoms in a molecule becomes large, the spacings between the energy levels become less and less, and in fact there is a tendency for the electronic energy levels to become grouped into bands.   When this occurs, it is no longer possible to use the spectroscopic methods and ideas in discussing the ionization and dissociation of such large molecular systems.   To deal with these cases and to enable the theoretical origins of the mass spectra of large molecules to be understood, the quasi-equilibrium theory of mass spectra was developed by Rosenstock, Wallenstein, Wahrhaftig, and Eyring [3, 167].

This theory assumes that the initial ionization process is "vertical" and that the parent molecular ion will, except in a few cases, possess low symmetry; thus all the low-lying electronic states will essentially form a continuum.   Radiationless transitions will then result in the transfer of electronic energy into vibrational energy in times comparable with the periods of nuclear vibrations.   It is assumed that the low-lying excited electronic states of the ions will not be repulsive states, and hence the parent molecular ion will not dissociate immediately on its formation, but will have a lifetime sufficiently long to allow the excess electronic energy to become randomly distributed over the whole ion as vibrational energy.

The rates of dissociation of the molecular ion in the many possible ways are determined by the probabilities that the energy randomly distributed over the molecular framework becomes concentrated in the particular modes required to give the several actuated complex configurations which may dissociate.   It is to be noted that rearrangements of the parent molecular ion or of any fragment ions can take place in a similar fashion. If the initial parent molecular ion possessed sufficient energy, the fragment ion may in turn have enough energy to undergo further decomposition.

The application of the absolute-reaction-rate theory [168] leads to the following expression for the unimolecular rate constant for the ionic dissociation process:

$$k(E) = \int_0^{E-\epsilon_0} \frac{1}{h} \frac{\rho^{\ddagger}(E, \epsilon_0, \epsilon_t)}{\rho(E)} \, d\epsilon_t \qquad (12.110)$$

where $\rho(E)$ is the density of energy levels for the system with total energy E, and $\rho^{\ddagger}(E, \epsilon_0, \epsilon_t)$ is the density of energy levels for the system in the activated complex with configuration energy ϵ_0 and translational energy ϵ_t along the reaction coordinate.   If it is assumed that the system can be treated as a collection of N harmonic oscillators, Eq. (12.110) can be simplified to

$$k(E) = \left(\frac{E - \epsilon_0}{E}\right)^{N-1} \frac{\prod\limits_{j=1}^{N} \nu_j}{\prod\limits_{i=1}^{N} \nu_i^{\ddagger}} \qquad (12.111)$$

In (12.111) the ν's are the vibrational frequencies of the molecular ion in its ground state ν_j and of the activated complex states $\nu_i^{\ddagger}$.   If the rate constants for all possible reactions were known as a function of the energy, the mass spectrum could be calculated as the integral over the energy of these parameters, multiplied by the probability function describing the amount of excess electronic energy given to the parent molecular ions in the initial ionizing process. Nothing exact is known about this latter probability function, so quite arbitrary distribution functions have to be used.

Quantitative calculations based on this theory have been carried out by a number of workers. Certain aspects of meta-stable ion transitions in the mass spectra of large molecules have been treated in terms of this theory by Rosenstock, Wahrhaftig, and Eyring [169]. These workers have shown that the spontaneous decomposition of large molecular ions could occur with a small distribution of half-lives, including a small range of values which would lead to the meta-stable transitions observed experimentally. For small molecules it is to be expected that this quasi-equilibrium model would be unsatisfactory. For these cases, e.g., methane, certain meta-stable transitions observed must be either collision-induced or the result of predissociation. Melton and Rosenstock [170] have studied meta-stable ion transitions and collision-induced dissociations in the mass spectra of n-butane and isobutane and have discussed these in terms of the above quasi-equilibrium theory of mass spectra. Later they reported work on collision-induced dissociations in the mass spectrum of methane [171]. Friedman, Long, and Wolfsberg [171a] have made similar calculations on various alcohols. Chupka [4] has made a quite extensive study of the effect of unimolecular decay kinetics on the interpretation of the appearance potentials of ions from large molecules. Further work has been carried out applying the theory to propane [3, 167, 173], simple esters [172], and mercaptans [174]. In general, the calculated mass spectra have at best been in only semi-quantitative agreement with the experimental data. Quite serious discrepancies were noted, particularly at low ionizing electron voltages [172]. These latter aspects have been fully discussed by Eyring and Wahrhaftig [173].

12.17. High-temperature Chemistry Studies

In recent years considerable interest has developed in the thermo-chemistry of substances at elevated temperatures. Mass-spectrometric methods have proved to be a particularly fruitful way of studying the thermal properties of chemical compounds in the region from 1000 to 2500°K. Furthermore, a great amount of interesting information has been amassed concerning the composition of inorganic materials in the vapor state at high temperatures. It has also been possible to measure accurately the vapor pressures of a large number of substances over a wide range of elevated temperatures. From these results it has frequently been possible to calculate accurate values for the heats of sublimation of the compounds and elements studied, as well as to determine values for certain bond-dissociation energies which could not readily be obtained by any other method. It has been possible, for example, to determine the latent heat of sublimation of carbon [175, 176] and to evaluate the bond-dissociation energies of the dimers of the elements in Group IB and IVB of the Periodic Table [177] and the dissociation energies of di-, tri-, and tetra-atomic intergroup IVB molecules [178]. The dissociation energies of AgAu, AgCu, and AuCu have also been determined [179].

In these studies the normal source of the mass spectrometric is modified so that a crude molecular beam of the vapor of the compound to be

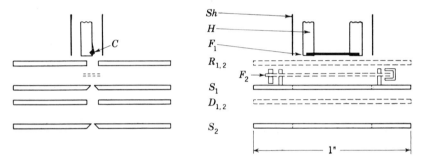

Fig. 12.22. Apparatus used by Honig to study the sublimation of involatile solids. Sh is a tantalum radiation shield; H, a molybdenum filament holder; C, molybdenum filament clips; F_1, a graphite filament; $R_{1,2}$, repeller plates; S_1, the plate of the first slit; $D_{1,2}$, deflector plates; S_2, the plate of the second slit; F_2, the electron filament. [*From R. E. Honig, J. Chem. Phys.*, **22**, 126 (1954).]

studied can be introduced to the ionization chamber. In the case of Honig's [175] experiments to determine the vapor pressure of carbon, a heated graphite filament was placed near the entrance to the ionization chamber, and the evaporating carbon atoms allowed to enter the chamber directly. The techniques for these studies were developed almost simultaneously by Honig [175] and by Chupka and Inghram [176]. Honig's apparatus is very simple but versatile (Fig. 12.22). The apparatus used by Chupka and Inghram is somewhat more elaborate and is applicable to a wide range of problems. The Knudsen effusion cell, which can be heated to the desired temperature, is mounted above the entrance to the ion source. Gaseous molecules leaving the surfaces in the Knudsen cell pass through several slits in the ion source into the ionization chamber. Positive ions produced by electron bombardment are then accelerated into the mass spectrometer for mass analysis. A shutter located between the effusion cell and the ion source makes it possible to control the number of molecules which enter the ion source from the effusion cell. When the slit in the shutter plate is aligned with the slits in the ion source and the orifice of the effusion cell, the maximum number of molecules pass through the ion-source slits. The crucible oven is heated by radiation from an electrically heated tungsten filament surrounded by several sheets of radiation shielding for uniform heating. The effusion cells consist of a metal crucible with liners of different refractory materials, platinum or stainless steel being used for the crucible material. The temperature of the cell was measured by a thermocouple or by an optical pyrometer. As high accuracy was required in measuring the ion currents resulting from the electron bombardment of the effusing gas, most of the mass spectrometers employed electron multipliers as detectors and often ion currents were measured with vibrating-reed electrometers. High sensitivities were thus obtained and are required, for frequently it is necessary to measure accurately small ion currents due to ions of polymeric species or multiply charged particles.

As we have already mentioned, one of the very interesting applications of the above methods was the determination of the latent heat of sublimation of graphite. Honig's [175] work, which was carried out using the rather simple techniques mentioned above, is based on the following principles. The Clapeyron equation relating the vapor pressure P of a solid with the latent heat of sublimation $L(T)$ is

$$\frac{dP}{P} = L(T)\frac{dT}{RT^2} \tag{12.112}$$

The latent heat $L(T)$ can be calculated at any temperature by the Kirchhoff equation. This equation can be integrated to give

$$R \ln P = \frac{L(T_1)}{T} + K \tag{12.113}$$

where T_1 is the mean temperature in the range studied. In Honig's experiments T_1 was approximately $2400°K$.

The pressure P can be related to the ion current in the mass spectrometer as follows. When positive ions are produced from neutral species by electron impact, the current at the collector I^+ is given by the relation

$$I^+ = \eta Q I(e^-) l n_i \tag{12.114}$$

where η = efficiency of ion collection, i.e., ratio of ions collected to ions formed in ionizing region

Q = ionization section, cm^2

$I(e^-)$ = electron-beam current, in units consistent with those of I^+

l = active path length of electrons, cm

n_i = concentration of neutral species in ionizing region, cm^3

Substituting for n_i from the ideal-gas equation, the following expression results for the pressure of the neutral species:

$$P = k_2 I^+ T \qquad (12.115)$$

For negative ions, which are emitted directly from the filament, pressure is related to the current density I^-/A' by the expression well known in gas kinetics,

$$P^- = \frac{I^-}{A'} (2\pi m k T)^{1/2} = k_3 I^- T^{1/2} \qquad (12.116)$$

Thus, if equilibrium may be assumed, the heat of sublimation of the neutral species is obtained by plots of $\ln (I^+ T)$ against $1/T$. It should be remembered that, in the case of graphite, the neutral species evaporating from a hot filament of this substance will contain C atoms and also C_2, C_3, . . . , C_n molecules. It may be noted that Drowart, Burns, De Maria, and Inghram (180) have detected the following ions: C^+, C_2^+, C_3^+, C_4^+, and C_5^+. The heats of sublimation of each species can be obtained from measurements of the ion currents I_{C^+}, $I_{C_2^+}$, $I_{C_n^+}$ as functions of the temperature of the emitting graphite filament.

The method used by Chupka and Inghram [176], while similar in principle, is perhaps more precise, for they calibrate their ion-beam intensities by using silver as a calibrating element. The vapor pressure of pure silver is known to a very high degree of accuracy. The determination of the sensitivity of the mass spectrometer is carried out by placing almost 10 mg of pure silver in the Knudsen cell and measuring the Ag^+ and Ag^{++} ion currents at $1250°K$. This method gave an average of about 6.0×10^{11} intensity units (arbitrary) of Ag^{107++} formed by 150 ev/ atm of Ag pressure inside the cell.

These two researches, namely, that by Honig and the one by Chupka and Inghram, gave, respectively, the values of 179 ± 10 kcal/mole and 170 ± 6 kcal/mole for the latent heat of sublimation of graphite.

The above methods, or essentially similar ones, have been used extensively in recent years by workers interested in determining the composition of the vapors of salts at high temperatures and in measuring bond-dissociation energies of oxides, halides, and intermetallic compounds. An early study of the evaporation products of the alkaline-earth oxides had been carried out by Pelchowitch [181]. This was repeated with their more sensitive apparatus by Inghram, Chupka, and Porter [182], who obtained the value of $D_0°(\text{BaO})$ of 5.65 ± 0.2 ev.

Inghram and his school have made some very important contributions in this field in the past several years. Their results have contributed much new and important information concerning the composition of the species in the vapor state of many high-melting compounds. Thus Berkowitz, Chupka, and Inghram [183], in studying the evaporation of

WO_3 from a Knudsen cell, found that the gaseous molecules W_3O_4, W_4O_{12}, and W_5O_{15} are formed. Later work [184] showed that the vapor subliming from V_2O_5 contained the species V_4O_{10}, V_6O_{14}, V_6O_{12}, with some evidence for V_4O_8 and V_2O_4. These studies, which included measuring the vapor pressure of VO as well as V_2O_5, led to the value of $D_0^{\circ}(VO) = 6.4$ ev and a value of 12.8 ev for the heat of atomization of VO_2. Similar work [185] on Ta_2O_5 indicated that the predominant species in the vapor were TaO and TaO_2. In this work the value of $D_0^{\circ}(T_2O)$ was found to be 8.4 ev and the heat of atomization of Ta_2O_5 to be 15.0 ev. Chupka, Berkowitz, and Giese [186] have studied the vaporization of BeO and found that the vapor consists predominantly of Be and O atoms with $(BeO)_3$ and $(BeO)_4$ molecules. Small amounts of the following molecules were detected: O_2, BeO, $(BeO)_2$, $(BeO)_5$, $(BeO)_6$, and, from the interaction with tungsten in the crucible, WO_2, WO_3, and $WO_x(BeO)_y$, where $x = 1$, 2 and $y = 1$, 2, 3. Thermodynamic treatment of their results yielded the value of 4.6 ev for $D_0^{\circ}(BeO)$. The value obtained for the heat of vaporization of the $(BeO)_4$ molecules suggested they had a cyclic structure. The vapors of molybdenum oxide were found to contain the species MoO_2, MoO_3, $(MoO_3)_2$, and $(MoO_3)_3$ [187]. In the gaseous products from the systems $MoAl_2O_3$ and $U-Al_2O_3$, the following species were found: MoO, MoO_2, MoO_3, WO, WO_2, WO_3, UO, UO_2, and UO_3 [188]. Studies on the vapor of Al_2O_3 led to the determination of the dissociation energies of AlO, Al_2O, and Al_2O_2 [189]. This school also studied NcO [190] and Cr_2O_3 [191] and obtained values for the dissociation energies of the following oxides: NcO, CrO, CrO_2, and CrO_3. In the $O(s)-O_2$ system they were able to determine the enthalpy change for the reaction $O(s)O_4 = O(s)O_3(g) + \frac{1}{2}O_2$ [192].

A detailed study of the gaseous species produced in the vaporization of potassium hydroxide in the temperature range 300 to 450°C was carried out by Porter and Schoonmaker [193]. These workers found that KOH vaporizes mainly as gaseous dimers. The heat of the reaction

$$2KOH(s) = K_2(OH)_2(g)$$

was found to be $\Delta H_{626}^{\circ} = 36 \pm 2$ kcal/mole of dimer. Later the same workers [194] extended their researches to include all the alkali hydroxides. Monomeric and dimeric species were detected in all cases except that of LiOH in the temperature range 650 to 900°K. A study of the heterogeneous reaction between $Li_2O(s)$ and $H_2O(g)$ between 1110 and 1400°K by Berkowitz, Meschi, and Chupka [195] showed that the major product was LiOH, with a smaller amount of $Li_2(OH)_2$ and a trace of $Li_3(OH)_3$. In the case of a corresponding study on $B_2O_3(s)$ and $H_2O(g)$, the main ions found were H_2O^+, HBO_2^+, $H_3BO_3^+$, and $(HBO_2)_3^+$ [196]. Porter and his students have reported interesting results on alkali fluoride vapors [197–199]. Similar studies have been reported by Chupka [200], who studied the vapors of the halides NaCl, KCl, KBr, and RbCl in the temperature range 700 to 900°K. For these four compounds he found that the most abundant ionic species were of the type M^+ and M_2X^+ in comparable amounts. It may be mentioned that perhaps the first observation of a complex ion in the vapors of mixed alkali halides was that of Hayden [201], who, during his mass-spectrometric study of the mixed vapors of KCl and NaCl, found the ion $NaKCl^+$. Chupka found small amounts of $M_3X_2^+$ ions, and in certain circumstances, negative ions of the type X^- and MX_2^- were detected. An attempt was made to detect the $K(H_2O)^+$ ion in equilibrium with water vapor and KCl at about 840°K, but positive identification of the ion could not be made. In an earlier paper Berkowitz and Chupka [202] studied in some detail the molecular species of alkali halide vapors in equilibrium with their respective condensed phases by mass-spectrometric analysis of the ions produced by electron impact. They found dimers, trimers, and, in the cases of LiF, LiCl, LiBr, and NaF, tetramers. The lithium halides exhibited more dimeric than monomeric species. Their results were in substantial agreement with earlier molecular-beam velocity-selector experiments of Miller and Kusch [203], who obtained a velocity profile which could not be attributed to simple diatomic species alone. They also substantiated earlier studies by Friedman [204], who studied mass-spectrometrically the ions resulting from the electron bombardment of lithium iodide molecules effusing from a crucible. He found that approximately 50% of the ion current was due to Li_2I^+, which was assumed to be formed by the reaction $(LiI)_2 + e = Li_2I^+ + I + 2e$. Similar studies by Schoonmaker and Porter [205] showed the existence of monomers, dimers, and trimers in all the alkali fluoride vapors. These authors were also able to determine the heats of dimerization of the mixed fluorides in the reaction $M_2F_2 + N_2F_2 = 2MNF_2$. A further study by the same authors of the vaporization of ferrous bromide led to the observation that the monomer is the predominant vapor species in the temperature range 620 to 665°K, but at the melting point the dimer concentration becomes significant. The heats of dimerization of the alkali chlorides LiCl, NaCl, KCl, RbCl, and CsCl have been determined mass-spectrometrically by Milne and Klein [206], who were also able to obtain a value for the heat of trimerization of LiCl. Molecular association in sodium cyanide vapor has been studied by Porter [214], who has detected the presence of $Na_2(CN)_2$ molecules at 1000°K. The mass-spectral investigation of the vapor of ethyl lithium has shown that the ions of the formulas $Li_nR_{n-1}^+$, where $n = 1, 2, 3, 4, 5,$ and 6, were observed [215], hexamer and tetramer species being the most predominant.

The methods which have been discussed above are also of considerable importance in studying the composition of metallic vapors and have been extensively used to derive values for the dissociation energies of diatomic molecules formed from metallic elements.   Drowart and Honig [207] studied the vapors of Cu, Hg, and Au, using C, Mo, Al_2O_3, and SiO_2 as crucible materials.   They also used silver to calibrate their mass spectrometer, and they mention that they were able to detect Ag_2^+ in silver vapor, which Chupka and Inghram [176] had failed to do.   They derived values for the dissociation energies of the diatomic molecules Cu_2, Ag_2, and Au_2.   Using the same methods, but BeO and graphite crucibles, they later [208] measured the vapor pressures of Ga and Ir over a range of temperatures, but were unable to detect sufficient quantities of the dimer ions of either Ga_2 or Ir_2 to enable values to be obtained for the dissociation energies of these molecules.   In a later paper [209], they give the results of their studies on the vapors of the elements in Group IB and IVB and have derived values for the dissociation energies of the dimers of their elements.   The values which they gave for the dissociation energies are $D_0^\circ(C_2) = 6.2$ ev, $D_0^\circ(Si_2) = 3.2$ ev, $D_0^\circ(Ge_2) = 2.8$ ev, $D_0^\circ(Sn_2) = 2.0$ ev, $D_0^\circ(Pb_2) = 1.0$ ev, $D_0^\circ(Cu_2) = 2.0$ ev, $D_0^\circ(Ag_2) = 1.6$ ev, and $D_0^\circ(Au_2) = 2.1$ g ev.   Drowart and Goldfinger [210] carried out similar studies for some of the elements in Groups III–V and VI.   They studied InF, InSb, and GaAs, as well as CdSe and CdTe.   In this work they observed the following ions, which are of some interest: P^+, P_2^+, and P_4^+ from InP; A_s^+, $A_{s_2}^+$, $A_{s_3}^+$, $A_{s_4}^+$, from GaAs.   They were able to calculate the following values for the dissociation energies $D_0^\circ(P—P) = 2.42$ ev, $D_0^\circ(A_{s_2}—A_{s_2}) = 3.11$ ev, and $D_0^\circ(Sb_2—Sb_2) = 2.63$ ev.   These values are in good agreement with later work from the same laboratory which was reported by Goldfinger and Jeunehomme [211], who studied the evaporation of InAs, GaSb, and ZnS.   In the case of ZnS, it was found that it decomposes when heated to near 1000°K, and Zn and S_2 evaporate in equivalent amounts.   The systems Pb-Se and Pb-Te have been studied by Porter [212], and the dissociation energies of the molecular PbSe and PbTe evaluated.   Drowart, De Maria, Boerboom, and Inghram [213] have studied the systems germanium-graphite and germanium-silicon-graphite by the same method.   Several gaseous molecules containing Ge, Si, and/or C were observed, as indicated by the ions $GeSi^+$, $GeSiC^+$, Ge_2Si^+, Ge_2SiC^+, and Ge_3Si^+, which were detected.   Their data enabled values for the heats of atomization of these molecules to be calculated.

REFERENCES

1. H. D. Hagstrum and J. T. Tate: *Phys. Rev.*, **59**, 343 (1941).
2. J. D. Morrison: *Rev. Pure Appl. Chem.* (Australia), **5**, 22 (1955).
3. H. M. Rosenstock, M. B. Wallenstein, A. L. Wahrhaftig, and H. Eyring: *Proc. Natl. Acad. Sci. U.S.*, **38**, 667 (1952).
4. W. A. Chupka: *J. Chem. Phys.*, **30**, 191 (1959).
5. O. A. Schaeffer and J. M. Hastings: *J. Chem. Phys.*, **18**, 1048 (1950).
6. D. P. Stevenson: *J. Am. Chem. Soc.*, **82**, 5961 (1960).

7. C. E. Moore: *Natl. Bur. Std. (U.S.) Circ.* 467, 1958.
8. H. D. Smyth: *Proc. Roy. Soc. (London)*, **A102**, 283 (1922).
9. J. W. Warren and C. A. McDowell: *Discussions Faraday Soc.*, **10**, 53 (1951).
10. P. T. Smith: *Phys. Rev.*, **36**, 1293 (1930).
11. R. E. Honig: *J. Chem. Soc.*, **16**, 105 (1948).
12. F. P. Lossing, A. W. Tickner, and W. A. Bryce: *J. Chem. Soc.*, **9**, 1254 (1951).
13. J. D. Morrison: *J. Chem. Phys.*, **19**, 1305 (1951).
14. J. D. Morrison and A. J. C. Nicholson: *J. Chem. Phys.*, **20**, 1021 (1952).
15. J. D. Morrison: *J. Chem. Phys.*, **21**, 1767 (1954).
16. J. D. Morrison: *J. Chem. Phys.*, **22**, 1219 (1954).
17. J. D. Morrison: *J. Chem. Phys.*, **21**, 2090 (1954).
18. G. R. Hercus and J. D. Morrison: *Rev. Sci. Instr.*, **23**, 118 (1952).
19. J. D. Morrison: *J. Chem. Phys.*, **22**, 1219 (1954).
20. J. D. Morrison: *J. Appl. Phys.*, **28**, 1409 (1957).
21. A. J. C. Nicholson: *J. Chem. Phys.*, **29**, 1312 (1958).
22. V. H. Dibeler and R. M. Reese: *J. Res. Natl. Bur. Std.*, **54**, 127 (1955).
23. D. P. Stevenson and J. A. Hipple: *Phys. Rev.*, **62**, 237 (1942).
24. M. G. Inghram and R. J. Hayden: A Handbook on Mass Spectrometry, *Natl. Acad. Sci.-Natl. Res. Council, Nucl. Sci. Ser., Rept.* 14, 1954.
25. J. D. Morrison and A. J. C. Nicholson: *J. Chem. Phys.*, **31**, 1320 (1959).
26. C. E. Melton: *Rev. Sci. Instr.*, **29**, 250 (1958).
27. C. Duval, P. LeGoff, and R. Valentine: *J. Chim. Phys.*, **53**, 369 (1956).
28. W. B. Nottingham: *Phys. Rev.*, **55**, 203 (1939).
29. E. M. Clarke: *Can. J. Phys.*, **32**, 764 (1954).
30. A. L. Hughes and J. H. McMillan: *Phys. Rev.*, **34**, 291 (1929).
31. P. Marmet and L. Kerwin: *Can. J. Phys.*, **38**, 787 (1960).
32. L. Kerwin and P. Marmet: *J. Appl. Phys.*, **31**, 2071 (1960).
33. C. Brion, D. C. Frost, and C. A. McDowell: unpublished work; see also P. Marmot and J. D. Morrison: *J. Chem. Phys.*, **36**, 1238 (1962).
34. R. E. Fox, W. M. Hickam, T. Kjeldaas, Jr., and D. J. Grove: *Phys. Rev.*, **84**, 859 (1951).
35. R. E. Fox, W. M. Hickam, D. J. Grove, and T. Kjeldaas, Jr.: *Rev. Sci. Instr.*, **12**, 1101 (1955).
36. G. C. Cloutier and H. I. Schiff: "Advances in Mass Spectrometry" (J. Waldron, ed.), Pergamon Press, New York, 1959, p. 473.
37. W. M. Hickam, R. E. Fox, and T. Kjeldaas, Jr.: *Phys. Rev.*, **96**, 63 (1954).
38. D. C. Frost and C. A. McDowell: *Proc. Roy. Soc. (London)*, **A230**, 227 (1955).
39. H. D. Hagstrum: *Rev. Mod. Phys.*, **23**, 185 (1951).
40. J. D. Craggs, C. A. McDowell, and J. W. Warren: *Trans. Faraday Soc.*, **48**, 1093 (1952).
41. W. M. Hickam and R. E. Fox: *J. Chem. Phys.*, **25**, 642 (1956).
42. D. C. Frost and C. A. McDowell: *J. Chem. Phys.*, **29**, 964 (1958).
43. D. C. Frost and C. A. McDowell: *Can. J. Chem.*, **38**, 407 (1960).
44. J. D. Morrison, H. Hurzeler, M. G. Inghram, and H. E. Stanton: *J. Chem. Phys.*, **33**, 821 (1960).
45. D. C. Frost and C. A. McDowell: *J. Chem. Phys.*, **29**, 503 (1958).
46. D. C. Frost and C. A. McDowell: *J. Am. Chem. Soc.*, **80**, 6183 (1958).
47. D. C. Frost and C. A. McDowell: unpublished experiments; see also *J. Chem. Phys.*, **29**, 1424 (1958).
48. W. W. Lozier: *Phys. Rev.*, **35**, 1285 (1930).
49. J. Marriott, R. Thorburn, and J. D. Craggs: *Proc. Phys. Soc. (London)*, **B67**, 437 (1954).

50. J. D. Craggs and B. A. Tozer: *Proc. Roy. Soc. (London)*, **A254**, 229 (1960).
51. J. D. Craggs and B. A. Tozer: *Proc. Roy. Soc. (London)*, **A247**, 337 (1958).
52. J. D. Craggs, R. Thorburn, and B. A. Tozer: *Proc. Roy. Soc. (London)*, **A240**, 473 (1957).
53. A. W. Petrocelli: M.Sc. Thesis, Providence College, Rhode Island, 1958.
54. P. L. Randolph and R. Geballe: *U.S. Office Ordnance Res. Tech. Rept.* 6, 1958.
55. K. Watanabe: *J. Chem. Phys.*, **22**, 1564 (1954).
56. K. Watanabe: *J. Chem. Phys.*, **26**, 542 (1957).
57. W. C. Walker and G. L. Weissler: *J. Chem. Phys.*, **23**, 1540 (1955).
58. A. N. Terenin and B. Popov: *Z. Sowjetunion*, **2**, 229 (1932).
59. F. P. Lossing and I. Tanaka: *J. Chem. Phys.*, **25**, 1031 (1956).
60. R. F. Herzog and F. F. Marmo: *J. Chem. Phys.*, **27**, 1202 (1957).
61. E. Schonheit: *Z. Physik*, **149**, 153 (1957).
62. W. M. Brubaker: quoted in G. L. Weissler, J. A. R. Sampson, M. Ogawa, and G. R. Cook, *J. Opt. Soc. Am.*, **49**, 338 (1959).
63. H. Hurzeler, M. G. Inghram, and J. D. Morrison: *J. Chem. Phys.*, **28**, 76 (1958).
64. M. Seya: *Sci. Light (Tokyo)*, **2**, 8 (1952).
65. T. Namioka: *Sci. Light (Tokyo)*, **3**, 15 (1954).
66. F. I. Vilesov, B. L. Kurbatov, and A. N. Terenin: *Dokl. Akad. Nauk S.S.S.R.*, **94**, 122 (1958).
67. G. L. Weissler, J. A. R. Sampson, M. Ogawa, and G. R. Cook: *J. Opt. Soc. Am.*, **49**, 338 (1959).
68. G. C. Eltenton: *J. Chem. Phys.*, **10**, 403 (1942).
69. J. A. Hipple and D. P. Stevenson: *Phys. Rev.*, **63**, 121 (1943).
70. F. P. Lossing and A. W. Tickner: *J. Chem. Phys.*, **20**, 207 (1952).
71. W. C. Price: *Chem. Rev.*, **41**, 257 (1947).
72. G. Herzberg and J. Shoosmith: *Nature*, **183**, 1801 (1959).
73. K. Watanabe: *J. Chem. Phys.*, **26**, 542 (1957).
74. J. T. Tate and P. J. Smith: *Phys. Rev.*, **39**, 270 (1932).
75. R. S. Mulliken: *Rev. Mod. Phys.*, **4**, 3 (1932).
76. R. S. Mulliken and D. S. Stevens: *Phys. Rev.*, **44**, 720 (1938).
77. A. G. Gaydon: "Dissociation Energies of Diatomic Molecules," Chapman & Hall, Ltd., London, 1953.
78. W. C. Price and G. Collins: *Phys. Rev.*, **48**, 714 (1935).
79. Y. Tanaka and T. Takamine: *Phys. Rev.*, **59**, 771 (1941).
80. R. E. Fox and W. H. Hickam: *J. Chem. Phys.*, **22**, 2059 (1954).
81. F. H. Dorman, J. D. Morrison, and A. J. C. Nicholson: *J. Chem. Phys.*, **32**, 378 (1960); **34**, 578 (1961).
82 P. Marmet and J. D. Morrison: *J. Chem. Phys.*, **35**, 746 (1961).
83. D. C. Frost and C. A. McDowell: *Proc. Roy. Soc. (London)*, **A241**, 194 (1957).
84. C. A. McDowell and D. C. Frost: *J. Chem. Phys.*, **24**, 173 (1956).
85. W. C. Price: *J. Chem. Phys.*, **4**, 539 (1936).
86. D. C. Frost and C. A. McDowell: *Can. J. Chem.*, **36**, 34 (1958).
87. C. A. McDowell: *J. Chem. Phys.*, **24**, 618 (1956).
88. W. C. Walker and G. L. Weissler: *J. Chem. Phys.*, **23**, 1540 (1958).
89. C. A. McDowell: *Ind. Chim. Belges*, **19**, 713 (1956).
90. M. E. Wacks and V. H. Dibeler: *J. Chem. Phys.*, **31**, 1557 (1959).
91. J. E. Lennard-Jones and G. G. Hall: *Discussions Faraday Soc.*, **10**, 8 (1951).
92. J. L. Franklin: *J. Chem. Phys.*, **22**, 1304 (1954).
93. D. P. Stevenson: *Symp. Mech. Homogeneous and Heterogeneous Hydrocarbon Reactions*, Reprint 29, ACS Meeting, Kansas City, 1954.
94. F. P. Lossing, P. Kebarle, and J. B. de Sousa: "Advances in Mass Spectrometry" (J. Waldron, ed.), Pergamon Press, New York, 1959, p. 431.

95. H. W. Massey and E. H. S. Burhop: "Electronic and Ionic Impact Phenomena," Oxford University Press, London, 1952.

96. D. R. Bates, A. Fundaminsky, J. W. Leech, and H. S. W. Massey: *Phil. Trans. Royal Soc. London*, **A243**, 93 (1950).

97. E. P. Wigner: *Phys. Rev.*, **73**, 1002 (1948).

98. S. Geltman: *Phys. Rev.*, **102**, 171 (1956).

99. S. Geltman: *Phys. Rev.*, **112**, 176 (1958).

100. W. M. Hickam, R. E. Fox, and T. Kjeldaas, Jr.: *Phys. Rev.*, **96**, 63 (1954).

100a. R. E. Fox: "Advances in Spectrometry" (J. Waldron, ed.), Pergamon Press, New York, 1959, p. 397; *J. Chem. Phys.*, **32**, 200 (1960).

101. F. H. Dorman, J. D. Morrison, and A. J. C. Nicholson: *J. Chem. Phys.*, **31**, 1335 (1959).

101a. F. H. Dorman and J. D. Morrison: *J. Chem. Phys.*, **34**, 1407 (1961); **35**, 575 (1961); **36**, 2808 (1962).

102. E. P. Wigner and E. E. Witmer: *Z. Physik*, **31**, 859 (1928).

103. R. S. Mulliken: *Phys. Rev.*, **43**, 279 (1933).

104. R. S. Mulliken: *J. Chem. Phys.*, **1**, 492 (1933); **3**, 517 (1935).

105. C. A. McDowell: *Trans. Faraday Soc.*, **50**, 423 (1954).

106. A. D. Liehr: *J. Chem. Phys.*, **27**, 476 (1957).

107. K. J. Laidler: *J. Chem. Phys.*, **22**, 1740 (1954).

108. C. A. McDowell: "Applied Mass Spectrometry," p. 129, The Institute of Petroleum, London, 1954.

109. G. Herzberg: "Infrared and Raman Spectra," p. 122, D. Van Nostrand Company, Princeton, N.J., 1945.

110. J. D. Morrison and H. E. Stanton: *J. Chem. Phys.*, **28**, 9 (1958).

111. H. W. Washburn and C. E. Berry: *Phys. Rev.*, **70**, 559 (1946).

112. R. Taubert: *Advan. Mass Spectrometry Proc.*, 1959, p. 489.

113. C. A. McDowell and J. W. Warren: see Ref. 9; *Trans. Faraday Soc.*, **48**, 1084 (1952).

114. R. I. Reed and W. Snedden: *Trans. Faraday Soc.*, **54**, 301 (1958).

115. F. P. Lossing, K. W. Ingold, and I. H. S. Henderson: *J. Chem. Phys.*, **22**, 621 (1954).

116. A. J. Langer, J. A. Hipple, and D. P. Stevenson: *J. Chem. Phys.*, **22**, 621 (1954).

117. E. W. B. Clarke and C. A. McDowell: *Proc. Chem. Soc.*, 1960, p. 69.

118. D. Oberghaus and R. Taubert: *Z. Physik. Chem. (Frankfurt)*, **4**, 264 (1955).

119. D. P. Stevenson: *Discussions Faraday Soc.*, **10**, 35 (1951).

120. J. F. Burns: *J. Chem. Phys.*, **23**, 1347 (1955).

121. D. C. Frost and C. A. McDowell: *Proc. Roy. Soc. (London)*, **A236**, 278 (1956).

122. R. J. Kandel: *J. Chem. Phys.*, **22**, 1496 (1954).

123. J. T. Herron and V. H. Dibeler: *J. Am. Chem. Soc.*, **82**, 1555 (1960).

124. L. Brewer, L. K. Templeton, and F. A. Jenkins: *J. Am. Chem. Soc.*, **73**, 1462 (1951).

125. J. T. Franklin, V. H. Dibeler, R. M. Reese, and M. Krauss: *J. Am. Chem. Soc.*, **80**, 298 (1958).

126. A. E. Douglas and G. Herzberg: *Can. J. Res.*, **A20**, 71 (1942).

127. P. T. Smith: *Phys. Rev.*, **51**, 663 (1937).

128. L. H. Long and R. G. W. Norrish: *Proc. Roy. Soc. (London)*, **A178**, 337 (1946).

129. L. H. Long: *Proc. Roy. Soc. (London)*, **A198**, 62 (1949).

130. F. H. Field and J. L. Franklin: "Electron Impact Phenomena and the Properties of Gaseous Ions," p. 152, Academic Press Inc., New York, 1957.

131. T. Cottrell: "The Strengths of Chemical Bonds," Butterworth & Co. (Publishers), Ltd., London, 1960.

132. A. L. Marshall and F. J. C. Norton: *J. Am. Chem. Soc.*, **72**, 2166 (1950).
133. L. Brewer, P. Gilles, and F. A. Jenkins: *J. Chem. Phys.*, **16**, 797 (1948).
134. R. E. Honig: *J. Chem. Phys.*, **21**, 573 (1953).
135. W. A. Chupka and M. G. Inghram: *J. Chem. Phys.*, **59**, 100 (1955).
136. R. I. Reed and W. Snedden: *Trans. Faraday Soc.*, **54**, 301 (1958).
137. R. I. Reed and W. Snedden: *Trans. Faraday Soc.*, **55**, 876 (1959).
138. D. C. Frost and C. A. McDowell: *J. Chem. Phys.*, **29**, 1424 (1958).
139. J. B. Farmer and F. P. Lossing: *Can. J. Chem.*, **33**, 861 (1955).
140. A. G. Harrison and F. P. Lossing: *J. Am. Chem. Soc.*, **82**, 819 (1960).
141. J. Collin and F. P. Lossing: *J. Am. Chem. Soc.*, **79**, 5848 (1959).
142. C. A. McDowell and B. C. Cox: *J. Chem. Phys.*, **20**, 1496 (1952).
143. R. R. Bernecker and F. A. Long: private communication.
144. L. M. Branscomb, D. C. Burch, S. J. Smith, and S. Geltman: *Phys. Rev.*, **111**, 504 (1958).
145. N. I. Yonov: *Compt. Rend. Acad. Sci. U.R.S.S.*, **28**, 512 (1940).
146. I. N. Bakalina and N. I. Yonov: *Dokl. Acad. Nauk S.S.S.R.*, **105**, 680 (1955).
147. T. L. Bailey: *J. Chem. Phys.*, **28**, 792 (1958).
148. I. N. Bakalina and N. I. Yonov: *Dokl. Acad. Nauk S.S.S.R.*, **116**, 41 (1957).
149. I. N. Bakalina and N. I. Yonov: *Zh. Fiz. Khim.*, **33**, 2063 (1959).
150. H. D. Pritchard: *Chem. Rev.*, **52**, 529 (1953).
151. J. A. Hipple, R. E. Fox, and E. U. Condon: *Phys. Rev.*, **69**, 347 (1946).
152. A. L. Wahrhaftig: *Advan. Mass Spectrometry Proc.*, 1959, p. 274.
153. A. Henglein: "Applied Mass Spectrometry," p. 158, The Institute of Petroleum, London, 1954.
154. D. O. Schissler, S. O. Thompson, and J. Turkevich: *Proc. Faraday Soc.*, **10**, 46 (1951).
155. D. I. Stevenson: *J. Chem. Phys.*, **19**, 17 (1951).
156. P. N. Rylander, S. Meyerson, and H. M. Grubb: *J. Am. Chem. Soc.*, **79**, 842 (1957).
157. S. Meyerson and P. N. Rylander: *J. Am. Chem. Soc.*, **79**, 1058 (1957).
158. S. Meyerson and P. N. Rylander: *J. Chem. Phys.*, **27**, 901 (1957).
159. J. H. Beynon, G. R. Lester, and A. E. Williams: *J. Am. Chem. Soc.*, **63**, 1861 (1959).
160. S. Meyerson, P. N. Rylander, E. L. Eliel, and J. D. McCollum: *J. Am. Chem. Soc.*, **81**, 2606 (1959).
161. S. Meyerson and P. N. Rylander: *J. Phys. Chem.*, **62**, 2 (1958).
162. S. Meyerson, J. D. McCollum, and P. N. Rylander: *J. Am. Chem. Soc.*, **83**, 1401 (1961).
163. F. W. McLafferty: *Anal. Chem.*, **31**, 82 (1959).
164. F. W. McLafferty: *Advan. Mass Spectrometry Proc.*, 1959, p. 355.
165. J. H. Beynon: *Advan. Mass Spectrometry Proc.*, 1959, p. 328.
166. J. H. Beynon: "Mass Spectrometry and Its Applications to Organic Chemistry," p. 259, Elsevier Publishing Company, Amsterdam, 1960.
167. A. Kropf, E. M. Eyring, A. L. Wahrhaftig, and H. Eyring: *J. Chem. Phys.*, **32**, 149 (1960).
168. S. Glasstone, K. J. Laidler, and H. Eyring: "The Theory of Rate Processes," McGraw-Hill Book Company, Inc., New York, 1941.
169. H. M. Rosenstock, A. L. Wahrhaftig, and H. Eyring: *J. Chem. Phys.*, **23**, 2200 (1955).
170. G. E. Melton and H. M. Rosenstock: *J. Chem. Phys.*, **26**, 568 (1957).
171. H. M. Rosenstock and C. E. Melton: *J. Chem. Phys.*, **26**, 314 (1957).
171a. L. Friedman, F. A. Long, and M. Wolfsberg: *J. Chem. Phys.*, **27**, 613 (1957).

172. A. B. King and F. A. Long: *J. Chem. Phys.*, **29**, 374 (1958).
173. E. M. Eyring and A. L. Wahrhaftig: *J. Chem. Phys.*, **34**, 23 (1961).
174. J. Collin: *Bull. Soc. Roy. Sci. Liége*, **25**, 520 (1956).
175. R. E. Honig: *J. Chem. Phys.*, **21**, 573 (1953).
176. W. A. Chupka and M. G. Inghram: *J. Phys. Chem.*, **59**, 100 (1955).
177. J. Drowart and R. E. Honig: *J. Chem. Phys.*, **61**, 980 (1959).
178. J. Drowart, G. de Maria, A. J. H. Boerboom, and M. G. Inghram: *J. Chem. Phys.*, **30**, 308 (1959).
179. M. Ackermann, F. E. Stafford, and J. Drowart: *J. Chem. Phys.*, **33**, 1784 (1960).
180. J. Drowart, R. P. Burns, G. de Maria, and M. G. Inghram: *J. Chem. Phys.*, **31**, 1131 (1959).
181. I. Pelchowitch: *Philips Res. Rept.*, **9**, 42 (1954).
182. M. G. Inghram, W. A. Chupka, and R. F. Porter: *J. Chem. Phys.*, **23**, 2159 (1955).
183. J. Berkowitz, W. A. Chupka, and M. G. Inghram: *J. Chem. Phys.*, **27**, 85 (1957).
184. J. Berkowitz, W. A. Chupka, and M. G. Inghram: *J. Chem. Phys.*, **27**, 87 (1957).
185. M. G. Inghram, W. A. Chupka, and J. Berkowitz: *J. Chem. Phys.*, **27**, 569 (1957).
186. W. A. Chupka, J. Berkowitz, and C. F. Giese: *J. Chem. Phys.*, **30**, 827 (1959).
187. R. P. Burns, G. de Maria, J. Drowart, and R. T. Grimley: *J. Chem. Phys.*, **32**, 1363 (1960).
188. G. de Maria, R. P. Burns, J. Drowart, and M. G. Inghram: *J. Chem. Phys.*, **32**, 1373 (1960).
189. J. Drowart, G. de Maria, R. P. Burns, and M. G. Inghram: *J. Chem. Phys.*, **32**, 1366 (1960).
190. R. T. Grimley, R. P. Burns, and M. G. Inghram: *J. Chem. Phys.*, **34**, 551 (1961).
191. R. T. Grimley, R. P. Burns, and M. G. Inghram: *J. Chem. Phys.*, **34**, 664 (1961).
192. R. T. Grimley, R. P. Burns, and M. G. Inghram: *J. Chem. Phys.*, **33**, 308 (1960).
193. R. F. Porter and R. C. Schoonmaker: *J. Phys. Chem.*, **62**, 234 (1958).
194. R. F. Porter and R. C. Schoonmaker: *J. Chem. Phys.*, **29**, 1070 (1958).
195. J. Berkowitz, D. J. Meschi, and W. A. Chupka: *J. Chem. Phys.*, **33**, 533, (1960).
196. D. J. Meschi, W. A. Chupka, and J. Berkowitz: *J. Chem. Phys.*, **33**, 530 (1960).
197. R. C. Schoonmaker and R. F. Porter: *J. Chem. Phys.*, **31**, 830 (1959).
198. R. F. Porter and E. E. Zeller: *J. Chem. Phys.*, **33**, 858 (1960).
199. R. F. Porter: *J. Chem. Phys.*, **33**, 950 (1960).
200. W. A. Chupka: *J. Chem. Phys.*, **30**, 458 (1959).
201. R. J. Hayden: *Phys. Rev.*, **74**, 651 (1948).
202. J. Berkowitz and W. A. Chupka: *J. Chem. Phys.*, **29**, 653 (1958).
203. R. C. Miller and P. Kusch: *J. Chem. Phys.*, **25**, 860 (1956).
204. L. Friedman: *J. Chem. Phys.*, **23**, 477 (1955).
205. R. C. Schoonmaker and R. F. Porter: *J. Chem. Phys.*, **30**, 283 (1959).
206. T. A. Milne and H. M. Klein: *J. Chem. Phys.*, **33**, 1628 (1960).
207. J. Drowart and R. E. Honig: *J. Chem. Phys.*, **25**, 581 (1956).
208. J. Drowart and R. E. Honig: *Bull. Soc. Chim. Belges*, **66**, 411 (1957).
209. J. Drowart and R. E. Honig: *J. Phys. Chem.*, **61**, 980 (1957).
210. J. Drowart and P. Goldfinger: *J. Chim. Phys.*, **55**, 721 (1958).
211. P. Goldfinger and M. Jeunehomme: "Advances in Mass Spectrometry" (J. Waldron, ed.), Pergamon Press, New York, 1959, p. 534.
212. R. F. Porter: *J. Chem. Phys.*, **34**, 583 (1961).
213. J. Drowart, G. de Maria, A. J. H. Boerboom, and M. G. Inghram: *J. Chem. Phys.*, **30**, 308 (1959).
214. R. F. Porter: *J. Chem. Phys.*, **35**, 318 (1961).

215. J. Berkowitz, D. A. Bafus, and T. L. Brown: *J. Phys. Chem.*, **65**, 1380 (1961).
216. D. C. Frost, D. Mak, and C. A. McDowell: *Can. J. Chem.*, **40**, 1064 (1962).
217. F. A. Elder, C. Giese, E. Steiner, and M. Inghram: *J. Chem. Phys.*, **36**, 3293 (1962).
218. F. H. Dorman and J. D. Morrison: *J. Chem. Phys.*, **34**, 578 (1961).
219. K. Biemann: "Mass Spectrometry," McGraw-Hill Book Company, Inc., New York, 1962.

13

Ion-Molecule Reactions

D. P. Stevenson

Shell Development Co.
Emeryville, California

13.1. Introduction

The knowledge that secondary processes involving reactions between primary ions and unionized gas take place in a mass spectroscope is as old as the art of mass spectroscopy. J. J. Thomson [1] noted the various effects of such secondary processes—the formation of new ionic species corresponding to no known molecular species that appeared as lines, as well as the appearance of diffuse lines, bands, and general background, with his original parabola mass spectroscope. In particular, Thomson observed a line corresponding to a particle of mass-charge ratio $(m/q) = 3$ in the products from a discharge in hydrogen. In 1916 Dempster [2] showed that this $m/q = 3$ particle was the triatomic hydrogen species H_3^+ and that it was truly the product of a secondary reaction. The identity of the $m/q = 3$ particle was independently confirmed by Aston [3] in 1920. On the basis of the results of their rather extensive investigations of the ionization of hydrogen by low-energy electrons, Smyth [4] and Hogness and Lunn [5] correctly suggested that the most probably mode of formation of this ionic allotrope of hydrogen is by the reaction

$$H_2^+ + H_2 = H_3^+ + H \qquad (13.1)$$

Reaction (13.1) is the prototype of the most extensively studied class of secondary ionic reaction that occurs in mass spectroscopes and that may be more generally written

$$X^+ + YH = XH^+ + Y \qquad (13.2)$$

The implicit knowledge of the existence of reactions of the form of (13.2) with YH equal to either H_2 or H_2O had important effects on the early mass-spectrographic investigations of the isotopic constitution of the elements.   Thus, for example, the neon isotope ^{21}Ne was not discovered until 1928 by Hogness and Kvalnes [6] because Aston [7] attributed the weak line at $m/q = 21$ that he saw on a 1920 plate of the neon mass spectrum to the presence of ^{20}NeH$^+$.   The difficulties these reactions gave in the detection of rare isotopes one mass unit heavier than an abundant isotope [8] were in part compensated by the additional lines, and thus members of multiplets, that could be used in the precision determination of the nuclear masses.   Thus the quadruplet H$_2$D$^+$, D$_2$$^+$, He$^+$, and ^{12}C^{3+} at $m/q = 4$ permits the simultaneous interrelating of the nuclear masses of the isotopes of hydrogen, helium, and carbon.

During the 1930s the primary interest of mass spectroscopists was in the study of the primary ionization process and its concomitant induced dissociations, and thus efforts were bent toward avoiding conditions that would lead to the observation of secondary processes.   These efforts were generally, but not completely, successful because of the improvements in high-vacuum techniques and the enormous improvements in the sensitivity of ion-detecting devices that permitted studies at the then attainable lower pressures.   The lack of complete success in the suppression of the secondary reactions is attested by the problems encountered by Bleakney and Gould [9] in the determination of low concentrations of HD in H$_2$, because of H$_3$$^+$ interference.   This particular problem remained unsolved even through the war years [10].

Until at least 1951 the only criterion used to distinguish primary ions from those arising in secondary processes was the different dependence of the intensity of a primary ion from that of a secondary ion on the pressure.   As is apparent from the equations defining the formation of primary (P_i^+) ions and secondary (S^+) ions,

$$P + e^- = P^+ + 2e^- \tag{13.3}$$
$$P^+ = P_i^+ + F_1 + \cdots$$
$$P_i^+ + \mathrm{R} = S^+ + T + \cdots \tag{13.4}$$

the intensity or current of primary ions will depend on the first power of the pressure of P, while the intensity or current of secondary ions will depend on the square of the total pressure.   Further, the only explicit means taken to avoid the appearance of secondary ions in the mass spectra was that of working at as low pressures as was compatible with accurate measurement of the intensities of the primary ions.   Hogness and Harkness [11], in their studies of the formation of I$_3$$^+$ and I$_3$$^-$ in the mass spectrum of iodine vapor, indicated an alternative, but much less convenient, means of distinguishing primary from secondary ions that may have found implicit use in the design of mass spectrometers for the study of primary processes.   This alternative procedure makes use of the fact that the relative intensities of primary ions do not depend on the length of their path through unionized gas, while the probability of collisions, and hence the probability of formation of secondary ions, increases with this path length, and hence that the intensity of a secondary ion relative to that of a primary ion will increase with the path length.

In 1951 Washburn and coworkers [12] described a third method for the suppression of secondary ions from mass spectra that is by far the most easily applied and now forms the most readily used criterion for distinguishing between primary and secondary ions in mass spectra. This third method makes use of the fact that the probability that a primary ion will suffer a reactive collision with an unionized gas molecule decreases with the residence time of the ion in the gas [12a]. Thus Washburn and coworkers found that, by increasing the ion-repeller potential that drives the ions from the ion source of a mass spectrometer, they were able to render negligible the interference of the hydronium ion H_3O^+ with the determination of HDO at its natural abundance in water. Although not recognized as such by them, this discovery of Washburn, Berry, and Hall forms the basis of the method of quantitative determination of the rates of such reactions as (13.2) or (13.4), which will be discussed in detail in Sec. 13.2.

Through the years prior to 1950, though there was little direct interest in the secondary reactions, new reactions of the type of (13.2), particularly, continued to be discovered. Thus, for example, Mann, Hustrulid, and Tate [13] showed the hydronium ion to be formed by the reaction

$$H_2O^+ + H_2O = H_3O^+ + OH \tag{13.5}$$

and Kondrat'ev and coworkers [14] found N_2H^+ and HCO^+ to be formed in N_2-H_2 and in H_2-CO mixtures, respectively. Among the more interesting reactions of type (13.2) was that discovered by G. C. Eltenton [15] in 1940, namely,

$$CH_4^+ + CH_4 = CH_5^+ + CH_3 \tag{13.6}$$

a reaction that was independently rediscovered by at least two other investigators [16, 17].

Because of the decline in interest in the use of the parabola mass spectroscope that accompanied the development of mass spectrographs and spectrometers, little was done to clarify the nature of the secondary reactions that give rise to the diffuse lines and bands found by Thomson [1] and Aston [18] until 1939, when Mattauch and Lichtblau [19] published the results of their thorough study of secondary rays in mass spectrographs. They showed that the "spurious" lines and bands arose from charge-exchange reactions [20] between primary ions and unionized gas molecules and from dissociation reactions of primary ions induced by collisions with unionized gas molecules that occur in the analyzer section of the mass spectrograph. This work of Mattauch and Lichtblau has been considerably extended by the use of a specially constructed parabola mass spectroscope by Henglein and Ewald [21]. Melton and coworkers [22] have undertaken an extensive investigation of the collision-induced dissociation reactions of the primary ions in the drift section between the last accelerating electrode and the first edge of the analyzing magnetic field of a sector-analyzer mass spectrometer.

Commencing in about 1950, various groups of investigators more or less independently started investigations of the various reactions that can and do take place in a mass-spectrometer ion source, in which knowledge of the nature of the reactions was the primary, rather than an incidental, objective of the study. Norton [23] made qualitative studies of the mass spectra of "high"-pressure mixtures of argon, oxygen, and mercury with hydrogen and found ArH^+, O_2H^+, and HgH^+, as well as evidence for Ar_2^+ and Hg_2^+, in the mass spectra. Hornbeck [24] showed that when the rare gases (R) are admitted to a mass-spectrometer ion source at relatively high pressures, reactions of the types

$$R^* + R = R_2^+ + e^- \tag{13.7}$$
and
$$R^+ + 2R = R_2^+ + R \tag{13.8}$$

where R^* is an excited rare-gas atom, take place. Tal'roze [17] started an extensive series of studies of reactions of the type of (13.2), where either X or YH is a hydrocarbon molecule, in the course of which he rediscovered reaction (13.6), as mentioned above, as well as new reactions like

$$C_3H_6^+ + C_3H_6 = C_3H_7^+ + C_3H_5 \tag{13.9}$$

Gutbier [25] and Stevenson and Schissler [26] independently undertook quantitative measurement of the rate or cross section of these secondary reactions that can be observed with a mass spectrometer, particularly reactions of the type of (13.2).

The fact that these secondary reactions can be readily observed in modern analytical mass spectrometers is now recognized as having considerable significance beyond the immediate field of mass spectrometry. The extraordinarily high rate that is required for such observability indicates that such reactions must play an important role in the primary initiation reactions of gaseous-radiation chemistry [27–30]. From the knowledge of the existence of a reaction like (13.2) with a very high specific rate, one can make deductions [31] concerning the relation that exists between the dissociation energies, $D(X—H^+)$ and $D(Y—H)$ that supplement and complement the results of the various other spectrometric and chemical methods [32].

A consequence of the recognition of these implications of the observability of ion-molecule reactions has been the development of a new field of mass-spectrometric study, namely, the mass spectra from high-pressure ion sources.   Special mass spectrometers with special differential pumping between ion source and ion analyzer have been constructed for this purpose.   Two such especially worth mention are those at the Oak Ridge National Laboratory and at the Humble Oil and Refining Co. Laboratory.   In the first of these [61], Melton and coworkers have employed alpha particles from Po^{208} as the ionizing agent instead of the more conventional thermionic electrons, and thus avoided the difficulties associated with the effects of high gas pressure on the emission characteristics of the cathode.   In the second of these [60], by use of high-speed pumps and special construction, Field and coworkers are able to maintain an ion-analyzer vacuum of less than 0.01 μ with an ion-source pressure as high as 600 μ and, as a consequence, have been able to observe reactions under conditions approaching those of conventional radiolyses of gases.

With these remarks we close the historical review of the subject.   In the following sections we shall consider the method and results of quantitative studies of the rates of the secondary reactions.   We shall then review the current status of the theory of the rates of these reactions with a comparison of experimental with theoretical results.   Finally, we shall conclude the chapter with a review of the applications in radiation chemistry and a critique of the determination of dissociation energies.

13.2. Rates: Their Measurement and the Results of Measurement

The basic measurement with a mass spectrometer is the resolved intensity or current of an ion of particular mass-charge ratio.   In the commonly used direction-focusing mass spectrometers of the Dempster type, only those ions that are formed in the ion source prior to acceleration to the mass-analyzing kinetic energy are observed as well-defined "peaks" of integral or small fraction ($\frac{1}{2}$, $\frac{1}{3}$, or $\frac{1}{4}$) of integral m/q.   In fact, it is a simple matter in these mass spectrometers, by the use of an appropriate retarding potential, to ensure that only ions formed in the ion source will be observed [33].   For primary ions the general nature of the relationships between the current of ions entering the ion accelerating and collimating system and the observed resolved current are reasonably well understood [34–36], and it is usually assumed that the slit that gives ions entry to the accelerating-collimating system collects all ions formed in the ion source equally efficiently [64].   In order to attempt an interpretation of intensities of secondary ions in terms of the rate of the reactions of primary ions in the ion source, it is necessary to assume that the overall efficiency of detection of secondary ions is the same as that of the primary ions.   Thus far the results of an investigation of the validity of this assumption have not appeared.   The quality of the agreement between the results of measurements in different laboratories in which different mass spectrometers were employed suggests that the assumption is not in error by more than a factor of 2, and probably considerably less.

As is discussed elsewhere in this volume, the primary ions are formed in the ion source in the more or less well defined plane of the ionizing electron beam located a distance l from the slit that gives entry to the ion accelerating-collimating system. The primary ions are continuously swept toward the exit slit as they are formed by a more or less uniform electric field E_r, produced by an adjustable difference in potential between

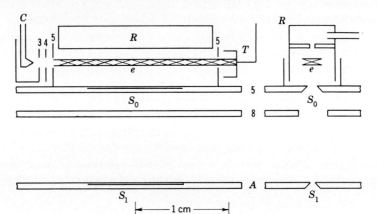

Fig. 13.1.   Geometry of an ion source. C = cathode; 3, 4 = electron-drawing-out and focusing electrodes; 5 = ion-chamber reference cathode; R = ion-repeller electrode and gas inlet; T = electron trap; S_0 = exit slit from ion chamber; 8 = ion focus electrode; S_1 = ion accelerating slit; crosshatched region e represents the approximate volume of the ionizing electron beam; $S_0 = 0.45_6$ mm; $S_1 = 0.20_3$ mm.   (*From Stevenson and Schissler* [26].)

the ion-repeller electrode and the exit slit.   (These statements can be visualized with the aid of Fig. 13.1, in which a scale drawing of a commonly used mass-spectrometer ion source is given.)   In traversing their paths from their plane of origin to the exit slit, the primary ions may undergo collision and reaction with unionized gas molecules, forming secondary ions.   In accordance with the implications of the preceding paragraph, we assume that such secondary ions follow, on the average, the paths that the corresponding primary ion would have followed. Since an extremely small fraction of the gas flowing continuously through the ion source is ionized,† we may take the concentration of gas in the

† For the ion source depicted in Fig. 13.1, under typical operating conditions, 1.5×10^{15} moles H_2 per second flow through the source to maintain a steady-state concentration equal to 2.5×10^{11} moles/cm³.   An ionizing electron current of 10 μ of 75 ev generates 2.8×10^9 primary ions (H_2^+) per second.   To maintain the same pressure in the ion source, 3.4×10^{14} argon atoms per second would be the required

source to be constant and equal to n_r. Then the attenuation of the primary current due to reaction to form secondary ions will be

$$i_p(l) = i_p(0) \exp (-ln_rQ) \qquad (13.10)$$

where Q, the phenomenological reaction cross section for the reaction

$$P^+ + R = S^+ + T + \cdots \qquad (13.11)$$

is the analog of the molal extinction coefficient in the familiar Beer-Lambert law for the attenuation of a photon beam traversing an absorbing medium. If the primary ions react to form but a single kind of secondary ion, the current of secondary ions generated will be

$$i_s(l) = i_p(0) - i_p(l) \qquad (13.12)$$

or, for the ratio of the secondary current to the primary current,

$$\frac{i_s(l)}{i_p(l)} = \exp (ln_rQ) - 1 \qquad (13.13)$$

and we shall associate the observed ratio of secondary-ion current to primary-ion current, i_s/i_p, with $i_s(l)/i_p(l)$.

Given a knowledge of l and n_r [65] corresponding to a measurement of the ratio i_s/i_p, Eq. (13.13) can be rewritten to permit the evaluation of Q, the measure of the rate of reaction (13.11). We write

$$Q = (ln_r)^{-1} \ln \left(1 + \frac{i_s}{i_p} \right) \qquad (13.14)$$

or, for $i_s/i_p \ll 1$,

$$Q = \frac{i_s}{ln_r i_p} \qquad (13.15)$$

with error less than 5% for $i_s/i_p \leq 0.1$.

The dependence of the reaction cross section Q for a number of reactions of the type of (13.2) on the magnitude of the ion-repeller field strength E_r and the temperature of the ion source, and thus that of the unionized gas, has been studied by Gutbier [25] and by Stevenson and Schissler [26]. In both of these studies sector-analyzer mass spectrometers were employed, but the instruments have ion sources of quite different design detail. The results of these independent studies are in complete agreement with regard to the functional dependence of Q upon E_r and T, namely, Q proportional to $(E_r)^{-\frac{1}{2}}$ and independent of the gas tempera-

---

flux, and the indicated ionizing electron current would yield 1×10^{10} ions per second. We see that, in general, less than a few hundredths of a per cent of the gas flowing through the ion source is ionized, and thus that neither ionization nor possible secondary reactions of the primary ions can change the concentration of the gas in the ion source.

ture.   These relations are illustrated in Figs. 13.2 to 13.4.   Furthermore, the actual values of the cross sections are also in reasonably good agreement.   The quality of the agreement in the absolute magnitudes of the cross sections found in these investigations is the only evidence for the validity of the assumption discussed above concerning the equality between the ratio of the intensity of the resolved ion currents of secondary to primary ions and the ratio of the currents of such ions generated in the ion source.   The quality of the agreement may be seen in Table 13.1,

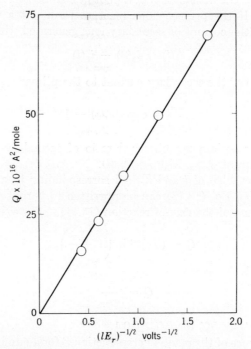

Fig. 13.2.   The dependence of the cross section of the reaction $D_2^+ + D_2 \rightarrow D_3^+ + D$ on the ion-repeller field strength.

showing the results of the two investigations referred to above, along with the values calculated from the simple classical theory that will be discussed in Sec. 13.3.

The experimental results shown in Table 13.1 are expressed as the specific bimolecular rate constants for the indicated chemical reactions. The conversion of the observed cross sections to specific rates was carried out by the dimensional argument of Stevenson and Schissler [26], which is justified by the theoretical treatment to be given in Sec. 13.3.   The relation of the observed reaction cross sections as a function of the ion-repeller field strength to the specific rate is obtained as follows.   The linear dependence of Q upon $(E_r)^{-\frac{1}{2}}$ that is demonstrated in Figs. 13.2

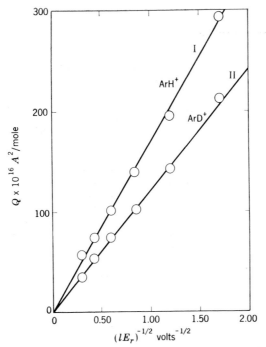

Fig. 13.3. The dependence of the cross section of the reactions (I) $Ar^+ + H_2 \rightarrow ArH^+ + H$, (II) $Ar^+ + D_2 \rightarrow ArD^+ + D$, on the ion-repeller field strength.

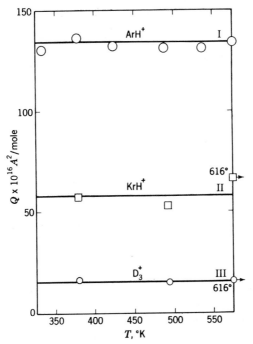

Fig. 13.4. Temperature independence of the reaction cross section for the reactions (I) $Ar^+ + H_2$; (II) $Kr^+ + H_2$; (III) $D_2^+ + D_2$. The ion-repeller field measure $(lE_r)^{-1/2} = 0.85_0 v^{-1/2}$ for (I) and (II), and $0.42_1 v^{-1/2}$ for (III).

TABLE 13.1. Observed and Calculated Specific Rates of Some Simple
Ion-Molecule Reactions

$k \times 10^9 \ \mathrm{cm^3 \ mole^{-1} \ sec^{-1}}$

Reaction	Gutbier [25]	Stevenson and Schissler [26]	Theory [39]
$H_2^+ + H_2$	2.9	......	2.05
$D_2^+ + D_2$	....	1.4_3	1.45
$He^+ + H_2$	0.62	......	1.78
$Ne^+ + H_2$	0.42	0.27	1.53
$Ar^+ + H_2$	1.6	1.68	1.50
$Ar^+ + D_2$	....	1.35	1.09
$Kr^+ + H_2$	0.64	0.49	1.47
$Kr^+ + D_2$	....	0.30	1.05
$N_2^+ + H_2$	2.2	......	1.44
$N_2^+ + D_2$	2.6	7.6	2.16
$CO^+ + D_2$	....	1.63	1.11
$CO_2^+ + H_2$	1.4	1.1_5	1.49
$Xe^+ + H_2$	....	<0.03	1.48

and 13.3 for the formation of D_3^+ and ArH^+, respectively, indicates that Eq. (13.15) can be written

$$i_s = \frac{i_p n_r l Q'}{E_r^{-\frac{1}{2}}} \tag{13.16}$$

Now the terminal kinetic energy with which the primary ions m_p/q reach the exit slit from the ion source is lqE_r, since the thermal kinetic energy with which the ion is formed will in general be negligibly small compared with that which it acquires from the electric field. For a uniformly accelerated ion, starting essentially from rest, this terminal kinetic energy will equal $2m_p\bar{v}^2$, where $\bar{v}$ is the (time) average speed with which the ion traversed its path to the exit slit. Thus Eq. (13.16) can be rewritten

$$i_s = i_p \left(\frac{l}{\bar{v}}\right) n_r Q' \left(\frac{lq}{2m_p}\right)^{\frac{1}{2}} \tag{13.17}$$

But $(l/\bar{v})$ is the mean residence time of the primary ions in the ion source, and the product of this quantity by the flux of primary ions i_p is the average concentration of primary ions in the ion source, and i_s is the flux, or rate of formation, of secondary ions. Thus $Q'(lq/2m_p)^{\frac{1}{2}}$ can be written equal to k_2, the equivalent of the bimolecular rate constant of reaction (13.11).

A question of no little importance that must be answered before a significant cross section or reaction rate can be computed is, which are the reactants? For example, in the case of the formation of the argonium ion ArH^+, are the reactants Ar^+ and H_2 or Ar and H_2^+? Several means are available for the determination of the ionic reactant when such ambiguity exists. The most direct is that of determining the appearance potential of the secondary ion and comparing this appearance potential with those of the various possible reactants. This method has been well demonstrated by Tal'roze and Frankevich [14] in the identification of the ionic reactants giving rise to H_3O^+ in mixtures of water vapor with ammonia, hydrogen sulfide, propane, and acetylene. In many cases it is not necessary actually to determine the appearance potential of the secondary ion; simply observing the dependence of the ratios i_s/i_{p_1} and i_s/i_{p_2}, where p_1 and p_2 are alternative possible primary ions, on the energy of the ionizing electrons will suffice. The ratio corresponding to the actual primary ion will be independent of the ionizing electron energy, while the ratio corresponding to the incorrect primary ion will either decrease sharply or increase sharply with decreasing low-ionizing electron energy, in accordance with whether the ionization potential of the erroneous primary ion is less than or greater than the ionization potential of the actual reactant primary ion [26]. As has been noted by Stevenson and Schissler [26], care must be taken in measurements at low-ionizing electron energies that the electron-beam space-charge neutralization of the ion-repeller field does not markedly distort the shape of the ionization efficiency curve of the secondary ion and thus lead to error in the appearance potential or erroneous shape of the i_s/i_p versus electron energy curve.

A second method of identifying the reactant ion that is of more limited applicability is that of comparing possible observed reaction rates with the upper limit that may be calculated from the classical theory of the rate of these reactions given below. As will be shown, the theoretical limit set to the specific rate of a reaction,

$$P^+ + R \xrightarrow{k} S^+ + T + \cdots \qquad (13.11)$$

is

$$k \leq k_t = 14.7_5 \times 10^{-10} \left(\frac{P_E}{\mu}\right)^{1/2} \quad \text{cm}^3 \text{ mole}^{-1} \text{ sec}^{-1} \quad (13.18)$$

where P_E = molecular refraction of reactant molecule R
μ = reduced mass of reactant pair $P^+ + R$, in amu
For alternative ionic reactants, P^+ or R^+, one will have alternative apparent rates k_1 and k_2 given by

$$k_1 = \frac{i_s}{i_p} \frac{1}{l n_r} \left(\frac{qE_r l}{600 m_p}\right)^{1/2}$$

and $\qquad k_2 = \frac{i_s}{i_p} \frac{1}{l n_p} \left(\frac{qE_r l}{600 m_r}\right)^{1/2}$ $\qquad (13.19)$

If an apparent rate k_1 or k_2 exceeds by an experimentally significant amount the corresponding theoretical limit to the rate k_{t1} or k_{t2} and the other apparent rate is less than or equal to its theoretical limit, one has strong presumptive evidence that the latter is that corresponding to the actual reaction [26].

A third method of identifying the reactants is directly related to one of the applications of observations on these reactions—that of limiting the magnitude of certain bond energies. As will be discussed below, it is essentially necessary that a reaction not be *endoergic* in order that its rate be sufficiently *great* to lead to observability in a mass-spectrometer ion source. Hence, if sufficient is known from other measurements to permit the calculation of the heats of possible alternative reactions, those that are endoergic may be eliminated from further consideration. This method is well illustrated in the work of Field and Lampe [37] in identifying "hydride transfer reactions" between hydrocarbon ions and molecules.

13.3. Theory of Ion-molecule Collisions and Reactions [38, 39]

The rate of reaction w of two structureless species P and R at a point may be expressed in terms of the velocity distribution functions $f_p(v_p)$ and $f_r(v_r)$ of the reactants and the microscopic reaction cross section $\sigma(g)$, where the v are the velocities of the particles and g is the relative velocity $|v_p - v_r|$ of the reactant pair. The expression for the rate is

$$w = \int_{-\infty}^{\infty} \int_{-\infty}^{\infty} f_p(v_p) f_r(v_r) g\sigma(g) \, dv_p \, dv_r \tag{13.20}$$

In the following paragraphs we shall outline the classical evaluation of w for the case of the reaction of mass-spectrometric interest,

$$P^+ + R \rightarrow S^+ + T + \cdots \tag{13.11}$$

Langevin [40] found the form of the classical collision cross section of a particle of charge q with a particle of polarizability α_r in his investigation of the mobility of ions in gases. His derivation of the collision cross section follows. The trajectories of a pair of particles moving under a mutual potential energy function $\varphi(r)$ can be expressed in terms of the impact parameter of the trajectory b and the initial relative velocity g in the form

$$\theta = -\int_{\infty}^{r} \frac{(b/r^2) \, dr}{\left(1 - \frac{2\varphi(r)}{\mu g^2} - \frac{b^2}{r^2}\right)^{\frac{1}{2}}} \tag{13.21}$$

where r, θ = plane polar coordinates of one particle referred to the other at the origin

μ = reduced mass of the pair

The impact parameter b is the distance of closest approach that would obtain for an encounter if the incident particle were not deflected by the interaction potential.

For a particle of unit charge $q = e$, incident on a neutral particle of polarizability α_r, the appropriate potential energy function is that for a charge-induced dipole interaction, namely,

$$\varphi(r) = -\frac{e^2\alpha}{2r^4} \tag{13.22}$$

Consideration of the conditions imposed on the denominator of Eq.

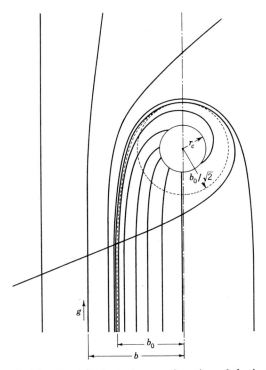

Fig. 13.5. A typical family of trajectories as a function of the impact parameter b. The dotted trajectory is the critical one for $b = b_0$ and approaches the circle $r = b_0/\sqrt{2}$. Only trajectories with $b < b_0$ will enter the reaction sphere if it has radius r_c less than this; that is, $r_c < b_0/\sqrt{2}$. Thus the cross section for entrance into the reaction sphere depends on b_0 but not, within limits, r_c. On the assumption that all molecules which enter the reaction sphere do react, the same may be said of the reaction cross section. These curves are a modified form of those given by Langevin [40].

(13.21) by the requirement of real trajectories shows that the trajectories are of two types: (1) open, those with impact parameters

$$b \geq b_0 = \left(\frac{4e^2\alpha_r}{\mu g^2}\right)^{\frac{1}{4}} \tag{13.23}$$

and distances of closest approach,

$$r_0 \geq \frac{b_0}{\sqrt{2}} \tag{13.24}$$

and (2) closed, those with impact parameters

$$b < b_0 \tag{13.25}$$

that pass through the origin. Typical trajectories are shown in Fig. 13.5, while the distance of closest approach is shown as a function of the impact parameter in Fig. 13.6.

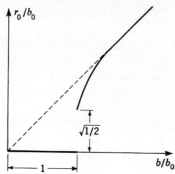

Fig. 13.6. Dependence of distance of closest approach r_0 upon the impact parameter b. The energy enters only through the critical impact parameter b_0, which is a scale factor.

We may define collisions as those encounters with closed trajectories, and this leads to the expression

$$\pi b_0{}^2 = \frac{2\pi e}{g} \left(\frac{\alpha_r}{\mu} \right)^{1/2} \tag{13.26}$$

as the classical collision cross section for encounters of P^+ with R. If the necessary and sufficient condition for reaction of P^+ with R is that their separation be less than $r_c < b_0/\sqrt{2}$, then the reaction cross section will be equal to the collision cross section

$$\sigma(g) = \frac{2\pi e}{g} \left(\frac{\alpha_r}{\mu} \right)^{1/2} \tag{13.27}$$

This assumption, that the only condition to be satisfied for reaction to occur is one of closeness of approach, eliminates any possible dependence of the reaction cross section on the energy distribution among the internal coordinates and any requirement of interaction between the internal coordinates of the reactants. This implies no activation-energy barrier to reaction, and thus that the reaction cannot be endoergic.

With this simple form for the reaction cross section, $\sigma(g)$, the expression for the point rate of reaction simplifies to

$$w = 2\pi e \sqrt{\frac{\alpha_r}{\mu}} \int_{-\infty}^{+\infty} \int_{-\infty}^{\infty} f_p(v_p) f_r(v_r)\, dv_p\, dv_r \tag{13.28}$$

The total rate of reaction W in the ion source of the mass spectrometer that we shall wish to associate with the generated current of secondary ions i_s will be

$$i_s = W = \int w \, d\tau \tag{13.29}$$

where $d\tau$ is an element of volume in the ion source, and the integration extends over the ion source.

The velocity distribution function for the neutral reactants $f_r(v_r)$ will be Maxwellian, but such will not be the case for the velocity distribution $f_p(v_p)$ of the ions formed from Maxwellian molecules in the presence of the ion-repeller field E_r. This latter distribution function is to be found by solution of the appropriate Boltzmann equation [39]. On the assumption that the electron beam is of zero thickness in the plane $z = 0$, at right angles to the direction of the uniform ion-repeller field E_r, the distribution function for the z (field direction) component of the velocity of the ions has been shown to be [39]

$$f_p(v_z, z) = v(v_z, z) \frac{P^+ m_p}{2kT} \frac{\exp\left[-(1/kT)(\tfrac{1}{2}mv_z{}^2 - eE_r z\right]}{(\pi/kT)^{\frac{1}{2}}(\tfrac{1}{2}mv_z{}^2 - eE_r z)} \tag{13.30}$$

where P^+ is the number of ionizations per unit area in the plane, $z = 0$, and $v(v_z, z)$ has one of the values

$$
\begin{aligned}
v &= 1 && \text{for } z < 0 \\
&= 2 &&
\begin{cases}
z > 0 \\
v_z > 0 \\
\tfrac{1}{2}mv_z{}^2 - eE_r z > 0
\end{cases} \\
&= 0 && v_z < 0 \\
&= 0 && \tfrac{1}{2}mv_z{}^2 - eE_r z < 0
\end{aligned}
$$

while the distribution function for the X and Y components of velocity, at right angles to the field E_r, remain Maxwellian.

Substitution in Eq. (13.29) and integration (over z from $-\infty$ to l) yields, for the total rate of reaction in the ionization chamber,

$$W = n_r i_p 2\pi e \sqrt{\frac{\alpha_r}{\mu}} \sqrt{\frac{2m_p}{eE_r l}} \tag{13.31}$$

where there has been introduced the further, quite accurate approximation that the energy $eE_r l$ that the ion acquires from the field is large compared with its initial thermal energy $\sim kT$. Identifying the rate of production of secondary ions, W, with their current i_s, we have, for Q, the phenomenological reaction cross section,

$$Q = 2\pi e \sqrt{\frac{\alpha_r}{\mu}} \sqrt{\frac{2m_p}{eE_r l}} \tag{13.32}$$

This form, inverse proportionality to the square root of the ion-repeller field strength, independence of temperature, and inverse dependence on the root-reduced mass of the reactant pair, agrees exactly with that found by Gutbier [25] and by Stevenson and Schissler [26] for a number of simple reactions indicated in the preceding section of this chapter.

In the absence of the electric field E_r and with neglect of the field due to the space charge of the electron beam, the velocity distribution functions for both the ionic and neutral reactants are Maxwellian and the rate of production of secondary ions per unit volume

$$w = n_p n_r g \sigma_{pr}(g) = n_p n_r 2\pi e \sqrt{\frac{\alpha_r}{\mu}} \qquad (13.33)$$

and the rate constant

$$k_t = 2\pi e \sqrt{\frac{\alpha_r}{\mu}} = 14.7_5 \times 10^{-10} \left(\frac{P_E}{\mu}\right)^{1/2} \quad \text{cm}^3 \text{ mole}^{-1} \text{ sec}^{-1} \qquad (13.34)$$

becomes

$$k_t = \left(\frac{eE_r l}{2m_p}\right)^{1/2} Q \qquad (13.35)$$

in terms of the phenomenological reaction cross section. This verifies the legitimacy of the dimensional argument given in the preceding section for relating Q to a bimolecular rate constant $Q'(lq/2m_p)^{1/2}$, in the notation of the preceding section [Eq. (13.17)].

It should be noted [39] that, for encounters of particles with large relative velocities g, the Langevin cross section $\sigma(g)$[Eq. (13.27)] becomes less than the hard-sphere collision cross section that can be taken to be approximately given by $\sigma_h = 4\pi(\alpha_r)^{2/3}$ [52]. Hence the relation between the bimolecular rate constant and the phenomenological reaction cross section [Eq. (13.35)] has not only the lower limiting condition ($kT/eE_r l$) $\ll$ 1, but also the upper limiting condition $\sigma(g) \geq \sigma_h$, which implies $eE_r l < \sim \frac{1}{8}e^2 \alpha_r^{-1/3}$.

The question of the error introduced by the assumption of zero thickness of the electron beam was investigated [39], and it was found that, provided the density of electrons is uniform through a region not thicker than $\pm 20\%(l)$ about its median plane, the error introduced in the above-given averaging over "z" is less than 0.5%.

Although many reactions have been found that show the simple dependence of reaction cross sections upon ion-repeller field strength here derived and indicated above, other reactions have been found for which this relation does not obtain. An example that has been studied in two laboratories with results in qualitative, if not quantitative, agreement is

$$CH_4^+ + CH_4 \rightarrow CH_5^+ + CH_3 \qquad (13.6)$$

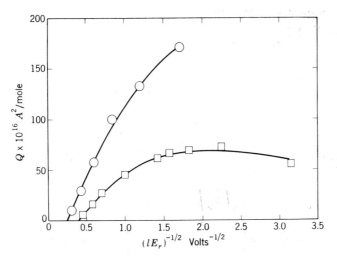

Fig. 13.7. The dependence of the cross section for the reaction $CH_4 + CH_4 \rightarrow CH_5^+ + CH_3$ on ion-repeller field strength. Illustrates the failure of the simple $Q \propto (lE_r)^{-1/2}$ law. ($\bigcirc$ from [27, 42, 47]; $\square$ from [30].)

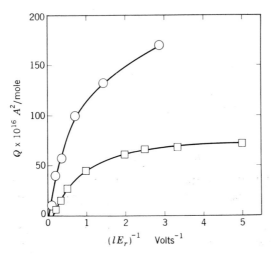

Fig. 13.8. Same as Fig. 13.7, but illustrating the failure of a proposed $Q \propto (lE_r)^{-1}$ law.

Reaction cross sections Q found by Schissler and Stevenson [27] and Field, Franklin, and Lampe [30] are shown as functions of $(lE_r)^{-1/2}$ in Fig. 13.7. The data of neither set of investigators can be fit by the classical form derived above. Incidentally, contrary to the assertion of the authors [30], these data cannot be interpreted as indicating the reaction cross section to be proportional to $(lE_r)^{-1}$. The failure of the $(lE_r)^{-1}$ law proposed by Field et al. [30] is shown in Fig. 13.8.

Recent studies by Hamill and coworkers [53–55] show, in a definitive manner, that the simple Eq. (13.32) is indeed not generally adequate to describe the ion-repeller field-strength dependence of the phenomenological reaction cross section observed mass-spectrometrically for reactions of complex substances, as was indicated by the earlier studies of the methane-ion reactions. If one takes into account the fact mentioned above, that for collisions involving relative velocities larger than g_k $\left[\text{of the order of } \left(\dfrac{1}{4}\dfrac{e^2\alpha_r^{-1/3}}{\mu}\right)^{1/2}\right]$ the collision cross section is relative-velocity-independent, it can be shown that $Q(E_r)$ takes the following forms:

$$Q(E_r) = 2\pi e \sqrt{\frac{\alpha_r}{\mu}} \sqrt{\frac{2m_p}{eE_r l}} \qquad eE_r l \leq \frac{2\pi^2 e^2 \alpha_r}{\sigma h^2} \qquad (13.32)$$

$$Q(E_r) = \sigma_h \left[1 + \frac{4\pi^2 e^2 \alpha_r m_p}{\sigma h^2 \mu} \frac{1}{eE_r l}\right] \qquad eE_r l \geq \frac{2\pi^2 e^2 \alpha_r}{\sigma h^2} \qquad (13.35a)$$

That is, for reactions of simple substances, $Q(E_r)$ would be expected to approach the finite limit σ_h as $(eE_r l)^{-1}$, as the ion-repeller field strength increases without limit. In the case of more complex substances, it would be expected that, in the case of higher-energy collisions, additional reactions (endoergic ones, for example) should become possible, and thus that it must be necessary to formulate the microscopic reaction cross section as the product of a collision cross section by a reaction probability that is a function of the relative kinetic energy, say, $\eta(g)$. If it is assumed that $(eE_r l) \gg kT$, so that the integrations over the velocity distribution functions simply give the factor $i_p n_r$, the total rate of reaction that we associate with the flux of secondary ions becomes

$$W = i_p n_r \int_0^l \eta(z)\sigma(z) \, dz \qquad (13.35b)$$

where $\eta(z)\sigma(z)$ is the reaction cross section in the plane z in the presence of an ion-repeller field E_r. The z dependence of the reaction cross section can be derived from an assumed g dependence by means of the relation between the z component of velocity of the ion, v_z, the ion-repeller field strength, and the z position of the ion, $\frac{1}{2}m_p v_z^2 = eE_r z$, if one makes the association $g \sim v_z$. This latter association is appropriate for the condition $eE_r l \gg kT$.

If one retains the hypothesis that the necessary and sufficient condition for reaction is that the separation of ion and neutral be less than $r_c \leq b_0/\sqrt{2}$ and assumes the reaction probability to be

$$\eta(z) = 1 \qquad \text{for } z_i \leq \frac{m_p g_i^2}{2eE_r}$$

$$\eta(z) = 0 \qquad \text{for } z_i \geq \frac{m_p g_i^2}{2eE_r} \qquad (13.35c)$$

then one finds, for $Q(E_r)$, the forms

$$Q(E_r) = 4\pi e \sqrt{\frac{m_p \alpha_r}{2\mu}} \sqrt{\frac{1}{eE_r l}} \qquad \text{for } z_i \geq l, \sigma(l) \geq \sigma_h$$

$$Q(E_r) = 2\pi e m_p g_i \sqrt{\frac{\alpha_r}{\mu} \frac{1}{eE_r l}} \qquad \text{for } z_i \leq l, \sigma(z_c) \geq \sigma_h \qquad (13.35d)$$

$$Q(E_r) = \frac{4\pi^2 e^2 m_p \alpha_r}{\mu \sigma_h} \frac{1}{eE_r l} + \frac{z_h - z_i}{l} \sigma_h \qquad \text{for } z_h \geq z_i, \text{ but } z_n \leq l, \text{ etc.}$$

where $\sigma(z_n) = \sigma_n$.

Boelrijk and Hamill [54] have obtained more complex expressions for $Q(E_r)$ similar to Eqs. (13.35d), by a different method. These authors abandon the hypothesis that ion-neutral separation less than r_c is the necessary and sufficient condition for reaction, and substitute the hypothesis that the microscopic reaction cross section is

$$\sigma'(g) = \eta_L(g)[\sigma(g) - \sigma_h] + \eta_h(g)\sigma_h \qquad (13.35e)$$

The physical interpretation of this essentially arbitrary hypothesis of Boelrijk and Hamill is that, for collisions of particular relative velocity g, the reaction probability is $\eta_L(g)$ for those with angular momenta within the range from $\sqrt{\mu\sigma_h}\,(g/\pi)$ to $\sqrt{\mu}\,b_0(g/\pi)$, and $\eta_h(g)$ for those with angular momenta less than $\sqrt{\mu\sigma_h}\,(g/\pi)$. They then show that, if the reaction probabilities are independent of g, the phenomenological reaction cross section takes the forms

$$Q(E_r) = \eta_L 2\pi e \sqrt{\frac{\alpha_r}{\mu}} \sqrt{\frac{2m_p}{eE_r l}} + (\eta_h - \eta_L)\sigma_h \qquad E_r \leq E_r^k$$

$$Q(E_r) = \eta_L \left(\frac{E_r^k}{E_r}\right)\left(2\pi e \sqrt{\frac{\alpha_r}{\mu}} \sqrt{\frac{2m_p}{eE_r l}} - \sigma_h\right) + \eta_h\sigma_h \qquad E_r \geq E_r^k \qquad (13.35f)$$

where E_r^k is the magnitude of the ion-repeller field that causes the Langevin cross section $\sigma(g)$ to equal the hard-sphere cross section σ_h at the exit slit from the ion source ($z = l$). It is further shown [53–55] that equations (13.35f) provide a quite satisfactory description of $Q(E_r)$ for a large number of reactions with the assumption that the reaction probabilities are approximately unity and that the σ_h have their usual kinetic-theory values.

We seriously question the utility of attempts to extend the "slow ion theory" [39] to the description of the reactions of fast ions by the introduction of arbitrary reaction probabilities such as those of Eqs. (13.35c) or by means of arbitrary modified reaction cross sections such as Eq. (13.35e). Such attempts are only subterfuges to circumvent the real problem, which is that of the manner in which the internal coordinates of the reactants enter into the rate-of-reaction expression. Possibly of paramount importance to this problem is the question of the exchange of energy between the relative motion of the reactants and their internal degrees of freedom. In the cases of certain exothermic reactions, Tal'roze and Frankevich [42] have shown that the excess energy is not evolved as kinetic energy of the reaction products, but is retained as internal excitation energy. This finding is perhaps also implicit in the results of Meisels, Hamill, and Williams [43], who showed that the charge-exchange reaction of Ar^+ with CH_4 leads primarily (within the error of their measurement) to dissociation of the methane ion to $CH_3^+ + H$. That such energy exchange can take place is implicit in the finding that endoergic reactions are observed with fast ions, as exemplified in the collision-induced dissociations [22]. In this connection the results of the extensive studies by Lindholm and coworkers of the nature of mass spectra that result from ions produced by impact of ions of known kinetic energy are undoubtedly pertinent [56–58]. On the other hand, it appears that, in the case of the $CH_4^+ + CD_4$ reaction involving relatively slow ions, the reactants essentially retain their identity in the reaction complex, since only single-atom exchange is observed [41].

13.4. Ion-molecule Reactions and Molecular Energetics

In Sec. 13.3 it has been shown that the frequency of collision between a charged particle e and a spherical molecule of polarizability α is

$$k_t = 2\pi e \left(\frac{\alpha}{\mu}\right)^{1/2} = 14.7_5 \times 10^{-10} \left(\frac{P_E}{\mu}\right)^{1/2} \quad \text{cm}^3 \text{ mole}^{-1} \text{ sec}^{-1} \quad (13.34)$$

where P_E = molecular refraction

μ = reduced mass of colliding pair in amu

Further, it has been shown that the rate of many observed ion-molecule reactions approximates this collision frequency and that these observed reactions have no temperature coefficient. Actually, consideration of the acceptable range in pressure regime of conventional (mass-spectrometer) ion sources in connection with the geometry and the dynamic range of the ion-detecting systems that are in use indicate that, for an ion-molecule reaction to be observable, it is indeed necessary that the rate be within a few orders of magnitudes of the collision frequency. In order for the collision efficiency to exceed 10^{-3}, it is necessary that the reaction have an activation energy less than 10 kcal/mole if we take $\sim 400°C$ as a reasonable upper limit to the operating temperature of the ion source. Hence, to an accuracy of about 5 kcal/mole, it may be asserted that a *necessary* condition for the observability of a secondary reaction in a mass spectrometer is that the reaction must *not* be *endoergic*.

Thus, if the reaction

$$X^+ + YH \rightarrow XH^+ + Y \tag{13.2}$$

is observed, we may write the inequalities

$$\Delta E \leq 0 \tag{13.36}$$

$$D(\text{H}\!-\!\text{X}^+) - D(\text{Y}\!-\!\text{H}) \geq 0 \tag{13.37}$$

and $\qquad P_A(\text{X}) \geq I^z(\text{H}) - I^z(\text{X}) + D(\text{Y}\!-\!\text{H}) \tag{13.38}$

where $P_A(\text{X})$, the proton affinity of X, is the bond-dissociation energy of XH^+ in the sense of the reaction

$$\text{XH}^+ \to \text{X} + \text{H}^+ \qquad P_A(\text{X}) = D(\text{X}\!-\!\text{H}^+) \tag{13.39}$$

Hence, depending on which of the dissociation energies, $D(\text{H}\!-\!\text{X}^+)$ [or $D(\text{X}\!-\!\text{H}^+)$] or $D(\text{Y}\!-\!\text{H})$ is known in addition to the ionization potentials of H and X, one may calculate an upper limit to $D(\text{Y}\!-\!\text{H})$ or a lower limit to $D(\text{X}\!-\!\text{H}^+)$ [or $D(\text{H}\!-\!\text{X}^+)$].

These consequences of the necessary condition for the mass-spectrometric observation of secondary reactions have been immediately recognized by all workers in the field. However, Tal'roze and Frankevich [14, 17] have taken the courageous position that, if a possible ion-molecule reaction is not endoergic, this will also be a *sufficient* condition for it to be observed in the mass spectrometer. This postulate is generically related to the hypothesis of Eyring, Hirschfelder, and Taylor [45] that the long-range charge-induced dipole force between an ion and a polarizable molecule will result in the absence of activation energy for an exothermic reaction. Tal'roze and Frankevich base their postulate on the argument that a large class of the ionic reactants, H_2^+, Ar^+, H_2O^+, etc., are odd electron systems, i.e., free radicals, and that it is well known that reactions involving free radicals with molecules usually have very small or zero activation energies [44]. On the basis of this hypothesis they propose that, in many cases, it should be possible to bracket an energetic quantity of interest within an upper and lower limit.

For example, consider the case of the dissociation energy $D(\text{H}\!-\!\text{X}^+)$. Suppose we look for reactions

$$\text{X}^+ + \text{Y}_i\!-\!\text{H} \to \text{XH}^+ + \text{Y}_i \tag{13.40}$$

where the Y_i are chosen so that

$$D(\text{Y}_1\!-\!\text{H}) < D(\text{Y}_2\!-\!\text{H}) < D(\text{Y}_3\!-\!\text{H}) \cdots < D(\text{Y}_n\!-\!\text{H}) \tag{13.41}$$

Now, according to the hypothesis that the nonendoergicness of the reaction is both a necessary and sufficient condition for its observation, we shall observe XH^+ for all Y_i for which

$$D(\text{H}\!-\!\text{X}^+) \geq D(\text{Y}_i\!-\!\text{H}) \tag{13.42}$$

and shall observe no reaction for

$$D(\text{H}\!-\!\text{X}^+) \leq D(\text{Y}_k\!-\!\text{H}) \tag{13.43}$$

If $i = j$ is the largest $D(Y_i\!-\!H)$ for which XH^+ is observed and $i = j + 1$ is the smallest $D(Y_i\!-\!H)$ for which XH^+ is not observed, then one would have $D(H\!-\!X^+)$ bounded as follows:

$$D(Y_i\!-\!H) \leq D(H\!-\!X^+) \leq D(Y_{i+1}\!-\!H) \qquad (13.44)$$

There are several criticisms of the sufficiency postulate of Tal'roze and Frankevich that suggest that limits of dissociation energies based on their hypothesis will be less reliable than those that are based on the definite necessity condition. These criticisms largely arise out of the fact that the sufficiency condition depends on a criterion of negative evidence, the failure to observe a reaction rather than the positive criterion of the observance of a reaction that is associated with the necessity condition. Failure to observe a secondary reaction in the mass spectrometer can arise from causes other than an activation energy, such as insufficient accuracy of measurement of ion currents or dynamic range of ion detection or the inability of a reaction of interest to compete with a more favorable reaction, i.e., an unfavorable steric factor.

The manner in which a more favorable reaction may mask the occurrence of an exothermic reaction is illustrated in the cases of the reactions of argon and krypton ions with methane. Since $D(CH_3\!-\!H)$ is about 0.06 ev less than $D(H\!-\!H)$, it is apparent, from the facts that the reactions

$$Ar^+ + H_2 \rightarrow ArH^+ + H \qquad (13.45)$$
$$Kr^+ + H_2 \rightarrow KrH^+ + H \qquad (13.46)$$

are observed, that the energy conditions for the reactions

$$Ar^+ + CH_4 \rightarrow ArH^+ + CH_3 \qquad (13.47)$$
$$Kr^+ + CH_4 \rightarrow KrH^+ + CH_3 \qquad (13.48)$$

are satisfied. Under conditions similar to those which permit the ready observation of reactions (13.45) and (13.46) in rare-gas–hydrogen mixtures, reactions (13.47) and (13.48) are not observed in rare-gas–methane mixtures. However, by working at pressures several orders of magnitude greater than can be obtained in conventional mass-spectrometer ion sources, Field, Head, and Franklin [59] have succeeded in observing the occurrence of reactions (13.47) and (13.48). The reason (13.47) and (13.48) have not been detected under more conventional conditions is that the most probable reactions that take place as a consequence of the rare-gas-ion (Ar^+ or Kr^+)–methane molecule collisions are charge exchange,

$$Ar^+ + CH_4 \rightarrow Ar + [CH_4^{+*} \rightarrow CH_3^+ + H \cdot \cdot \cdot] \qquad (13.49)$$
$$Kr^+ + CH_4 \rightarrow Kr + [CH_4^+ + CH_4^{+*} \rightarrow CH_3^+ + H] \qquad (13.50)$$

Field et al. [59] showed the cross section for reactions (13.49) and (13.50) to be two orders of magnitude greater than those of (13.47) and (13.48).

A final evaluation of the reliability of the postulate of Tal'roze and

Frankevich must await the time when we have a better appreciation of the factors that govern the occurrence of these reactions.

13.5. Radiation-chemistry Implications

The very large specific rate required of an ion-molecule reaction to admit its observation with a mass spectrometer ($k > 10^{-11}$ cm^3/mole^{-1} sec^{-1}) ensures that any secondary reaction observed with a mass spectrometer must be considered in the formulation of the mechanism of the radiolysis of systems with the same components. Thompson and Schaeffer [48], for example, have demonstrated the importance of the reactions

$$H_2^+ + H_2 \to H_3^+ + H \qquad (13.1)$$
$$H_3^+ + Kr \,(or\ Xe) \to H_2 + H + Kr^+ \,(or\ Xe^+) \qquad (13.51)$$

in accounting for the large M/N of the alpha particle–induced H_2-D_2 exchange reaction and its inhibition by traces of either of the two rare gases krypton or xenon. Similarly, Meisels, Hamill, and Williams [43] and Lampe [49] have discussed the significance of the reactions

$$CH_4^+ + CH_4 \to CH_5^+ + CH_3 \qquad (13.52)$$
$$CH_3^+ + CH_4 \to C_2H_5^+ + H_2 \qquad (13.53)$$

in the X-ray and electron-induced radiolysis of methane, and the former authors have further shown the reactions

$$Ar^+ + CH_4 \to Ar + CH_3^+ + H \qquad (13.54)$$
$$Kr^+ + CH_4 \to Kr + CH_4^+ \qquad (13.55)$$

to be important in the different modifying effects of these two rare gases on the methane radiolysis.

Because of the importance of these secondary reactions that can be observed with a mass spectrometer to the formulation of the mechanism of radiolyses, and their further importance as sources of interference in conventional analytical applications of the mass spectrometer when extreme sensitivity of detection is of interest, it appears desirable to include in this chapter a brief survey of the kinds of reactions that have been observed. No attempt is made in the following paragraph to provide an exhaustive summary of all reactions that have been reported.

The major portion of the reactions that have been observed involve hydrogen transfer—as proton, as atom, as hydride, or ambiguous. Examples of established proton transfer reactions are

$$H_2^+ + O_2 \to HO_2^+ + H \qquad \text{(Refs. 23, 26)}$$
$$(13.56)$$

$$H_2S^+ + H_2O \to H_3O^+ + HS \qquad \text{(Ref. 42)}$$
$$(13.57)$$

$$CH_2OHCHOH^+ + CH_2OHCHOHCH_2OH \to C_3H_9O_3^+ + C_2H_4O \qquad \text{(Ref. 46)}$$
$$(13.58)$$

Hydrogen-atom transfer reactions that have been observed include

$$R^+(R=He, Ne, Ar, Kr) + H_2 \rightarrow RH^+ + H \qquad \text{(Refs. 23, 25, 26)} \qquad (13.59)$$
$$H_2O^+ + C_2H_2 \rightarrow H_3O^+ + C_2H \qquad \text{(Ref. 42)} \qquad (13.60)$$
$$H_2O^+ + NH_3 \rightarrow H_3O^+ + NH_2 \qquad \text{(Ref. 42)} \qquad (13.61)$$
$$N_2^+ + H_2 \rightarrow N_2H^+ + H \qquad \text{(Refs. 14, 25, 26)} \qquad (13.62)$$

Two examples of the recently discovered hydride transfer reaction are [37]

$$C_2H_3^+ + C_2H_6 \rightarrow C_2H_4 + C_2H_5^+ \qquad (13.63)$$
$$C_3H_5^+ + neo\text{-}C_5H_{12} \rightarrow C_3H_6 + C_5H_{11}^+ \qquad (13.64)$$

In the ambiguous class of hydrogen transfer reactions there are

$$H_2^+ + H_2 \rightarrow H_3^+ + H \qquad (13.65)$$
$$C_3H_6^+ + C_3H_6 \rightarrow C_3H_7^+ + C_3H_5 \qquad \text{(Ref. 17)} \qquad (13.66)$$
$$i\text{-}C_4H_8^+ + i\text{-}C_4H_8 \rightarrow C_4H_9^+ + C_4H_7 \qquad \text{(Ref. 17)} \qquad (13.67)$$
$$CH_3OH^+ + CH_3OH \rightarrow CH_3OH_2^+ + CH_3O \text{ and } CH_2OH \qquad \text{(Ref. 46)} \qquad (13.68)$$

In hydrocarbons an enormous number of complex reactions can take place. Examples are

$$CH_3^+ + CH_4 \rightarrow C_2H_5^+ + H_2 \qquad \text{(Ref. 27)} \qquad (13.69)$$
$$C_2H_4^+ + C_2H_4 \rightarrow C_3H_5^+ + CH_3 \qquad \text{(Refs. 17, 27, 30)} \qquad (13.70)$$

The last example is of particular interest in that other products are observed from the same reactants, $C_4H_7^+ + H$, to indicate but one additional pair. The relative quantities of the various products seem to bear definite relationship to the relative quantities of the same ionic products observed among the primary dissociation products formed from, in this case, butylene ions generated by single-electron impact. This suggests the interpretation [30] that the $C_2H_4^+ + C_2H_4$ yield, as an activated complex in their reaction, an ion similar to the excited butylene ions formed by electron impact from butylene itself. This interpretation is reasonable even though the dimer ion, $C_4H_8^+$, is not observed, unless the total pressure is so great that three-body collisions can result in stabilization of the dimer ion [60]. In this connection, Chang, Yang, and Wagner [50] have found it necessary to assume such an ion-molecule dimerization reaction in the radiolysis of 1-hexene and 1-octene in the liquid phase. The excitation energy with which such dimer ions would be formed could presumably be readily dissipated in the condensed phase. Further, in the hypothesis of association of ion and molecule it is of interest to note that Hamill and coworkers [51] have recently reported observing such association products in the "high-pressure" mass spectra of alkyl iodides, and Henglein has found similar association products in the "high-pressure" mass spectra of both acrylonitrile and benzene [65].

In conclusion, the reader's attention is directed to two recent mutually complimentary reviews of the subject of ion-molecule reactions. The more comprehensive of these reviews is that of Pahl [62], which discusses such other means of studying ion-molecule reactions as microwave discharges and mass spectrometers with field-emission ion sources in addition to conventional or semiconventional mass-spectrometric methods. The second review is that of Durup [63], which is particularly concerned with the implications for radiation chemistry. This review of Durup contains a comprehensive compilation of reactions observed prior to 1960.

REFERENCES

1. J. J. Thomson: "Rays of Positive Electricity," p. 32, Longmans, Green & Co., Ltd., London, 1933.
2. A. J. Dempster: *Phil. Mag.*, **31,** 438 (1916).
3. F. W. Aston: "Mass Spectra and Isotopes," 2d ed., p. 52, Edward Arnold Publishers) Ltd., London, 1942.
4. H. D. Smyth: *Phys. Rev.*, **25,** 452 (1925).
5. T. R. Hogness and E. G. Lunn: *ibid.*, **26,** 44 (1925).
6. T. R. Hogness and H. M. Kvalnes: *ibid.*, **32,** 942 (1928).
7. Ref. 3, p. 134.
8. See, for example, in Ref. 3, Copper and Zinc.
9. W. Bleakney: *Phys. Rev.*, **41,** 32 (1932); also W. Bleakney and A. J. Gould: *Phys. Rev.*, **44,** 265 (1933).
10. A. O. Nier: *Rev. Sci. Instr.*, **18,** 398 (1947).
11. T. R. Hogness and R. W. Harkness: *Phys. Rev.*, **32,** 784 (1928).
12. H. W. Washburn, C. E. Berry, and L. G. Hall: Mass Spectrometry in Physics Research, no. 20, *Natl. Bur. Std. (U.S.) Circ.* 522, 1953.
12a. This principle had been previously considered and described by R. Wertzler and J. F. Kinder in a Consolidated Engineering Corporation Mass Spectrometry Group Report, 1948, of limited circulation.
13. M. M. Mann, A. Hustrulid, and J. T. Tate: *Phys. Rev.*, **58,** 340 (1940).
14. Quoted without reference by V. L. Tal'roze and E. L. Frankevich: *Dokl. Akad. Nauk S.S.S.R.*, **111,** 376 (1956); see also G. E. Eltenton: *Nature*, **141,** 975 (1938).
15. G. C. Eltenton: Monthly Reports of Shell Development Co., Emeryville, Calif., April, 1940.
16. A. O. Nier: 1940–1946, private communication to D. P. Stevenson.
17. V. L. Tal'roze: Dissertation, Academy of Science of the U.S.S.R., Institute of Chemistry and Physics, 1952; also V. L. Tal'roze and A. K. Loubimova, *Dokl. Akad. Nauk* S.S.S.R., **86,** 909 (1952).
18. Ref. 3, p. 21.
19. J. Mattauch and H. Lichtblau: *Phys. Z.*, **40,** 16 (1939).
20. H. S. W. Massey and E. H. S. Burhop: "Electronic and Ionic Impact Phenomena," chap. 8, Oxford University Press, London, 1952.
21. A. Henglein and H. Ewald: Ref. 12, no. 28; also A. Henglein: *Z. Naturforsch.*, **7A,** 165 (1952).
22. H. M. Rosenstock and C. E. Melton: *J. Chem. Phys.*, **26,** 314 (1957); also C. E. Melton and H. M. Rosenstock: *ibid.*, **26,** 568 (1957); C. E. Melton, M. M. Bretscher, and R. Balddock, *ibid.*, **26,** 1302 (1957); C. E. Melton, and C. F. Wells: *ibid.*, **27,** 1132 (1957).
23. F. J. Norton: Ref. 12, no. 27.

24. J. A. Hornbeck: *Phys. Rev.*, **80**, 297 (1950); **84**, 615 (1951); also J. A. Hornbeck and J. P. Molnar: *ibid.*, **84**, 621 (1951).
25. H. Gutbier: *Z. Naturforsch.*, **A12**, 499 (1957).
26. D. P. Stevenson and D. O. Schissler: *J. Chem. Phys.*, **23**, 1353 (1955); **29**, 282 (1958).
27. D. O. Schissler and D. P. Stevenson: *ibid.*, **24**, 926 (1956).
28. D. P. Stevenson: *J. Phys. Chem.*, **61**, 1453 (1957).
29. G. G. Meisels, W. H. Hamill, and R. R. Williams: *J. Chem. Phys.*, **25**, 790 (1956).
30. F. Field, J. Franklin, and F. Lampe: *J. Am. Chem. Soc.*, **79**, 2419 (1957).
31. See Refs. 14 and 17.
32. M. Szwarc: *Chem. Rev.*, **47**, 75 (1950).
33. J. A. Hipple and E. U. Condon: *Phys. Rev.*, **68**, 54 (1945); also J. A. Hipple, R. E. Fox, and E. U. Condon: *ibid.*, **69**, 347 (1946).
34. N. Coggeshall: *J. Chem. Phys.*, vol. 12 (1944).
35. H. W. Washburn and C. E. Berry: *Phys. Rev.*, **70**, 559 (1946); also C. E. Berry, *ibid.*, **78**, 597 (1950).
36. Discussed in Chapters 4 and 5 of this volume.
37. F. H. Field and F. W. Lampe: *J. Am. Chem. Soc.*, **80**, 5587 (1958).
38. E. Vogt and G. H. Wannier: *Phys. Rev.*, **95**, 1190 (1954).
39. G. Gioumousis and D. P. Stevenson: *J. Chem. Phys.*, **29**, 294 (1958).
40. P. Langevin: *Ann. Chim. Phys.*, **5**, 245 (1905).
41. C. D. Wagner, P. A. Wadsworth, and D. P. Stevenson: *J. Chem. Phys.*, **28**, 517 (1958).
42. V. L. Tal'roze and E. L. Frankevich: *Trans. First All-Union* (S.S.S.R.) *Conf. Radiation Chem.*, 1957, pp. 13-18.
43. G. G. Meisels, W. H. Hamill, and R. R. Williams: *J. Phys. Chem.*, **61**, 1456 (1957).
44. N. N. Semenov: "Some Problems in Chemical Kinetics and Reactivity" (transl. M. Boudart), Princeton University Press, Princeton, N.J., 1958.
45. H. Eyring, J. O. Hirschfelder, and H. S. Taylor: *J. Chem. Phys.*, **4**, 479 (1936).
46. Author's (D. P. Stevenson's) unpublished observations.
47. V. L. Tal'roze and E. L. Frankevich: *J. Am. Chem. Soc.*, **80**, 2344 (1958).
48. S. O. Thompson and O. A. Schaeffer: *ibid.*, **80**, 553 (1958); *Radiation Res.*, **10**, 671 (1959).
49. F. W. Lampe: *J. Am. Chem. Soc.*, **79**, 1055 (1957); **83**, 3559 (1961); *Radiation Res.*, **10**, 691 (1959).
50. P. C. Chang, N. C. Yang, and C. D. Wagner: *J. Am. Chem. Soc.*, **81**, 2060 (1959).
51. R. P. Pottie, R. Barker, and W. H. Hamill: *Radiation Res.*, **10**, 664 (1959).
52. J. H. Jeans: "The Dynamical Theory of Gases," p. 332, Dover Publications, Inc., New York, 1954.
53. R. P. Pottie, A. J. Lorquett, and W. H. Hamill: *J. Am. Chem. Soc.*, **84**, 529 (1962).
54. N. Boelrijk and W. H. Hamill: *ibid.*, **84**, 730 (1962).
55. L. P. Theard and W. H. Hamill: *ibid.*, **84**, 1134 (1962).
56. E. Lindholm: *Z. Naturforsch.*, **9a**, 535 (1954).
57. H. von Koch and E. Lindholm: *Arkiv Fysik*, **19**, 123 (1960).
58. P. Wilmenius and E. Lindholm: *ibid.*, **21**, 97 (1961).
59. F. H. Field, H. N. Head, and J. L. Franklin: *J. Am. Chem. Soc.*, **84**, 1118 (1962).
60. F. H. Field: *J. Am. Chem. Soc.*, **83**, 1523 (1961). In this and subsequent papers [*ibid.*, **83**, 3555 (1961) and Ref. 59], a variety of reactions are described, the consequences of ternary and higher-order processes that become observable when the ion-source pressure is increased to the order of 100 μ, i.e., by a factor of 100 over the normal upper limit to ion-source pressure.

61. G. E. Wells and C. E. Melton: *Rev. Sci. Instr.*, **28**, 1065 (1957); see also P. Rudolph and C. E. Melton: *J. Chem. Phys.*, **29**, 968 (1958); **30**, 847 (1959); *J. Am. Chem. Soc.*, **63**, 916 (1959).
62. M. Pahl: *Ergeb. Exakt. Naturw.*, **34**, 182 (1962).
63. J. Durup: "Les reactions entre ions positifs et molecules en phase gaseuse," Gauthier-Villars, Paris, 1960.
64. N. D. Coggeshall: *J. Chem. Phys.*, **36**, 1640 (1962). An excellent account of the efficiency of ion sources and the limitations on the assumption stated in the text.
65. A. Henglein: *Z. Naturforsch.*, **17a**, 44 (1962).

Name Index

Subject Index

Absolute-reaction-rate theory, 575
Adiabatic and vertical ionization potentials, nitrogen, table, 544
for O_2 and $O_2{}^+$, table, 544
Alkali fluoride vapors, study of existence of monomers, dimers, and trimers in, 581
Alphatron, 311
Analysis, chemical, 334–370
computational methods, 347
volatile inorganic and semiorganic, 363
functional group and structure, 355
by mass spectrometry (see Chemical analysis by mass spectrometry)
Appearance-potential equation, 553
Appearance potentials, 506
determination, experimental methods, 513
by filament of thoriated iridium or oxide cathode of BaO/SrO, 520
ionization efficiency curves, 514
instrument for rapid determination, 518
ionization potentials of some organic compounds, table, 517
second differential ionization curve, 519
second differentials $\Delta^2 I/\Delta V^2$ by special electronic circuits, 518
semilogarithmic, by Morrison, 516
by tungsten and other filaments, 520–521
vanishing-current, for evaluating ionization potentials, 514
by photoionization, 536
of free radicals, 542

Appearance potentials, of negative ions, 525
attachment peak shape, 531
electron-capture process, 526
RPD method for studying, 530
electron energy-scale calibration, 528
kinetic energy of fragment ions, 534
Lozier apparatus, 530, 534
molecular resonance-capture process, 527
Argonne instrument, 242
Aromatic hydrocarbons, excited states, 545
Aston's mass spectrograph, 9, 203
Atomic masses, determination, 47
Atoms produced by action of electrical discharges, 476

Bainbridge double-focusing mass spectrograph, 228
Bainbridge's apparatus, 204
Bainbridge-Jordan double-focusing mass spectrograph, 230
Beam-half-width method, 558
Bendix time-of-flight mass spectrometer, 502
Benzenium ions, 573
Berry's formula, 168
Bond-dissociation energies, 553
of AgAu, AgCu, AuCu, study, 577
appearance-potential equation, 553
derived from complex dissociation processes, 558
of di-, tri-, tetra-atomic intergroup IVB molecules, study, 577
of diatomic molecules Cu_2, Ag_2, Au_2, study, 582